understanding human communication

understanding human communication

FIFTEENTH EDITION

Ronald B. Adler
SANTA BARBARA CITY COLLEGE

George Rodman
BROOKLYN COLLEGE, CITY UNIVERSITY OF NEW YORK

Athena du Pré
UNIVERSITY OF WEST FLORIDA

Barbara Cook Overton
LOUISIANA STATE UNIVERSITY

OXFORD
UNIVERSITY PRESS

Oxford University Press is a department of the University of Oxford.
It furthers the University's objective of excellence in research, scholarship,
and education by publishing worldwide. Oxford is a registered trade mark
of Oxford University Press in the UK and in certain other countries.

Published in the United States of America by Oxford University Press
198 Madison Avenue, New York, NY 10016, United States of America.

© 2024, 2020, 2017, 2014, 2012, 2009, 2006 by Oxford University Press

For titles covered by Section 112 of the US Higher Education Opportunity
Act, please visit www.oup.com/us/he for the latest information about
pricing and alternate formats.

All rights reserved. No part of this publication may be reproduced,
stored in a retrieval system, or transmitted, in any form or by any means,
without the prior permission in writing of Oxford University Press,
or as expressly permitted by law, by license or under terms agreed with
the appropriate reprographics rights organization. Inquiries concerning
reproduction outside the scope of the above should be sent to the Rights
Department, Oxford University Press, at the address above.

You must not circulate this work in any other form
and you must impose this same condition on any acquirer

CIP data is on file at the Library of Congress.

Library of Congress Control Number: 2022951024

ISBN 9780197615638

9 8 7 6 5 4 3 2 1

Printed by Quad/Graphics, Inc., Mexico

Brief Contents

Preface *xvii*   Acknowledgments *xxv*   About the Authors *xxix*

PART ONE FUNDAMENTALS OF HUMAN COMMUNICATION

- **1** Communication: What and Why 3
- **2** Communicating with Social Media 21
- **3** The Self, Perception, and Communication 49
- **4** Communication and Culture 75

PART TWO COMMUNICATION ELEMENTS

- **5** Language 105
- **6** Listening 129
- **7** Nonverbal Communication 161

PART THREE INTERPERSONAL COMMUNICATION

- **8** Understanding Interpersonal Communication 193
- **9** Managing Conflict 227

PART FOUR COMMUNICATING IN GROUPS, TEAMS, AND ORGANIZATIONS

- **10** Communicating for Career Success 257
- **11** Teamwork and Leadership 289

PART FIVE PUBLIC COMMUNICATION

- **12** Preparing and Presenting Your Speech 323
- **13** Speech Organization and Support 351
- **14** Informative Speaking 383
- **15** Persuasive Speaking 413

Notes *N–1*   Glossary *G–1*   Credits *C–1*   Index *I–1*

Contents

Preface *xvii*
Acknowledgments *xxv*
About the Authors *xxix*

PART ONE FUNDAMENTALS OF HUMAN COMMUNICATION

1 Communication: What and Why 3

1.1 Characteristics of Communication 4
Defining Communication 4
Transactional Communication Model 5

1.2 Types of Communication 8
Intrapersonal Communication 8
Dyadic Communication 8
Interpersonal Communication 8
Small-Group Communication 9
Organizational Communication 9
Public Communication 9
Mass Communication 10
Social Media Communication 10

1.3 Communication Competence 11
There's No "Ideal" Way to Communicate 11
Competence Is Situational 11
Competence Is Relational 12
Competent Communicators Are Empathic 12
Competence Can Be Learned 12
Competence Requires Effort 13
Competent Communicators Self-Monitor 13
Competent Communicators Are Committed 13

1.4 Misconceptions About Communication 15
Myth: Communication Requires Complete Understanding 15
Myth: Communication Can Solve All Problems 15
Myth: Communication Is Good 15
Myth: Meanings Are in Words 16
Myth: Communication Is Simple 16
Myth: More Communication Is Always Better 16

Making the Grade 17

Key Terms 18

Public Speaking Practice 18

Activities 18

FEATURES
■ Communication Strategies: *Maintaining a Healthy Relationship with Social Media* 10
■ Understanding Your Communication: *What Type of Communicator Are You?* 14

2 Communicating with Social Media 21

2.1 The Roles of Social and Mass Media 22
Characteristics of Social Media 23
Social Media Uses and Gratifications 24
Masspersonal Communication 24

2.2 Mediated Versus Face-to-Face Communication 27
Message Richness 27
Synchronicity 29
Permanence 29

2.3 Benefits and Drawbacks of Social Media 30
Benefits of Social Media 30
Drawbacks of Social Media 34

2.4 Influences on Mediated Communication 40
Gender **40**
Age **40**

2.5 Communicating Competently with Social Media 41
Maintaining Positive Relationships **42**
Protecting Yourself **43**

Making the Grade 45

Key Terms 46

Public Speaking Practice 46

Activities 46

FEATURES
Understanding Your Communication: *What Type of Social Media Communicator Are You?* 26
Communication Strategies: *Using LinkedIn for Career Success* 32
Communication Strategies: *Evaluating (Mis)information* 38

3 The Self, Perception, and Communication 49

3.1 Communication and the Self 50
Self-Concept **50**
Self-Esteem **51**
Personality **52**
Reflected Appraisal **52**
Social Comparison **52**
Self-Fulfilling Prophecies **53**

3.2 Perceiving Others 56
Selection **56**
Organization **57**
Interpretation **57**

3.3 Problematic Perceptual Tendencies 58
Categorizing People **58**
Clinging to First Impressions **59**
Paying More Attention to Negative Impressions than to Positive Ones **60**
Judging Yourself More Charitably than You Judge Others **61**
Overgeneralizing **61**
Gravitating to the Familiar **61**

3.4 Perceptual Skill Builders 62
Emotional Intelligence **62**
Perception Checking **63**

3.5 Communication and Identity Management 65
Public and Private Selves **65**
Facework **66**
Why Manage Identities? **66**
Identity Management and Honesty **67**
Characteristics of Identity Management **68**

Making the Grade 71

Key Terms 72

Public Speaking Practice 72

Activities 72

FEATURES
Communication Strategies: *Keeping It Real on Social Media* 54
Communication Strategies: *Ways to Reverse Self-Defeating Thinking* 56
Communication Strategies: *Focus on Individuality Rather than Stereotypes* 59
Understanding Your Communication: *How Emotionally Intelligent Are You?* 64
Communication Strategies: *Humblebragging* 67
Communication Strategies: *Work Lessons from Undercover Boss* 70

4 Communication and Culture 75

4.1 Understanding Cultures and Cocultures 76
Salience 76
In-Group and Out-Group 77

4.2 How Cultural Values and Norms Shape Communication 79
Individualism and Collectivism 80
High and Low Cultural Context 82
Uncertainty Avoidance 83
Power Distance 84
Talk and Silence 86

4.3 Cocultures' Influence on Communication 86
Intersectionality Theory 87
Race and Ethnicity 87
Sex and Gender 90
Religion 92
Disability 92
Political Viewpoints 94
Age/Generation 95

4.4 Becoming an Effective Intercultural Communicator 96
Contact with a Diverse Array of People 97
Tolerance for Ambiguity 98
Open-Mindedness 98
Knowledge and Skill 99
Patience and Perseverance 99

Making the Grade 101

Key Terms 102

Public Speaking Practice 103

Activity 103

FEATURES
Understanding Your Communication: *How Much Do You Know About Other Cultures?* 78
Communication Strategies: *Talking About Race* 89
Communication Strategies: *Communicating Respectfully About Gender* 91
Communication Strategies: *Discussing Politics Responsibly on Social Media* 95
Communication Strategies: *Coping with Culture Shock* 100

PART TWO COMMUNICATION ELEMENTS

5 Language 105

5.1 The Nature of Language 106
Language Is Symbolic 106
Words Have Varying Interpretations 107
Meaning Is Negotiated 107
Language Is Governed by Rules 108

5.2 The Power of Language 110
Gender References 110
Names 111
Accents 112
Assertive and Collaborative Language 112

5.3 Language and Misunderstandings 113
Abstract Language 115
Equivocal Language 115
Relative Words 116
Slang 116
Jargon 116
Euphemisms 116

5.4 Troublesome Language 117
Confusion About Facts, Inferences, and Opinions 118

Lies and Evasions **119**
Emotive Language **121**
Microaggressions **121**

Making the Grade **124**

Key Terms **126**

Public Speaking Practice **126**

Activities **126**

FEATURES

Understanding Your Communication: *How Do You Use Language?* **109**

Communication Strategies: *Balancing Assertive and Collaborative Language* **114**

Communication Strategies: *Expressing Yourself Clearly* **117**

Communication Strategies: *Distinguishing Between Facts and Opinions* **119**

Communication Strategies: *Engaging in Microresistance* **122**

 Listening **129**

6.1 The Value of Listening **130**

6.2 Misconceptions About Listening **132**
Myth: Listening and Hearing Are the Same Thing **132**
Myth: Listening Is a Natural Process **134**
Myth: All Listeners Receive the Same Message **135**
Myth: People Have One Listening "Style" **135**
Myth: Women Are More Supportive Listeners Than Men **135**
Myth: The Majority of Listening Happens Offline **136**

6.3 Overcoming Challenges to Effective Listening **137**
Message Overload **137**
Noise **137**
Cultural Differences **138**

6.4 Faulty Listening Habits **140**
Pretending to Listen **141**
Tuning In and Out **141**
Missing the Underlying Point **141**
Dividing Attention **141**
Being Self-Centered **142**
Talking More Than Listening **142**
Avoiding the Issue **142**
Being Defensive **142**

6.5 Listening to Connect and Support **144**
Allow Enough Time **145**
Be Sensitive to Personal and Situational Factors **145**

Ask Questions **146**
Listen for Unexpressed Thoughts and Feelings **146**
Encourage Further Comments **147**
Reflect Back the Speaker's Thoughts **147**
Consider the Pros and Cons When Analyzing **149**
Reserve Judgment, Except in Rare Cases **149**
Think Twice Before Offering Advice or Solutions **149**
Offer Comfort, If Appropriate **150**

6.6 Listening to Learn, Analyze, and Critique **151**
Task-Oriented Listening **151**
Analytical Listening **154**
Critical Listening **156**

Making the Grade **157**

Key Terms **158**

Public Speaking Practice **159**

Activities **159**

FEATURES

Understanding Your Communication: *What Are Your Listening Strengths?* **133**

Communication Strategies: *Listening in a Virtual Space* **139**

Communication Strategies: *Active Listening* **143**

Communication Strategies: *When Is a Question Not a Question?* **153**

7   Nonverbal Communication   161

7.1   Characteristics of Nonverbal Communication   163
- Nonverbal Communication Is Unavoidable   163
- Nonverbal Communication Is Ambiguous   164
- Nonverbal Cues Convey Emotion   167
- Nonverbal Cues Help Manage Identities   168
- Nonverbal Cues Affect Relationships   168

7.2   Functions of Nonverbal Communication   169
- Repeating   170
- Substituting   170
- Complementing   170
- Accenting   170
- Regulating   170
- Contradicting   172
- Deceiving   172

7.3   Types of Nonverbal Communication   175
- Body Movements   175
- Voice   178
- Appearance   179
- Touch   182
- Space   183
- Time   185

7.4   Influences on Nonverbal Communication   186
- Culture   186
- Gender   187

Making the Grade   189

Key Terms   190

Public Speaking Practice   190

Activities   191

FEATURES
- Communication Strategies: *Managing Nonverbal Communication at Work*   165
- Understanding Your Communication: *How Worldly Are Your Nonverbal Communication Skills?*   171
- Communication Strategies: *Deception Detection Hacks*   174

PART THREE   INTERPERSONAL COMMUNICATION

8   Understanding Interpersonal Communication   193

8.1   Characteristics of Interpersonal Communication   194
- What Makes Communication Interpersonal?   194
- Content and Relational Messages   196

8.2   Interpersonal Relationship Building   197
- How People Choose Relational Partners   197
- Metacommunication   199
- Self-Disclosure   200
- Interpersonal Communication Online   202

8.3   Communicating with Friends and Family   205
- Friendships Have Unique Qualities   205
- Friendships Develop with Communication   206
- Friendships Can Build Bridges   209
- Family Relationships   211

8.4   Communicating with Romantic Partners   213
- Stages of Romantic Relationships   214
- Love Languages   218

8.5   Relational Dialectics   220
- Connection Versus Autonomy   220
- Openness Versus Privacy   221
- Predictability Versus Novelty   221

Making the Grade 223

Key Terms 225

Public Speaking Practice 225

Activities 225

FEATURES

▪ Communication Strategies: *Questions to Ask Yourself Before Self-Disclosing* 203

▪ Communication Strategies: *How to Be a Good Friend* 207

▪ Understanding Your Communication: *What Kind of Friendship Do You Have?* 208

▪ Communication Strategies: *How to Make Friends with a Wide Range of People* 210

▪ Communication Strategies: *Strengthening Family Ties* 213

▪ Communication Strategies: *Meeting an Online Date for the First Time* 216

▪ Understanding Your Communication: *What Is Your Love Language?* 219

▪ Communication Strategies: *Managing Dialectical Tensions* 222

Managing Conflict 227

9.1 Understanding Interpersonal Conflict 228
Expressed Struggle 229
Interdependence 229
Perceived Incompatible Goals 229
Perceived Scarce Resources 230

9.2 Communication Climates 231
Confirming and Disconfirming Messages 231
How Communication Climates Develop 235

9.3 Conflict Communication Styles 237
Nonassertiveness 237
Indirect Communication 238
Passive Aggression 239
Direct Aggression 240
Assertiveness 240

9.4 Negotiation Strategies 245
Win–Lose 245
Lose–Lose 246
Compromise 247
Win–Win 247

9.5 Cultural Approaches to Conflict Communication 251
Individualism and Collectivism 251

High and Low Context 251
Emotional Expressiveness 252

Making the Grade 253

Key Terms 254

Public Speaking Practice 255

Activities 255

FEATURES

▪ Communication Strategies: *Managing Conflict in Online Classes and Teams* 230

▪ Communication Strategies: *Rules for Fighting Fair* 232

▪ Understanding Your Communication: *What's the Forecast for Your Communication Climate?* 236

▪ Communication Strategies: *Dealing with Sexual Harassment* 239

▪ Communication Strategies: *Protecting Yourself from an Abusive Partner* 241

▪ Understanding Your Communication: *How Assertive Are You?* 244

▪ Communication Strategies: *Negotiating with a Bully* 246

PART FOUR COMMUNICATING IN GROUPS, TEAMS, AND ORGANIZATIONS

10 Communicating for Career Success 257

- **10.1** Communication Skills Are Essential 258
- **10.2** Setting the Stage for Career Success 259
  - Developing a Good Reputation 259
  - Managing Your Online Identity 259
  - Cultivating a Professional Network 261
- **10.3** Preparing Job Search Materials 263
  - Create a Portfolio of Your Work 264
  - Write a Confidence-Inspiring Cover Letter 265
  - Construct a High-Quality Resume 265
  - Follow Application Instructions 269
  - Keep Organized Records of Your Interactions 269
- **10.4** Taking Part in a Job Interview 269
  - Preparing for an Interview 269
  - Participating in a Job Interview 272
- **10.5** Adapting to a New Work Environment 277
  - Culture in the Workplace 279
  - Patterns of Interaction 280
  - Communication and Workplace Etiquette 280
  - Working Remotely 283

Making the Grade 285

Key Terms 286

Public Speaking Practice 287

Activities 287

FEATURES

- Communication Strategies: *Building a Career-Enhancing Network* 263
- Communication Strategies: *Answering "What Is Your Greatest Weakness?"* 270
- Communication Strategies: *Creating a Job Interview Presentation* 272
- Communication Strategies: *Responding to Common Interview Questions* 275
- Communication Strategies: *Interviewing by Phone or Video* 278

11 Teamwork and Leadership 289

- **11.1** Communicating Well as a Follower 290
  - Be Proactive 291
  - Seek Feedback 291
  - Support Others 291
  - If Something Isn't Right, Speak Up 291
  - Handle Challenges Calmly 292
- **11.2** Communicating in Groups and Teams 294
  - What Makes a Group a Team? 294
  - Motivational Factors 295
  - Rules and Norms in Small Groups 295
  - Individual Roles 297
- **11.3** Making the Most of Group Interaction 298
  - Recognize Stages of Team Development 299
  - Enhance Cohesiveness 300
  - Manage Meetings Well 301
  - Use Meeting Technology Effectively 303
  - Use Discussion Formats Strategically 305
- **11.4** Group Problem Solving 307
  - Advantages of Group Problem Solving 307
  - A Structured Problem–Solving Approach 308
- **11.5** Communicating Effectively as a Leader 311
  - Leadership Can Be Learned 311
  - Power Comes in Many Forms 312
  - Leadership Approaches Vary 313
  - Good Leadership Is Situational 314
  - Transformational Leadership 315

11.6 Leaving a Job Graciously 316

Making the Grade 318

Key Terms 320

Public Speaking Practice 321

Activities 321

FEATURES

Understanding Your Communication: *How Good a Follower Are You?* 290

Communication Strategies: *Working with a Difficult Boss* 293

Communication Strategies: *Getting Slackers to Do Their Share* 296

Communication Strategies: *Dealing with Difficult Team Members* 300

Communication Strategies: *Making the Most of a Brainstorming Session* 305

Communication Strategies: *Maximizing the Effectiveness of Multicultural Teams* 308

Communication Strategies: *Ways to Reach a Group Decision* 310

Communication Strategies: *Demonstrating Your Leadership Potential* 312

Understanding Your Communication: *What's Your Leadership Style?* 317

PART FIVE PUBLIC COMMUNICATION

12 Preparing and Presenting Your Speech 323

12.1 Getting Started 324
- Choosing Your Topic 325
- Defining Your Purpose 325
- Writing a Purpose Statement 325
- Stating Your Thesis 326

12.2 Analyzing the Speaking Situation 327
- The Listeners 327
- The Occasion 331

12.3 Gathering Information 332
- Online Research 332
- Library Research 333
- Interviewing 334
- Survey Research 334

12.4 Managing Communication Apprehension 335
- Facilitative and Debilitative Communication Apprehension 335
- Sources of Debilitative Communication Apprehension 336
- Overcoming Debilitative Communication Apprehension 338

12.5 Presenting Your Speech 339
- Choosing an Effective Type of Delivery 339
- Practicing Your Speech 340

12.6 Guidelines for Delivery 341
- Visual Aspects of Delivery 341
- Auditory Aspects of Delivery 342

12.7 Sample Speech 344

Making the Grade 347

Key Terms 348

Public Speaking Practice 348

Activities 348

FEATURES

Communication Strategies: *Adapting With Integrity* 329

Communication Strategies: *Evaluating Websites* 333

Understanding Your Communication: *Speech Anxiety Symptoms* 339

Communication Strategies: *Practicing Your Presentation* 340

13 Speech Organization and Support 351

- **13.1** Building Your Speech 352
  - Your Preliminary Notes 352
  - Your Working Outline 354
  - Your Formal Outline 355
  - Your Full-Sentence Outline 356
  - Your Speaking Notes 356
- **13.2** Principles of Outlining 359
  - Standard Symbols 359
  - Standard Format 359
  - The Rule of Division 359
  - The Rule of Parallel Wording 360
- **13.3** Organizing Your Outline into a Logical Pattern 360
  - Time Patterns 360
  - Space Patterns 361
  - Topic Patterns 361
  - Problem-Solution Patterns 362
  - Cause-Effect Patterns 362
  - Monroe's Motivated Sequence 363
- **13.4** Beginnings, Endings, and Transitions 364
  - The Introduction 364
  - The Conclusion 366
  - Transitions 368
- **13.5** Supporting Material 368
  - Functions of Supporting Material 368
  - Types of Supporting Material 371
  - Styles of Support: Narration Versus Citation 374
  - Plagiarism Versus Originality 374
- **13.6** Sample Speech 375
  - Speech Outline 375
  - Annotated Bibliography 376

Making the Grade 379

Key Terms 380

Public Speaking Practice 381

Activities 381

FEATURES

- Communication Strategies: *Building a Full-Sentence Speech Outline* 357
- Understanding Your Communication: *Main Points and Subpoints* 363
- Communication Strategies: *Effective Conclusions* 367
- Communication Strategies: *Organizing Business Presentations* 369

14 Informative Speaking 383

- **14.1** Types of Informative Speaking 386
  - By Content 386
  - By Purpose 386
- **14.2** Informative Versus Persuasive Topics 387
  - Type of Topic 387
  - Speech Purpose 387
- **14.3** Techniques of Informative Speaking 388
  - Define a Specific Informative Purpose 388
  - Create Information Hunger 389
  - Make It Easy to Listen 389
  - Use Clear, Simple Language 390
  - Use a Clear Organization and Structure 391
- **14.4** Using Supporting Material Effectively 392
  - Emphasizing Important Points 392
  - Generating Audience Involvement 393
  - Using Visual Aids 395
  - Using Presentation Software 400
  - Alternative Media for Presenting Graphics 400
  - Rules for Using Visual Aids 401

14.5 Sample Speech 402

Making the Grade 410

Key Terms 411

Public Speaking Practice 411

Activities 411

FEATURES

Understanding Your Communication: *Are You Overloaded?* 385

Communication Strategies: *Techniques of Informative Speaking* 390

Communication Strategies: *The Pros and Cons of Presentation Software* 401

15 Persuasive Speaking 413

15.1 Characteristics of Persuasion 414
- Persuasion Is Not Coercive 415
- Persuasion Is Usually Incremental 415
- Persuasion Is Interactive 416
- Persuasion Can Be Ethical 416

15.2 Categorizing Persuasive Attempts 418
- By Type of Proposition 418
- By Desired Outcome 419
- By Directness of Approach 420
- By Type of Appeal: Aristotle's Ethos, Pathos, and Logos 421

15.3 Creating a Persuasive Message 422
- Set a Clear Persuasive Purpose 422
- Structure the Message Carefully 423
- Use Solid Evidence 425
- Avoid Fallacies 426

15.4 Adapting to the Audience 428
- Establish Common Ground 429
- Organize According to the Expected Response 429
- Neutralize Potential Hostility 429

15.5 Building Credibility as a Speaker 430
- Competence 431
- Character 431
- Charisma 432

15.6 Sample Speech 433

Making the Grade 437

Key Terms 438

Public Speaking Practice 438

Activities 439

FEATURES

Communication Strategies: *You Versus the Experts* 418

Communication Strategies: *Recognizing Cultural Differences in Persuasion* 427

Understanding Your Communication: *Persuasive Speech* 430

Communication Strategies: *Persuasion in the World of Sales* 432

Notes *N–1*
Glossary *G–1*
Credits *C–1*
Index *I–1*

Preface

"We are stronger when we listen, and smarter when we share."
—Rania Al-Abdullah

Effective communication is about sharing ideas and listening to others. It's a simple concept, but not always as easy as it might seem. Effective communicators appreciate that understanding diverse viewpoints is an important way to build relationships and solve problems—with friends, family, coworkers, and strangers. Understanding others is also the key to presenting yourself effectively and changing hearts and minds.

The conviction that communication can honor differences and inspire collaboration is central to the 15th edition of *Understanding Human Communication*. It features new and expanded coverage of diversity, civil discourse, social media and virtual interactions, and collaborative workplace communication.

This edition builds on the approach that has served more than 1 million students over four decades. Rather than focusing solely on either skills or scholarship, *Understanding Human Communication* embraces the idea that each enhances the other. Content is well researched, up-to-date, and clear without being overly simplistic. Real-life examples and engaging images make concepts interesting and relevant to students' lives.

Approach

Understanding Human Communication introduces students to the academic study of communication and to skills that will benefit them in a wide array of life experiences. To see how well this edition succeeds, we invite you to flip to any page and ask three questions: *Is the content important? Is the explanation clear? Is it useful?*

New to this Edition

Changes in this edition reflect the changing world that *Understanding Human Communication* seeks to explain. Updated graphics, photos, examples, and research are included throughout. Following is a description of new content in key areas.

Communicating About Race, Gender, Disability, and Culture

Chapter 4 focuses exclusively on culture and communication, with updated explorations of intersectionality theory, diverse gender identities, cultural humility, and ways to talk respectfully about race and disabilities. The discussion of culture extends throughout the book, with topics such as the impact of nonbinary pronouns (Chapter 5), gender and nonverbal communication (Chapter 7), intergroup contact hypothesis (Chapter 8), and cultural influences on conflict communication (Chapter 9). The public speaking chapters take up these issues in analyses of sample speeches dealing with issues such as students' mental health (Chapter 12), the impact of hearing disabilities (Chapter 13), and power dynamics in cultural differences (Chapter 15).

Emphasis on Civil Discourse

This edition expands upon a long-standing commitment by the authors to help communicators understand and learn from one another. New content includes discussions about how to spot misinformation on social media (Chapter 2), avoid stereotypical thinking and snap judgments (Chapters 3 and 6), consider the hurtful impact of microaggressions and tactfully engage in microresistance (Chapter 5), distinguish between facts and opinions (Chapters 5 and 15), befriend a diverse array of people (Chapter 8), stop sexual harassment and abuse (Chapter 9), respond to bullies (Chapter 9), and engage in fair and open-minded debates (Chapters 6, 9, and 15).

Expanded Coverage of Social Media and Virtual Communication

Chapter 2 is dedicated entirely to mediated communication, both in close relationships and with a wider audience via social media. In addition, readers will find information about online communication throughout the book, such as how to maintain a healthy relationship with social media (Chapters 1 and 6), avoid unrealistic social comparisons to social media content (Chapter 3), present a genuine self (Chapter 3), recognize social media snarks (Chapter 4), listen online and in person (Chapter 6), negotiate what constitutes digital infidelity with romantic partners (Chapter 8), and manage conflict online (Chapter 9). The public speaking chapters examine the role that online media can serve in terms of in-depth research and inspiration (Chapter 12), as well as the dangers of plagiarism (Chapter 13) and misinformation from that form of research (Chapter 14).

Strategies for Career Success

Chapters 10 and 11 focus on communication skills for the workplace. Chapter 10 presents new information on creating a professional portfolio, negotiating a job offer, learning about prospective employers, asking questions during a job interview, practicing workplace etiquette, and working from home. Chapter 11 begins with an expanded segment on the contributions of good followers. It also includes a new unit on participating effectively in virtual team meetings. Career-relevant communication strategies throughout the book also include networking on social media (Chapter 2), politely bragging during job interviews (Chapter 3) managing conflict in team meetings (Chapter 9), listening in virtual meetings (Chapter 6), making the most of nonverbal communication at work (Chapter 7), rethinking decisions in the workplace (Chapter 14), and using persuasion techniques in sales (Chapter 15).

Chapter Content

Understanding Human Communication reflects an integrated approach to developing communication skills and awareness in a wide variety of situations.

Part One, Fundamentals of Human Communication

The first four chapters establish a foundation for studying communication in today's high-tech, multicultural environment. Real-life examples provide inspiration and guidance, underscoring the relevance of communication in daily life.

- Chapter 1 (Communication: What and Why) defines communication and dispels some common myths about the process, such as the assumption that there is one correct way to interpret a message.
- Chapter 2 (Communicating with Social Media) considers the unique qualities of mediated communication as well as strategies for using social media to enhance understanding and build relationships while avoiding social isolation, deception, and loss of privacy.
- Chapter 3 (The Self, Perception, and Communication) describes how communication influences the way people think about themselves and others. It calls attention to perceptual habits that can lead to unfair stereotypes and distrust, and instead presents ways to enhance emotional intelligence and open-mindedness. The chapter includes new features on helicopter and bulldozer parenting and how to reverse self-defeating thinking.
- Chapter 4 (Communication and Culture) helps readers consider how cultural perspectives shape who they are and how they communicate. It also encourages them to look beyond their own experiences and learn about other cultures. The chapter includes strategies for talking respectfully about race, political differences, gender, and disabilities.

Part Two, Communication Elements

This section explores three key building blocks of communication—language, listening, and nonverbal cues. By the end of Chapter 7, students should be equipped to overcome common barriers to good listening and to appreciate the layers of meaning made possible by language-based and nonverbal cues.

- Chapter 5 (Language) emphasizes the potent and evolving nature of language. It explores the impact of words—including names, pronouns, and slang. Readers will consider how to balance assertive and collaborative language, distinguish between facts and opinions, and respond effectively to microaggressions. The segment on lies and evasions, previously in Chapter 8, now appears in this chapter.
- Chapter 6 (Listening) introduces readers to the concept of social listening and offers tips for listening in high-tech environments. A new feature on active listening invites readers to practice the verbal and nonverbal behaviors associated with person-centered listening.
- Chapter 7 (Nonverbal Communication) contains an expanded and updated discussion of gender influences, with gender nonconforming and nonbinary perspectives included. New research and examples encourage readers to consider what clothing, body art, fidgeting, makeup, and facial expressions communicate to others.

Part Three, Interpersonal Communication

A focus on close relationships reveals the rewards and challenges of sharing personal information, developing friendships and romantic partnerships, and nurturing family ties. An entire chapter on conflict management provides useful skills and reminds readers that, when managed sensitively, conflict can lead to positive outcomes.

- Chapter 8 (Understanding Interpersonal Communication) focuses on how people use communication to form (and sometimes to leave) close relationships. It considers the impact that metacommunication, self-disclosure, and online communication have on relationships of all types. Units are also devoted to the unique qualities of family communication, friendships, and romantic relationships.
- Chapter 9 (Managing Conflict) explores productive ways to address differences and solve problems. It helps readers distinguish between confirming and disconfirming messages and consider whether a relationship is in an upward or downward spiral. The chapter includes new strategies for managing conflict in online classes and virtual teams. It also presents code words and gestures people can use to discreetly call for help if they are in dangerous situations.

Part Four, Communicating in Groups, Teams, and Organizations

Communication is essential to career success, and employers know it. Readers will learn how to use communication to excel as job candidates, new employees, team members, and leaders.

- Chapter 10 (Communicating for Career Success) presents communication strategies for landing a desirable position and adapting to a new job, whether that involves working in person or remotely. The chapter includes new information on sharing professional credentials in an environment that frequently includes online applications and virtual interviews. It includes a sample cover letter and resume and suggestions for responding to common interview questions.
- Chapter 11 (Teamwork and Leadership) honors the contributions of followers and team members and presents communication strategies to excel in those roles. The chapter also considers how to communicate successfully as a leader, with an emphasis on collaborative and transformational approaches. Coverage of power, previously part of Chapter 10, has been moved here within the discussion of leadership.

Part Five, Public Communication

The public speaking section (Chapters 12 through 15) continues to guide students through the traditional time-tested process for developing effective in-person presentations, while incorporating new research, examples, and techniques to update and enhance that process. Sample speeches and excerpts, especially, provide scrutiny into important communication issues.

- Chapter 12 (Preparing and Presenting Your Speech) provides a close reading of a sample speech by a student about her own mental health challenges to encourage students to tell their own stories and tackle their own challenges through public speaking.

- Chapter 13 (Speech Organization and Support) examines one student's journey into the world of diversity as he explores the process of learning American Sign Language.
- Chapter 14 (Informative Speaking) explores a presentation by a super-professor that will encourage students to overcome the problems of communication overload and disinformation.
- Chapter 15 (Persuasive Speaking) explores one student's awakening as to how to help heal a politically toxic environment.

Learning Tools

Understanding Human Communication stimulates critical thinking and high-impact learning experiences.

- **Learning objectives** provide a clear map of what students should learn and where to find that material. Objectives correspond to major headings in each chapter and coordinate with end-of-chapter summaries and activities.
- **"Communication Strategies"** features throughout the book provide handy information and tips to help students build their communication skills. Topics include how to maintain a healthy relationship with social media (Chapter 1), identify misinformation online (Chapter 2), cope with culture shock (Chapter 4), practice active listening (Chapter 6), detect deception (Chapter 7), manage dialectical tensions (Chapter 8), fight fair in an argument (Chapter 9), build a career-enhancing network (Chapter 10), get slackers to do their share (Chapter 11), adapt a message with integrity (Chapter 12), organize a message for maximum impact (Chapter 13), use presentation software effectively (Chapter 14), and recognize the impact of culture in persuasive messages (Chapter 14).
- **"Understanding Your Communication"** features are introspective quizzes that invite students to evaluate and improve their communication skills. Topics include communication style (Chapter 1), communicating with social media (Chapter 2), emotional intelligence (Chapter 3), intercultural sensitivity (Chapter 4), listening styles (Chapter 6), nonverbal communication (Chapter 7), friendship types (Chapter 8), interpersonal communication climates (Chapter 9), follower and leadership styles (Chapter 11), communication anxiety (Chapter 12), and message organization (Chapter 13).
- **Photo captions** highlight the relevance of content in the book and end with critical-thinking questions that invite readers to apply the information.
- **Key terms** are boldfaced on first use and highlighted in the margins.
- **Information and activities** at the end of every chapter help students review and apply the information they have learned.
    o A **Making the Grade** section helps students test and deepen their mastery of the material. Organized by learning objective, this section summarizes key points from the text and presents questions and prompts to help students understand and apply the material.

- **Public Speaking Practice** prompts provide confidence-building opportunities to get students speaking in class before undertaking formal presentations.
- **Activities** help students apply the material to their everyday lives. Additional activities are available in the Instructor's Manual (*The Complete Guide to Teaching Communication*) at www.oup.com/he/adler-uhc15e.

Oxford Insight

Understanding Human Communication is available in **Oxford Insight**, which delivers the book's trusted and student-friendly content within a powerful, data-driven learning experience. Oxford Insight provides access to the e-book, multimedia resources, assignable/gradable activities and exercises, and analytics on student achievement and progress. As students work through the course material, Oxford Insight automatically sets personalized learning paths for them, based on their specific performance.

Developed with applied social, motivational, and personalized learning research, **Oxford Insight** enables instructors to deliver an immersive experience that empowers students by actively engaging them with assigned reading. This approach, paired with real-time, actionable data about student performance, helps instructors ensure that all students are best supported along their unique learning paths.

With Oxford Insight, instructors can:

- Assign auto-scored multiple-choice, fill-in, and other machine-gradable questions
- Score specific items (including open-ended questions) with feedback
- Export grades
- Establish a course roster and add/drop students
- Share courses and resources with students and faculty
- Sync real-time assignments with LMS/VLE gradebooks
- Author new content and/or customize the publisher-provided content

For more information on how *Understanding Human Communication* powered by **Oxford Insight** can enrich the teaching and learning experience in your course, please visit oxfordinsight.oup.com or contact your Oxford University Press representative.

Oxford Learning Link

Oxford Learning Link (OLL) www.oup.com/he/adler-uhc15e is a convenient, instructor-focused website that provides access to up-to-date teaching resources for this text, while guaranteeing the security of grade-significant resources. In addition, it allows OUP to keep instructors informed when new content becomes available. The following items are available on the OLL:

- *The Complete Guide to Teaching Communication*, written by coauthors Athena du Pré and Barbara Cook Overton, provides a complete syllabus, teaching tips, preparation checklists, grab-and-go lesson plans, high-impact learning activities, handouts, links to relevant video clips, and coordinating PowerPoint lecture slides.

- A comprehensive **Test Bank** includes 60 exam questions per chapter in multiple-choice, short-answer, and essay formats. The questions have been revised for this edition, are labeled according to difficulty, and include the page reference and chapter section where the answers may be found.
- **PowerPoint slides** include key concepts, video clips, discussion questions, and other elements to engage students. They correspond to content in the lesson plans, making them ready to use and fully editable so that preparing for class is faster and easier than ever.

Acknowledgments

Anyone involved with creating a textbook knows that success isn't possible without the contributions of many people. We owe a debt to our colleagues. Thanks yet again to Russ Proctor, University of Northern Kentucky, for sharing his work and insights, and to those whose reviews helped shape this edition:

Edward T. Arke	*Messiah University*
Lois K. Deerberg	*Kirkwood Community College*
Jennifer Millspaugh Gray	*Dallas College*
James Michael Hinson	*Tarleton State University*
Ronald Hochstatter	*McLennan Community College*
James Michael Hinson	*Tarleton State University*
Michele Mahi	*Leeward Community College*
Catherine Nichole Morelock	*Texas Tech University*
Jessica Martin	*Portland Community College*
Jonathan Riehl	*Wake Technical Community College*
Charlotte Toguchi	*Kapiolani Community College*

We also continue to be grateful to the many educators whose reviews of previous editions continue to bring value to this book: **Theresa Albury**, Miami Dade College; **Marcee Andersen**, Tidewater Community College; **Deanna Armentrout**, West Virginia University; **Manuel G. Avilés-Santiago**, Arizona State University; **Miki Bacino-Thiessen**, Rock Valley College; **Marie Baker-Ohler**, Northern Arizona University; **Kimberly Batty-Herbert**, South Florida Community College; **Mark Bergmooser**, Monroe County Community College; **Pete Bicak**, SUNY Rockland; **Brett N. Billman**, Bowling Green State University; **Shepherd Bliss**, Sonoma State University; **Jaime Bochantin**, University of North Carolina, Charlotte; **Beth Bryant**, Northern Virginia Community College, Loudoun; **Jo-Anne Bryant**, Troy State University–Montgomery; **Adam Burke**, Hawaii Pacific University; **Ironda Joyce Campbell**, Pierpont Community and Technical College; **Patricia Carr Connell**, Gadsden State Community College; **Cheryl Chambers**, Mississippi State University; **Kelly Crue**, Saint Cloud Technical & Community College; **Dee Ann Curry**, McMurry University; **Amber Davies-Sloan,** Yavapai College; **Sherry L. Dean**, Richland College–Dallas County Community College District; **Heather Dorsey**, University of Minnesota; **Rebecca A. Ellison**, Jefferson College; **Gary G. Fallon**, Broward Community College and Miami International University of Art and Design; **Amber N. Finn**, Texas Christian University; **Lisa Fitzgerald**, Austin Community College; **David Flatley**,

Central Carolina Community College; **Sarah Fogle**, Embry-Riddle Aeronautical University; **Cole Franklin**, East Texas Baptist University; **Mikako Garard**, Santa Barbara City College; **Karley Goen**, Tarleton State University; **Samantha Gonzalez**, University of Hartford; **Betsy Gordon**, McKendree University; **Sharon Grice**, Kirkwood Community College–Cedar Rapids; **Donna L. Halper**, Lesley University; **Lysia Hand**, Phoenix College; **Andrew Herrmann**, East Tennessee State University; **Deborah Hill**, Sauk Valley Community College; **Lisa Katrina Hill**, Harrisburg Area Community College–Gettysburg Campus; **Brittany Hochstaetter**, Wake Technical Community College; **Emily Holler**, Kennesaw State University; **Milton Hunt**, Austin Community College; **Tricia Hylton**, Seneca College; **Elaine Jansky**, Northwest Vista College; **Maria Jaskot-Inclan**, Wilbur Wright College; **Angela King**, Cape Cod Community College; **Kimberly Kline**, University of Texas at San Antonio; **Carol Knudson**, Gateway Tech College–Kenosha; **Kara Laskowski**, Shippensburg University of Pennsylvania; **Jennifer Lehtinen**, State University of New York at Orange; **Amy K. Lenoce**, Naugatuck Valley Community College; **Allyn Lueders**, East Texas Baptist University; **Kurt Lindemann**, San Diego State University; **Judy Litterst**, St. Cloud State College; **Natashia Lopez-Gomez**, Notre Dame De Namur University; **Brett Maddex**, St. Petersburg College and Harrisburg Area Community College; **Jennifer McCullough**, Kent State University; **Anne McIntosh**, Central Piedmont Community College; **Bruce C. McKinney**, University of North Carolina–Wilmington; **Denise Menchacha**, Northeast Lakeview College; **Brenda Meyer**, Anoka Ramsey Community College–Cambridge; **Jim Mignerey**, St. Petersburg College; **Jennifer Millspaugh**, Richland College/Grayson College; **Randy Mueller**, Gateway Technical College, Kenosha; **Kimberly M. Myers**, Manchester College and Indiana University–Purdue University Fort Wayne; **Gregg Nelson**, Chippewa Valley Technical College, River Falls; **Emily Normand**, Lewis University; **Kim P. Nyman**, Collin College; **Catriona O'Curry**, Bellevue Community College; **Emily Osbun-Bermes**, Indiana University–Purdue University at Fort Wayne; **Christopher Palmi**, Lewis University; **Doug Parry**, University of Alaska at Anchorage; **Daniel M. Paulnock**, Saint Paul College; **Cheryl Pawlowski**, University of Northern Colorado; **Stacey A. Peterson**, Notre Dame of Maryland University; **Kelly Aikin Petkus**, Austin Community College–Cypress Creek; **Evelyn Plummer**, Seton Hall University; **Russell F. Proctor**, Northern Kentucky University; **Shannon Proctor**, Highline Community College; **Robert Pucci**, SUNY Ulster; **Terry Quinn**, Gateway Technical College, Kenosha; **Leslie Ramos Salazar**, West Texas A&M University; **Elizabeth Ribarsky**, University of Illinois at Springfield; **Delwin E. Richey**, Tarleton State University; **Charles Roberts**, East Tennessee State University; **Dan Robinette**, Eastern Kentucky University; **B. Hannah Rockwell**, Loyola University Chicago; **Dan Rogers**, Cedar Valley College; **Theresa Rogers**, Baltimore City Community College, Liberty; **Michele Russell**, Northern Virginia Community College; **John H. Saunders**, University of Central Arkansas; **Gerald Gregory Scanlon**, Colorado Mountain College; **David Schneider**, Saginaw Valley State University; **Sara Shippey**, Austin Community College; **Cady Short-Thompson**, Northern Kentucky University; **Kim G. Smith**, Bishop State Community College; **Karen Solliday**, Gateway Technical College; **Patricia Spence**, Richland Community College; **Sarah Stout**, Kellogg Community College; **Linda H. Straubel**, Embry-Riddle University; **Don Taylor**, Blue Ridge Community College; **Raymond D. Taylor**, Blue Ridge Community College; **Cornelius Tyson**, Central Connecticut State University; **Curt VanGeison**, St. Charles Community College; **Lori E. Vela**, Austin Community College; **Robert W. Wawee**, The University of Houston–Downtown; **Kathy Wenell-Nesbit**, Chippewa Valley Technical College; **Shawnalee Whitney**, University of Alaska, Anchorage; **Karin Wilking**, Northwest Vista College; **Princess Williams**, Suffolk County Community College; **Rebecca Wolniewicz**, Southwestern College; **Archie Wortham**, Northeast Lakeview College; **Yingfan Zhang**, Suffolk County Community College; and **Jason Ziebart**, Central Carolina Community College.

Many thanks are also due to colleagues who developed and refined elements of the digital program and instructor resources:

Barbara Cook Overton (contributor)	Instructor's Manual and PowerPoints
Rachel Resnik, Elmhurst University	Test Bank
Ellen Bremen, Highline College	Review, Matching, and Chapter Quizzes

In an age when publishing is becoming increasingly corporate, impersonal, and sales driven, we continue to be grateful for the privilege and pleasure of working with the professionals at the venerable not-for-profit Oxford University Press. They blend the best old-school practices with cutting-edge thinking.

Our editor, Steve Helba, was a heroic advocate and mentor in the creation of this book. Development editor Lauren Mine embodies the best of all the communication skills we describe here—a great listener, writer, friend, and coach. And Associate Editor Alyssa Quinones has kept us organized in an exceptionally chaotic world. For that, we are forever grateful.

Our heartfelt thanks also go out to the following team members whose work defines this edition: Jaime Burns, Portfolio Manager; Maeve O'Brien, Assistant Content Editor; Melissa Yanuzzi, Senior Production Editor; Michele Laseau, Lead Designer; Laura Ewen, Marketing Manager; Kaylee Williams, Marketing Assistant; Karissa Venne, Digital Content Producer Lead; Nicolas Wehmeier, Digital Project Manager; and Michael Quilligan, Senior Media Editor.

Finally, as always, we thank our partners Sherri, Linda, Grant, and Clayton for their good-natured understanding and support while we've worked on this edition for more than a year. When it comes to communication, they continue to be the best judges of whether we practice what we preach.

Ron Adler
George Rodman
Athena du Pré
Barbara Cook Overton

About the Authors

Ronald B. Adler is Professor of Communication, Emeritus, at Santa Barbara City College. He is coauthor of *Interplay: The Process of Interpersonal Communication*; *Essential Communication*; *Looking Out, Looking In*; and *Communicating at Work: Principles and Practices for Business and the Professions*.

George Rodman is Professor in the Department of Television, Radio and Emerging Media at Brooklyn College, City University of New York, where he founded the graduate media studies program. He is the author of *Mass Media in a Changing World*; *Making Sense of Media*, and several books on public speaking, as well as the coauthor of *Essential Communication*.

Athena du Pré is Distinguished University Professor of Communication at the University of West Florida. She is the author of *Communicating About Health: Current Issues and Perspectives* and coauthor of *Essential Communication*, as well as other books, journal articles, and chapters on communicating effectively.

Barbara Cook Overton has a doctorate in health communication from Louisiana State University and a Master of Fine Arts in media production from the University of New Orleans. She is the author of *Unintended Consequences of Electronic Medical Records: An Emergency Room Ethnography* and coauthor of *Communicating About Health: Current Issues and Perspectives*.

understanding human communication

Communication: What and Why

CHAPTER OUTLINE

1.1 Characteristics of Communication 4
Defining Communication
Transactional Communication Model

1.2 Types of Communication 8
Intrapersonal Communication
Dyadic Communication
Interpersonal Communication
Small-Group Communication
Organizational Communication
Public Communication
Mass Communication
Social Media Communication

1.3 Communication Competence 11
There's No "Ideal" Way to Communicate
Competence Is Situational
Competence Is Relational
Competent Communicators Are Empathic
Competence Can Be Learned
Competence Requires Effort
Competent Communicators Self-Monitor
Competent Communicators Are Committed

1.4 Misconceptions About Communication 15
Myth: Communication Requires Complete Understanding
Myth: Communication Can Solve All Problems
Myth: Communication Is Good
Myth: Meanings Are in Words
Myth: Communication Is Simple
Myth: More Communication Is Always Better

MAKING THE GRADE 17

KEY TERMS 18

PUBLIC SPEAKING PRACTICE 18

ACTIVITIES 18

LEARNING OBJECTIVES

1.1
Define communication and explain its essential characteristics.

1.2
Distinguish between types of communication in a variety of contexts.

1.3
Analyze elements of effective versus ineffective communication.

1.4
Explain how misconceptions about communication can create problems.

"Communication is like a dance," says Oprah Winfrey. When people are in sync, things flow smoothly. But not everyone hears the same song. The best communicators, Winfrey says, embrace those differences and keep dancing together, even when missteps occur.[1]

Winfrey, who has a degree in communication, is famous for her talk show, acting career, production studio, broadcast network, books, and philanthropic efforts. She is one of the richest women in the United States. Nevertheless, what makes her a good communicator, she says, is the recognition that, at her core, she is no different from anyone else. "All of us are seeking the same thing," she says, "Everybody wants to fulfill the highest, truest expression of yourself. . . . My understanding of that has allowed me to . . . reach everyone," she says.[2]

After interviewing more than 37,000 people during her career, Winfrey says she has learned something from every one of them. "I know one thing for sure," she says, "Great communication begins with connection."[3]

You may take a different path than Winfrey, but regardless of what you do, communication is sure to play a central role in your life. The average person spends about 70 percent of their waking hours communicating.[4] Effective communication can help you present a positive identity to others, establish and maintain healthy relationships, gain support for your ideas, work well with others, and attain career success.

This book explores the many ways you can use communication to be more successful in your personal and professional life. The emphasis is on respectful discourse that can help people honor diversity and consider different perspectives with an open mind. This foundational chapter explains what communication is, the forms it takes, what qualifies as effective communication, and some common myths about the process.

1.1 Characteristics of Communication

What does it mean to communicate? This section explores defining characteristics and the transactional model of communication.

Defining Communication

Communication is the process of creating meaning through symbolic interaction. This definition highlights five fundamental qualities.

Communication Is a Process People often think about communication as if it occurs in discrete, individual acts. They might say, "I told them the news" or "We had a talk." But what seems to be an isolated experience is actually part of a larger process. Imagine that a friend says you look "fabulous." Your interpretation of that will depend on a range of factors: What have others said about your appearance in the past? How do you feel about the way you look? Is your friend prone to sarcasm? Metaphorically speaking, communication isn't a series of episodes pasted together like photographs in a scrapbook. Instead, it's more like an ongoing movie in which meaning emerges based on interrelated experiences from the past and present.

Communication Is Symbolic Words are **symbols** in that they represent people, things, ideas, and events—making it possible for people to communicate about them. One feature of symbols is their arbitrary nature. For example, there's no logical reason

---

communication The process of creating meaning through symbolic interaction.

symbol An arbitrary sign used to represent a thing, person, idea, or event in ways that make communication possible.

why the letters in the word *book* should stand for what you're reading now. Speakers of Spanish call it a *libro*, and Germans call it a *buch*. Even in English, another term would work just as well. The same applies to nonverbal behavior. When you tell a joke, your friend's eye roll might signify amusement or annoyance. You understand each other to the extent that you interpret it the same way.

Communication Is Collaborative Communication isn't something you do *to* others. Rather, it's collaborative—something you do *with* them. As psychologist Kenneth Gergen points out, "one cannot be 'attractive' without others who are attracted, a 'leader' without others willing to follow, or a 'loving person' without others to affirm with appreciation."[5] Because communication is collaborative, it's a mistake to suggest that any one individual is responsible for its success or failure. If a communication episode is disappointing, it's usually better to ask, "How did *we* handle this situation, and what can *we* do to make it better?"

Messages Are Conveyed via Channels A communication **channel** is the method by which a message is transmitted. Channels include face-to-face contact and mechanisms such as computers, phones, and other devices. The channel you use can make a big difference in the effect of a message. For example, if you wanted to say "I love you," a generic e-card probably wouldn't have the same effect as a handwritten note. Likewise, saying "I love you" for the first time on the phone might be a different experience from saying the words in person.

> **channel** The medium through which a message passes from one person to another.

Digital communication technology—which emerged in the 1990s and continues rapidly to evolve—means that you can use channels beyond the dreams of earlier generations. This presents many advantages. After all, imagine (or remember) how life would be different without a mobile device in the following situations: you become separated from your friends at a large outdoor event, your car stalls on a deserted roadway, or you need to get an urgent message to someone who doesn't have a land line. Thanks to technology, those situations probably don't present dilemmas for you. You can reach most people at any time regardless of where they are.

Your dilemma may be that almost anyone can reach out to *you*. "I just can't keep up!!" laments one social media enthusiast, who explains, "Between my instagram messages, facebook messages, patreon messages, and emails . . . I just literally do not have the time to respond to every message that comes my way."[6] We'll talk more about the advantages and challenges of today's abundant communication networks in Chapter 2 and throughout the book.

Communication Is Irreversible John E. Gardner once said that "life is art without an eraser." The same can be said of communication. At times, you've probably wished you could take back words you've said or actions you've taken. But once it's out there, you typically can't unsend or unreceive a message. This is especially true on social media, where a careless comment, photo, or video can haunt you forever. It's always wise to think before you speak, write, or post a message.

Transactional Communication Model

Our basic definition of communication and its characteristics are useful, but they only begin to describe the process we will examine throughout this book. One way to deepen your understanding of communication is to look at a model depicting what happens when two or more people interact.

The **transactional communication model** proposes that communicators co-create meaning while simultaneously sending and receiving messages within the

> **transactional communication model** Proposes that communicators co-create meaning while simultaneously exchanging messages within the context of social, relational, and cultural expectations.

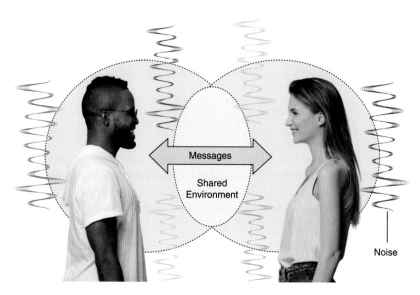

FIGURE 1.1 Transactional Model of Communication

context of social, relational, and cultural expectations.[7] A closer look at each component of the model (Figure 1.1) helps to explain.

Communicators Imagine making eye contact with a friend walking toward you. From the moment you are aware of each other, the two of you are communicating, even without trying. Perhaps his arms are folded over his chest, leading you to think he's upset or cold. At the same time, your body language suggests how you are feeling. As you get closer, your friend launches into an account of something that just happened to him. You are mostly silent, but you continue to convey information via your facial expressions, posture, and so on. As this example shows, people involved in personal communication continually and simultaneously send and receive messages. For that reason, the transactional model refers to people as *communicators*, rather than labeling one person a sender and the other a receiver.

message Information shared (intentionally or not) via words and nonverbal behaviors.

Messages In the transactional model, **messages** represent information shared (intentionally or not) between people via words or nonverbal cues. An old adage about communication is that you cannot *not* communicate. For example, your facial expressions, gestures, postures, and vocal tones may offer information to others even when you aren't aware of them. As you have certainly experienced, these messages can be powerful. If you yawn or look at the ceiling during an important conversation, the other person may assume (rightly or wrongly) that you disagree or aren't paying attention.

noise External, physiological, and psychological distractions that interfere with the accurate transmission and reception of a message.

Noise Many factors can get in the way of clear communication. Scholars use the term **noise** to refer to any force that interferes with the accurate reception of a message. Three types of noise can disrupt communication—external, physiological, and psychological.

- *External noise* (also called *physical noise*) includes factors outside of a person that are distracting or make hearing difficult. For example, a text message that catches your eye might divert your attention during a lecture, or a lawnmower might make it difficult to hear a professor's remarks clearly.

- *Physiological noise* involves biological factors that interfere with accurate reception. You might find it difficult to listen well if you are tired, sick, or hungry.

- *Psychological noise* refers to thoughts and feelings that interfere with the ability to express or understand a message accurately. If you come home annoyed or preoccupied, it will probably be harder to give the people you live with the attention they deserve.

An awareness of noise underscores that it's important to choose the right time and place for important conversations—and when you're part of those conversations, to exert the self-discipline necessary to overcome distractions. Chapter 6 focuses on the skills involved in being a good listener.

Co-Created Meaning A communication episode is co-created in that all participants help to shape it, and the meaning that emerges is a result of their collaborative effort. Here are a few examples:

- You pour your heart out to an attentive listener, and in the process, arrive at a new understanding of your situation.
- Members of a team brainstorm solutions to the parking problem on campus, coming up with ideas none of them had thought of before.
- When you discuss politics with your best friend, the conversation spirals into an argument about what is "right" and "wrong," even though neither of you intended to frame your position in those terms.

Co-creation is involved any time people engage in personal communication. That helps to explain why talking about a subject with someone is often different than simply *thinking* about it—and why conversations don't always go the way you have planned.

Context Communicators occupy **environments**—fields of experience that influence how people interpret others' behavior. In communication terminology, *environment* refers not only to a physical location but also to the personal experiences and cultural backgrounds that participants bring to a conversation. It's easy to imagine how your position on economic issues might differ depending on whether you are struggling financially or are well off, and how your thoughts on immigration reform might depend on how long your family has lived in this country.

Figure 1.1 shows that the environments of person A and person B overlap. This area represents the background that the communicators have in common. The overlap may be quite large if they belong to the same groups, have had similar life experiences, and share many of the same opinions. By contrast, communication may be challenging if participants' shared environment is small. That doesn't mean you should avoid it. Communicating with people who are different from you—perhaps of different ages, socioeconomic status, abilities, or ethnicities—can be rewarding. In Chapter 4, we discuss the advantages and skills involved in intercultural communication, and information throughout this book offers ways to bridge the gaps that separate each of us to a greater or lesser degree.

As you read about the transactional model, it may have occurred to you that it doesn't describe every type of communication. For example, mass communicated messages (such as what you see in the news) may flow in a one-way manner, thus aren't transactional. Swapping texts with a friend may involve noise, environment, and context—but not a simultaneous exchange of information—making the process transactional in some ways but not in others. Although the model doesn't apply perfectly to every situation, understanding it will help you appreciate key elements of communication.

> **environment** Both the physical setting in which communication occurs and the personal perspectives of the parties involved.

1.2 Types of Communication

Della and Honya met via a direct message on Instagram and then communicated for a year online and over the phone before they met in person. During that time, Della wondered, "Will that online friendship be the same in real life?"[8] It's a good question. Each method of communicating has unique qualities. In this section, we consider seven communication contexts, each with its own characteristics, advantages, and challenges.

Intrapersonal Communication

Intrapersonal communication involves "communicating with oneself."[9] It involves the mental voice that replays conversations, rehearses what you might say to others, and so on. Take a moment and listen to your own self-talk before reading on. It may have said something like, "What inner voice? I don't have any inner voice!" That's the "sound" of your thinking.

Intrapersonal communication affects almost every type of interaction. A conversation with a new acquaintance is shaped by your self-talk ("I'm making a fool out of myself" or "She likes me!"). Take this idea further by imagining your thoughts in each of the following situations:

- You're planning to approach a stranger you would like to get to know better.
- You pause for a minute and look at the audience before beginning a 10-minute speech.
- The boss yawns while you're asking for a raise.
- A friend seems irritated lately, and you're not sure whether you are responsible.

> **intrapersonal communication** Communication that occurs as internal dialogue.
>
> **dyadic communication** Message exchange between two people.
>
> **interpersonal communication** Two-way interactions between people who share emotional closeness and treat each other as unique individuals.

You'll read more about self-talk later. Much of Chapter 3 deals with how what you think shapes how you relate to others, and part of Chapter 14 explains how the right kind of intrapersonal communication can minimize anxiety when delivering a speech.

Dyadic Communication

Social scientists use the term **dyadic communication** to describe two-person interactions. They can occur when people talk in person or via electronic means (text, DM, or phone call). Dyadic conversations are the most common type of personal communication. Even communication within larger groups (such as classrooms, parties, and work environments) often consists of multiple dyadic encounters. Some people assume that dyadic communication is identical to interpersonal communication, but that's not always true.

Interpersonal Communication

Communication qualifies as **interpersonal** when the people involved treat each other as unique individuals. This helps to explain why dyadic communication isn't always interpersonal. You probably communicate one on

Source: © The New Yorker Collection 1984 Warren Miller from cartoonbank.com. ALL RIGHTS RESERVED.

one with many people (e.g., new acquaintances, strangers, salespeople) in ways that are informational, but not personal in any sense. And interpersonal communication can involve several people at a time, as when you engage in conversation with a group of close friends. Chapters 7 through 9 explore interpersonal communication in detail.

Small-Group Communication

Small-group communication occurs when the number of individuals is small enough that each person can interact with all of the other members. Small groups are a fixture of everyday life. Your family is an example. So are an athletic team, several students working on a project, and colleagues in different locations who work closely together online.

In a small group, the majority of members can overrule or put pressure (either deliberately or unconsciously) on the rest of the members. And for better and worse, groups often take risks that members wouldn't take if they were alone or in a dyad. On the plus side, groups can be effective at generating creative solutions since there are more people from whom to draw ideas. Groups are so important to career success that Chapter 11 focuses extensively on them.

Organizational Communication

Compared to small groups, organizations are larger and more permanent. **Organizational communication** occurs when their members work collectively to achieve goals. Organizations come in many forms, such as business (a corporation), nonprofit (the Humane Society or Habitat for Humanity), political (a government or political action group), health (a hospital or medical office), and recreational (a YMCA or sports league).

In organizations, members usually occupy specific roles (such as sales associate, general manager, corporate trainer) that influence who communicates with whom and how. Culture also plays a role. Each organization develops its own culture, which can be useful to consider when you apply for jobs and communicate in the workplace, the focus of Chapters 10 and 11.

Public Communication

Public communication occurs when one or more people deliver remarks to others who act as an audience. Even when audience members have the chance to ask questions and post comments (in person or online), the speakers are still mostly in control of the message. Public speakers usually have more opportunity to plan and structure their remarks than do

> **small-group communication** Communication that occurs within a group in which every member can participate actively with the other members.

> **organizational communication** Interaction among members of a relatively large, permanent structure (such as a nonprofit agency or business) in order to pursue shared goals.

> **public communication** Communication that occurs when a group becomes too large for all members to contribute; characterized by an unequal amount of speaking and by limited verbal feedback.

The television series *The Morning Show* depicts characters navigating organizational communication issues, such as teamwork and interpersonal rivalries, while simultaneously creating messages to be mass communicated to the fictional audience of their daybreak news program.

In what circumstances are you involved with more than one type of communication at the same time?

communicators in smaller settings. The last four chapters of this book describe the steps you can take to prepare and deliver effective speeches.

Mass Communication

Mass communication consists of messages that are transmitted to large, widespread audiences via electronic and print media such as websites, magazines, television, radio, and some blogs. Many messages sent via mass communication channels are developed, or at least financed, by large organizations such as advertisers and movie studios.

Social Media Communication

Social media use doesn't fit neatly within the groupings we've just described. A tweet or post might reach your closest friends *and* thousands of people you don't know. This can be inspiring, as when a grocery clerk's kindness to a teenager with autism

> **mass communication**
> The transmission of messages to large, usually widespread audiences via TV, internet, movies, magazines, and other forms of mass media.

COMMUNICATION STRATEGIES

Maintaining a Healthy Relationship with Social Media

There's a lot to adore about social media. It offers opportunities to connect with people all over the world.[12] The content can entertain, inspire, and educate. But the love affair isn't without a few rough spots. Screen time can be distracting and addictive. All in all, the best way to describe the public's love affair with social media might be "It's complicated." Here are some tips from the experts on keeping the relationship healthy.

Focus on the Positive

You probably wouldn't choose to befriend someone who's negative and hypercritical. Don't tolerate that on social media either. Instead, "follow inspiring, uplifting and empowering accounts that make you feel fabulous," advises tech writer Lucy Greenwell.[13] At the same time, make sure your own posts and comments are respectful toward others.

Decide When to Scroll and When to Stop

Social media is often called a time suck. You intend to check in for a moment, "but all of a sudden that 30 seconds became 30 minutes," reflects writer Bo Manry.[14]

It adds up. On average, teenagers in the United States spend more than 7 hours a day on their phones, compared to 1 hour a day doing homework.[15,16,17] To achieve a balance between screen time and other activities, set your phone aside during particular occasions. For example, you might declare classes, meetings, and meals to be device-free times. It's also a good idea to turn off notifications and to activate the "do not disturb" mode when you go to bed.

Be Present in the Moment

"Have you ever been in a room full of people but felt that no one was paying attention to anyone?" asks writer Julie Hambleton.[18] Social media allows you to attend a meeting or party at the same time you're scanning the news, catching up with friends, and posting thoughts about your latest find on Amazon. Multitasking has never been so easy. But never forget: Some of life's greatest pleasures involve pausing to look around, smell the air, enjoy a friend's laugh, be fully present with others, and simply enjoy a calmness of mind.[19] That beautiful rose you almost passed up? Stop and smell it. (You can post a photo of it later.)

became a viral video.[10] The clip was viewed by tens of thousands of people and became the focus of stories carried by CNN, ABC News, *The Washington Post*, and other news outlets. More attention isn't always better, however. UC Berkeley graduate student Connor Riley learned this the hard way when she tweeted "Cisco just offered me a job! Now I have to weigh the utility of a fatty paycheck against the daily commute to San Jose and hating the work." Company officials saw the tweet and withdrew their offer.[11]

Social media plays such an important role in today's communication environment that Chapter 2 is devoted entirely to the subject. Throughout the book, you'll also find guidance on discussing politics respectfully on social media (Chapter 4), distinguishing between facts and opinions (Chapter 5), avoiding social media distractions to be a better listener (Chapter 6), using your online identity to help with a job search (Chapter 10), and more.

As useful as social media is, it can inspire a love–hate relationship. The feature "Maintaining a Healthy Relationship with Social Media" offers some tips to maximize its positive aspects and minimize the negative.

1.3 Communication Competence

Most scholars agree that **communication competence** involves achieving one's goals in a manner that, ideally, maintains or enhances the relationship in which it occurs.[20,21] Is it effective communication if you win an argument but damage a valued relationship in the process? Probably not. Here are eight other maxims about communication competence.

> **communication competence** The ability to achieve one's goals through communication and, ideally, maintain healthy relationships.

There's No "Ideal" Way to Communicate

Some successful communicators are serious, while others are lighthearted. Some are talkative, while others are quiet. Some are straightforward, while others are subtle. Just as beautiful art takes many forms, there are many kinds of competent communicators. Others may inspire you to improve your communication skills, but ultimately, it's important to find approaches that suit you.

Many less-than-competent communicators are easy to spot by their limited approaches. Perhaps they are chronic jokers, relentlessly argumentative, or nearly always quiet. Like a piano player who knows only one tune or a chef who can prepare only a few dishes, these people rely on a small range of communication strategies again and again, whether or not they are successful. By contrast, competent communicators have a wide repertoire from which to draw, and they choose the most appropriate behavior for a given situation.

Competence Is Situational

It's a mistake to think that anyone either is or isn't a competent communicator. Individuals are better at communicating in some situations than in others, and competence is always a matter of degree. You might be quite skillful socializing at a party but less successful talking with professors during office hours. And no matter how good a communicator is, there's always more to learn. It's an overgeneralization to say, in a moment of distress, "I'm a terrible communicator!" It

Oprah Winfrey, who is often in the public eye, says she can relax and vent with her longtime best friend Gayle King.

How are you different with your best friends than you are in other situations?

would be more accurate to say, "I didn't handle this situation very well, even though I'm better in others."

Competence Is Relational

Because communication is something we do *with* others, behavior that's competent in one relationship or culture isn't necessarily effective or appropriate in others. For example, your friends might consider it fun to tease each other by trading insults, but the same comments directed at coworkers or classmates might offend. The challenge can be even greater when people are from different cultures and use words differently. If you tell someone from the United Kingdom that you packed two pairs of pants, they may think you odd. To them, *trousers* are what you wear on the outside and *pants* are what's underneath.[22] To their way of thinking, you just told them how many pairs of underwear are in your suitcase.

Competent Communicators Are Empathic

You have the best chance of developing an effective message when you understand the other person's point of view. And because people don't always express their thoughts and feelings clearly, the ability to *imagine* how an issue might look to someone else is an important skill.

Cognitive complexity is the ability to understand issues from a variety of perspectives. For instance, imagine that your longtime friend seems angry with you. Rather than jumping to a conclusion, you might consider a range of possibilities: Is your friend offended by something you have done? Did something upsetting happen earlier in the day? Or perhaps nothing is wrong, and you're just being overly sensitive. Listening is also important in such situations. It helps you understand others and judge how they are responding to your messages. Because empathy is such an important element of communicative competence, parts of Chapters 3 and 6 are devoted to the topic. Throughout the book, you will learn strategies for communicating about sex and gender, race, and disabilities (Chapter 4), as well as ways to avoid and respond to microaggressions (Chapter 5).

Competence Can Be Learned

There are numerous ways to build communication competence. You can boost your skills by taking classes such as the one you are in now, trying out new communication strategies, and observing what works well (and not so well) for you and the people around you.

Competence Requires Effort

Knowing *how* to communicate effectively is necessary, but it takes more than knowledge to be competent. Hard work is also an ingredient. For example, one study revealed that, even when college students were capable of paying attention to important conversational cues, they often didn't put in the effort to do so, perhaps because they were preoccupied or distracted.[23]

Developing competence takes time. Simply reading about communication skills in the following chapters doesn't guarantee that you'll put those skills to use right off the bat. As with any other skill—playing a musical instrument or learning a sport, for example—the road to competence in communication can be a long one. You can expect that your first efforts at communicating differently will be awkward. After some practice you should become more skillful, although you will still have to think about new ways of speaking or listening. Finally, after repeating new skills again and again, you may find you can perform them without much conscious thought.

Competent Communicators Self-Monitor

Self-monitoring involves paying close attention to situational cues and adapting your behavior accordingly. High self-monitors ask themselves questions such as "What's expected in this situation?" and "What's likely to please (or offend) the people around me?" Low self-monitors are driven by internal rather than external cues. For example, if a low self-monitor feels upset, they are more likely to raise their voice than to grin and bear it, whatever the situation may be.

> **self-monitoring** Paying close attention to situational cues and adapting one's behavior accordingly.

High self-monitors are more likely than low self-monitors to develop relationships with many types of people.[24] Being highly adaptive can lead to problems, however. High self-monitors may be inclined to overuse social media, perhaps because they're concerned with how they appear to others.[25] And people may wonder how high self-monitors really feel, since their behavior tends to vary by situation. You'll read more about self-monitoring in Chapter 3.

Competent Communicators Are Committed

One feature that distinguishes effective communication in almost any context is commitment. People who are emotionally committed to a relationship are more

Source: CALVIN and HOBBES © 1994 Watterson. Distributed by UNIVERSAL PRESS SYNDICATE. Reprinted with permission. All rights reserved.

UNDERSTANDING YOUR COMMUNICATION

What Type of Communicator Are You?

Answer the following questions for insight about your approach as a communicator.

1. You're puzzled when a friend says, "The complex houses married and single soldiers and their families."[27] What are you most likely to do?
   a. Tune out and hope your friend changes topics soon.
   b. Declare, "You're not making any sense."
   c. Ask questions to be sure you understand what your friend means.
   d. Nod as if you understand, even if you don't.

2. You're working frantically to meet a project deadline when your phone rings. It's your roommate, who immediately launches into a long, involved story. What are you most likely to do?
   a. Pretend to listen while you continue to work on your project.
   b. Interrupt to say, "I don't have time for this now."
   c. Listen for a few minutes and then say, "I'd like to hear more about this, but can I call you back later?"
   d. Listen and ask questions so you don't hurt your friend's feelings.

3. You are assigned to a task force to consider how to let new students know about a fitness program on campus. Which of the following are you most likely to do during task force meetings?
   a. Talk in a quiet voice to the person next to you.
   b. Express frustration if meetings aren't productive.
   c. Ask questions and take notes.
   d. Spend most of your time listening quietly.

4. Your family is celebrating your brother's high school graduation at dinner. What are you most likely to do at the table?
   a. Skip dessert so you can leave early.
   b. Keep your cell phone handy so you won't miss anything your friends post.
   c. Give your undivided attention as your brother talks about his big day.
   d. Paste a smile on your face and make the best of the situation, even if you feel bored.

INTERPRETING YOUR RESPONSES

For insight about your communication style, see which of the following best describes your answers. (More than one may apply.)

Distracted Communicator

If you answered "a" to two or more questions, you have a tendency to disengage. Perhaps you are shy, introverted, or easily distracted. You needn't change your personality, but you're likely to build stronger relationships if you strive to be more attentive and proactive. Active listening tips in Chapter 6 may be helpful.

Impatient Communicator

If you answered "b" more than once, you tend to be a straight-talker who doesn't like delays or ambiguity. Honesty can be a virtue, but be careful not to overdo it. Your tendency to "tell it like it is" may come off as bossy or domineering at times. The perception-checking technique in Chapter 3 offers a good way to balance your desire for the truth with concern for other people's feelings.

Tactful Communicator

If you answered "c" multiple times, you are able to balance assertiveness with good listening skills. Your willingness to actively engage with people is an asset. Use tips throughout the book to enhance your already-strong communication skills.

Accommodator

If you answered "d" two or more times, you tend to put others' needs ahead of your own. People probably appreciate your listening skills but wish you would speak up more. Saying what you feel and sharing your ideas can be an asset both personally and professionally. Tips in Chapters 11 and 15 may help you become more assertive and confident without losing your thoughtful consideration for others.

likely than others to talk about difficult subjects and to share personal information about themselves, which can strengthen relationships and contribute to a heightened sense of well-being.²⁶ Take the self-assessment quiz "What Type of Communicator Are You?" for insight about your communication strengths and challenges.

1.4 Misconceptions About Communication

Having spent time talking about what communication is, we should identify some things it is not.

Myth: Communication Requires Complete Understanding

Most people operate on the flawed assumption that the goal of all communication is to maximize understanding between communicators. In fact, there are times when complete comprehension isn't the primary goal. For example, when people in the United States ask social questions such as "How's it going?" they typically aren't asking for genuine details about how the other person is doing. Beyond formalities, competent communicators may also deliberately create ambiguous messages. For example, consider what you might say if someone you care about asks a personal question that you don't want to answer such as "Do you think I'm attractive?" or "Is anything bothering you?"

Myth: Communication Can Solve All Problems

"If I could just communicate better . . ." is the sad refrain of many unhappy people who believe that, if they could express themselves more effectively, their relationships would improve. This is sometimes true, but it's an exaggeration to say that communicating—even communicating clearly—is a guaranteed cure-all.

Myth: Communication Is Good

In truth, communication is neither good nor bad in itself. Rather, its value comes from the way it's used. Communication can be a tool for expressing warm feelings and useful facts, but under different circumstances words and actions can cause both physical and emotional pain.

"My wife understands me."

Source: Cartoonbank.com

Communication isn't a cure-all for problems. Sometimes people need time and space to think things through on their own.

In what circumstances would you like a break from communicating with others?

Myth: Meanings Are in Words

People often interpret the same words differently. You may have learned the hard way that words such as *freedom* and *control* can mean different things to different people. And new words continually emerge. If *periodt* appears at the end of a written comment, some people may think it's a misspelling, whereas others will assume that the writer feels strongly about the topic, as in, "I am staying home, periodt."

Myth: Communication Is Simple

Most people assume that communication is an aptitude that people develop without the need for training. After all, you've been swapping ideas with others since early childhood, and some people communicate well without ever taking a class on the subject. However, it's a mistake to assume that communication comes easily. Communication skills are a lot like athletic ability: Everyone can learn to be more effective with training and practice, and those who are talented can always become better.

Myth: More Communication Is Always Better

Although it's true that not communicating enough is a mistake, there are situations when *too much* communication is ill advised. Sometimes people go over the same ground again and again and again. You've probably had the experience of "talking yourself into a hole"—making a bad situation worse by pursuing it too far. There are even times when *no communication* is the best course. Any good salesperson will tell you that it's often best to stop talking and let the customer think about the product. And when two people are angry and hurt, they may say things they don't mean and will later regret. One key to successful communication, then, is to share an adequate amount of information in a skillful manner when the time is right.

This chapter began by describing Oprah Winfrey, who is famous for giving others a voice. "Oprah Winfrey is a world-renowned communicator and talented conversationalist, but her greatest quality is her ability to listen," proclaims media writer Colin Baker.[28] That isn't always easy for her or anyone else. As Winfrey puts it, in the dance of communication, people sometimes get tangled up and fall down. When that happens, she encourages, "extend a hand of connection and understanding to help your partner to [their] feet. . . . Then once you're face-to-face, offer three of the most important words any of us can ever receive: 'I hear you.'"[29]

MAKING THE GRADE

OBJECTIVE 1.1 **Define communication and explain its essential characteristics.**

- *Communication* is the process of creating meaning through symbolic interaction.
- Communication is achieved collaboratively with others by conveying messages through channels.
- The transactional model of communication proposes that communicators co-create meaning while simultaneously sending and receiving messages with the context of social, relational, and cultural expectations.
- Communication is influenced by three types of noise.
    > Describe a real-life example that illustrates how communication is process oriented, relational, and symbolic.
    > Have you ever overreacted to something someone said, only to realize later that you misinterpreted that person's intent? How can the knowledge that communication is an ongoing process help you put things in perspective?
    > Pause for a moment to consider what is on your mind. How might these thoughts create physiological noise that could interfere with your ability to listen closely? Brainstorm some techniques people might use to listen carefully rather than being distracted by their own thoughts.

OBJECTIVE 1.2 **Distinguish between communication in a variety of contexts.**

- Communication operates in many contexts, including intrapersonal, dyadic, interpersonal, small group, organizational, public, mass media, and social media.
    > Provide examples of communication in three of the contexts just listed.
    > In what way would you most like to improve your communication relevant to interpersonal communication? To small-group communication? To public speaking?
    > Overall, what goals would you most like to set for yourself as a communicator? Which contexts will your goals involve?
    > Imagine you woke up this morning to find that the internet and cell phones no longer worked. In what ways would your daily life be different without them?

OBJECTIVE 1.3 **Analyze elements of effective versus ineffective communication.**

- Communication competence is situational and relational in nature, and it can be learned.
- Competent communicators are able to choose from and perform appropriately using a wide range of behaviors, taking into account the perspectives of others.
    > Who are the most competent communicators you know? How does their behavior reflect the elements of competence described here?
    > Rate yourself from 1 to 10 in terms of how well your communication reflects each of the following: empathy, self-monitoring, and commitment. Which of these are you good at? Which might you improve and how?
    > Imagine that someone you know often tells jokes that you find offensive. Reflect on which of the following are you most likely to do: (a) say nothing, (b) ask a third party to talk to the jokester, (c) hint at your discomfort, (d) use humor to comment on your friend's insensitivity, or (e) express your discomfort in a straightforward way and ask your friend stop. Do you feel the strategies you use would be effective?

OBJECTIVE 1.4 Explain how misconceptions about communication can create problems.

- Communication doesn't always require complete understanding.
- Communication is not always a positive factor that will solve every problem.
- Meanings are in people, not in words.
- Communication is neither simple nor easy.
- More communication is not always better.

> Describe a time when a misunderstanding led to a mistake or a comical situation. How did interpretations of the initial message(s) differ?

> Which misconception(s) have been most problematic in your life? Give examples of each.

> How can you avoid succumbing to the misconceptions listed in this chapter? Give examples of how you could communicate more effectively.

KEY TERMS

channel p. 5
communication p. 4
communication competence p. 11
dyadic communication p. 8
environment p. 7
intrapersonal communication p. 8
interpersonal communication p. 8
mass communication p. 10
message p. 6
noise p. 6
organizational communication p. 9
public communication p. 9
self-monitoring p. 13
small-group communication p. 9
symbol p. 4
transactional communication model p. 5

PUBLIC SPEAKING PRACTICE

Describe a time when a listener's reaction led you to change your communication approach. What about the listener's feedback stood out to you? How did you change your communication as a result? Create a brief oral presentation in which you explain this encounter in terms of the transactional model of communication.

ACTIVITIES

1. Observe your interactions for one day. Record every occasion in which you are involved in communication as it is defined in this chapter. Based on your findings, answer the following questions:

   - What percentage of your waking day is involved in communication?
   - What percentage of time do you spend communicating in the following contexts: intrapersonal, dyadic, small group, and public?
   - What percentage of your communication is devoted to satisfying each of the following types of needs: physical, identity, social, and practical? (Note that you might try to satisfy more than one type at a time.)

   Based on your analysis, describe 5 to 10 ways you would like to communicate more effectively. For each item on your list of goals, describe who is involved (e.g., "my boss," "people I meet at parties") and how you would like to communicate differently (e.g., "act less defensively when criticized," "speak up more instead of waiting for them to approach me"). Use this list to focus your studies as you read the remainder of this book.

2. Construct a diary of the ways you use social media in a three-day period. For each instance when you use social media (email, a social networking website, phone, Twitter, etc.), describe:

   - The kind(s) of social media you use
   - The nature of the communication (e.g., "DM'd a friend on Instagram," "Reminded roommate to pick up dinner on the way home")
   - The type of need you are trying to satisfy (information, relational, identity, entertainment)

   Based on your observations, describe the types of media you use most often and the importance of social media in satisfying your communication needs.

3. Read the following scenarios and consider in each case whether you think it would be best to (a) say

what you are thinking, (b) listen, or (c) remove yourself from the situation. Explain your reasons.

- A close relational partner of yours was hurt by something you said. You have apologized, but they still want to talk about the issue.
- A classmate is considering a job offer that would require leaving school and moving to another city. You think it's a bad idea, but your classmate seems excited about it.
- Your ex calls at all hours of the night wanting to "talk," which usually means asking you over and over why the relationship ended.
- A coworker confides in you about a family issue that has them worried and upset.

What moral obligation, if any, do you have to express your thoughts and feelings in each of these situations? Which communication might have influenced your approach before reading this section?

Communicating with Social Media

CHAPTER OUTLINE

2.1 The Roles of Social and Mass Media 22
Characteristics of Social Media
Social Media Uses and Gratifications
Masspersonal Communication

2.2 Mediated Versus Face-to-Face Communication 27
Message Richness
Synchronicity
Permanence

2.3 Benefits and Drawbacks of Social Media 30
Benefits of Social Media
Drawbacks of Social Media

2.4 Influences on Mediated Communication 40
Gender
Age

2.5 Communicating Competently with Social Media 41
Maintaining Positive Relationships
Protecting Yourself

MAKING THE GRADE 45
KEY TERMS 46
PUBLIC SPEAKING PRACTICE 46
ACTIVITIES 46

LEARNING OBJECTIVES

2.1
Identify the distinguishing characteristics of social media, which combine elements of mass and personal communication.

2.2
Outline cases for using social media versus face-to-face communication, noting important differences.

2.3
Identify the benefits and drawbacks of mediated communication, and use this knowledge to maximize effectiveness and minimize the potential for harm.

2.4
Describe how factors such as gender and age can influence social media use.

2.5
Explain how to use social media competently to maintain positive relationships, minimize understandings, and protect yourself.

Christina Farr suspected she was spending too much time on social media. Her worst fears were confirmed after she checked her phone's activity log. "I spent more than five hours on Instagram in a single week. Five hours!"[1] Farr concedes that five hours may not seem like much, but, as she wrote for CNBC News, "I would have guessed an hour or two."[2] The realization that Farr could have spent those five hours reading a book, hanging out with friends, or learning a new language motivated her to delete Instagram and Facebook from her phone. She's not the only one. People are quitting social media in droves. And many of them are happier for it.[3]

This comes on the heels of a few no good, very bad years for social media. Companies such as Meta/Facebook[4] and Twitter have, according to Farr, "faced a reckoning,"[5] with reports surfacing about a number of ills ranging from social media's role in election interference,[6] to inciting violence culminating in an attack on the U.S. Capitol,[7,8] to spreading COVID-19 misinformation.[9] On top of that were leaked documents showing that Facebook's executives knew Instagram harmed teen girls.[10] With 64 percent of Americans in agreement that social media is a divisive force,[11] it's no wonder hashtags like #DeleteFacebook have trended over the last few years.[12]

Social media isn't all bad, however. As this chapter demonstrates, social media can be a powerful tool that helps people connect and communicate. And, yes, it can wreak havoc and cause considerable harm if it isn't used responsibly. The goal of this chapter is to help you understand how social media affects your life, your relationships, and how you communicate. (Even if you don't use social media much, you will find that its influence is inescapable). You'll begin by learning about social media's unique characteristics, its function as a type of masspersonal communication, and how it differs from face-to-face interaction. You'll then read about its benefits and drawbacks, how gender and age influence users' experiences, and tips for communicating competently and responsibly with social media.

2.1 The Roles of Social and Mass Media

What do you think of when you hear *media*? As you read in Chapter 1, the term refers to channels through which messages flow. (The singular form is *medium*.) Print media (such as newspapers and magazines) have been around for centuries. Electronic media (telephone, radio, and television) are more recent. And social media has emerged in the last 50 years.

Chapter 1 defined *mass communication* as messages transmitted to large, widespread audiences. Until the early 21st century, creators of mass media were almost always professional **gatekeepers**, such as the heads of motion picture studios, newspaper reporters, and television news directors, who controlled what information the public received. Audiences had little direct influence on the content of public messages.

Corporate mass media giants such as Fox News and CNBC still exert a powerful influence by framing information, shaping public discourse, and reinforcing people's beliefs. In the entertainment world, content producers such as Netflix, HBO, and Disney+ shape and reflect public tastes. But today, not all media are aimed at mass audiences or are controlled by corporate gatekeepers. **Social media** are dynamic websites and applications that enable individual users to create and share content

---

gatekeeper A person who acts as a hub for receiving messages and sharing (or not sharing) them with others.

social media Dynamic websites and applications that enable individual users to create and share content or to participate in personal networking.

or to participate in social networking. (Although the term *social media* is technically plural, we refer to it as singular, as do most people in everyday conversation.) If you tweet, post photos, or follow users on platforms such as TikTok, Instagram, YouTube, or Facebook, you're using social media. If you use LinkedIn, XING, or other professional networking sites, you're harnessing social media for career success. With the growth of social media, the number of content producers and channels available for reaching large audiences has increased. As a result, the agenda-setting power of mass media has diminished.[13]

Baby Yoda became one of the last decade's biggest pop culture phenomena. The popular character from Disney+'s hit show *The Mandalorian* spawned countless memes, and is just one example of mass media's power to influence public taste. The many memes and GIFs created and shared by social media users on platforms like Twitter and Instagram highlight the interplay between mass and social media.

Characteristics of Social Media

Social media is similar to other channels of communication in some respects and different in others. This section considers some distinguishing qualities. As you read, think about how these characteristics align with your own communication practices.

User-Generated Content Social media provides channels for individual users—not just big organizations—to share content. Go to YouTube's homepage or open TikTok and you'll see a plethora of videos made by regular people whose only tools were smartphone cameras and easy-to-use editing apps. Some of these videos are silly ("Baby Elephant Throwing a Tantrum"). Others can be helpful ("How to Change a Flat Tire Like a Boss" or "Warning Signs of Alcoholism"). Before social media, material like this was difficult or impossible for everyday people to create and share.

Variable Audience Sizes Voicemails and text messages target small numbers of people—or even a single receiver. Mass media, by contrast, target very large audiences. The audience size of social media varies, depending on the platform (whether you are using Snapchat or Instagram, for example), how many "followers" or "friends" you have, and your privacy settings. Your social media posts could be seen by a few folks . . . or by a few thousand. A post you intend for just a couple of people might be forwarded, shared, and reposted to countless others.

Interactivity The recipients of your social media messages can—and often do—talk back. Social media platforms make it easy for users to engage with one another. Most platforms offer the option to like, comment, and share. Because engagement is so easy, there are more than 5 billion likes and 7 million comments posted to Facebook every day.[14] Many social media applications also have direct messaging capabilities that allow users to privately text back and forth, share photos, and stream video. These two-way interactions underscore the difference between traditional media, in which communication is essentially one way, and far more synergistic web-based social media.

Collaboration The video-sharing platform TikTok was the most downloaded app in the U.S. in 2020.[15] It now has more than 1 billion users worldwide.[16] One of the

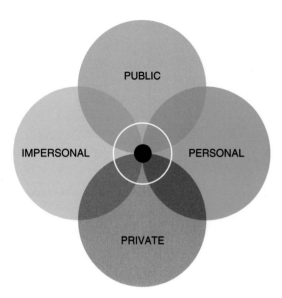

FIGURE 2.1 Overlapping Boundaries Between Different Types of Media

uses and gratification theory An approach used to understand why and how people actively seek out specific media to satisfy their particular needs, including providing information, facilitating personal relationships, defining personal identity, and entertainment.

masspersonal communication Messages that, to varying degrees, combine elements of mass and interpersonal communication.

reasons the platform is so popular is its emphasis on collaboration:[17] TikTok makes it easy for users to remix and add on to each other's videos. One of the biggest TikTok collaborations to date involved thousands of users worldwide who crafted a musical adaptation of the Disney Pixar film *Ratatouille*.[18] Besides being a viral hit, "Ratatouille: The TikTok Musical" helped raise more than $2 million for out-of-work theater professionals during the COVID-19 pandemic.[19] And TikTok isn't the only social media platform that encourages collaboration. Others—such as YouTube, Drooble, The Dots, and Vimeo—also facilitate creative teamwork.

Social Media Uses and Gratifications

Uses and gratification theory helps communication scholars understand why and how people use media. The theory suggests that, rather than passively consuming information, people actively seek out specific media to satisfy a variety of needs.[20] You probably don't watch television just because a TV screen happens to be in the room with you (although sometimes you might). More often, you choose to watch TV for a reason—maybe you want to see the news or catch your favorite sitcom. In this example, you are using a specific medium (i.e., TV) to satisfy your need for information or entertainment.

The same is true for social media. People may go on TikTok because they're bored or they want to see what their friends are posting. Researchers have identified several main reasons why most people use social media.[21] Social interaction is at the top of the list, followed by information seeking, passing the time, and entertainment. A single platform could ostensibly satisfy all these needs. For example, Meta's Facebook lets you simultaneously chat with friends in real time, check the location and hours of a business via its Facebook profile, scroll through your feed, and watch an assortment of videos and entertaining posts.

While many platforms such as Facebook and Instagram are one-stop shops, most social media platforms prioritize specific kinds of content. Acknowledging that there is some overlap between these categories, Table 2.1 outlines some primary types. As you look over the table, think about which platforms and content types you would use to scout for a job, relieve boredom, catch up with friends, or satisfy other needs.

Masspersonal Communication

With the advent of social media, the boundaries between interpersonal and mass communication have blurred. **Masspersonal communication** is a term some scholars use to describe the overlap between personal and public communication.[22] Figure 2.1 illustrates the fuzzy boundaries where some communication channels meet.

Consider, for example, video sharing services such as YouTube and TikTok. They provide a way for individuals to publish content (e.g., a graduation or a baby's first birthday party) to a small number of people. But some videos go viral, receiving thousands or even millions of hits. Why certain videos go viral is often a mystery. When 37-year-old TikTok user Nathan Apodaca posted a video of himself skateboarding down the street while drinking cranberry juice and singing along to a Fleetwood Mac song, he never imagined the video would rack up more than 70 million views in two

TABLE 2.1
Common Types of Social Media Content

CONTENT TYPE	EXAMPLE
Social network	Facebook, WT Social
Business network	LinkedIn, XING, Upstream
Blog	WordPress, Tumblr, Open Diary
Microblog	Twitter, Tumblr
Messaging	WhatsApp, Snapchat, Telegram
Discussion forum	Reddit, Digg
Advocacy	Care2, MoveOn, Wiser.org
Photo sharing	Instagram, Flickr, Imgur
Video sharing	TikTok, YouTube, Vimeo
Social bookmarking	Pinterest, Dribble, Flipboard
Product and service reviews	Yelp, TripAdvisor
Social shopping	Etsy, Facebook Shops
Goods and services trading	Couchsurfing, TaskRabbit
Crowdsourcing	Quora, InnoCentive, Wikipedia
Dating	Tinder, OKCupid, Bumble
Collaboration	The Dots, Drooble
Social gaming	Cellufun, Twitch, Discord
Virtual world	Second Life, Minecraft

months' time.[23] The video's popularity changed Apodaca's life. He went from working in a warehouse to starring alongside Snoop Dogg in commercials and music videos.

Not all viral videos have such happy outcomes. Brianna Ramirez lost her job after creating and posting a TikTok video titled "Exposing Panera." It showed her at work preparing mac and cheese by dropping a frozen packet into boiling water. Without her knowledge, the video was retweeted and was viewed more than 12 million times.[24]

Ramirez's experience underscores the fuzzy boundary between personal and mass media. While she may have intended her original post for a relatively small audience of TikTok followers, the video was reposted and shared beyond her control. Ramirez's ordeal is an important reminder that once you post something on social media, it has a life of its own.

Until somewhat recently, most of the videos found online and on social media were prerecorded. These days, however, most platforms have livestreaming capabilities, which means that people can broadcast content in real time. Livestreaming

UNDERSTANDING YOUR COMMUNICATION

What Type of Social Media Communicator Are You?

Answer the following questions for insight about your social media communication style.

1. When you and your friends go on a deep-sea fishing trip, you are nauseated the entire time. What are you most likely to do?
    a. Post selfies in which you pretend you're having a great time.
    b. Tweet about how much you detest the ocean and warn others never to take a similar trip.
    c. Post pictures of your friends boozing it up on deck. So funny!
    d. Set social media aside and try to be fully present with your friends, even if the circumstances aren't ideal.
    e. Concentrate on making it back to shore. You'll think about social media later. Maybe.

2. Imagine that you work for an outdoor furniture company that encourages you to promote products via your personal social media accounts. What are you most likely to do?
    a. Post pictures of yourself using the furniture around the pool, at the beach, at the park, and so on.
    b. Use social media to tell all your friends and followers about the benefits of your company's products compared to others.
    c. Post a message such as, "I need to boost my sales numbers. Place an order, please!"
    d. Share promotional messages only with people you think will be interested in the products.
    e. Decline to use your social media in this way, reasoning that personal and business interests are best kept separate.

3. Which of the following best describes your social media presence?
    a. I like to post pictures that show my activities and hobbies—my morning workout, the delicious dish I had for lunch, cute pictures of my pet, and so on.
    b. I mostly use social media to spread the word about issues I think are important, such as elections and social justice.
    c. My posts reflect how I'm feeling that day—whether it's happy, sad, frustrated, or excited.
    d. I strive for a balance between how much I "talk" and how much I "listen" to others on social media.
    e. I don't have much of an online presence.

4. Which of the following best describes you?
    a. I have more friends and followers on social media than most people do.
    b. I gravitate to people on social media whose opinions are similar to mine.
    c. I sometimes cringe when I look back at things I have shared on social media.
    d. I use social media mostly to stay in touch with people I care about.
    e. I wonder how people find the time to keep up with social media.

INTERPRETING YOUR RESPONSES

For insight about your communication style, see which of the following best describes you. More than one may apply.

Publicist

If you answered "a" more than once, you realize the image-enhancing potential of social media. This can help you present yourself and others in a favorable light, as a good publicist would. Just be sure that you are genuine in what you post and that you don't get too caught up in pleasing others. Publicists sometimes struggle to balance superficial relationships online with close relationships in person.

Advocate

If you answered "b" multiple times, you often use social media to garner support for ideas and issues. At their best, advocates inspire others; at their worst, they can come off as bossy or domineering. The trick is to stand up for what you believe, but also listen to others and keep an open mind.

> **Emoter**
>
> If you answered "c" two or more times, you tend to post emotional messages. This can help you connect with others and build strong relationships. It can also get you in trouble. Emoters are sometimes hyperpersonal online in ways they may regret later. Avoid this pitfall by using the asynchronous nature of social media to your advantage. Speak from the heart, but pause before posting.
>
> **Empath**
>
> If "d" answers best describe you, you use social media to foster connections with others. You are likely to be an empathic listener who considers how your messages will be received, both in person and online. The segment on mindful listening in Chapter 6 will probably resonate with you.
>
> **Abstainer**
>
> If you saw yourself reflected mostly in "e" options, you aren't likely to annoy your online friends with minutia or self-serving messages. Don't abstain from social media entirely, however. There are advantages to being on board. For example, Chapter 10 presents strategies for building an online presence that will impress potential employers and others in the career world.

took off during the coronavirus pandemic. Between January and August 2020, the number of livestreams grew by 1,500 percent.[25] Social distancing measures meant more people than ever used livestreams to connect with others. The relative ease of "going live" also helps social media users connect with large audiences. Compared with prerecorded videos, livestreams generate six times as many comments and likes.[26]

Livestreaming presents little or no chance to control unexpected content. Content creators have been embarrassed or banned for using profanity and for inadvertently showing nudity or violence. And it pays to be skeptical about what seems real, even in a livestream. Twitch user Angel Hamilton (known to gamers as zilianOP) collected viewer donations after presenting himself as paralyzed from the waist down—until he forgot to turn off his camera at the end of a livestream. Viewers saw him rise from his wheelchair and walk away.[27]

2.2 Mediated Versus Face-to-Face Communication

Mediated communication is facilitated by technology. Social media is one type of mediated communication. Others are emails, text messages, direct messages (DMs), and telephone conversations. In some ways, mediated and face-to-face communication are quite similar. Both include senders, receivers, channels, feedback, and so on. Both are used to satisfy physical, identity, social, and practical needs, as described in Chapter 1. Despite these similarities, the two forms of communication differ in some important ways.

Message Richness

Social scientists use the term **richness** to describe the degree to which nonverbal cues can clarify a verbal message.[28] As Chapter 7 explains in detail, face-to-face communication is rich because it abounds with nonverbal cues that hint at what others mean and feel.[29] By comparison, message **leanness** is the lack of nonverbal cues. Mediated channels are much leaner than in-person communication.

> **richness** The degree to which nonverbal cues clarify and/or reinforce a verbal message.

> **leanness** The lack of nonverbal cues to clarify a message.

Because of their high potential for ambiguity, mediated messages are often subject to misinterpretation.

When have your mediated messages ever been misunderstood? What were the consequences?

To appreciate how message richness varies by medium, imagine you haven't heard from a friend in several weeks and decide to ask, "Are you okay?" Your friend replies, "I'm fine." You could probably tell a great deal more about what your friend really means from a face-to-face response than from a DM, text, or phone call. That's because in-person interactions involve a richer array of cues, such as facial expressions and vocal tone. By contrast, DMs and text messages are lean because they contain only words and possibly emojis. A phone conversation—containing vocal cues but no visual ones—would probably fall somewhere in between.

Ambiguity in Mediated Messages Because most mediated messages are leaner than the face-to-face variety, they can be harder to interpret with confidence. For example, irony and attempts at humor can be misunderstood. As a receiver, it's important to clarify your interpretation of a message before jumping to conclusions. As a sender, think about how to send clear messages (as much as that is possible) so that you aren't misunderstood. Adding phrases such as "just kidding" or an emoji such as 😊 to your posts and comments can help your lean messages become richer.

Identity Management via Mediated Channels The leanness of mediated messages presents another challenge. Without nonverbal cues, people can create idealized—and sometimes unrealistic—images of one another. As you'll read in Chapters 3 and 5, the absence of nonverbal cues can help communicators strategically manage their identities. Online, you don't have to worry about bad breath, blemishes, or getting tongue tied. Such conditions encourage what Joseph Walther[30] calls **hyperpersonal communication**, messages that accelerate the discussion of personal topics and relational development beyond what normally happens face-to-face. Young communicators tend to self-disclose at higher rates and share more emotions online than they do in person, often leading to a hastened (and perhaps premature) sense of relational intimacy.[31] This accelerated disclosure may explain why communicators who meet online sometimes have difficulty shifting to face-to-face relationships.[32]

Richer doesn't always mean better. When you want people to focus on what you're saying rather than on your appearance, leaner communication can be the way to go. Researchers in one study found that text-only online messages sometimes bring people closer because they minimize the perception of differences due to gender, social class, race or ethnicity, and age.[33]

Choosing a Communication Channel Social scientists use the term **polymediation** to address the range of options available to communicators when they are choosing how to engage with others.[34] For example, you could send a message through Snapchat, Instagram, or WhatsApp, or you could send an email or text. The channel you choose can make a big difference in how a message is received and understood. Evidence suggests that design students looking for jobs benefit from posting their photographs or drawings in addition to creating resumes, because words alone don't showcase their talents. In other circumstances, you might choose to not use words at all. *Transmediation* refers to recasting a message from one medium (e.g., written or spoken language) into other media (e.g., music, art).[35] If you've ever expressed

hyperpersonal communication The phenomenon in which digital interaction creates deeper relationships than those which arise through face-to-face interaction.

polymediation The range of communication channel options available to communicators.

regrets or sympathy by sending flowers rather than speaking or writing, you've practiced transmediation.

Synchronicity

Communication that occurs in real time, such as in-person talks or phone conversations, is **synchronous**. By contrast, **asynchronous communication** occurs when there's a lag between receiving and responding to messages. Voice mail messages are asynchronous. So are "snail mail" letters, emails, and Tweets.

When you respond to asynchronous messages, you have more time to consider your wording. You might even ask others for advice about what to say. But taking too long to respond to a text or DM can send a message, whether you mean to or not ("She hasn't texted me back. She must be mad."). And unlike most messages that are conveyed face-to-face or over the phone, ignoring asynchronous messages is pretty easy. No response is sometimes the best response, especially when you receive unsolicited and/or inappropriate texts or DMs.

synchronous communication Communication that occurs in real time.

asynchronous communication Communication that occurs when there is a lag between receiving and responding to a message.

Permanence

What happens in a face-to-face conversation is transitory. By contrast, the text and video you send via mediated channels can be stored indefinitely and forwarded to others. Sometimes permanence is useful.[36] You might want a record documenting your boss's permission to take a day off or a VRBO confirmation for the vacation rental you reserved online.

In other cases, though, permanence can work against you. It's bad enough to blurt out a private thought or lash out in person, but at least there's no visible record of your indiscretion. By contrast, a regrettable text message, Tweet, or post can be archived virtually forever. Even worse, it can be retrieved and forwarded in countless ways. It could even go viral. That's what happened to Naomi H. After landing a coveted internship at NASA, she tweeted, "EVERYONE SHUT THE F— UP. I GOT ACCEPTED FOR A NASA INTERNSHIP."[37] When a Twitter user commented "Language" beneath her Tweet, Naomi H replied, "Suck my d— and balls I'm working at NASA."[38] The Twitter user on the receiving end of Naomi H's rant turned out to be a member of the National Space Council, the organization that oversees NASA. Screenshots of their exchange went viral. As you might have guessed, Naomi H lost the internship.

Some mediated platforms, unlike Twitter, are designed to prevent message permanence. Snapchat, on which content typically disappears within a few seconds of being seen, is one of the most popular of these time-limited messaging apps.[39] The ephemeral nature of Snapchat encourages less inhibited communication, such as flirting and sexting (i.e., sending sexually explicit texts or images). Because of this, Snapchat generates higher

A single indiscretion could come back to haunt you years later.

What can you do to avoid online embarrassments?

rates of partner jealousy compared with more permanent social media platforms like Facebook.[40] One study found that among the 74 percent of American adults who say they've sexted, 38 percent of them did it over Snapchat.[41] It's no wonder Snapchat makes people jealous! Believing their risqué photos and flirty messages will disappear emboldens some people who would not otherwise sext to do so. However, the ability to take screen shots makes it a risky bet that photos and messages you want to disappear will truly vanish forever. Given risks like these, the best advice is to take the same approach with mediated messages that you do when communicating in person: Think twice before sharing something you might later regret.

2.3 Benefits and Drawbacks of Social Media

By now it's clear that social media can be a boon for connecting with others. You can also see that using these communication channels presents some risks. In the following section, you will take a closer look at the benefits and risks of social media.

Benefits of Social Media

Steve Jobs, the legendary cofounder of Apple Computer, once suggested that personal computers should be renamed "*inter*personal computers."[42] He had a point. Technology has great potential to bring people together and enhance the quality of their relationships. This is especially true when it comes to social media. During the coronavirus pandemic, for instance, social media was invaluable in helping family and friends stay connected while they limited their face-to-face interactions. Even before the pandemic, more than 80 percent of U.S. teens said social media helped them feel more connected to their friends.[43]

Romance Social media has also made it easier than ever to make romantic connections. In fact, social media in general and Tinder in particular occupies "a crucial place in dating habits around the world,"[44] especially among younger adults. Compared with older generations, adults under 30 are far less likely to go to bars, nightclubs, and other places to meet potential romantic partners. That doesn't mean they're not going out though. Tinder users go on about 1 million dates a week.[45]

One way to connect online is to "slide into" someone's DMs (i.e., send them a direct message on their social media). Two of the three Jonas Brothers (Nick and Joe) met their wives (Priyanka Chopra and Sophie Turner) via DMs on Twitter and Instagram, respectively.[46] Plenty of non-celebrities meet their spouses on social media too. About a quarter of married couples surveyed in 2019 met their spouses on a dating app.[47] They may be onto something. Couples who meet online stay together about as much as those who meet in person, tend to transition to marriage more quickly, and on average have happier marriages than couples who meet in person.[48]

Mediated channels can also sustain romance when partners must be apart because of jobs, school, or other factors. Text messages, video calls, DMs, and social media can help couples bridge geographic gaps.[49] But social media use is not always relationship friendly. As you will read in a moment, some types of online communication can jeopardize close relationships.

Using social media to connect with other people, whether for romance or friendship, has certain advantages. To start, social media (especially dating apps) reduces the stress of reaching out—no more working up the courage to approach someone in person just to be told they're not available or not interested. Ouch! Because people's profiles generally let you know what they're looking for (e.g., friendship, a long-term relationship, or a casual fling), you can approach with more confidence. Also, social media lets you cut to the chase: If certain attitudes or behaviors are "deal breakers" (e.g., political views or smoking), this information can usually be sussed out before a lot of time is invested (or wasted) getting to know someone.

Nowadays, couples are considerably more likely to meet online than to meet any other way.

How might beginning a relationship online differ from beginning one in person?

Friendship In the jargon of social science, mediated channels offer "low-friction opportunities" to create and maintain close relationships.[50] About two-thirds of teens say they have made new friends through social media sites, and 9 in 10 say they keep in touch with established friends via technology.[51]

College students cite "keeping in touch" as one of the main reasons for using social media.[52] The masspersonal nature of social media platforms makes them a logical choice for keeping in touch. Imagine that you want to share news about landing a good job. You can inform everyone with a single Tweet, and people can easily add comments (such as "Congratulations!" or "Well done!"). In this way, social media provides more maintenance "bang" for the message-sending "buck."[53] (Social media may have even helped you find and land a good job. See the feature "Using LinkedIn for Career Success" for more.)

Social Support Social media can also play a role in sustaining your overall social network. For example, Facebook groups and subreddits (special interest forms on Reddit) create virtual communities among people who would otherwise be strangers.[59] Whether you're a fan of Premiere League soccer, an avid environmentalist, or a devotee of punk rock, you can find like-minded people on social media. These virtual community members often provide social support for each other.

Communicating via social media can be especially helpful for people who are shy[60] or find it difficult to get out and about.[61] Mediated relationships can help alleviate feelings of loneliness, particularly for older adults.[62] Electronic communication isn't a replacement for the face-to-face variety, but it expands the world for those seeking connection beyond the people they already know.

Before social media, finding support for personal problems usually meant reaching out to friends, family members, or trusted members of the local community (e.g., teachers and religious leaders). Those personal contacts are still important, but social media now provides an alternative source of support for matters ranging from depression[63] to marital problems,[64] substance abuse,[65] suicide prevention,[66] and coping with senseless acts of violence.[67] Sometimes social media is the primary source of support for members of marginalized communities, such as transgender teens[68] and persons who are morbidly obese.[69]

COMMUNICATION STRATEGIES

Using LinkedIn for Career Success

With more than 800 million members, LinkedIn is the go-to social networking tool for advancing your career. If you are looking for a job, LinkedIn can match you with recruiters who are searching for someone with your skill set. With more than 57 million companies posting 15 million job openings on the platform each year,[54] you should probably consider joining if you are not already a user.

You can also use LinkedIn to build your personal brand, presenting yourself the way you want others to see you. LinkedIn is especially useful for expanding your professional network and connecting you with people who can help you succeed. Here are some tips for using LinkedIn effectively.

Write a Compelling Summary

In 2,000 characters or less, make users want to read about your education, professional history, skills, and accomplishments. This summary is likely to be the first section users see, so it's essential that you call attention to your qualifications and what sets you apart. Proofread everything you write. A single error can demolish your credibility.

Use Key Words in Your Profile

Your LinkedIn profile functions as an online resume. When writing your profile, think about what skills, experience, and personal traits you want to highlight as well as the key words users are likely to search for. Look at job descriptions and other profiles as points of reference. Your profile can also serve as a portfolio of your work. When appropriate, incorporate multimedia samples to showcase your skills and accomplishments.

Keep It Professional

Make your profile public so anyone can view it, and be sure to include a headshot that looks professional. Profiles with photos get about seven times more visits than those without pictures.[55] Don't use a selfie though! Instead, have someone take a picture of you in a well-lit space. Make sure you're dressed for success—business attire is best. You'll learn more about the importance of clothing in Chapter 7.

Seek (and Provide) Positive Recommendations

Post endorsements from professors, colleagues, supervisors, or clients. It's fine to ask them to highlight qualities you want to promote: your attitude, skills, achievements, professionalism, and ethics. Consider paying it forward by recommending others, especially when users may find your perspective valuable.

Expand and Engage Your Professional Network

Use LinkedIn to connect with people who can enhance your career success. Sending personalized invitations to connect are often more successful than sending the default invitation. And while it's important to have contacts, don't mistake quantity for quality. You may impress some people if you have thousands of contacts, but others might think you are superficial.

Rather than focus on the numbers, researchers suggest you focus on engagement. Regular, frequent participation in discussions is more strongly associated with career benefits than the size of a person's network.[56] Make comments on posts, share articles important in your field, and join groups.[57]

LinkedIn is an indispensable tool for many users, helping grow professional networks and advance careers. With 87 percent of job recruiters consulting LinkedIn during the hiring process,[58] it's important for job seekers to have a LinkedIn presence. Follow the aforementioned tips and you'll be on your way to leveraging LinkedIn's power. It may even help you land your dream job!

When people are asked why they turn to social media for support, a common response is that they feel more comfortable talking with like-minded others with whom they have few formal ties—particularly when the issues are embarrassing or stigma laden.[70] The words of one recovering alcoholic demonstrate the value of social support. After reaching out for help,

> I began to get emails, phone calls, text messages, Tweets and other digital notes from people around the world. Some offered kind words. Some offered support. Many people shared their own stories of addiction. In my darkest times, these notes would come. And always, without question, they pulled me back from the brink.[71]

During the coronavirus pandemic, millions of people on lockdown would have been unable to give or receive social support without social media.

How did the pandemic change your social media use?

Because online support groups are relatively anonymous and the participants are similar, they can offer help in ways that make strangers seem like close friends.

Advocacy and Fundraising The "mass" dimension of social media has dramatically increased the power of individuals and informal groups to change society via communication. Consider the effects of the many movements that have spread via social media, including School Strike 4 Climate. This particular movement began in 2018 when 15-year-old activist Greta Thunberg skipped school to protest government leaders' failure to address global warming. Her Tweets and Instagram posts were shared widely, prompting students around the world to go on strike as well. A million and a half people had joined the movement by 2019, staging Friday protests using the hashtag #fridaysforfuture.[72]

Hashtags link social media content to specific themes, posts, and comments. Like keywords, hashtags are searchable, which makes it easy for people to find and contribute to topics and conversations they find interesting. Hashtags can also be followed. This means everything posted under a certain hashtag will show up in your feed, like #fridaysforfuture, even if you don't know or follow the people making the posts. Because hashtags aggregate content from disparate users, social media helps activists develop a collective identity. Researchers stress the importance of a cohesive, shared identity if a social movement is going to be effective.[73]

Beyond advocacy, social media also provides a way to raise funds quickly in support of a candidate or cause. You may remember the Ice Bucket challenge. Back in 2014, the social media phenomenon raised $115 million for amyotrophic lateral sclerosis (ALS) research. Spurred on by hashtags such as #ALSIceBucketChallenge and #IceBucketChallenge, it was among the first fundraising challenges to go viral.[74] Another example is the crowdfunding platform GoFundMe, which has facilitated over $9 billion in donations for a wide range of causes—personal, professional, and political.

Drawbacks of Social Media

Despite all its advantages, social media isn't a replacement for face-to-face interaction. Along with the potential benefits, social media can have a dark side.[75] Understanding the potential drawbacks can help you guard against them.

Superficial Relationships Social scientists have concluded that most people can only sustain about 150 relationships.[76] (That figure has been termed "Dunbar's number" in recognition of Oxford University anthropologist Robin Dunbar, who established it.) If you're lucky, you have an inner circle of five "core" people and an additional layer of 10 or 15 close friends and family members.[77] Beyond that lies a circle of roughly 35 reasonably strong contacts.[78] That leaves about 100 more people to round out your group of meaningful connections. You almost certainly don't have the time or energy to sustain relationships with many more people.

Dunbar's number is much smaller than the array of "friends" that many people boast on social media sites. You've probably seen Facebook or Instagram accounts with thousands of friends or followers. Dunbar explored the discrepancy between "true" and mediated friends by studying the online interactions of those who have thousands of friends and those who have far fewer friends.[79] He discovered that, regardless of how many online friends users had, they only maintained relationships with the same number of people—roughly 15.

A large number of online "friendships" can actually yield diminishing returns. You may impress others if you have a few hundred Facebook friends, but research shows that as that number doubles or triples, you're likely to be perceived as shallow or desperate.[80] Some scholars have suggested that seeking an unrealistically large number of social media friends might be compensation for low self-esteem.[81]

Mental Health Issues Researchers have known for some time that heavy internet use is linked with conditions such as depression, loneliness, and social anxiety.[82] The same goes for social media. Additionally, too much time on social media can lead to sleep deprivation[83] and poor academic performance,[84] lower self-esteem and body satisfaction,[85] and altered perceptions of what constitutes an ideal body.[86]

Time spent looking at others' seemingly "perfect" social media profiles, bios, and pictures can make viewers feel inferior by comparison.[87] Of course, a lot of the "perfect" pictures making people feel bad about themselves are heavily edited. Many social media apps have features that erase wrinkles, banish blemishes, whiten teeth, and contour faces. A consequence is that in addition to comparing ourselves to seemingly perfect others, we're now comparing ourselves to ourselves! Some people have actually gotten plastic surgery to more closely resemble Snapchat versions of themselves.[88]

Another condition affecting social media users is called *Facebook Addiction Disorder* (FAD), although it can apply to all forms of social media. It's estimated that about 210 million people are addicted to the internet and social media.[89] Symptoms include depression, anxiety, insomnia, and stress.[90] Therapists specializing in addiction explain

"It says no one really knows who he is, but that he's got 400,000 followers on Twitter."

Source: Cartoonist Group

that receiving likes and positive comments activates the brain's reward center and provides an exhilarating but transient boost in self-esteem.[91] For people with FAD, getting that feel-good boost requires ever increasing amounts of time on social media.

For some people, social media can lead to social isolation. Loneliness is correlated with what social scientists call a *preference for online social interaction*.[92] Some users prefer mediated interactions, which can detract from in-person relationships, resulting in feelings of loneliness.[93] In turn, feelings of loneliness can drive people to seek online connections even more. It can be a vicious cycle.

A few factors help explain why online communication can crowd out face-to-face interaction. One involves social skills—or more accurately, a lack of them. People who are nervous or anxious about communicating in person may feel more comfortable doing so through mediated channels[94] that allow them to edit their thoughts and transmit them when and how they want. They can also craft online identities in an attempt to seem more attractive or interesting.

The bad news is that, whereas positive interactions on social media can help lonely and socially anxious people feel good about themselves, disconfirming feedback can make them feel worse.[95]

Plastic surgeons use the term "Snapchat dysmorphia" to describe social media users who are preoccupied with perceived flaws in their physical appearance. Severe emotional distress and lower quality of life are common among sufferers who obsess over their looks and overuse photo editing apps.

Do you use filters or edit photos of yourself that you post to social media? What are some advantages and drawbacks of using photo editing apps?

Harm to Relationships Earlier in the chapter, you read about ways social media can bring people together and help them stay connected, but research shows social media can just as easily wreak havoc on relationships. Scholars know from survey data that about a quarter of people are bothered by how much time their significant others spend on social media.[96] About the same number say their romantic partners' social media interactions have made them jealous. As people spend more time on social media, their partners often report less interpersonal intimacy[97] and more relational conflict (this is often due to jealousy, spying, and sometimes infidelity[98]). Feeling jealous may lead partners to spend more time on social media themselves,[99] which can create even more jealousy in the relationship.

It is important to note that *people* cause relationship problems; social media in and of itself is just a tool. Granted, it's a powerful tool that enables and, at times, encourages questionable behavior (e.g., sexting over Snapchat). Understanding the risks may help you use social media more responsibly, thereby enhancing your mental health and benefiting rather than harming your relationships.

Deception There's a pretty good chance that you will be catfished at some point. Almost certainly, someone you know will be catfished. Maybe someone you know *is* a catfish?

"Catfishing" is deceiving others using fake online personas. Nev Schulman coined the term in his 2010 documentary of the same name. In the film, Schulman discovers that the woman he fell for online was not who she claimed to be.[100]

In another high-profile catfishing case, a Notre Dame football star, Manti Te'O, captured the nation's attention when just moments before a big game, he learned that both his grandmother and his girlfriend had died—his grandmother was real, his girlfriend wasn't. Te'O had been in a romantic relationship with "Lenna Kekua" via Facebook posts, text messages, and phone calls over a two-year period.[101,102] For Te'O, it was true love, but for Ronaiah Tuiasosopo, the man behind the hoax, it was a cruel game.[103] When Tuiasosopo decided to end things, he gave Te'O's "girlfriend" terminal cancer (he even went as far as staging a funeral).

Not all online deceptions play out as spectacularly as those experienced by Schulman and Te'O. And not all deceptions qualify as catfishing (you don't need a fake account to fool people). Online misrepresentation is common. That's not surprising since it's possible to craft deceptive messages that wouldn't be possible in person. For instance, online dating profiles might use edited photos or contain outright lies underreporting someone's weight or overreporting their height.[104] Sometimes people declare they are single when they're actually in a romantic relationship, and others salt their LinkedIn profiles with jobs they never held. Given the unreliable nature of online self-characterizations, it's probably a good idea to view them with at least a little skepticism.

Here are a few tips to help you detect catfishing:

- If someone you've been corresponding with for a while won't Zoom, Skype, or Facetime with you, be wary.
- Early in the relationship, a catfish may declare feelings of love and/or sexual attraction. Flattery lowers your guard, making you more susceptible to manipulation.
- A catfish may talk about high-paying jobs or powerful connections to impress you, or they may recount hardships and setbacks to win your sympathy. If their story sounds fishy to you, consider curtailing your interactions.
- If you suspect a catfish, check their social media accounts. If they haven't posted much or have few friends, their accounts might be fake.
- Reverse-search photos shared with you to see if they are associated with other people's names or social media accounts.

online surveillance Discreet monitoring of the social media presence of unknowing targets.

cyberstalking Ongoing obsessive and/or malicious monitoring of the social media presence of a person.

cyberbullying A malicious act in which one or more parties aggressively harasses a victim online, often in public forums.

Stalking and Harassment You probably have searched the internet to find out more about an intriguing stranger, a former friend, or a romantic partner. **Online surveillance** is a discreet way of monitoring the social media presence of unknowing targets.

Although occasional online surveillance is relatively harmless, it's a problem if it escalates into an unhealthy obsession or full-blown **cyberstalking**.[105,106] In most cases, cyberstalkers are males monitoring their female exes.[107] People who discover that they're being cyberstalked can suffer the same types of mental and emotional trauma experienced in offline stalking. If you believe you're under unwanted surveillance by someone you know, you can alert legal authorities and victim assistance professionals. You also might want to consider getting off social media until you feel safe again.[108]

Another form of online harassment is **cyberbullying**, a malicious act in which one or more parties aggressively harass a victim online, often in public forums.[109] Cyberbullies can post hateful messages on social media, circulate false rumors, send disparaging texts and emails, and distribute photos of their victims without consent. Cyberbullying has been linked to a variety of negative consequences, including poor academic performance, depression, withdrawal, psychosomatic pain, drug and alcohol abuse, and even suicide.[110] Middle school is the peak period for cyberbullying (nearly 6 in 10 teens are victims[111]), but it can start as early as grade school and continue into college.[112] Around 40 percent of adults say they have been bullied online.[113]

An alarming number of adults have endured a particularly insidious form of cyberbullying—*revenge porn*, posting sexually explicit images of people without their permission. Survey data show that 1 in 8 social media users have been targets of revenge porn; about 75 percent of victims are women.[114]

Non-consensual sharing of content can have disastrous consequences, especially when the content is intimate or sexual in nature. Consider Noelle Martin's story. When Martin was 18, she discovered that some of her social media pictures had been copied, edited to appear pornographic, and then uploaded to thousands of porn sites and forums. Worse yet, many of the convincingly real-looking pictures bore her full name and address! Even though Martin's original images were not sexually explicit, she worried the realistic-looking fakes would damage her online reputation and threaten her future employment prospects.[115]

Most of the time, revenge porn involves real images (e.g., "nudes" intended for romantic partners). Sharing nude images is a routine part of courtship for many couples, but, in the heat of the moment, people sometimes fail to consider what will happen to those images if the relationship ends.[116] Many victims of revenge porn suffer from posttraumatic stress disorder, anxiety, and depression, and have suicidal thoughts.[117] In some tragic cases, victims have taken their lives.[118] Even without such dire consequences, it's not hard to imagine the unpleasant ramifications of a private photo or text going public.

Misinformation Close to 70 percent of Americans say they get news from social media.[119] Facebook, Twitter, and Reddit are the most popular platforms overall. The problem is that a lot of so-called "news" on social media isn't actually news. Misinformation and conspiracy theories are often packaged to resemble credible reports.[120] For instance, a news item you see on social media might use the same fonts and branding as *The New York Times* or CNN, making it seem legitimate.

When people share junk news items within their social networks, they unwittingly spread misinformation. The items most often shared are usually negative and/or about controversial topics.[121] This is problematic because people tend to trust the information their friends and peers share with them.[122] During the coronavirus pandemic, researchers discovered that COVID-19 misinformation was shared at higher rates than news from credible health sources.[123] In this instance, junk news spread on social media posed a significant risk to public health.[124,125] Other studies have found similar patterns play out in the political sphere, especially during election years when junk news (usually consisting of polarizing content and conspiracy theories) is tweeted more often than news from credible sources.[126]

If you or someone you know has been a victim of revenge porn, StopNCII.org can help. It's a free service run by Revenge Porn Helpline that helps identify and remove nonconsensual intimate images from the internet.

COMMUNICATION STRATEGIES

Evaluating (Mis)information

You probably wouldn't believe headlines like "Rock & Roll Hall of Fame Under Pressure to Return Looted Ancient Mesopotamian Stratocaster" or "Adele Postpones Vegas Residency Rather Than Give Up Seat at Hot Slot Machine." Only the most gullible readers would fall for these stories, which were posted by a news parody site called *The Onion*. (For the record, Stratocaster guitars weren't around 3,000 years ago and Adele's Vegas residency was delayed because of COVID-19). If all fake news were so easy to spot, misinformation would be a nonissue. Instead, as one American lawmaker described it, "misinformation in the modern age is maybe the greatest threat that we face."[128]

In an environment where anybody can publish claims, no matter how bogus, it's vitally important to separate solid information from information that is either intentionally or unintentionally inaccurate or misleading. These tips can help you think critically about content you encounter online and on social media.

Google It

If you run across a news item on social media that gives you pause, google it. Are several other reputable news agencies also covering the story? If so, it is (or was) likely legitimate. You could be seeing old news however, so check the dates.

Stories published years ago are regularly recirculated online as if they are new. Sometimes this is done deliberately to mislead readers, many of whom may not notice publication dates before reading or sharing old stories.[129] Taken out of context, old articles can fuel discord. For example, old news about gasoline and a note written in Arabic found near Notre Dame cathedral was widely circulated on Facebook after fire destroyed much of the cathedral in 2019. The articles, however, were three years old and unrelated to the fire. They were resurfaced primarily to ignite anti-Muslim sentiments.[130] Do your due diligence and check a story's date before you click "share."

Consider the Source

Many false news stories are hard to spot because they come from sites that disguise themselves as legitimate. For example, a site branded abcnews.com.co could easily be confused for the real *ABC News* (the actual URL for *ABC News* is abcnews.go.com). An authentic-appearing *Boston Tribune* website whose "contact us" page lists only a Gmail address should strike you as suspicious. As should misspellings, grammatical errors, words in all caps, and lots of exclamation points ("THIS IS NOT A HOAXE!").

Some fake sites are convincingly real, though. At least a thousand websites in the United States masquerade as local news media with credible-sounding names like *East Michigan News* and *Grand Canyon Times*.[131] Funded by political interest groups, these sites publish biased, agenda-driven stories (some are completely fabricated while others misrepresent actual events). Dead giveaways include featured articles attributed to bogus news services with names like Metric Media News Service, Local Labs News Service, and Franklin Archer, as well as numerous links to (fake) local news outlets.[132] Another clue to watch for is a "contact us" page with no physical address or phone number.

Check the Author

Another tell-tale sign of a fake story is often the byline. Sometimes a search of the writer's name will reveal that they have dubious or nonexistent credentials. Another tipoff is seeing the same writer's name on most or all of the stories on a particular site.

It's also worth checking into the writer's sources. Are the people interviewed qualified to speak on the subject at hand? Look up their credentials. Does the writer only quote one person? That's a red flag.

Consult Fact Checkers

Plenty of services can help validate or debunk stories that appear in your news feed. Among the best are FactCheck.org, Snopes.com, the *Washington Post* Fact Checker, and PolitiFact.com.

Do a Reverse Image Search

Seeing is believing, right? Well, not necessarily. These days, photo manipulation is common. (If you have ever used a Snapchat filter, you know how easy it is to alter an image.) If you suspect a photo in a news story has been doctored or used out of context, do a reverse image search on Google. You'll be able to see if the photo appears elsewhere and ways in which it may have been altered. You can conduct reverse image searches on TinEye and REVEye as well (these sites come in handy if you're trying to uncover catfish or fake social media accounts).

Watch for Deepfakes

Deepfakes are videos made using a type of artificial intelligence called deep learning (hence the name). Simple versions can be made using apps such as Reface and Zao that let you insert yourself into music videos, movies, and TV shows by mapping your face onto someone else's body. Convincing looking and sounding deepfakes require more sophisticated software and processing power (google "Jon Snow apologies for Season Eight of *Game of Thrones*" for an example). By combining elements of real and computer-generated footage, deepfakes run the gamut from fake porn to political leaders hurling insults at one another. The real worry is that plausible deepfakes may influence voters, impact economies, and sow division.[133] If you suspect a video is a deepfake, there are a few things to watch for: flickering near the edge of transposed faces, reflections in the eyes which are inconsistent or irregular, and digital artifacts over fine or detailed items (such as strands of hair or jewelry).[134]

Check for Biases

Recognizing your own lack of objectivity can be difficult. What's known as confirmation bias leads people to trust information that reinforces their existing beliefs and discount information that doesn't. Ask yourself how someone less disposed to your belief system would react to information you accept uncritically.

Go Directly to Credible News Sources

Rather than rely on social media to stay abreast of current events, get your news straight from the source. The Associated Press and Reuters are good sources for trustworthy, impartial news. ABC, NBC, CBS, and *The Wall Street Journal* rank among the most trusted news sources in the United States.[135] AllSides is also a good source. It presents news from left, centrist, and right political perspectives giving you a more complete picture of the issues.

Just remember that not everything you see on a cable news channel is news. Shows like *Anderson Cooper 360* on CNN and *Tucker Carlson Tonight* on Fox do not offer objective reflections of the day's news, only the hosts' opinions.

Discerning fact from fiction can be really hard, especially when so much misinformation is so expertly crafted. Following these tips will help.

A consequence of getting news from social media is that you probably see a narrow range of topics and points of view. This is because most people's social media networks are composed of friends and followers with whom they share similar beliefs and values. On top of that, Meta (Facebook's parent company) organizes users' feeds by giving priority to content posted by their friends rather than to news posted by journalists or news media.[127] These factors can contribute to the formation of ideological echo chambers fueling political and social polarization. The feature "Evaluating (Mis)information" offers tips to help you evaluate the quality and accuracy of information you find online, particularly on social media.

Pink slime journalism describes false, agenda-driven stories published by partisan media companies masquerading as local news. It gets its name from filler used in processed meats—a pink paste made of finely ground animal byproducts.

2.4 Influences on Mediated Communication

Who you are can affect the way you use social media and other forms of mediated communication. Factors such as race, socioeconomic status, and educational attainment play a role, but two of the strongest influences are gender and age.

Gender

As in face-to-face interaction, gender influences how people communicate using social media[136,137]—from their motivations for going online and the platforms they prefer down to the words they post. For instance, when compared with men, women tend to seek out and give social support on social networking sites in far greater numbers.[138] They are also more likely to exhibit signs of social media addiction.[139] Men, on the other hand, are more apt to become gaming addicts.[140] Men use Reddit and LinkedIn more than women do, while women outnumber men on Facebook and Instagram.[141,142] In general, men use more profanity than women do and are more likely to make object references (talking about things rather than people). Women use more emotion-related words, personal pronouns, verbs, and hedge phrases ("I think," "I feel").

For women, nonbinary individuals, and members of marginalized communities, social media is an important way to engage in public discourse. The downside is that they are often harassed, bullied, and threatened, especially when their social media posts contradict normative power structures that have traditionally privileged White, heterosexual male perspectives.[143,144] Even benign content can invite scrutiny if it's posted by a woman. For example, women regularly receive more disparaging comments on YouTube than men do, even when the videos they post are similar.[145] Negative comments targeting women YouTubers often include threats of physical and sexual violence.[146] It's probably no surprise then that numerous studies find women and girls suffer a disproportionate share of social media's ill effects: depression, anxiety, poor body image, and low self-esteem.[147,148] A 2021 press leak revealed that Facebook's own in-house researchers concluded that using Instagram harms teen girls.[149]

When it comes to gender equality online, social media can be a double-edged sword. It causes many women and nonbinary people harm while at the same time buffering against the negative effects of harassment. This is the case when people connect with and receive support from like-minded others. As you read earlier in the chapter, enabling social support is one of social media's benefits.

Age

As you have likely observed, different generations prefer using different communication channels. If you were born after the early 1990s, using social media probably feels as natural as breathing. It's a different story for people who grew up in a world without internet and smartphones, technology that most take for granted today.[150] Older communicators are more likely to prefer phone conversations, and teens are more likely to text and consider phone calls annoying or intrusive.[151,152,153] Contemporary parent–child arguments may include questions such as "Why don't you just call?" followed by "Why don't you just text me back?"

Although younger communicators use social media more than older communicators, the gap is narrowing. The number of older adults who use Facebook has more than doubled since 2012.[154] At the same time, Facebook's popularity with younger

Source: Rina Piccolo Cartoon used with permission of Rina Piccolo and the Cartoonist Group. All rights reserved.

users has waned—Millennials use Instagram and Snapchat more than all other social media platforms,[155] while Gen Zers favor YouTube and TikTok.[156]

How people use punctuation in their mediated messages varies with age, and that can sometimes cause confusion. When using social media, older communicators tend to use formal punctuation rules they learned in school, which gives them a distinct "digital accent." One example involves the use or nonuse of punctuation such as periods.[157] Whereas older communicators may commonly use periods, lifelong digital communicators may interpret periods in texts, tweets, and DMs as a sign of negative feelings.[158] Other language mechanics in texts can also signal age, as one person says of her dad: "[he] signs his texts 'ILY, Daddy,' as if I didn't know who was texting me in the first place." ILY as an acronym for "I love you" is not a shortcut younger people would use.[159]

2.5 Communicating Competently with Social Media

Perhaps you've found yourself in situations like these:

- You want to bring up a delicate issue with a friend, family member, or colleague. You aren't sure whether to do so in person, on the phone, or through text.
- You're enjoying a film at the theater—until another moviegoer starts livestreaming.
- A friend posts a picture of you online that you would rather others not see.
- Someone you care about is spending too much time on social media sites and having fewer face-to-face interactions.
- You receive so many comments and DMs that you have trouble responding to all of them and staying on top of your school and work deadlines.

None of these situations would have existed a generation ago. They highlight the need for a set of social agreements that go beyond the general rules of communicative competence outlined in this chapter. Although the guidelines offered here won't cover every situation involving social media, they can help you avoid some problems and deal with others that are bound to arise.

Maintaining Positive Relationships

Goodwill and respect are important in both face-to-face and mediated relationships. But when it comes to building and maintaining relationships, the leanness of online messages can muddy the waters. It's important to be a competent communicator online and off. Here are a few suggestions for helping make the most of social media exchanges.

Don't assume you understand. Before jumping to conclusions about the meaning of a message or the intentions of the sender, consider alternate interpretations. Instead of interpreting ambiguous messages in the worst possible way, start by taking other people's words at face value, assuming that what they *actually* say is sincere. Ask the other person for clarification. The skill of perception checking (Chapter 3) provides a template: "When you posted a barfing emoji after my comment, I thought you were mocking what I said. But maybe you meant it as a joke. I'm confused—what *did* you mean?"

Seek common ground. Start discussions by looking for beliefs you share instead of focusing on differences. You may be surprised to find that disagreements aren't as stark as they first seem. If you're debating politics, for example, try to find areas where you agree with others—making the world better for the next generation, seeking fairness, reducing economic inequality, and so on. Even if you still disagree, you can learn by exploring the thinking behind positions that differ from your own. In some cases, however, it may be best to walk away from debates on social media, especially if comments become abusive or threatening.

disinhibition The tendency to transmit messages without considering their consequences.

Keep your tone civil. If you've ever shot back a nasty reply to a text or DM or posted a mean comment, you know that it's easy to behave badly when the recipient of your message isn't right in front of you. This explains why abuse is more likely in mediated channels than in face-to-face contact.[160] According to researchers, lack of eye contact is the number-one factor contributing to online hostility and disinhibition[161] (**disinhibition** is the academic term for transmitting messages without considering their consequences). When people attack others through online channels it's called **trolling**.

trolling Attacking others via online channels.

The temptation to troll can be strong, especially when you're angry, but psychologist Maria Konnikova has this advice: Be like Lincoln.[162] When Abraham Lincoln was angry, he channeled his negative emotions into what he called a "hot letter" and then he set it aside. Never mailed nor read by the addressees, the hot letters provided catharsis without the repercussions of confrontation. Lincoln vented his anger, cooled off, and moved on. Rather than putting quill to paper, you might try composing a hot email or text instead. Just don't hit "send."

Respect others' need for undivided attention. If you have tried having a conversation with someone who paid more attention to their phone than to

Hostile, abusive language, usually sent anonymously, is a sad fact of online life. When you encounter these sorts of messages, the best strategy is to not respond.

How would you rate your own online civility using the tips outlined in this section?

you, then you know how infuriating phubbing can be. The term "**phubbing**" blends "phone" and "snubbing." It describes the act of snubbing people in one's physical presence by paying attention to one's phone instead of talking with others.[163] Phubbing often leads to feelings of exclusion, resentment, anger, and jealousy as well as diminished trust and lower relational satisfaction. It happens with alarming regularity: Half of people in committed relationships say they are often or sometimes phubbed by their partners.[164]

Chapter 6 has plenty to say about the challenges of listening effectively when you're multitasking. Even if you think you can interact with others while dealing with mobile devices, it's important to realize they may think you are being rude. As one person put it, "While a quick log-on may seem, to the user, a harmless break, others in the room perceive it as a silent dismissal. It announces: 'I'm not interested.'"[165] Social media analyst Sherry Turkle suggests creating "device-free zones." By committing to set devices aside at the dinner table, in the car, or in the living room, you can carry on a conversation without distractions.[166]

Using mobile devices in social settings can be distracting and annoying, but sometimes it's unavoidable. Perhaps you are expecting an important call or someone is trying to reach you in an emergency. If you must check your devices, do so discreetly. If you need to take a call or respond to a message, politely excuse yourself. When possible, make calls in private. Use headphones or earbuds (if you have them), and lower your screen's brightness to avoid disturbing others.

> **phubbing** A mixture of the words *phoning* and *snubbing*, used to describe episodes in which people pay attention to their devices rather than to the people around them.

Protecting Yourself

Respecting others and avoiding misunderstandings are only part of competent online communication. You also need to keep yourself safe. The following sections offer tips on how to do so.

Be Safe. As a rule, don't post information on social media that you wouldn't tell a stranger on the street. The safest bet is to assume that your posts can be seen by unintended recipients, some of whom you may not know or trust. There are a few things you can do to protect yourself online:

- Consider changing your default privacy setting to "Friends Only." This affords you *some* privacy, but remember "friends" can share your posts with others.
- Review your friends list on a regular basis so you know who you are sharing information with. Consider "unfriending" strangers.
- If a stranger sends you a friend request, the safest thing you can do is ignore it.
- Be careful about sharing your birthdate and where you were born. Too much information on your profile can make you an easy target for identity thieves.
- If you're being bullied online, keep copies of harassing messages. Block the offender. If the harassment continues or if you feel unsafe, contact law enforcement.

Be Careful What You Post. You don't want regrettable social media posts coming back to haunt you. As a cautionary tale, consider the case of 10 students whose college acceptance offers were rescinded after university officials discovered offensive content that they posted in a private Facebook group.[167] In another case, a student's admission was denied because of racist comments he posted years earlier.[168] At least 12 universities rescinded offers in 2020 because of students' social media posts.[169] And don't forget about Naomi H. whose profanity-laced Tweet cost her a NASA internship.

A social media "detox" can help you feel happier and more connected to the world around you. Experts recommend taking regular breaks, even if it's only for a few hours at a time.

How much time do you spend on social media each day? How might you otherwise use that time?

Don't Believe Everything You See. People tend to believe that what others communicate is true. That's because people *are* honest, most of the time.[170] The "default to truth" mindset means people generally aren't looking for deception. This may help explain how Manti Te'O was catfished for two years!

In addition to catfishing, you are likely to encounter junk news as well. Social media is a breeding ground for misinformation, so you should evaluate the news items you read carefully. Refer back to the feature "Evaluating (Mis)information" for a refresher on evaluating information you find online and on social media.

Limit the Amount of Time You Spend Online. It's easy to make a case that many relationships are better because of social media. And as you've already read in this chapter, some research supports this position. But even with all the benefits of social media, your own experience probably supports research saying that too much time online is unhealthy. By some estimates, most people swipe, stroke, tap, or click using their smartphones hundreds or even thousands of times a day.[171] Although social media may tune us in to people who are not nearby, it can rob us of presence in the current moment.

One way to reveal the impact of social media on your life is to try a "detox" in which you stay off your devices for a period of time. Evidence from one experiment showed that people who quit using social media for a month spent less time using the internet and more time socializing with family and friends. They also reported feeling better overall.[172]

If you are worried about your social media use:

- Keep track of the amount of time you spend online so you can accurately assess whether it's too much.
- Plan to spend a limited amount of time each day on social media and see if you can stick to your plan.
- Don't reach for a device every time you get a free moment. Instead, take stock of what you are seeing, feeling, smelling, and hearing. It will make you more attentive to the people and things around you.
- Make a list of problems in your life that may have occurred because of your social media use.
- Share your feelings in person every so often. Rather than posting or tweeting your thoughts, consider sharing them face to face.
- If you do not feel able to change your behavior on your own, seek the help of a counselor or therapist.

Taking regular breaks from social media can be invigorating. Christina Farr, whom you read about at the start of the chapter, said that within a few weeks of abstaining from social media she started to feel happier and lighter.[173] And with all that extra time on her hands, she began volunteering for a crisis support center and taking French lessons. Farr recommends everyone try detoxing from social media (at least once). Valuable advice for using social media comes from Meta technology executive Andrew Bosworth: Think of it as sugar—best enjoyed in moderation.[174]

MAKING THE GRADE

OBJECTIVE 2.1 Identify the distinguishing characteristics of social media, which combine elements of mass and personal communication.

- Social media is distinguished from other forms of communication by user-generated content, variable audience size, and interactivity.
- Uses and gratifications theory explains that social media serve the following functions: social interaction, providing information, passing the time, and entertainment.
- Social media can be "masspersonal," combining elements of both mass and interpersonal communication.
    > How does your use of social media reflect the characteristics of user-generated content, variable audience size, interactivity, and collaboration?
    > For a 1-week period, describe how your use of social media reflects the following uses and gratifications: social interaction, providing information, passing the time, and entertainment. What other functions does using social media serve?
    > To what extent is your social media use "masspersonal" in nature?

OBJECTIVE 2.2 Outline cases for using social media versus face-to-face communication, noting important differences.

- Mediated communication is less rich than face-to-face interaction. For that reason, it is more prone to ambiguity. When asynchronous, mediated communication is easier to use for managing one's identity. The permanent nature of mediated content means that regrettable behavior can have long-lasting repercussions.
- Paradoxically, mediated communication can be hyperpersonal, accelerating the discussion of personal topics and relational development beyond what normally happens in face-to-face interaction.
- Keeping these characteristics in mind can help communicators decide which communication channel is most appropriate and effective in a given situation.
    > Apply the information in this section to a representative sample of your communication, explaining in each case what communication channel is most effective.
    > Describe a situation in which choosing a different communication channel might have led to better results.

OBJECTIVE 2.3 Identify the benefits and drawbacks of mediated communication, and use this knowledge to maximize effectiveness and minimize the potential for harm.

- Use of social media has the potential for great benefit, including sparking romance, facilitating friendships, and providing opportunities to connect with others you might not see face to face. In addition, using social media can help sustain and enrich personal relationships, provide social support, and promote social advocacy and fundraising.

- The drawbacks of social media use include potential harm to offline relationships and increased risk of depression, loneliness, and social anxiety. It's more difficult to detect deception in mediated relationships, and social media can provide an avenue for abusive behavior including stalking, harassment, and cyberbullying. Social media also enables misinformation to proliferate, which is often difficult to detect.

  > How is your life richer from using social media? How much do the downsides described in this section affect your personal and relational well-being?

OBJECTIVE 2.4 Describe how factors such as gender and age can influence social media use.

- Social media content, platform preference, word choice, and likelihood of harassment vary by gender.
- People raised with the internet differ significantly in their orientation toward social media than do those who learned online communication as adults. Even for "digital natives," the content of mediated communication changes according to age.

  > Interview a digital immigrant who lived in the world before social media. How different has the digital immigrant's life been in terms of the benefits and risks outlined in this section?

  > Use the material in this section to compare your use of social media with the use of classmates or others with different gender orientations. Based on your findings and the information in this section, how can you adapt your communication when interacting with people from different demographic groups?

OBJECTIVE 2.5 Explain how to use social media competently to maintain positive relationships, minimize misunderstandings, and protect yourself.

- Maintaining positive relationships online comes from using a civil tone, preserving privacy boundaries, respecting others' needs for undivided attention, and being mindful of bystanders.
- Self-protection includes using mobile devices safely, being careful about what you post, being a critical consumer of online information, and balancing mediated and face time.

  > Ask several of your social media contacts to evaluate how well you follow the guidelines for maintaining positive relationships. Based on the feedback you receive, how can you become a more competent user of social media?

  > Use the guidelines for staying safe to evaluate how well you protect yourself when using social media. What are your strengths? In what areas do you need to improve?

KEY TERMS

asynchronous communication p. 29
cyberbullying p. 36
cyberstalking p. 36
disinhibition p. 42
gatekeeper p. 22
hyperpersonal communication p. 28
leanness p. 27
masspersonal communication p. 24
online surveillance p. 36
phubbing p. 43
polymediation p. 28
richness p. 27
social media p. 22
synchronous communication p. 29
trolling p. 42
uses and gratification theory p. 24

PUBLIC SPEAKING PRACTICE

Describe a time when you posted something to social media that you later regretted posting. What happened? How did the post affect you and/or others? Prepare to share your experiences in a brief oral presentation.

ACTIVITIES

1. Come up with a list of do's and don'ts for communicating on social media.

   - What social media practices would you list as rude or ineffective. Why?
   - What social media practices would you encourage? Why?
   - What do you think would happen if everyone followed your advice?

2. Write down how much time you think you spend on social media. Check your phone to see how much time you *actually* spend on social media apps each day.
   - Did you accurately estimate your social media time?
   - Which social media app do you spend the most time using?
   - How could you otherwise use the time you spend on social media? Come up with a list of other activities you can engage in instead of going online.

3. Study the LinkedIn profile of someone you consider to be highly successful. Based on what you see:
   - How would you describe that person's brand?
   - What are the person's main accomplishments?
   - What about the profile is most impressive to you?
   - What additions or revisions might make the profile better?

4. Take a social media break for a few days. Note how unplugging makes you feel.

The Self, Perception, and Communication

CHAPTER OUTLINE

🔍 3.1 Communication and the Self 50
Self-Concept
Self-Esteem
Personality
Reflected Appraisal
Social Comparison
Self-Fulfilling Prophecies

🔍 3.2 Perceiving Others 56
Selection
Organization
Interpretation

🔍 3.3 Problematic Perceptual Tendencies 58
Categorizing People
Clinging to First Impressions
Paying More Attention to Negative Impressions than to Positive Ones
Judging Yourself More Charitably than You Judge Others
Overgeneralizing
Gravitating to the Familiar

🔍 3.4 Perceptual Skill Builders 62
Emotional Intelligence
Perception Checking

🔍 3.5 Communication and Identity Management 65
Public and Private Selves
Facework
Why Manage Identities?
Identity Management and Honesty
Characteristics of Identity Management

MAKING THE GRADE 71

KEY TERMS 72

PUBLIC SPEAKING PRACTICE 72

ACTIVITIES 72

LEARNING OBJECTIVES

3.1 Describe how communication influences individuals' self-concept and self-esteem.

3.2 Explain the three steps people follow while forming perceptions of others.

3.3 Evaluate the impact of perceptual tendencies that lead to misconceived notions about people.

3.4 Apply communication skills related to emotional intelligence and perception checking.

3.5 Identify how identity management operates in both face-to-face and online communication.

Moniya was a quiet, socially awkward child who largely kept to himself. He was a mediocre student and inept at sports. Adulthood was not easy for him, either. He was arrested at least 13 times and served multiple prison sentences.

What words come to mind after reading this description? Perhaps you feel that Moniya was misunderstood, immoral, or irresponsible. We form impressions of people based on how they behave and communicate with others. That's unavoidable. But our perceptions are based on incomplete information.

Moniya was a childhood nickname. You probably know him as Mahatma Gandhi, the civil rights leader who was willing to be jailed for his beliefs.[1] Ultimately, his words and nonviolent approach attracted followers, helped India gain independence, and inspired Nelson Mandela, Martin Luther King Jr., the Dalai Lama, Malala Yousafzai, and many others.

In this chapter, we explore how communication shapes the way people understand themselves and others. We face up to some bad habits that lead to distorted and unfair assessments of others, and then practice techniques to be more emotionally intelligent and inquisitive. The chapter concludes with a section on identity management in person and via social media.

3.1 Communication and the Self

Let's start with you. Who are you? How did you come to view yourself this way? How does the way you see yourself shape your communication with others? We'll consider these types of questions as we explore the idea of self-concept.

Self-Concept

self-concept A set of largely stable perceptions about oneself.

The **self-concept** is a set of largely stable perceptions individuals have of themselves. You might imagine the self-concept as a mental mirror that reflects how you view yourself. It shows what is unique about you and what makes you both similar to and different from others. The picture may involve your gender identity, age, religion, and occupation. It's also likely to include your physical features, emotional states, talents, likes and dislikes, values, and roles.

Take a few minutes to list as many ways as you can to identify who you are. Try to include all the major characteristics that describe you, including:

- Groups with whom you identify (e.g., southerners, musicians, business majors)
- Your common moods or feelings
- Your appearance and physical condition
- Your career goals
- Your social traits
- Talents you possess or lack
- Your race and ethnicity
- Your intellectual capacity
- Your gender identity
- Your belief systems (religion, philosophy)
- Your social roles

Of course, to make this self-portrait even close to complete, your list would have to be hundreds—or even thousands—of words long. And not all items on your list are equally important to you. For example, you might define yourself primarily by your social roles (parent, veteran), culture (Mexican American, Chinese), or beliefs (libertarian, feminist). Others might define themselves more in terms of physical qualities (tall, Deaf), or accomplishments and skills (athletic, scholar).

Self-Esteem

An important element of the self-concept is **self-esteem**, which involves evaluations of self-worth. If your self-concept includes being athletic or tall, your self-esteem indicates how you feel about these qualities: "I'm glad that I am athletic" or "I'm worried about being so tall." There's a powerful link between communication and self-esteem. It's probably no surprise that people who have close, supportive interactions with others are more likely to have high self-esteem.[2,3] And the same principle works in reverse. People with high self-esteem are more likely than others to take a chance on starting new relationships[4] and showing affection to others,[5] which can enhance their self-esteem even more.

Researchers have found that people who feel good about themselves are more likely than others to believe and enjoy compliments.[6] They're also more resilient in the face of criticism and even cyberbullying. For example, individuals with healthy self-esteem are more likely than those with low self-esteem to report bullying and to see bullies as immature and eager to prove their own status.[7] That's not to say that

self-esteem The part of the self-concept that involves evaluations of self-worth.

The musical group BTS is famous for promoting confidence and self-love, as in their song "Epiphany," which features the chorus "I'm the one I should love in this world, shining me, precious soul of mine.... Not so perfect but so beautiful, I'm the one I should love."

Under what conditions is your self-esteem the highest? When is it the lowest?

bullying is okay or can always be shrugged off. It does suggest, however, that being silent or self-critical can make unkind comments feel even worse.

Despite its obvious benefits, self-esteem doesn't guarantee success in personal and professional relationships. People with an exaggerated sense of self-worth may mistakenly *think* they make a great impression, even though the reactions of others don't always match this belief. It's easy to see how people with an inflated sense of self-worth could irritate others by coming across as condescending know-it-alls.

Personality

Take another look at the list of terms you used to describe yourself. You'll almost certainly find some that reflect your **personality**—characteristic ways you think and behave across a variety of situations. Researchers estimate that people inherit about 40 percent of their personality.[8] For example, if your parents or grandparents were shy, extroverted, or anxious, you may have similar traits.[9]

> **personality** Characteristic ways that a person tends to think and behave in a variety of situations.

That's far from the whole story, though. You may have a disposition toward some personality traits, but you can do a great deal to control how you actually communicate. To test this idea, researchers evaluated 143 university students in terms of introversion, autonomy, emotional stability, and other attributes. Then they followed the students as they completed two communication courses. To the researchers' surprise, personality traits did not predict mastery of communication skills. For example, introverted and extroverted students learned to be equally effective at asking questions, showing empathy, and telling stories. The researchers concluded that, while particular communication skills may come easier for some than others, personality needn't be a factor in choosing one's path in life.[10]

Throughout this book you'll learn about communication skills that you can build into your repertoire if you are willing to practice. We begin by considering how communication contributes to the self-concept.

Reflected Appraisal

The term **reflected appraisal** describes the influence of others on one's self-concept. If people treat you as if you're smart, good looking, goofy, funny, athletic, lazy, or whatever, those assessments are likely to affect how you think about yourself. While it's unwise to let others' opinions control or demoralize you, no one is entirely immune to social influence. "Yes, I still care" what people think, admits blogger Aushaf Widisto, "because I need affection, I need approval. . . . The primal need for group inclusion still applies."[11]

> **reflected appraisal** The influence of others on one's self-concept.

Social Comparison

Your self-concept is shaped by **social comparison**—the process of evaluating yourself as compared to others.[12] Are you attractive? Successful? Intelligent? It depends on who you measure yourself against.

> **social comparison** Evaluating oneself in comparison to others.

People sometimes make *downward* social comparisons. Perhaps you feel good about earning a C compared to classmates who made lower grades. Or you may think "I'm compassionate" when you're the only one who stops to help a person in need.

More often, people focus *upward*—comparing themselves to those who seem better in some way.[13] That's not necessarily bad. Deciding to exercise more because you were inspired by a physically fit celebrity can be beneficial. Likewise, friendly competition with peers you admire can motivate you to try harder. There are three potential problems with upward social comparisons, however:

- *Comparisons can be unrealistic.* For example, most people on Instagram don't actually look that way in real life. Filters, lighting, computer touch-ups, and exotic backdrops make people and their lives look more appealing than they are.[14] Closer to home, you can bet that the beautiful person in your history class has as many worries and insecurities as anyone else.

Make-up and so-called "catfish filters" make it possible for people to look very different on social media than they do in real life.

Do you ever wish you looked more like the "people" you see online?

- *Comparing up can be a distraction.* No matter how much you have going for you, you can find someone who seems—at least on the surface—to be richer, happier, more successful, better looking, or smarter than you. Obsessing about those people can distract you from appreciating and nurturing your own merits.

- *Feeling down about yourself is counterproductive.* People who feel they don't measure up are likely to feel inferior, guilty, envious, defensive, depressed, and disappointed with themselves.[15,16] In addition to making then miserable, a negative outlook can make negative outcomes more likely, as you'll see in a moment.

Technology offers around-the-clock opportunities for social comparisons. The feature "Keeping It Real on Social Media" offers some tips for making the most of networking apps and presenting yourself authentically in the process.

Self-Fulfilling Prophecies

The self-concept is reinforced by expectations and assumptions. If you consider yourself a good student but you do poorly on a test, you might think, "I had an off day." But if you consider yourself a poor student, you might conclude, "I'm really bad at this subject," in line with your self-expectations.

There's even more to the story. Expectations don't only affect how you interpret events after they have happened. You may actually be *causing* things to happen as expected. A **self-fulfilling prophecy** occurs when a person's expectation of an outcome and their subsequent behavior increase the chances that the outcome will occur.

One type of self-fulfilling prophecy occurs when a person's expectations influence their *own* behavior. Imagine your friend goes to a party and is convinced they will have a terrible time. While there, they complain about the temperature and the music, sit alone, and brush off attempts at conversation. Sensing that your friend is annoyed or uninterested, other people leave them alone, which reinforces your

self-fulfilling prophecy A prediction or expectation of an event that makes the outcome more likely to occur than would otherwise have been the case.

COMMUNICATION STRATEGIES

Keeping It Real on Social Media

Newly wed to the love of her life, Rachel Leonard could relax on her front porch and enjoy a beautiful view of the Blue Ridge Mountains, all the while looking forward to the birth of their first child. To be more accurate, *virtual* Rachel had all those things.

Rachel's social media posts included happy wedding pictures, gorgeous mountain scenes, and pregnancy updates. But real-life Rachel was grappling with a difficult pregnancy, and a growing realization that she had married the wrong person. And the beautiful scenery? The mountain view straight ahead *was* gorgeous, "but if you looked to the left, you could see this huge factory," she admits, adding, "Of course, I didn't take [or post] pictures of the factory because why would you do that?"[17]

Rachel faced a common dilemma rooted in self-concept, communication, and perception. She wanted to present herself favorably to others. At the same time, she craved the genuine approval of people who understood and accepted her as she was. Concerns such as these are central to the communication choices people make, especially on social media. Here are two tips that may help.

Get to Know Your Goals and Priorities

Your social media use can fuel introspection about who you are and want to be. If your posts are mostly about having fun with friends, perhaps relationships are especially important to you. Or maybe you respond enthusiastically when content shows people mastering new skills or visiting exotic locations. Once you identify some priorities, you might create a journal or vision board to help you focus on them in real life. Social media is a glamorized version of reality, but it can provide inspiration for developing the genuine you.

Dare to Be Unique

Some people won't like it if you act different or look different than everyone else. As author Brené Brown points out, there will always be cowards in the "cheap seats" who are more concerned with criticizing others than working on themselves.[18] That's okay, because being genuine has its own rewards. People who present an authentic version of themselves on social media score higher than average in terms of life satisfaction and positivity.[19] When Amanda McElvey posted an honest description of having a bad day, she says she was touched by the number of supportive comments from people who cared and related to her situation. Celebrities such as Lizzo, Sarah Hyland, and Sam Smith have earned followers by posting unfiltered photos of themselves in everyday situations.

Ultimately, confidence arises from a sense of being accepted for the genuine you. College students who accept their own strengths and weaknesses are more likely than others to show their true selves on social media.[20] Consequently, they enjoy the security of knowing that others like them for who they really are, imperfections and all. As one social media analyst puts it, stop *trophy hunting*—trying to find that perfect picture or story that will play well on social media—and enjoy your life.[21] Share what happens naturally, not what you have manufactured to impress others.

---

friend's impression that the party is boring, which makes them seem even less approachable, and so on. It may be obvious to you (but not to your friend) that their behavior all but guarantees that the party will be as disappointing as they expected. A similar pattern may occur if you're so nervous during a job interview that you answer questions poorly—which makes you even more nervous. On a positive note, if you meet someone you have admired from afar, you might focus on their appealing qualities, concluding they are as wonderful as you expected. In each of these cases, the outcome happened at least in part because of the expectation that it would.

Another type of self-fulfilling prophecy occurs when *other people's* expectations influence your actions.[22] Perhaps an early mentor said you were good (or bad) at sports. You may behave in ways that support that assessment of your abilities, even if you don't realize you're doing so. This principle was demonstrated in a classic experiment.[23] Researchers told teachers that 20 percent of the children in a certain elementary school showed unusually high potential for intellectual growth. The names of the students were actually drawn at random, but eight months later, the children who had been indicated as "unusually gifted" showed significantly greater gains on test scores than the other children. Directly or indirectly, the teachers had communicated the message "I think you're bright" to the selected students. This message had affected the students' self-concepts, which ultimately affected their performance.

As you might expect, parents' expectations have a strong influence. Students whose parents believe they are academically talented tend to perform better than those whose parents have less confidence in them.[24] Unfortunately, low expectations can be self-fulfilling as well. College students with "helicopter" (always hovering) or "bulldozer" (obstacle removing) parents typically adjust more slowly to college life, and they often experience more anxiety and depression than others.[25] As a former dean at Stanford University puts it:

> What parents don't realize when they do the homework, or registration, or email, is what they're really doing is telling their child. . . . "Hey kid, I don't think you're going to be successful at this task, so I need to do it for you."[26]

"I don't sing because I am happy. I am happy because I sing."

Source: Edward Frascino The New Yorker Collection/The Cartoon Bank

It's an ironic twist of self-fulfilling prophecies that fearing a bad outcome can increase the chances that it will occur. When parents assume their youngsters are incapable, they tend to communicate that skepticism and do things for them—thereby short-circuiting opportunities for the children to develop confidence, experience, and skills.

There's a caveat: Although self-fulfilling prophecies can be powerful, expecting a particular outcome doesn't always bring it about. Children benefit from their parents' high opinion of them to an extent, but if expectations are *too* high, the children may feel anxious and discouraged instead.[27] Likewise, believing you'll do well in a job interview when you're not qualified for the position is unrealistic. And there will be people you don't like and occasions you won't enjoy, no matter what your attitude.

Even if you are surrounded by supportive people, it's easy to give more weight to unfavorable assessments than to positive ones.[28] The feature "Ways to Reverse Self-Defeating Thinking" offers experts' suggestions for turning that mindset around.

COMMUNICATION STRATEGIES

Ways to Reverse Self-Defeating Thinking

"You have been criticizing yourself for years and it hasn't worked," observes author Louise Hay, who suggests, "Try approving of yourself and see what happens."[29] It's often easier to be our own critics than our own supporters. But it's possible to make the shift. Here are some strategies to try.

Write a New Script for Your Inner Critic

"I'm not very good at this." "I'll never be able to...." Self-critical thoughts such as these can become self-fulfilling. A better alternative is to replace overly negative beliefs with more accurate and less discouraging ones, such as "I'm not very good at this yet" and "I've overcome more difficult challenges than this." Motivational speaker Mel Robbins recommends, "If you wouldn't say it to a friend, don't say it to yourself."[30]

Remember That a Lot of What You See Isn't Real

Compared to much of what you see on social media, normal life (and normal bodies) can feel substandard. But that's not a fair comparison. Stop and consider, "Is what I'm seeing realistic?" You might search "Instagram vs. reality" or #socialmediavsreality to see the difference between what people post and how they really look. (While you're browsing, avoid critiquing people's actual appearance. Real is okay!)

Steer Clear of Downers

Regularly exposing yourself to people, images, and messages that make you feel inferior is self-defeating. Stress specialist Susan Biali Haas asks, "Are there certain activities, such as strolling through a high-end shopping center, or driving through an expensive neighborhood, that frequently make you discontented with your life?" If so, she says, direct your energy into activities that have more value for you.[31]

Set Realistic Goals and Take Small Steps

To improve what's most important to you, experts suggest that you start with a realistic goal in mind, then make a sustainable plan with mini-deadlines and milestones to measure your progress. Stay focused but be patient. "Learn to take baby steps," advises consultant Royale Scuderi, noting that "the happiest and most successful people will tell you that they have achieved their level of life and work success by taking small steps and making one positive choice after another."[32]

3.2 Perceiving Others

Blogger Tiffany Tan has a good friend she describes as creative and talented. Her friend has autism. "That might make it seem like she's different," says Tiffany, "but let's be real, we're all different."[33] Tan's attitude illustrates the importance of **perception**, the way people regard others and the world around them. Whereas some people judge or avoid those who don't fit a particular mold, Tan embraces the notion that everyone has unique qualities.

One of the most powerful functions of communication is to find common ground with others. In this section, we consider the role that perception plays in that process. Typically, people form perceptions of others in three stages: selection, organization, and interpretation.

perception A process in which people reach conclusions about others and the world around them.

Selection

Psychologist William James described an infant's world as "one great blooming, buzzing confusion."[34] Babies are bombarded by unfamiliar stimuli nearly all the time.

You probably feel the same way when you move to a new city, take on a new position at work, or enter a room full of people you don't know. But even a fairly ordinary day is filled with more stimuli than you can possibly process. Thus, the first step in perception is **selection**, paying attention to some stimuli while ignoring others. Here are some key factors that influence that process.

Stimuli are likely to attract your attention if they are *intense* or *novel*. Anything that's louder, larger, or brighter than its surroundings tends to stand out. And people notice contrasts and changes more than predictable occurrences. That's why you may take a person's consistently wonderful qualities for granted. Online commentator Megan Elford reflects that she never appreciated her mother's habit of offering frank, unsolicited advice until her mom was gone. "She was far more wise than I gave her credit for," Elford says in retrospect.[35]

Selective perception is also affected by your *goals* and *emotional state*. On the lookout for romance? Attractive people may catch your eye more than normal. Adjusting to parenthood? Your baby's cooing probably captivates your attention. Day to day, your mood also makes some stimuli more noticeable than others. In one study, researchers computer enhanced a photo to include facial expressions suggesting both happiness and sadness. People in a good mood who viewed the photo typically thought the model was happy, whereas people who felt sad assumed that the model was too.[36] A similar process of selective attention happens in relationships. If you're happy with your partner, you're likely to focus on their good qualities, but if you're angry or discontent, you may fixate on things about them that annoy you.[37]

selection The perceptual act of attending to some stimuli in the environment and ignoring others.

Organization

In the **organization** phase of perception, you mentally arrange elements within your awareness. Central to this process is the *figure–ground principle*, which states that the mind tends to perceive some stimuli as primary (figure) and others as backdrop (ground). When you look at Figure 3.1, do you see an apple core or two faces in profile? Your answer represents the figure that initially attracted your attention. You also experience a figure–ground effect when you add a new word to your vocabulary and it suddenly seems to pop up in books and conversations. In reality, the word was probably always there, but you overlooked it (treated it as ground).

When you focus on the apple core in Figure 3.1, the faces largely fade from your awareness and vice versa. It can be hard to focus on multiple things at a time, but with effort you can change your perspective. Try this with people. If you typically concentrate on how slowly a colleague accomplishes tasks, try focusing on the outcome of their efforts. If you tend to categorize people by age, gender, skin color, or size, challenge yourself to notice different aspects of them instead. The way you mentally organize stimuli can make you aware of qualities in people you might otherwise overlook.

organization The perceptual process of mentally grouping stimuli into patterns.

FIGURE 3.1 The Figure–Ground Principle. Which do you see first: an apple core or two faces? While you focus on one part of the image, you perceive it as the *figure* and the rest as the *ground*.

Interpretation

Because the brain tends to process information quickly, it may seem that you select, organize, and interpret stimuli simultaneously. But if you could create a slow motion replay of perception, you would be able to distinguish these as separate functions.

To illustrate, imagine that you look across a crowded room and someone you know winks at you. Amid everything else going on, you might, however momentarily, *select* that to focus on. The wink in itself is ambiguous, so you mentally group (*organize*) stimuli that seem relevant to it. Perhaps the person is also nodding their head in your direction and gazing steadily at you. You will also figure in other factors—such as the occasion, the nature of your relationship, the time, the place, and so on. Based on all of that, you are likely to form an **interpretation**—that is, assign meaning to what you are perceiving. Your interpretation might be "They think I'm cute" or "This is my cue to rescue them from a tedious conversation." Interpretations are an integral part of understanding others, but they are fallible. Perhaps your friend just has something in their eye. Ultimately, perception is an ongoing process, not a clear-cut path to certainty.

> **interpretation** The perceptual process of attaching meaning to stimuli that have previously been selected and organized.

Although we have talked about selection, organization, and interpretation as a linear sequence, they can actually happen in any order. A babysitter's past *interpretation* ("little Jason is a troublemaker") can influence what the babysitter pays *selective* attention to in the future, and how they *organize* available information. For example, if there's a mess, the babysitter may key in on that and assume that Jason was involved in creating it. As with all communication, perception occurs within a complex interplay of internal and external factors.

3.3 Problematic Perceptual Tendencies

Everyday activities can be humiliating for Rachel Hoge. By all accounts, she is highly accomplished—having written for the *Washington Post*, edited 200 books, supervised a staff of 40 people, and more. But because Hoge stutters, strangers often assume that she's up to no good. A bank teller refused to offer her service when she couldn't quickly say her name, instead getting stuck on "RRRRRRRRRR . . ." A police officer assumed she was drunk when she stammered while trying to explain where she was going. "Thankfully, not all strangers are quick to judge," Hoge writes. "Sometimes I'll meet someone who regards me with patience and not pity, with respect and not condemnation . . . but they are few and far between."[38]

In this section, we consider perceptual tendencies that often result in misconceived notions about people. Some of the greatest obstacles to understanding arise from **attribution**—the process of attaching meaning to behavior. People make attributions when they assume that Hoge's speech patterns indicate that she's lying. As you will see, attributions are problematic when they're based on social labels, snap judgments, and a tendency to assume the worst or best about others. Following are six perceptual tendencies that can lead to inaccurate attributions—and to communication problems.

> **attribution** The process of attaching meaning to behavior.

Categorizing People

The basis for unfair attributions can be laid before people even meet. **Stereotypes** are widely held but oversimplified or inaccurate ideas about social groups. Even seemingly positive stereotypes (such as the myth "All Asians are good at math") can be demoralizing in that they overlook a person's unique qualities and impose irrelevant expectations on them.

> **stereotype** A widely held but oversimplified or inaccurate idea about a group of people.

Negative stereotypes can undermine people's self-confidence and make them feel that they must work overtime to disprove the assumptions.[39] While earning a Ph.D. in chemistry, Stephanie Santos-Díaz says she felt responsible for exemplifying the potential of an underrepresented group. She reflects that even well-intentioned statements such as "You have to finish" amped up the pressure.[40]

Another problem with stereotypes is a double-bind that punishes people for following them *and* for breaking them. Kara Johnson says that, if she acts in keeping with stereotypes about Black people, she risks being typified as "hood," "ghetto," or "ratchet."[41] But if she doesn't, she's criticized as being "not Black enough," or an "Oreo" (Black on the outside, white on the inside).[42] Johnson says she frequently reminds herself that stereotypes are inaccurate and unfair by remembering the words of a friend: "I am Black. So any way I choose to act is 'acting Black.'"[43]

Stereotyping is a mental shortcut that gives the illusion of understanding others rather than taking the effort to actually get to know them. The feature "Focus on Individuality Rather than Stereotypes" presents some ways to set aside lazy thinking and appreciate people for who they really are.

Clinging to First Impressions

Even if you haven't categorized someone in advance, there's a tendency to do so within a few minutes of encountering them. Given limited information, it's easy to conclude that "This person seems cheerful," or "They're awfully arrogant." Problems arise when these first impressions are inaccurate but you cling to them, ignoring

---

COMMUNICATION STRATEGIES

Focus on Individuality Rather than Stereotypes

You have probably felt the sting of stereotypical thinking yourself—when someone assumed that you aren't a great student because you play sports, or that you don't understand politics because you're young. Unfortunately, stereotypes are suggested by media representations, things people say, and selective perception. Here are some ways to resist their influence.

Don't Judge by Appearances

A person's most noticeable characteristics aren't usually the most important things about them. For example, Vincent Pisano leads a normal life, but strangers tend to categorize him as helpless as soon as they notice that he has cerebral palsy, a condition that sometimes causes involuntary movements of the body. "If I am at a store, attendants give me extra attention because they don't believe I can find anything on my own or know what I am looking for," Pisano says. "If I am with a female, people are amazed, almost like, 'what does she want to do with him?'" Pisano's request is simple: "Treat me normal. . . . [W]e were just born differently."[44]

Consider the Assumptions People Make About You

Think about the stereotypes others might impose on you—perhaps based on your physical appearance or the groups with which you identify. It's probably clear that these generalizations don't accurately capture who you are. Recognizing that can help you avoid prejudging others.

Get to Know People

One of the best ways to dispel a stereotype is firsthand contact. Strike up a conversation with someone you might otherwise categorize as different. Treat them like you would a friend. For example, if the person is transgender, "you don't have to talk about 'trans stuff,'" says Matt Kailey. Instead, "talk about sports, the weather, taxes, television shows, movies, music—you're likely to find something you have in common."[45]

The Disney movie *Encanto* has been praised for portraying Latin people in a positive manner, in contrast to negative stereotypes.

What stereotypes do people often make about a group to which you belong? How do those stereotypes affect the way people communicate with you and about you?

conflicting information. Suppose, for instance, that you mention your new neighbor to a friend. "Oh, I know him," your friend replies. "He seems nice at first, but it's all an act." Perhaps this appraisal is off base, but it will probably influence the way you respond to the neighbor. Your response may in turn influence your neighbor's behavior, creating a negative self-fulfilling prophecy.

Snap judgments are particularly likely in text-based communication, which often presents few cues other than writing style. A team of researchers sent out emails that ended with either "Thanks!" "Best," "Thank you," or no closing salutation. Recipients judged "Thanks!" to be less professional when the email was sent by a woman than by a man,[46] perhaps because the exclamation mark triggered a negative stereotype of women as emotionally effusive.

Paying More Attention to Negative Impressions than to Positive Ones

Consider a time when you received feedback about your contributions to a team project. Even if 9 out of 10 comments were positive, the negative one probably hit you hardest and stayed with you longest. **Negativity bias** is the tendency to focus more on negative impressions than on positive ones.[47,48] Scientists speculate that it may have evolved as a survival advantage. Spotting the one threatening element in the environment may ultimately be more important than focusing on all the safe ones.[49] But this tendency can skew judgments, affecting careers and relationships. Potential employers may rule out otherwise qualified job candidates after finding a negative post on social media.[50] And if a friend says something that hurts your feelings, you may be tempted to focus on that statement rather than the dozens of supportive things your friend has said.

negativity bias The perceptual tendency to focus more on negative indicators than on positive ones.

One lesson here is to pause and reflect on the big picture. Just as you would hate to be judged for one statement or mistake, avoid judging yourself or others based on isolated incidents.

Judging Yourself More Charitably than You Judge Others

Although we're quick to be critical of others, we tend to judge ourselves in generous terms. Social scientists call this tendency *fundamental attribution error*, or simply the **self-serving bias**.[51] When others suffer, we often blame the problem on their personal qualities. By contrast, when *we* suffer, we find explanations outside ourselves. Consider a few examples:

- When someone else botches a job, you might think they didn't try hard enough. When you make a mistake, you blame unclear directions or inadequate time.
- When someone else lashes out angrily, you may say they are moody or oversensitive. When you blow off steam, you point to the pressure you are under.
- When others don't reply to your text or email, you might assume they are inconsiderate, disrespectful, or unprofessional. When you don't reply, you may say you were too busy, you didn't see the message, or it didn't seem necessary to reply.

self-serving bias The tendency to judge other harshly but cast oneself in a favorable light.

As these examples show, uncharitable attitudes toward others affect communication. Your harsh opinions can lead to judgmental messages, which trigger defensive responses. At the same time, you may be defensive when people question your behavior—which can keep you from trying to improve.

Overgeneralizing

People tend to generalize based on a single positive or negative trait or experience. When someone has one positive quality, you might unduly assume other positive qualities—a bias that scholars call the **halo effect**.[52] For example, people often suppose that physically attractive people are more intelligent than others, even when they are not.[53] The converse of the halo effect is known as the **horns effect**—perceiving others in an unfairly negative light on the basis of a single negative trait or experience.[54]

halo/horns effect A form of bias that overgeneralizes positive or negative traits.

Gravitating to the Familiar

People tend to favor characteristics and ideas that are similar to their own. In one study involving social media, participants were more likely to rate a profiled person as likeable if they perceived common interests and shared group membership, such as attending the same university.[55]

A preference for similarities often leads people to mistakenly project their own attitudes and ideas onto others. For example, you might assume that an off-color joke won't offend a friend, but it does. Others don't always think or feel the way we do, and assuming similarities can lead to problems.

When you become aware of differences, resist the temptation to distance yourself or demean the other person. Competent communicators are able to talk respectfully about viewpoints that differ from their own. As one blogger puts it, "The overall objective of expressing your views is supposed to be to encourage conversation and gain/provide new perspectives. It is not to demean and disprove."[56]

3.4 Perceptual Skill Builders

Having focused on perceptual tendencies that can lead to unfair perceptions of others, let's consider how to be open-minded, deliberate, and attentive in our efforts to understand and be understood.

Emotional Intelligence

When Rafael wasn't given the promotion he wanted, he threw his laptop to the floor and stormed out of the office, yelling, "I quit!" The odds are that Rafael's emotional outburst made the people around him glad that he wasn't going to be on their team any longer.

Rafael seems to lack **emotional intelligence (EI)**—the ability to understand and manage one's own emotions and deal effectively with the emotions of others. The idea was made famous by psychologist Daniel Goleman,[57] who proposes that EI has five dimensions.

Self-awareness involves understanding your own feelings. This can be harder than it sounds. For one thing, people often experience a mixture of emotions. When something great happens to your best friend, you might feel happy and jealous. During a breakup, you might feel both grief and relief. Another impediment is that people tend to describe emotions in vague terms such as sad, mad, and happy. To better understand (and express) how you feel, psychologist Lisa Feldman Barrett suggests more nuanced terms such as remorseful, melancholy, dejected, aggravated, ecstatic, thrilled, and blissful.[58] Try it yourself. The next time you feel *afraid*, challenge yourself to consider whether words such as *terrified*, *suspicious*, *anxious*, or *nervous* more accurately describe your feelings.

Self-regulation involves managing emotions effectively. People who lack self-regulation may lose their temper, say things they wish they hadn't, or be plagued by remorse or envy. Self-regulation doesn't mean stifling or ignoring emotions. Instead, it stems from a deliberate effort to express them effectively. Imagine if Rafael had taken time to cool off. He might have been able to express himself more clearly and initiate a productive dialogue about the issue. At the very least, he would have known that he communicated as effectively as possible.

Internal motivation involves finding the inner strength and determination to accomplish important goals. Consider success as a student. You may assume that people who do well in school are smarter than others, but research shows their success often depends more on confidence and perseverance.[59] You might be able to escalate your internal motivation by celebrating incremental gains, taking breaks, giving yourself pep talks, and sharing your feelings with a trusted friend.

Empathy involves the willingness and ability to experience things from another person's point of view.[60] Here are some ways to improve your ability to be empathic:[61]

> **emotional intelligence (EI)** A person's ability to understand and manage their own emotions and to deal effectively with the emotions of others.

"How would you feel if the mouse did that to you?"

Source: William Steig The New Yorker Collection/The Cartoon Bank

- Set aside your own opinions and suspend judgment of the other person. Try to understand as clearly as possible what they are *thinking*.
- In addition to focusing on the person's thoughts, try to understand their *feelings*.
- Show genuine concern for the other person. Sincere statements such as "I can understand why you are upset" can encourage understanding and open communication.

Some people seem to have an inborn capacity for empathy, but most can develop their capacity by being aware and attentive. Children whose parents encourage empathy for others are less likely than other children to engage in bullying and cyberbullying.[62] Empathy is so important that some medical schools have begun considering applicants' capacity for empathy when making admissions decisions.[63]

Social skills involve the ability to communicate effectively with others. Some questions to ask yourself include: Should I talk or listen right now? Is this a good time to bring up a touchy subject? How much personal information about myself should I share? Insights throughout this book are designed to help.

Emotional intelligence offers many benefits. Travis Bradberry, who wrote *Emotional Intelligence 2.0*, observes that emotionally intelligent people don't get stressed as easily as others, are better at sharing their feelings, have a larger vocabulary of emotional terms, more easily forgive themselves and others for mistakes, and are misunderstood less often.[64] What's more, organizations led by people with high EI are usually more successful than others, partly because members are self-aware, exercise emotional control, and are good at understanding how coworkers and customers feel.[65] Keep reading for tips about a process that can help you build empathy and an "How Emotionally Intelligent Are You?" quiz.

Perception Checking

After spending weeks creating a project proposal, Anastasia was dismayed when her boss left the room halfway through her presentation. She wondered to herself: Was the boss disappointed in my presentation? Was she called away to an emergency? What should I think about this?

Perception checking is a three-part communication technique for referencing a particular occurrence and asking for feedback about it rather than jumping to conclusions. Because the goal of perception checking is mutual understanding, it's a cooperative approach that minimizes defensiveness and displays respect. Here are the three steps involved.

The first is to *reference a specific behavior*, as in, "I noticed you left the room during my presentation." It's important to avoid assumptions and value judgements, and you shouldn't presume that you know how the other person is feeling. For example, it's not fair to say, "It was inconsiderate of you to walk out" or "I can tell you didn't like my ideas." Such statements are likely to cause defensiveness, and they signal that you have already made up your mind without trying to understand the other person's feelings.

The second step is to *offer two options* that may be true. This opens the way for conversation and makes it clear that you don't presume to know the other person's motives. For example, you might say, "When you left, I wasn't sure if something important came up or if you were disappointed in my presentation."

perception checking
A three-part method for verifying the accuracy of interpretations, including an objective description of the behavior, two possible interpretations, and a request for confirmation of the interpretations.

UNDERSTANDING YOUR COMMUNICATION

How Emotionally Intelligent Are You?

Answer the following questions for insights about how EI influences you as a communicator.

1. A friend says something that hurts your feelings. What are you most likely to say?
   a. "That is so insensitive! I can't believe you just said that!"
   b. "I feel hurt by what you said."
   c. "That makes me feel bad. Tell me why you feel that way."
   d. Say nothing. You'll get over it.

2. It's Monday morning and you feel great. What are you most likely to do?
   a. Take the day off. This feeling is too good to waste at work.
   b. Announce to everyone at work, "I feel like a million bucks!"
   c. Channel your positive energy into being a great team member.
   d. Set your emotions aside and get to work. You'll enjoy yourself later.

3. Your usually talkative roommate is quiet today and seems to be looking out the window rather than focusing on the book he's trying to read for school. What are you most likely to do?
   a. Tell him, "Focus! That book's not going to read itself."
   b. Say that you understand because you've had a hard day too.
   c. Ask if anything is bothering him and then listen attentively to what he says.
   d. Give him some space. He's probably just tired.

4. The grade on your research paper is not as high as you had hoped. How are you most likely to respond?
   a. Fume about what an idiot the professor is.
   b. Post on social media that you are sad and discouraged today.
   c. Go over the paper carefully to learn what you might do better next time.
   d. Tell yourself, "What's done is done" and try to forget about it.

INTERPRETING YOUR RESPONSES

For insight about your emotional intelligence, see which of the following best describes your answers. (More than one may apply.)

Emotionally Spontaneous

If you answered "a" to two or more questions, you tend to display your emotions expressively and spontaneously. This can be a bonus, but be careful not to let feelings get the best of you. Your unfiltered declarations may sometimes offend or overwhelm others, and they may prevent you from focusing on what other people are thinking and feeling. Suggestions for perception checking and self-monitoring in this chapter may help you strengthen the empathy and self-regulation components of EI.

Emotionally Self-Aware

If you answered "b" two or more times, you tend to be aware of your emotions and express them tactfully. You score relatively high in terms of emotional intelligence. Just be careful to pair your self-awareness with active interest in others. You may feel impatient with people who are not as emotionally aware as you are. Stay tuned for listening tips and strategies in Chapter 6.

EI Champion

If you answered "c" to two or more questions, you balance awareness of your own emotions with concern for how other people feel. Your willingness to be self-reflective and a good listener will take you far. Communication strategies throughout this book provide opportunities to build on your already strong EI.

Emotion-Avoidant

If you answered "d" two or more times, you tend to downplay emotions—yours and other people's. While this may prevent you from overreacting to situations, it may also make it difficult to build mutually satisfying relationships and to harness the benefits of well-managed emotions. You may sometimes feel that others are taking advantage of you when they actually don't know how you feel. The tips for self-disclosure in Chapter 8 may be especially useful to you.

The final step is simple but important. Simply *ask the other person* to tell you what they were (or are) feeling. You might say "Can you tell me what was going on?"

Here are some examples of perception checking in different situations that include all three steps:

"When you slammed the door [behavior], I wasn't sure whether you were mad at me [first interpretation] or just in a rush [second interpretation]." Are we good, or do you want to talk [request for clarification]?"

"You haven't laughed much in the last couple of days [behavior]. I wonder whether something's bothering you [first interpretation] or whether you're just feeling quiet [second interpretation]. What's up [request for clarification]?"

As you can see, perception checking takes a respectful approach that implies "I know I'm not qualified to understand your feelings without some help." Of course, it can succeed only if your nonverbal behavior reflects the open-mindedness of your words. An accusing tone of voice or a hostile glare will suggest that you have already made up your mind about the other person's intentions no matter what you say.

3.5 Communication and Identity Management

During her honeymoon, Jenny Parrish photographed a wineglass, never dreaming that anyone would know she snapped the picture while naked. But when she posted the image on Instagram, an alert viewer messaged her "I don't know if you care but you can see yourself in the reflection." Parrish quickly removed the photo and replied, "I care so much. I literally care so much, thank you."[66]

We *do* care how others see us. In fact, people are engaged in **identity management**—the communication strategies people use to influence how others see them—almost constantly. In this section, we consider how people use communication to create the "selves" that they share with the world.

> **identity management** Strategies used by communicators to influence the way others view them.

Public and Private Selves

So far we have referred to the "self" as if each of us has only one identity. In truth, every person has many selves, some private and others public. Often these selves are quite different.

Your **perceived self** is the person you believe yourself to be in moments of honest self-examination. The perceived self is private in the sense that you're not likely to reveal all of it to another person. There are probably some elements of yourself there that you wouldn't disclose to many people, and some that you don't share with anyone. For example, you might be reluctant to share some feelings about your appearance ("I think I'm rather unattractive"), your intelligence ("I'm not as smart as I wish I were"), your goals ("The most important thing to me is becoming rich"), or your motives ("I don't really like to meet new people").

> **perceived self** The person we believe ourselves to be in moments of candor.

In contrast to the perceived self, the **presenting self** is a public image—the way you want to appear to others. In most cases, a person seeks to create a presenting self that is a socially approved image: diligent student, loving partner, conscientious

> **presenting self** The image a person presents to others.

worker, loyal friend, and so on. Social norms often create a gap between the perceived and presenting selves. For example, you may present yourself as more confident than you feel.

Facework

Sociologist Erving Goffman used the word **face** to describe the presenting self, and **facework** (synonymous with identity management) to describe the verbal and nonverbal ways people try to maintain a positive image.[67] Goffman argued that each person can be viewed as a kind of playwright who creates roles they want others to believe, and as a performer who acts out those roles. Depending on the circumstances, you may behave in ways that suggest to others that you are nice, competent, or intelligent, for example.

It may seem logical that people would strive for the best face possible—the "nicest person in the world" or the "most talented artist in town." Two factors discourage this, however. One is an embarrassing loss of face if you can't live up to that image. The other is the way your face goals make others feel. You've probably known people who acted as if they were "the best," with the unwelcome implication that others were inferior to them. All in all, although everyone wants to be viewed in positive terms, it's face-saving not to overdo it.

Why Manage Identities?

People engage in identity management to show consideration for others and to recruit assistance in reaching their own goals. One example is being polite. Good manners essentially help others maintain a positive identity. Growing up, you probably learned manners for various occasions: meeting strangers, attending school, going to religious services, and so on. Young children who haven't learned all the do's and don'ts of polite society often embarrass their parents by behaving inappropriately ("Mommy, why does that person walk funny?"). As we mature, we're likely to realize that it's important to make others feel valued and that doing so increases the chances that they will think highly of us in return.

Identity management can also help you accomplish personal goals. You might, for example, dress up for a visit to traffic court in the hope that your front (responsible citizen) will convince the judge to treat you sympathetically. You might act more friendly and lively than you feel on meeting a new person so that you'll appear likable. You might smile to show the attractive stranger at a party that you'd like to get better acquainted. In situations like these, you aren't being deceptive as much as putting your best foot forward.

Identity management is critical to career success. It would be impossible to keep a job, for example, without meeting certain expectations. Salespeople are obliged to treat customers with courtesy. Employees should appear reasonably respectful when talking to the boss. Some forms of clothing would be considered outrageous at work. By agreeing to take on a job, you sign an unwritten contract that you will present a certain face at work, whether or not that face reflects the way you feel at a particular moment.

A common dilemma in identity management is reconciling competing expectations. For example, it may seem rude to brag about yourself, but in situations such

---

face The socially approved identity that a communicator tries to present.

facework Verbal and nonverbal behavior designed to create and maintain a communicator's face and the face of others; synonymous with identity management.

COMMUNICATION STRATEGIES

Humblebragging

Life is hard for humblebraggers. Every golden moment inspires an ostensibly modest complaint about a dazzling accomplishment. Can't you just feel the envy—and anguish—of their social media followers?

> Another raise. I'm turning into the one-percenter I used to mock in college. So embarrassed!
>
> I show up for the interview with a hangover and they still pick me from dozens of other candidates. Like really?
>
> I'm bummed. My new corner office is 100 feet from the board room where I meet VIP clients. Not fair!

After reading even a few humblebrags, you probably agree they're annoying. A Harvard study confirms that trying to pass off a brag as a complaint usually doesn't fool anyone.[68] After reading humblebraggers' online posts, participants in the study gave them low marks in terms of likability, sincerity, and competence.

Despite the blatant faux modesty of humblebrags, sometimes it's necessary to self-promote, especially in the world of work. People were humblebragging in job interviews long before the term was invented: "My greatest weakness? It's probably that I'm a perfectionist." Self-serving comments like this sound phony. But are they worse than honest confessions such as, "I'm not very organized" or "I'm not a team player"? How can you share your accomplishments without being self-defeating on one hand or boastful on the other?

In the second part of their humblebragging study, the Harvard team put those questions to the test. The researchers asked college students to respond in writing to the classic job interview question: "What is your biggest weakness?" Trained evaluators then evaluated how likely they would be to hire the students based on their responses. In the end, humblebraggers ("I find myself doing a lot of favors for others") were less likely to be hired than those who revealed honest weaknesses ("I sometimes tend to procrastinate").

Should you humblebrag in job interviews? Probably not. An honest account of your talents and challenges is more likely to be appreciated.

---

as job interviews, you may miss out if you don't. The feature "Humblebragging" explores that dilemma.

Identity Management and Honesty

Managing identities doesn't necessarily make you phony or a liar. It's difficult—even impossible—not to create impressions. After all, you have to send some sort of message. If you don't act in a friendly manner when meeting a stranger, you have to act aloof, indifferent, hostile, or in some other manner. If you don't act businesslike, you have to behave in an alternative way: casual, silly, or whatever. You wouldn't act the same way with strangers as you do with close friends, for example, or talk to a two-year-old the same way you talk to a peer.

Each of us has a repertoire of faces—a cast of characters—and part of being a competent communicator is choosing the best role for the situation. In countless situations every day, you have a choice about how to act. When meeting a new roommate, you face the challenge of deciding how much to reveal about yourself and how soon: Is it appropriate to say right off the bat that you snore? That you're overcoming a painful breakup? That you really hope the two of you will become good friends?

It's an oversimplification to say there is only one honest way to behave in a particular circumstance. Instead, impression management involves deciding which face—which part of yourself—to reveal. In Chapter 8 we'll talk more about the rewards and risks of disclosing personal information.

Characteristics of Identity Management

Now that you have a sense of what identity management is, let's look at four key characteristics of the process.

In the TV show *Never Have I Ever*, the lead character Devi Vishwakumar is an Indian-American teenager who navigates the different expectations of her immigrant mother and her classmates in a San Fernando Valley high school.

Have you ever felt that changes in your life (such as, a new neighborhood, school, or set of friends) challenged or changed your social identity? If so, how did you reconcile the differences?

People Have Multiple Identities In the course of even a single day, most people take on a variety of roles: respectful student, joking friend, friendly neighbor, and helpful worker, to suggest just a few. In his Netflix special, comedian W. Kamau Bell humorously presents some of his different identities. As one observer describes it:

> Sometimes [Bell is] speaking as a parent, who has to go camping because his kids enjoy camping. Sometimes he's speaking as an African-American, who, for ancestral reasons, doesn't see the appeal of camping ("sleeping outdoors *on purpose?*"). Sometimes—as in a story about having been asked his weight before boarding a small aircraft—he's speaking as "a man, a heterosexual, cisgender *Dad* man." (Hence: "I have no idea how much I weigh.") [69]

You may even play a variety of roles with the same person. With your parents, for instance, perhaps you acted as a responsible adult sometimes ("You can trust me with the car!") and at other times as a helpless child ("I can't find my socks!"). Perhaps on birthdays or holidays you were a dedicated family member, but at other times you played the role of rebel.

The ability to construct multiple identities is one element of communication competence. We recall a colleague who was also minister of a Southern Baptist congregation. On campus his manner of speaking was typically professorial, but a visit to hear him preach one Sunday revealed a speaker whose style was much more animated and theatrical, reflecting his identity in that context.

Identity Management Is Collaborative As people perform like actors trying to create a persona (character), their "audience" is made up of other actors who are trying to create their own personas. Identity-related communication is a kind of theater in which people collaborate with other actors to improvise scenes.

At other times, people are active agents in helping one another save face. For example, what might you do if someone arrives at a party with their fly unzipped? If you know the person well, you might point it out so they can avoid further embarrassment. Or you might pretend you don't notice. Either way, you're engaged in a cooperative effort to help that person save face, just as you hope others will help you.

Communication is central to the process. Imagine that your roommate has left dirty dishes in the sink and you're annoyed. You might exclaim, "You act like a child. Learn to clean up after yourself!" That would almost certainly challenge the identity your roommate hopes to accomplish—and your identity as a kind person. Imagine rewriting the scene in a way that allows both of you to save face. You might say, "I noticed you left dishes in the sink this afternoon," to which your roommate might reply, "I'm sorry! It took me longer than I expected to finish an assignment and I had to rush to class. I promise it won't happen again." In this version, both you and your friend accept each other's bids for identity as basically thoughtful people. As a result, the conversation runs smoothly.

The point here is that virtually all conversations provide an arena in which communicators construct their identities in response to the behavior of others. As you read in Chapter 1, communication isn't made up of discrete events that can be separated from one another. Instead, what happens at one moment is influenced by what each party brings to the interaction and by what happened in their relationship up to that point.

Identity Management Can Be Deliberate or Unconscious Sometimes people act largely out of an unconscious sense of what's appropriate. Without stopping to think about it, you might say "Hi" to people you pass on the street or answer the phone at work with a script such as "Good afternoon. How can I help you?"

There are occasions, however, in which you're probably highly aware of managing your identity, as when you're on a job interview or a first date. **Frame switching** involves adopting different perspectives based on the cultures and situations in which you find yourself.[70] You probably frame switch when you talk to your friends versus your grandparents, or your colleagues at work versus your best friend. People who aren't good at frame switching tend to say things that are inappropriate for the situation and offend others by being more aggressive, casual, or standoffish than others expect them to be.

Identity management can be especially hard work for people who operate in two or more cultures. A Filipino American man describes the duality of his work and family life this way: "At my first job I learned I had to be very competitive and fighting with my other coworkers for raises all the time, which I was not ready for—being brought up as nice and quiet in my family."[71]

The feature "Work Lessons from *Undercover Boss*" offers some tips for frame switching in professional environments.

People Differ in Their Degree of Identity Management Some people are more aware of their identity management behavior than others. So-called **high self-monitors** pay close attention to their own behavior and to others' reactions, adjusting their communication to create the desired impression. By contrast, **low self-monitors** express what they are thinking and feeling without much attention

frame switching
Adopting the perspectives of different cultures.

high self-monitors
People who pay close attention to their own behavior and to others' reactions, adjusting their communication to create the desired impression.

low self-monitors
People who express what they are thinking and feeling without much attention to the impression their behavior creates.

COMMUNICATION STRATEGIES

Work Lessons from *Undercover Boss*

In the television show *Undercover Boss,* frontline employees are asked to train a "new colleague" who is actually a high-level executive in disguise. The process presents a number of tips you can use at work to create a favorable impression with peers, subordinates, and managers.

Imagine that Every Coworker Is the CEO

Your behavior may never be broadcast on TV, but word does get around. To build a favorable reputation, behave as if the CEO can see and hear everything you do on the job. Strive for excellence, display a good attitude, and don't complain or overshare.

Be Patient with Yourself and Others

A comical part of nearly every *Undercover Boss* episode is the CEO clumsily trying to master new tasks. There is truth in that. Even if you've been on the job for a while, it's hard to learn new things. Show your character by handling learning experiences well. "Take in as much as you can by observing, asking, and taking notes," suggests a team of career advisors, adding, "For now, be ok with the fact that you don't know everything."[72] When the tables are turned and you're training others, be patient and willing to repeat yourself a few times.

Flip the Script and Act Like a Leader

Leadership coach Tom Triumph suggests an alternative plotline: Rather than disguising a CEO as a frontline worker, be a frontline worker who thinks like a CEO.[73] If you were the boss, what changes would you implement to make the organization more successful? What might you do to help team members feel supported and appreciated? Suggesting answers to questions such as these can demonstrate your investment in the organization and your leadership potential. One executive reflects that "employees who came to me and said, 'I have five new ideas I think would be fantastic for our products and our department and company'" went to the top of the "promotion" list when opportunities arose.[74]

to the impression their behavior creates.[75] There are advantages and disadvantages to both approaches.

High self-monitors are generally good actors and good "people readers" who can act interested when bored, or friendly when they feel quite the opposite. This allows them to handle social situations smoothly, often putting others at ease and getting a desired reaction from them. For example, high self-monitors tend to post pictures and messages on Facebook that make them seem especially outgoing, which correlates to a higher-than-average number of "likes."[76] The downside is that they are unlikely to experience events completely because a portion of their attention is always devoted to viewing the situation from a detached position.

People who score low on the self-monitoring scale live life quite differently from their more self-conscious counterparts. They have a simpler, more focused idea of who they are and who they want to be. Low self-monitors are likely to have a more limited repertoire of behaviors, so they act in more or less the same way regardless of the situation. This means that they are easy to read. "What you see is what you get" might be their motto. However, a lack of flexibility may sometimes cause them to seem awkward or tactless.

It's probably clear that neither extremely high nor low self-monitoring is the ideal. There are some situations in which paying attention to yourself and adapting your behavior can be useful, but sometimes, being yourself without conforming to expectations is a better approach. If following others' wishes contradicts with what you believe is right, going against the flow can be act of courage. In keeping with the notion of communicative competence outlined in Chapter 1, flexibility and good judgment are essential to successful communication.

MAKING THE GRADE

OBJECTIVE 3.1 Describe how communication influences individuals' self-concept and self-esteem.

- The self-concept is a set of largely stable perceptions about oneself.
- Self-esteem involves evaluations of self-worth.
- Although some personality characteristics are innate, communication and cultural/social factors also help shape the self-concept.
- Self-concept is influenced by reflected appraisal, social comparison, and self-fulfilling prophecies.
  > Name at least four factors that influence a person's self-concept. Explain the role of communication relevant to each of these factors.
  > How might your self-concept affect both your academic performance and your communication patterns?
  > Describe a relationship or event that had a powerful impact on your self-concept. What role did communication play?
  > Present a scenario in which a self-fulfilling prophecy might help someone achieve an important goal.

OBJECTIVE 3.2 Explain the three steps people follow while forming perceptions of others.

- Perception is a multistage process that includes selection, organization, and interpretation of information.

  > Describe an experience in which you met someone new. How did your impressions of that person evolve in the context of the perceptual phases of selection, organization, and interpretation?

OBJECTIVE 3.3 Evaluate the impact of perceptual tendencies that lead to misconceived notions about people.

- Perceptual tendencies and errors can affect the way we view and communicate with others.
- Stereotypes can be demoralizing in that they overlook a person's unique qualities and impose irrelevant expectations on them.
- First impressions tend to linger, even when they are inaccurate.
- A negativity bias is the tendency to focus on negative impressions.
- People tend to judge themselves more favorably than they judge others.
- When a person makes positive or negative generalizations about another on the basis on limited information it is known as a halo or horns effect.
- People tend to favor characteristics and ideas that are similar to their own.
  > What first impressions do you think people form about you? Are they accurate or not? Explain how.

- Describe a time when you changed your opinion of someone. Looking back, what perceptual tendencies were involved in your initial assessment of the person?
- What advice do you have for avoiding hurtful and unfair judgments about people?

OBJECTIVE 3.4 Apply communication skills related to emotional intelligence and perception checking.

- Emotional intelligence involves the ability to understand one's own emotions and the emotions of others, regulate expression of emotion, maintain internal motivation, and engage in effective communication with others.
- Perception checking is a three-step technique for increasing the accuracy of perceptions and for increasing empathy.
  - Describe a person you know who exhibits qualities of emotional intelligence. Describe someone who doesn't. How is their communication different?
  - Explain the three steps involved in perception checking and provide an example that illustrates all three.

OBJECTIVE 3.5 Identify how identity management operates in both face-to-face and online communication.

- Identity management consists of strategic communication designed to influence others' perceptions of an individual.
- Identity management is collaborative.
- Identity management is meant to respect others and to reach personal goals.
- Although identity management might seem manipulative, it can be an authentic form of communication. Because each person has a variety of faces that they can present, choosing which one to present is not necessarily being dishonest.
  - Explain how people use facework and frame switching to manage their private and public identities.

KEY TERMS

attribution p. 58
emotional intelligence (EI) p. 62
face p. 66
facework p. 66
frame switching p. 69
halo/horns effect p. 61
high self-monitors p. 69
identity management p. 65
interpretation p. 58
low self-monitors p. 69
negativity bias p. 60
organization p. 57
perceived self p. 65
perception p. 56
perception checking p. 63
personality p. 52
presenting self p. 65
reflected appraisal p. 52
social comparison p. 52
selection p. 57
self-concept p. 50
self-esteem p. 51
self-fulfilling prophecy p. 53
self-serving bias p. 61
stereotype p. 58

PUBLIC SPEAKING PRACTICE

Introduce yourself, including some of your interests and passions. Describe at least three different identities you present in various situations or relationships.

ACTIVITIES

1. For one day, keep a log of the identities you notice yourself creating in different situations: at school and at work, and with strangers, various family members, and different friends. For each identity:

a. Describe the persona you are trying to project (e.g., "responsible son or daughter," "laid-back friend," "attentive student").

b. Explain how you communicate to promote this identity. What kinds of things do you say (or not say)? How do you act?

2. Look up someone on social media and evaluate how genuine and realistic their posts seem to be. What do you admire about this person? If they were to post more realistic images of themselves and described their worries occasionally, how might your opinion of them change?

Communication and Culture

CHAPTER OUTLINE

4.1 Understanding Cultures and Cocultures 76
Salience
In-Group and Out-Group

4.2 How Cultural Values and Norms Shape Communication 79
Individualism and Collectivism
High and Low Cultural Context
Uncertainty Avoidance
Power Distance
Talk and Silence

4.3 Cocultures' Influence on Communication 86
Intersectionality Theory
Race and Ethnicity
Sex and Gender
Religion
Disability
Political Viewpoints
Age/Generation

4.4 Becoming an Effective Intercultural Communicator 96
Contact with a Diverse Array of People
Tolerance for Ambiguity
Open-Mindedness
Knowledge and Skill
Patience and Perseverance

MAKING THE GRADE 101

KEY TERMS 102

PUBLIC SPEAKING PRACTICE 103

ACTIVITY 103

LEARNING OBJECTIVES

4.1
Analyze the influence of cultures and cocultures on communication, including the concepts of salience and group membership.

4.2
Distinguish among the following cultural norms and values, and explain how they influence communication: individualism and collectivism, high and low context, uncertainty avoidance, power distance, and talk and silence.

4.3
Evaluate how factors such as race and ethnicity, sex and gender, religion, disability, political viewpoints, and age influence communication.

4.4
Implement strategies to increase intercultural and cocultural communication competence, such as increasing contact with diverse others, and being open-minded, knowledgeable, and patient.

CHAPTER 4 Communication and Culture

Many surprises were in store for Edit Vasadi when she moved to the United States after living in Serbia, Hungary, and Italy. One discovery was how friendly Americans are. They "make you feel like you really matter" by smiling, waving, and asking questions, Vasadi declares. Responding to these overtures can be tricky, however. Unaccustomed to strangers asking, "What have you been doing today?" or "How are you?", Vasadi says she often wonders whether to give an honest answer or a quick reply.[1]

Throughout the chapter, we'll learn from experts and from Vasadi and others who have navigated cultural differences. The odds are that you have engaged in intercultural communication yourself. It isn't necessary to be a world traveler to encounter people from different cultures. Social media and the internet make communicating around the world as easy as talking to a neighbor, and society is more diverse than ever.

This chapter explores what **culture** is, and when it does (and doesn't) affect communication. We'll take an imaginary trip around the world to see how people communicate in different cultures. Closer to home, we'll explore how factors such as **race**, gender, and age constitute cocultures, and then conclude with communication tips for learning about new perspectives. Along the way, you'll practice communication strategies for talking about race, communicating with people who have disabilities, discussing politics respectfully, and surviving culture shock.

> **culture** The language, values, beliefs, traditions, and customs shared by a group of people.

> **race** A construct originally created to explain differences between people whose ancestors originated in different regions of the world—Africa, Asia, Europe, and so on.

4.1 Understanding Cultures and Cocultures

Even everyday encounters can be puzzling when different cultures are involved. As we use the term, culture is "the language, values, beliefs, traditions, and customs people share and learn."[2] Edit Vasadi's observations relate to intercultural expectations about greeting behaviors. When people from different countries are involved, even fairly routine exchanges can be perplexing. But differences exists *within* a society as well. A **coculture** is a group that is part of an encompassing culture. The children of immigrants, for example, might be immersed in American culture while still identifying with the customs of their parents' homeland. Examples of other cocultures include:

> **coculture** A group that is part of an encompassing culture.

- Race and ethnicity (e.g., Black, Latino, Asian)
- Sex and gender (e.g., male, female, trans)
- Religion (e.g., Mormon, Muslim)
- Physical disability (e.g., wheelchair user, Deaf person)
- Political viewpoints (e.g., liberal, conservative)
- Age (e.g., child, teen, older adult)

You may identify with one or more of these groups and many others. Each is likely to have its own communication style, and vocabulary.

Salience

Depending on the situation, cultural differences may seem nearly insurmountable or practically nonexistent. Intercultural and intergroup communication—at least as the terms are used here—don't always occur when people from different backgrounds interact. Those backgrounds must have a significant impact on the exchange before we can say that culture has made a difference. The term **salience** describes how much weight people attach to cultural characteristics in a particular situation. Consider a few examples in which culture is more or less salient:

- A group of preschool children is playing together in a park. These three-year-olds don't recognize that their parents come from different countries, or even that they speak different languages. At this point we wouldn't say that intercultural or intergroup communication is taking place. Only when cultural factors (such as diet or parental discipline) become salient might the children think of one another as different.

- Members of a school athletic team—some Asian American, some Black, some Latina, and some European American—are intent on winning the league championship. During a game, cultural distinctions aren't salient. They may or may not seem important later, when the team talks about the game and what it means to win or lose.

- A husband and wife were raised in different religious traditions. Most of the time their religious heritages make little difference and the partners view themselves as a unified couple. Every so often, however—perhaps during the holidays or when meeting members of each other's family—the different backgrounds are more salient. At those times the partners may feel quite different from each other culturally.

These examples illustrate that intercultural communication is contextual. Even if people are from different backgrounds, they may communicate in much the same way depending on what they are doing.

salience How much weight people attach to cultural characteristics in a particular situation.

in-group Members of a social group with which a person identifies.

out-group People who do not belong to the social groups with which a person identifies.

In-Group and Out-Group

If you consider everyone you know, it's probably obvious that you identify more closely with some than with others. Social scientists use the term **in-group** to describe people you consider to be similar to you and with whom you have an emotional connection. The term **out-group** describes people you view as different and with whom you have little or no sense of affiliation.[3] Your sense of group membership may differ by situation. At home, your family members may feel in-group to you, but at a rock concert, your friends may feel more in-group than your family members.

Perceived differences can discourage intergroup communication. For example, one study revealed that students studying abroad were often not included in informal social gather-

Tina Fey, who produced *Mean Girls* the musical, has pointed out that Cinderella's stepsisters were the original mean girls. In the fairy tale, as in the *Mean Girls* movies and play, a clique with social privileges excludes others from joining the group.

When do you feel in-group? Under what circumstances do you feel excluded?

UNDERSTANDING YOUR COMMUNICATION

How Much Do You Know About Other Cultures?

Answer the following questions to test your knowledge about what is culturally appropriate around the world.

1. Japanese visitors are in town. You've heard that Japanese custom involves gift-giving. What should you know about this practice?
   a. It's important that gifts be expensive and of the finest quality.
   b. Avoid gifts that come in threes, as in three flowers or three candies.
   c. It's preferable to sign the accompanying card in green ink rather than black.
   d. It is not customary to wrap gifts in Japan.

2. You are interacting with a person who is Deaf and who uses an interpreter. What should you do?
   a. Address your comments to the interpreter, then look at the Deaf person to see how they react.
   b. Maintain eye contact with the Deaf person rather than the interpreter.
   c. Offer to communicate in written form so that the interpreter will be unnecessary.
   d. Speak very slowly and exaggerate the movements your mouth makes.

3. While traveling in China, you should be aware of which rule of dining etiquette?
   a. It's considered rude to leave food on your plate.
   b. You should put your drinking glass on your plate when you finish eating.
   c. Cloth napkins are just for show. Use a paper napkin to wipe your mouth.
   d. Avoid sticking your chopsticks upright in your food when you are not using them.

4. You are meeting with a group of Arab business people for the first time. What should you know?
   a. They favor greetings that involve kissing on each cheek.
   b. It's polite to say no if an Arab host offers you coffee or tea.
   c. Men tend to be touch avoidant and to stand at least 3 feet from one another during conversations.
   d. They consider the left hand unsanitary and hold eating utensils only with their right hands.

INTERPRETING YOUR RESPONSES

Read the following explanations to see how many questions you got right.

Question 1

Gift-giving is an important ritual in Japan, but gifts needn't be extravagant or expensive. The number 3 is fine, but avoid gifts that involve 4 or 9, as these numbers rhyme with the Japanese words for "death" and "suffering," respectively, and are considered unlucky.[11] Black is associated with death or bad luck, so green ink, which symbolizes good luck, is preferred.[12] Gift wrapping is expected and is even considered an art form. The correct answer is c.

Question 2

The short answer is: Treat Deaf people with the same courtesy as anyone else, which means maintaining eye contact and focusing on them. If the circumstances may make it difficult for the interpreter to see clearly, make arrangements in advance so that is not an issue.[13] The correct answer is b.

Question 3

Cultures vary in terms of whether it is rude to eat everything or rude not to. In China, leaving a little food on your plate lets your hosts know that they provided plentifully for you. However, sticking your chopsticks upright in your food evokes thoughts of funerals, where it's customary to place a stick of lighted incense upright in a container of rice.[14] The correct answer is d.

Question 4

Members of Arab cultures may kiss on each cheek, but usually only with people they already know well. A handshake is more appropriate for an introductory business meeting.[15] It's polite to accept a host's offer of coffee or tea. Men tend to speak at close distances (less than 3 feet) unless the conversation involves a woman, in which case it is typically considered rude to touch or crowd her.[16] It's considered unclean to eat with one's left hand (even if you are left handed),[17] harkening back to days when the left hand was used for personal hygiene. The correct answer is d.

ings among host-country students. As a result, they tended to feel lonely and isolated, learned less about the host culture than they might have, and associated more with students from their home countries than those from the host country.[4] A sense of isolation can also occur closer to home, especially for people who are considered out-group in many contexts. Music producer and singer Zhu says, "My whole life I have been the 'Other.' I'm not white enough to be accepted by the whites and not Chinese enough to be accepted by the Chinese."[5]

The good news is that, when people from different backgrounds get to know one another well, they often find that their common interests outweigh their differences.[6] And those who take part in intergroup friendships tend to experience greater well-being and less discrimination than individuals with fewer cross-group friendships.[7] This is especially true if the friends trust one another and make an effort to understand their different perspectives. Danny Richardson (who is White) and Erin "Big Debo" Ridgeway (who is Black) talk about friendship, racism, and hilarious life events in their podcast "My Black Friend."[8] "The main thing I learned through our friendship is to just listen," Richardson says. Even then, he advises, "stop trying to assume you know what everyone is going through. You may hear their experiences, but that doesn't mean you know what it's like to actually experience them."[9]

Although it's eye-opening to look at cultural differences, there are sometimes greater differences *within* cultures than *between* them. Within every culture, members display a wide range of communication styles. For example, Jaclyn Samson, whose parents are from the Philippines, is frustrated when people typecast her in ways that don't reflect who she really is. In school "people saw me as the quiet Asian girl in class who was good at math and kept to herself," she says, when actually "I was outgoing and talkative. . . . I had a ton of friends, and to be quite honest, I was terrible at math."[10] She makes a good point: Cultural differences are generalizations—broad patterns that do not apply to every member of a group.

Take the "How Much Do You Know About Other Cultures" quiz to gauge your knowledge of communication norms around the world. (Don't be discouraged if you don't make 100 percent. The important point is to be curious and eager to learn.)

4.2 How Cultural Values and Norms Shape Communication

Growing up in the Netherlands, Daniëlle didn't anticipate that she would fall in love with a Sudanese man. However, she and her now husband, Hussam, connected right away. "Even though we were from different continents, we had an insane amount of things in common," she says. "We both loved to read the same books and liked playing around with graphic design. We understood each other."[18]

Even so, Daniëlle was initially nervous about getting to know Hussam based on stereotypes she had heard about Arab men. Over time, she learned a valuable lesson. "We are not the stereotypes people have about us," she says. "We are all just people, with differences and similarities, strengths and weaknesses, habits and customs."

One way to reduce the uncertainty of communicating with people from different cultures is to better understand diverse norms and values. Here is a look at six patterns that help distinguish cultures around the world. Unless people are aware of

individualistic culture A group in which members focus on the value and welfare of individual members more than on the group as a whole.

collectivistic culture A group in which members focus on the welfare of the group as a whole more than on individual identity.

these cultural differences, they may consider others to be unusual—or even rude—without realizing that seemingly odd behavior comes from following different sets of beliefs and unwritten rules about the "proper" way to communicate.

Individualism and Collectivism

Members of **individualistic cultures**—such as the United States, Canada, and Great Britain—tend to regard people as unique and independent. Idioms such as "Be your own person" and "Stand up for yourself" emphasize individuality. Conversely, communicators in **collectivistic cultures**—including those in China, Korea, and Japan—typically put more emphasis on membership in groups, such as one's extended family, community, or workplace.[19,20] Popular sayings in Japan include "Different body, same mind" and "Child of a frog is a frog," which emphasize shared values and characteristics.

Following are six differences in the ways members of individualistic and collectivistic cultures tend to approach communication.

Names When asked to identify themselves, individualistic Americans, Canadians, Australians, and Europeans usually respond with their given name and then their surname. But collectivistic Asians do it the other way around. They're likely to begin with their family name and provide their given name second[21] because their primary emphasis is on group membership rather than individual identity. It's easy to imagine the confusion created when Western paperwork asks for a person's "first name" and "last name," assuming that everyone orders names the same way.

Pronouns When American-born Ann Babe moved to South Korea to teach English, she was confused when a colleague said, "Our husband is also a teacher." Then Babe came to understand that South Koreans' use of pronouns reflects a cultural emphasis on collectivism. They use *our* and *my* (also *we* and *I*) interchangeably but prefer the collective terms because they seem less egocentric. That helped her understand why South Korean students were initially put off by English, which sounded selfish to them. As one student put it, everything in English is "my, my, my" and "me, me, me."[22]

Direct or Indirect Speech Members of individualistic cultures are relatively tolerant of conflicts and tend to use a direct, solution-oriented approach. In Hungary, where individualism is highly valued, the locals pride themselves on being straight talkers who don't shy away from sensitive topics.[23] "Hungarians are self-expressed and to-the-point," says a Californian who married a woman from Hungary and moved there, adding, "If someone has the

Elon Musk and Grimes have named their children Exa Dark Sideræl and X Æ A-12 to reflect terms from supercomputing and space exploration.

Why did your parents give you the name they did? How does your name influence your life?

slightest problem with something, they're going to let you know.... That's just the way it is here. Don't take it personally—tempers flare, decibels rise."[24]

By contrast, members of collectivistic cultures are less direct, often placing greater emphasis on harmony and group interests. "Everyone is soooo polite," observes Spike Daeley of his time in Japan, saying, "Two of the most common phrases you'll hear in Japanese are 'sumimasen' and 'gomenasai.' They mean 'excuse me' and "I'm sorry,' respectively."[25] The emphasis there is on maintaining good relations and not causing offense.

Competition or Collaboration Individualistic cultures are characterized by self-reliance and competition. A website for international students explains, "Americans thrive on competition. From a young age, children are encouraged to work hard and try their best.... Even Girl Scouts vie to sell the most cookies."[26]

By contrast, collectivistic cultures are more focused on collaboration and group effort.[27] As a result, members are often adept at seeing others' point of view. When researchers studied people in 63 countries, they found that those who grew up in collectivistic cultures were typically more empathic and more likely to volunteer than were people from individualistic cultures.[28]

There are pros and cons to both perspectives. Competition can be fun and motivational. However, it may also damage relationships and create hostile environments.[29] Cooperation can enhance relationship development, trust, and open communication. At the same time, it may discourage people from challenging each other's opinions and reporting bad behavior.[30]

Humility or Self-Promotion In some cultures, it's considered polite to downplay one's accomplishments. To investigate, researchers asked female college students in Iran to answer questions about the quality of instruction at the school.[31] The real purpose of the interviews, however, was to see how the women would respond when the researchers complimented them. When a research assistant told a student, "Your shoes are really beautiful," the student replied, "This is not so stylish." Another time, after a researcher complimented a participant's phone, the student said, "You can have it." The lead researcher reflects that, in the collectivistic Iranian culture, these responses are courteous ways to demonstrate modesty and to avoid implying that you are better than anyone else. (To members of the Iranian culture, it's clear that "you can have it" is most often a polite response to a compliment, not an actual invitation to assume ownership.)

By contrast, individualistic Americans are relatively comfortable talking about their personal accomplishments and even saying they are "the best" at something. This doesn't mean they are more self-absorbed than people in other cultures. From an American's perspective, sharing news of an accomplishment can be a means of including others in a happy life event. For example, if an American student says "I made an A on test!" a friend may say, "That's great! Let's celebrate." In other circumstances, "boasting" can be a form of verbal play, as when one person tells another "My team will crush yours in the big game" even if that is very unlikely. These behaviors are not usually considered rude. In fact, self-promotion is expected, and even respected, in an individualistic culture. An employee or job candidate who downplays their accomplishments may be seen as unconfident or ill prepared.

Communication Apprehension Cultural differences can also affect the level of anxiety people feel when communicating. In societies in which the desire to conform is great, people admire those who are reserved and quiet while they communicate with people outside of their intimate circle. Since standing out from the crowd is frowned upon, it's understandable that people in those cultures tend to be especially apprehensive about public speaking.[32] By contrast, people in more individualistic cultures typically revere those who are comfortable being in the limelight and aspire to be like them. That's not to say that collectivistic people are all shy, or that individualistic people are fearless when it comes to public speaking. It's more accurate to say that cultures emphasize and admire some ways of communicating more than others.

Table 4.1 illustrates some of the differences between individualistic and collectivistic cultures.[33,34,35]

High and Low Cultural Context

Cultures also differ in how much they rely on words versus situational cues.[36,37] Members of **low-context cultures** mostly use *language* to express thoughts, feelings, and ideas. They prefer specific, straight-forward communication. Mainstream cultures in the United States, Canada, northern Europe, and Israel fall toward the low-context end of the scale. Edit Vasadi says she's amused by signs in the United States that tell people not to leave their children unattended, how to wash their hands, and so on.[38] It may not seem necessary to post messages such as these, but members of low-context cultures tend to spell things out, even when the meaning seems rather obvious.

By contrast, members of **high-context cultures** rely heavily on unspoken and situational cues. As the term suggests, they interpret meaning based more on the *context* in which a message is delivered—the nonverbal behaviors of the speaker, the history of the relationship, and the general social rules that govern interaction between people, and so on. They are typically attentive listeners who place a higher value on social harmony than on direct communication.[39] People in most Asian, Latin, and Middle Eastern cultures fit this pattern. Let's explore two communication activities that illustrate the difference between high and low context.

low-context culture A group that uses language primarily to express thoughts, feelings, and ideas as directly as possible.

high-context culture A group that relies heavily on subtle, often nonverbal cues to convey meaning and maintain social harmony.

TABLE 4.1

Individualistic Versus Collectivistic Cultures

INDIVIDUALISTIC CULTURES	COLLECTIVISTIC CULTURES
People are likely to emphasize aspects of themselves that are unique and independent.	People are likely to emphasize their membership in particular groups.
Cultural members are relatively comfortable with conflict and are likely to address it directly.	Cultural members emphasize harmony and address conflict indirectly.
A high value is placed on competition and self-promotion.	Empathy, humility, and cooperation are highly valued.
Those who call attention to themselves are admired.	Those who are quiet are admired.

Doing Business Imagine a meeting that involves professionals from around the world. Members of low-context cultures will probably want to "get down to business" quickly by talking about the details of a deal or an idea. They tend to be verbal, direct, and results oriented. Talking for very long about other matters, such as personal information, can seem to them like a waste of time.[40]

By contrast, "in countries such as Mexico, conversations are first and foremost an opportunity to enhance the relationship," observe international business consultants Melissa Hahn and Andy Molinsky.[41] From a high-context perspective, focusing on business without first developing a relationship can feel dismissive and disrespectful. It implies that the other person has no value. And because members of high-context cultures focus on nonverbal cues and what *isn't* said, they may wonder how they are supposed to trust or understand the other person without getting to know them first.

Managing Conflict Members of low-context cultures are likely to state their concerns and complaints up front, whereas people raised in high-context cultures usually hint at them.[42,43] An exchange student gives an example:

> Suppose a guy feels bad about his roommate eating his snacks. If he is Chinese, he may try to hide his food secretly or choose a certain time to say, "My snacks run out so fast, I think I need to buy more next time." Before this, he also may think about whether his roommate would hate him if he says something wrong. But Americans may point out directly that someone has been eating their food.

The roommate from China may feel that it's obvious, based on the situation and his indirect statement, that he's upset about the roommate eating his food. But an American—who may expect a friend to say outright if they are upset—may miss the point entirely.

In summary, it's easy to see how the clash between directness and indirectness can present challenges. To members of high-context cultures, communicators with a low-context style can appear overly talkative, lacking in subtlety, and redundant. To people from low-context backgrounds, high-context communicators often seem evasive, or even dishonest. As with all cultural mores, no approach is inherently better than another. The point is to understand each other better by recognizing differences, and ideally, finding ways to meet everyone's goals. Table 4.2 summarizes some key differences in how people from low- and high-context cultures use language.

Uncertainty Avoidance

Uncertainty may be universal, but cultures have different ways of coping with it. **Uncertainty avoidance** reflects the degree to which members of a culture feel threatened by ambiguous situations and try to avoid them.[44] As a group, residents of some countries (including Australia, Great Britain, Denmark, Sweden, and the United States) tend to embrace change, whereas others (such as natives of Ecuador, Indonesia, Iran, and Turkey) are more likely to find new or ambiguous situations uncomfortable.[45]

In cultures that avoid uncertainty, definitive information is highly valued. For example, people from Turkey—who tend to have low tolerance for ambiguity—may be frustrated when their physicians take a "wait and see" approach or seem unsure about what's wrong with them. A Turkish immigrant in the Netherlands told researchers that it's unsatisfying when her doctor asks her questions rather than

> **uncertainty avoidance**
> The cultural tendency to seek stability and honor tradition instead of welcoming risk, uncertainty, and change.

TABLE 4.2

High- and Low-Context Communication

LOW CONTEXT	HIGH CONTEXT
The majority of information is carried in explicit verbal messages, with less focus on the situational context.	Information is carried in contextual cues such as time, place, relationship, and situation. There is less reliance on explicit verbal messages.
Business people typically prefer to "get down to business" quickly and may be impatient with conversations that don't seem results oriented.	Business people usually spend time getting to know one another and developing trust before they talk specifically about the business at hand.
Conflict is often addressed directly and verbally.	People are likely to manage conflict indirectly, through hints and nonverbal cues.

focusing on observable indicators of her condition. "Sometimes I think: what kind of question is this?" she said, exclaiming, "All these stupid questions. Do some research!"[46]

The misunderstandings that arise in relation to ambiguity can be striking. A person who stands out as different or who expresses ideas that challenge the status quo may be considered untrustworthy or even dangerous by people who avoid uncertainty, but may be seen as a visionary and an inspiration by those who are comfortable with change. The take-away lesson is that, people from different cultures are likely to have varied reactions to ambiguous information and to communicate differently about it. (See Table 4.3 for a summary of differences.)

Power Distance

American Kate Sweetman remembers being bewildered when an Indonesian coworker with great ideas often failed to speak up in meetings.[47] The quiet colleague's

TABLE 4.3

Differences Between Low and High Uncertainty Avoidance

LOW UNCERTAINTY AVOIDANCE	HIGH UNCERTAINTY AVOIDANCE
Uncertainty in life is accepted and even welcomed.	Uncertainty is unsettling and can feel threatening.
People may be willing to accept that a clear answer is not possible.	Clear, well-supported explanations are expected.
Those who "march to the beat of their own drum" or "forge new paths" are often admired.	People and ideas that don't fit the norm may be regarded with suspicion.

reluctance is easier to understand if you know that she grew up in a culture that observes high **power distance**, which is the perceived gap between those with substantial power and resources and those with less.

In cultures with high power distance, it's considered natural and respectful to treat some people as more important than others. Examples include Malaysia, India, Singapore, and Thailand.[48] Traditionalists in those countries are likely to expect obedience and to honor the opinions of authority figures without question. For example, medical patients in Pakistan, Japan, and Thailand seldom question their physicians' advice or pose questions because they think it might seem disrespectful.[49] In Japan, new work associates exchange business cards immediately, which helps establish everyone's relative status. The oldest or highest ranking person receives the deepest bows from others, the best seat, the most deferential treatment, and so on. This treatment is not regarded as elitist or disrespectful. Indeed, treating a high-status person the same as everyone else would seem rude.

At the other end of the spectrum are cultures with low power distance, in which people believe in minimizing the difference between various social classes. They tend to subscribe to the egalitarian belief that one person is as good as another regardless of their station in life—rich, poor, educated, or uneducated. Argentina, Denmark, Colombia, and Finland are among the most egalitarian countries.[50] Most cultures in the United States and Canada value equality, even though that ideal is not always perfectly enacted. Americans often call supervisors by their first names and engage in brainstorming sessions and even debates with people of higher status.

The potential for misunderstanding is great. In the workplace, employees who are accustomed to high power distance may consider it inappropriate to call attention to themselves or voice their own opinions (especially if they differ from those of people higher up the corporate ladder). The reverse is true in cultures with low power distance, where employees may feel unappreciated when they aren't consulted or when their opinions are ignored. If a manager is accustomed to high power distance, they may give high marks to a deferential employee who communicates submissively toward them. But if the leader is accustomed to low power distance, they may assume that quiet and nonassertive employees lack initiative and creativity—traits that help people gain promotions in their culture. (See Table 4.4 for an overview of low and high power distance cultures).

> **power distance** The perceived gap between those with substantial power and resources and those with less.

TABLE 4.4
Differences Between Low and High Power Distance

LOW POWER DISTANCE	HIGH POWER DISTANCE
It's assumed that all people are created equal.	It's accepted that some people are more important than others.
People who question authorities and the status quo are often considered visionary.	People are expected to follow authority figures' decisions without question.
It's usually permissible to communicate with authority figures in a friendly manner.	It's usually required that people treat authority figures with deference and social distance.

Talk and Silence

Beliefs about the very value of talk differ from one culture to another.[51] Members of Western cultures tend to view talk as desirable and use it for social purposes as well as to perform tasks. Silence can feel embarrassing and awkward in these cultures. It's likely to be interpreted as lack of interest, or as hostility, anxiety, shyness, or interpersonal incompatibility.

On the other hand, silence is valued in Asian cultures. For thousands of years, many cultures have favored silence over unrestrained verbal expression of thoughts and feelings. Taoist sayings propose that "In much talk there is great weariness" and "One who speaks does not know; one who knows does not speak." Unlike most Westerners, who tend to find silence uncomfortable, Japanese and Chinese traditionalists more often believe that remaining quiet is proper when there is nothing to be said. A talkative person is often considered a show-off or a fake.

Sometimes silence says more than words.

Describe the last time you were with a group of people in silence. How did it feel?

Members of some Native American communities revere silence as well. For example, traditional members of Western Apache tribes maintain silence when others lose their temper. As one member explained, "When someone gets mad at you and starts yelling, then just don't do anything to make him get worse."[52] Apache also consider that silence has a comforting value. The idea is that words are often unnecessary in periods of grief, and it's comforting to have loved ones present without the pressure to maintain conversations with them.

It's easy to see how these different views about speech and silence can lead to communication problems when people from different cultures meet. Both the "talkative" Westerner and the "silent" Asian and Native American are behaving in ways they believe are proper, yet each may view the other with disapproval and mistrust. Only when they recognize the different standards of behavior can they adapt to one another, or at least understand and respect their differences.

4.3 Cocultures' Influence on Communication

Much of how individuals view themselves and how they relate to others grows from their cultural and cocultural identity—the groups with which they identify. Where do you come from? What's your ethnicity? Your religion? Your gender? Your age? All of these group identities are important, but their significance is different for every person.

Intersectionality Theory

Imagine that you're an Italian American who uses a wheelchair. It would be preposterous for someone to assume that they understand your life because they have a friend who is Italian and another who uses a wheelchair.

Intersectionality theory observes that each person experiences life at the juncture of multiple identities that give rise to a unique perspective and collection of experiences.[53,54] The theory recognizes that it's a very different experience to be an Asian gay man than an Amish gay man, a Latino custodian than a Latino CEO, and so on.

Some people experience life at the intersection of multiple identities that put them at a disadvantage socially. For example, in the United States, women make about 18 percent less money than men, and Black men make about 13 percent less than White men.[55,56] So you might expect Black women's salaries to be in the same range, somewhere between 13 percent and 18 percent less. But they're not. They average 35 percent less than men overall—far lower than the average pay for women *and* for Black men.[57,58] The issue isn't that Black women are unqualified. Females earn about 61 percent of master's degrees conferred in the United States,[59] and nearly 70 percent of Black students pursuing advanced degrees in the United States are female.[60] The explanation, experts say, has more to do with cultural assumptions and communication:

- Because women of color (Asian, Black, and Latina) comprise only 3 percent of top-level executives and 9 percent of senior managers in the United States,[61] opportunities to develop role models and mentors are relatively scarce.

- Because women of color aren't well represented in leadership roles, people may have a hard time recognizing their potential and envisioning them in high-paying positions.

- Women have traditionally been taught that it's inappropriate to be hard bargainers on their own behalf.

The good news is that you can use communication to help minimize such gaps. No matter what your race or gender, if you can offer career advice and networking assistance, consider reaching out. "I've always imagined how much more confident and inspired I would feel knowing that I could identify with people at the top," says journalist Rebecca Stevens A.[62] If you have a place at the table during hiring decisions, point out that judgments such as "this candidate doesn't fit our culture" rule out diverse candidates who may have a lot to offer.

Here we look at some—though by no means all—of the factors that help shape cultural identity, and hence the way people perceive and communicate with others. As you read on, think about other cocultures that might be added to this list. With intersectional theory in mind, treat the descriptions not as a formula for understanding particular individuals, but as a way of exploring qualities that contribute in countless ways to creating cultural diversity.

> **intersectionality** The idea that people are influenced in unique ways by the complex overlap and interaction of multiple identities.

Race and Ethnicity

The notion of race was created hundreds of years ago to reflect differences between people whose ancestors originated in different regions of the world—Africa, Asia, Europe, and so on. Modern scientists acknowledge that, although there are some genetic differences between people with different heritage, they mostly involve

Tennis champion Naomi Osaka has been open about her experiences as a person of Haitian and Japanese descent.

How do the racial and ethnic groups to which you belong affect the way you communicate with others?

superficial qualities such as hair color and texture, skin color, and the size and shape of facial features. Consequently, race is not a reliable indicator of individual differences. As one analyst puts it, "There is less to race than meets the eye. . . . Knowing someone's skin color doesn't necessarily tell you anything."[63]

There are several reasons why race has little use in explaining individual differences. Most obviously, racial features are often misinterpreted. A well-traveled friend of ours from Latin America says that she is often mistaken as Italian, Indian, Spanish, or Native American.

More importantly, there is more genetic variation within races than between them. For example, some people with Asian ancestry are short, but others are tall. Some have sunny dispositions, while others are more stern. Some are terrific athletes, while others were born clumsy. The same applies to people from every background. Even within a physically recognizable population, personal experience plays a far greater role than superficial characteristics such as skin color. As you read in Chapter 3, stereotyping is usually a mistake.

Ethnicity is a social construct that refers to the degree to which a person identifies with a particular group, usually on the basis of nationality, culture, religion, or some other perspective. Ethnicity isn't tied to physical indicators. For example, people from around the world who speak Spanish may look very different from one another but experience a sense of shared identity to some degree. A person may have physical characteristics that appear Asian, but may identify more strongly as a Presbyterian or a member of the working class. Intersectionality theory draws attention to the fact that people have many overlapping identities.

Identifying with more than one group can be a bonus. Research indicates that people who do are usually more comfortable establishing relationships with a diverse array of people, which increases their options for friendships, romantic partners, and professional colleagues.[64]

Although there is no significant biological basis for race and ethnicity, their impact is powerful at a social level. About 10,000 racial slurs a day appear on Twitter alone.[65] More than half of Black, Latino, and Asian teens in the United States say they encounter demeaning comments about racial and ethnic groups online.[66] At the same time, social media can be a platform for people to speak out about discrimination. More than 1 in 10 social media users in the United States say they have changed their views (one way or another) about race relations based on content they have seen posted.[67] See the feature "Talking About Race" for strategies to discuss this important topic in a respectful and open-minded way.

ethnicity A social construct that refers to the degree to which a person identifies with a particular group, usually on the basis of nationality, culture, religion, or some other perspective.

COMMUNICATION STRATEGIES

Talking About Race

Talking about race is uncomfortable, reflects writer Matt Vasilogambros, "and when we do talk about race, it's usually with people who look like us."[68] Despite the challenges, respectful conversations about race are essential to improve understanding and establish equal opportunities. The suggestions here can help you engage constructively with people whose experiences are different from your own.

Expect Strong Emotions

There will almost certainly be emotional heat in conversations about race.[69] It can help to cultivate what writer David Campt calls "mindful courage"—a commitment to listen respectfully, exercising curiosity and humility.[70]

Put Yourself in the Other Person's Shoes

"You may think that someone is making a mountain out of a molehill," says Ijeoma Oluo, author of *So You Want to Talk About Race*, "but when it comes to race, actual mountains are indeed made of countless molehills stacked on top of each other."[71] It's easy to assume that racism doesn't exist if you haven't been the target of it. Interracial conversations are one way to gain insights.

Don't Debate

Journalist and author Renni Eddo-Lodge says she's tired of talking to White people about race because, all too often, many of them "try to interrupt, itching to talk over you but not to really listen, because they need to let you know that you've got it all wrong."[72] If you are White, you can avoid this frustrating pattern by focusing your efforts on listening and seeking to understand the experiences of others.

Learn and Apologize, If Appropriate

One stumbling block to open communication about race is a fear of saying the wrong thing or being misunderstood. Even with the best motives, people sometimes blunder and cause pain. The real test of good intentions is to learn from those mistakes and do better in the future. Invite people around you to let you know if your words or actions cause pain. And if they point something out, listen, thank them, apologize, and avoid making the same mistake again. If you mean it sincerely, you might say, "Now that you've explained, I can see how my remarks were hurtful. I'm really sorry. I learned a lesson here."

Be an Ally

If you hear someone make a hurtful statement (whether or not it's directed at you), a good approach is to remain calm and point out why the person's statement is problematic. For example, if someone says "Black people are always late," Oluo suggests that "you can definitely say, 'Hey, that's racist' but you can also add, 'and it contributes to false beliefs about black workers that keep them from even being interviewed for jobs.'" The value of this communication approach, says Oluo, is that it calls attention to the harmful effects of racism and reduces the chance the speaker will say, "It's not that big of a deal, don't be so sensitive."[73]

Don't Force the Issue

If you haven't experienced racism, you may have a genuine desire to learn from people who have, but it isn't fair to insist that they tell you about it. Talking about such a sensitive subject requires a great deal of emotional labor. As writer John Meta puts it, "When you ask a Black person to teach you how to be a better white person, you are scratching a wound."[74] Asking questions can be relatively easy and passive, but Meta points out that answering them can be painful and exhausting. White people can educate themselves about racism via books, lectures, blogs, and other resources. When people of color are asked questions they would rather not answer, they can suggest resources such as *So You Want to Talk About Race*,[75] *White Fragility*,[76] *Uncomfortable Conversations with a Black Man*,[77] and *Why Are All the Black Kids Sitting Together in the Cafeteria?*[78] Helpful communication strategies are also provided in this book—such as how to avoid and respond to microaggressions (Chapter 5), be a better listener (Chapter 6), and develop friendships with a diverse array of people (Chapter 9).

If all of this feels overwhelming, take heart. Even Ijeoma Oluo, who wrote a book about the subject, finds it difficult to talk about social differences. "I'd love to not talk about race ever again," she attests. Yet she keeps the dialogue open because she believes that communication is the key to bridging differences and facilitating change. "Our humanity is worth a little discomfort," she says, 'it's actually worth a *lot* of discomfort.'"[79]

Sex and Gender

At different times, Paula Stone Williams has experienced the world as a man and as a woman. "The differences are massive," Williams says.[80] Men's clothing is more comfortable, and the pockets are larger. But the biggest difference, she says, is how people communicate with her. "Apparently, since I became a female, I have become stupid," she says wryly. People are bossier toward her and more likely to question her intelligence since the gender transition.

Williams's story involves the difference between sex and gender. Although people tend to use the terms interchangeably, they actually mean different things. **Sex** is a biological category. A person's sex might be classified as male or female based on physiological features. But physical attributes and hormone levels make biological sex a more complicated formula than you might think. Everyone's body produces some estrogen and some testosterone, which are sex-related hormones that influence how the body and mind develop.[81] Biologically speaking, no one is entirely masculine or feminine,[82] so it's unrealistic to expect people to observe strict gender differences in the way they communicate.

In contrast to sex, **gender** is a socially constructed set of expectations about what it means to be masculine, feminine, and so on. Expectations about gender are often communicated the moment people are born—or even before, with the selection of a name, clothing, toys, and nursery décor—and may be renegotiated through their lives.

Gender Identities   We live in a gender-expansive time, meaning that society recognizes a wide range of identities.[83] Chapter 5 explores how language evolves to describe a changing world and our place in it. For now, let's consider a few of the terms used to describe gender identities:

- *Cisgender* describes a person whose current gender is the same as the sex attributed to them at birth. For example, a person who was categorized as female at birth and whose identity is female is considered a cis woman.
- *Transgender* individuals' gender is different from the biological sex attributed to them at birth. Paula Stone Williams says she felt like a girl from a young age, although society viewed her as male until she came out as a transgender woman.
- *Nonbinary* and *genderqueer* describe those who do not identify according to a dichotomous notion of male or female. Suzannah Weiss, who identifies as nonbinary, explains: "Growing up, I never felt people were *wrong* when they called me a woman, but it felt like a label imposed on me rather than one that fit.... I learned about non-binary identity, and that did fit."[84]
- *Agender* individuals do not identify in terms of gender.[85] Nessie Avery describes what it feels like to be agender: "I'm not a woman. I'm not a man. I'm not somewhere between the two. I'm just me, and I don't fit in any of the boxes."[86]

This list is by no means complete. By some accounts, there are 64 terms to describe gender, and the list is growing.[87] The feature "Communicating Respectfully About Gender" presents some thoughtful ways to honor people's gender identity.

Lewis Hancox, a filmmaker and comedian who is transgender, reflects on people's reactions to diverse gender identities. Some are supportive. "Other people may

---

sex A biological category such as male or female.

gender A socially constructed set of expectations about what it means to be masculine, feminine, and intermediate points along a masculine-feminine continuum.

> **COMMUNICATION STRATEGIES**
>
> ## Communicating Respectfully About Gender
>
> Here are a few ways you can demonstrate to others that you honor their gender identity.
>
> **Use Preferred Pronouns and Share Yours**
>
> "You can't always know what someone's gender pronoun is by looking at them," says a spokesperson for the Lionel Cantú Queer Center at UC Santa Cruz.[89] Therefore, it's useful to pay attention to people's preferences, share your own, and do everything possible to use preferred pronouns. Many individuals now include their pronouns (such as *he/him*, *she/her*, or *they/them*) next to their name in video conferencing meetings and in their email signatures. It's also fine to ask, "What pronouns do you use?"[90]
>
> **Avoid Deadnaming**
>
> Calling someone by a previous name after they have transitioned to one that suits them better is known as deadnaming, and it can be hurtful. "Any time I hear my deadname, it strikes this feeling in me that, oh yeah, other people will never see me as the person I see myself as, and that's really hard," says Hanley Smith, who is trans.[91]
>
> **If You Mess Up, Apologize and Correct Yourself**
>
> If you accidentally misgender someone—as by calling them by a deadname or unpreferred pronoun—say you're sorry and make the correction, even if they aren't present. For example, you might say, "I'm sorry. I mistakenly referred to Tracy as she. I meant to say he." Your sincere desire to get it right is likely to be appreciated.

think you're confused," Hancox observes. No matter your gender, he says, the best policy is to love and accept yourself, connect with people who are supportive, have patience with those who don't immediately know how to respond, and keep in mind all the qualities that make you unique and special as an individual. "It took me a long time to realize that being transgender didn't make me any less of a guy, or more importantly, any less of a person," Hancox says. "We're all different in our own right and we should embrace those differences."[88]

Social Expectations Although gender roles are evolving, traditional beliefs are still present in many ways. For example, people who frequently watch crime shows and sporting events on TV are more likely than others to believe that men should be physically strong, socially dominant, and traditionally unfeminine.[92] The same pattern is evident in those who regularly play video games with violent content.[93] People who subscribe to such beliefs are likely to consider it acceptable for men to communicate assertively and to compete for promotions in the workplace, but consider it inappropriate for women to display the same behaviors.[94]

Because gender roles are learned, they can change. Some people feel they should. Actor Justin Baldoni, for one, says he's tired of social ideas about masculinity that devalue women and limit his experience of the world. "It is exhausting trying to be 'man enough' for everyone all the time," Baldoni says. He rejects the notions that masculine is the opposite of feminine and that "boys are strong and girls are weak."[95]

Sheryl Sandberg, a former Meta/Facebook executive and author of *Lean In*, is another advocate for change. She calls attention to a "broken rung" in the corporate ladder that often keeps qualified women from advancing.[96] Part of the issue involves the way people think about and talk about gender. A study of 132 companies revealed that women are more likely than men to be labeled "bossy," "intimidating," and "aggressive" when they ask for a raise.[97] With the conviction that words matter, a range of celebrities—including Beyoncé, Condoleezza Rice, Jennifer Garner, Jimmie Johnson, and others—are part of the "Ban Bossy" movement that urges people to stop calling girls and women "bossy" when they are assertive.[98]

Religion

Religion often shapes how and when people communicate. It may influence what they eat and drink, what holidays they observe, what type of services they attend, and much more.[99] In some regions of the United States, religion is so integral to social life that newcomers are often asked, "What church do you attend?"

Unfortunately, religious differences are sometimes used as the basis for harsh judgment of out-group members. Peace-loving people who are Muslim and who live in the United States have often been singled out and vilified by people who stereotype them as terrorists. Yasmin Hussein, who works at the Arab American Institute, reflects:

> Many Muslims and individuals of other faiths who were thought to be Muslim have been attacked physically and verbally. Young children have been bullied at schools, others told to go back home and social media has become at times (a lot of the time) an ugly place to be on.[100]

In an effort to dispel unrealistic stereotypes, tens of thousands of Muslims have joined the #NotInMyName social media movement in which they denounce terrorist groups such as ISIS and condemn violence in the name of the religion.

Although religion can be a touchy topic, it's an important one in some situations. For example, you may wish to know about a romantic partner's spiritual beliefs, or they may be curious about yours. If you're the one asking, relationship therapist Becca Hirsch recommends simply saying, "Do you have any religious or spiritual beliefs?" and then listening attentively. When sharing your beliefs, Hirsch suggests presenting them in a clear and succinct way that isn't designed to convert the other person or assert that your perspective is better. You can decide for yourself if religious differences are a deal-breaker concerning a long-term relationship. But even if they are, Hirsch says, be curious, open, and respectful in the way you communicate about them.[101]

Religion can be a sensitive subject in the workplace. When researchers in the United States interviewed hundreds of people of many different faiths, they found that many of them avoid discussing their beliefs at work because the topic feels overly personal or they fear it will create opportunities for discrimination.[102] Some make casual references to religion ("Oh, did you know I'm Jewish?") if the topic seems relevant or if it's necessary to request religious holidays off. Of the relatively small number of people who talk about religious views with coworkers, most said they proceed cautiously to make sure others are receptive and interested first.

Disability

"When my mom first told me I had Down syndrome, I worried that people might think I wasn't as smart as they were, or that I talked or looked different," remembers Melissa Riggio. "But having Down syndrome is what makes me 'me.' And I'm proud of who I am. I'm a hard worker, a good person, and I care about my friends."[103] Riggio's comments are a good reminder that people with disabilities are mostly like everyone else—except that people often treat them as if they are different or helpless. As in any group, there's no absolute consensus among people with disabilities on how to communicate respectfully and accurately, but here are some overall general best practices.

Act the Way You Normally Do It's common courtesy to look people in the eye, engage with them, and treat them in an age-appropriate manner. Don't treat people with disabilities any differently. Speak directly to the person, whether or not they have someone with them, and whether or not they have difficulty responding.

Shaquem Griffin is the first player with one hand to be drafted to play in the National Football League. His hand was amputated when he was four because of a painful condition he was born with. Griffin has said that his mother encouraged him that he could accomplish anything.

Who encourages you when things are hard? What is the best thing they can say and do?

Phoebe is a bright, outgoing 11-year-old. Symptoms of cerebral palsy make it necessary for her to use a wheelchair and to speak with the use of an electronic device. She loves math, swimming, and skiing, and hopes for a career in television. Despite Phoebe's age and abilities, strangers tend to use baby talk when they address her—or they speak *about* her as if she isn't present, as in asking her parents, "How old is your daughter?" It's hard to imagine an 11-year-old who would appreciate being treated that way, and Phoebe is no different. "I am clever. I am just like you," she says, "You might meet me or someone else with a physical disability. Don't treat us like a baby or ignore us. Talk to me."[104]

Be Cautious About Supposed Euphemisms Terms such as "differently abled," "challenged," and "special" are typically not preferred.[105,106] Emily Ladau of the Center for Disability Rights explains that it can feel demeaning "when non-disabled people try to dance around the world 'disabled'" as if it is a "negative quality or derogatory word, when in fact, disabled is what I am."[107]

Avoid Judgments Don't use language that implies that people with disabilities should be pitied ("victim" or "sufferer") or regarded as special ("heroic" or "inspiring"). Also avoid saying that someone is "wheelchair-bound." Instead, say that they use a wheelchair.[108] As Dot Nary puts it, "I personally am not 'bound' by my wheelchair. It is a very liberating device that allows me to work, play, maintain a household, connect with family and friends, and 'have a life.'"[109]

Adapt but Don't Overreact Small communication adaptations can be helpful. For example, when meeting a person who is visually impaired it's courteous to identify yourself and others who may be with you. And when communicating with a person who is hearing impaired, make eye contact and make sure they can see your mouth while you speak. But don't overcompensate. Most of the time, shouting, speaking slowly, or making exaggerated movements is unnecessary and makes the encounter awkward. "If we miss something, we might ask you to say it again or say it slowly or in a different way," says Charlie Raine, who is Deaf. But for the most part, she urges, "just speak naturally."[110]

Don't Presume That People Need Help If someone is laboring to reach an item that's within your reach, they'll probably appreciate it if you offer to help, whether they have a disability or not. But nobody likes it when others presume that they are helpless or force "assistance" on them without asking. Devarshi Lodhia, a university student who has one hand, says that it's frustrating and embarrassing when people insist on helping him lock up his bike or when restaurant servers cut up his food before bringing it to him.[111] Likewise, people in wheelchairs don't appreciate it when others move or push them without permission. Recognize that a wheelchair is part of the user's personal body space, and except for rare instances, they are probably able to use it without assistance.

Don't Be Nosy "Would you ask a stranger why they wear glasses? Nope," says Barbara Twardowski. But people often ask her why she uses a wheelchair. "This line of questioning quickly becomes far more personal that I am comfortable discussing with a stranger," she says, suggesting, "Instead, ask me if I like to cook, then tell me about your killer recipe for guacamole." An exception, Twardowski says, is the genuine curiosity of children. "I understand how direct they can be, but there's no need for you to be embarrassed or correct them for it."[112]

Relax Don't be embarrassed if you happen to use common expressions such as "See you later" or "Did you hear about that?" that seem to relate to a person's disability.

Melissa Riggio, whose story we began with, sums it up this way: "I can't change that I have Down syndrome, but one thing I would change is how people think of me.... Judge me as a whole person, not just the person you see. Treat me with respect, and accept me for who I am. Most important, just be my friend. After all, I would do the same for you."[113]

Political Viewpoints

As one analyst quipped, "Liberals like to have fun. Conservatives like to have fun. Liberals and Conservatives like to have fun with each other .... unless they're talking about politics."[114] Political discussions have become one of the most contentious examples of intercultural communication in the United States.

It's no secret that politics can push emotional hot buttons. Nearly two-thirds of Americans say they avoid talking about the subject for fear of offending others or losing their jobs.[115] Online, about half of the people who have experienced harassment say it had to do with their political beliefs.[116] And about the same percentage of social media users say they're tired of seeing political messages online, mostly because there seems to be little room for a respectful exchange of ideas.[117]

Part of the problem is an effort by some people to inflame online discussions. **Social media trolls** are individuals whose principal goal is to disrupt public discourse by posting false claims and prejudiced remarks, usually anonymously. In a similar but more personal way, **social media snarks** post insulting comments about others to get a rise out of them. You can minimize the hurtful impact of these saboteurs by following guidance in the feature "Discussing Politics Responsibly on Social Media."

---

social media trolls Individuals whose principal goal is to disrupt public discourse by posting false claims and prejudiced remarks, usually anonymously.

social media snarks People who post insulting comments about others.

COMMUNICATION STRATEGIES

Discussing Politics Responsibly on Social Media

If you've ever been tempted to block, unfriend, or unfollow someone on social media because their political comments make your blood boil, you appreciate how emotional the subject can be. "Debates about politics aren't just about issues," observes leadership expert Andrew Blotky, "they are now almost always about character, values, and our understanding of both independence and community."[118] It's a true test to listen and learn when such heartfelt issues are involved. Experts' tips for engaging in responsible political discourse on social media can help.[119]

Don't Assume Everything You See is True

Sometimes it's difficult to know which sources to trust. Check to see if multiple sources with different perspectives report the same information.

Don't Be a Troll or Snark

It's okay to disagree respectfully, but never post messages that insult or belittle others. As one observer puts it, "Disagreement is perfectly normal," but using those differences to bash or inflame others is unacceptable.[120]

Look for Common Ground

"Give people the space to explain themselves," suggests Liz Joyner, an advocate for bipartisan problem solving.[121] You may find that you disagree on policies but agree on key principles, such as treating people with respect, reducing crime, and providing economic opportunities.

Ask Open-Ended Questions

Avoid loaded and leading questions, such as "How can you possibly overlook all the evidence on climate change?"[122] and "Don't you agree that the current minimum wage is more than fair?" Instead, engage in sincere curiosity by asking questions like "What experiences have shaped your thinking on this issue?"[123]

Be Open Minded to Differing Opinions

Resist the temptation to block responsible messages that differ from your own. You'll never learn anything new if you aren't willing to hear about different ways of thinking.

Age/Generation

We tend to think of getting older as a purely biological process. But age-related communication reflects culture at least as much as biology. In many ways, people learn how to "do" being various ages—how to dress, how to talk, and what not to say and do—in the same way they learn how to play other roles in their lives, such as student, parent, or employee.

Older and Younger Perspectives Communicators often underestimate older adults' capabilities. Even though gray or thinning hair and wrinkles don't signify diminished cognitive capacity, they may be interpreted that way—with powerful consequences. People who believe older adults have trouble communicating are less likely to interact with them. When they do, they tend to engage in "elderspeak"—speaking LOUDLY and s-l-o-w-l-y to them and using simple words and repetition.[124] Even when these speech styles are well intentioned, they can have harmful effects. Older adults who are treated as less capable than their peers tend to perceive *themselve*s as older and less capable.[125,126] And objecting to ageist treatment presents seniors with a dilemma:

Speaking up can be taken as a sign of being cranky or bitter, reinforcing the stereotype that older adults are curmudgeons. Researchers have found that people of different generations can often identify common interests by sharing their answers to questions such as: Who are some famous people you look up to and admire? What do (or did) you like and dislike about being a student? and What are your dreams and goals?[127]

Teens and young adults typically feel intense pressure, both internally and from people around them, to establish their identity and prove themselves.[128] At the same time, adolescents typically experience what psychologists call a **personal fable** (the sense that they are different from everybody else) and an **imaginary audience** (a heightened self-consciousness that makes it seem as if people are always observing and judging them).[129] These characterize a natural stage of development, but they lead to some classic communication challenges. For one, teens often feel that their parents and others can't understand them because their situations are different and unique. Couple this with teens' belief that others are being overly attentive and critical of them and you have a recipe for conflict and frustration. Parents may be baffled that their "extensive experience" and "good advice" are summarily rejected. And young people may wonder why adults butt into their affairs with "overly critical judgments" and "irrelevant advice."

Age and Workplace Communication A lot has been written about generational differences between coworkers. But mounting evidence suggests that, when adults of different ages work together, they have a surprising amount in common. For example, in a study of Baby Boomers (born 1946–1964), Generation Xers (born 1965–1980), and Millennials (born 1981–1994), researchers found that members of all three generations typically rate self-development and personal growth as top priorities at work.[130] Slight differences *are* evident in confidence levels and desire for support, but those probably reflect their amount of work experience more than anything else. As you might predict, in one study of nurses, Baby Boomers reported higher levels of confidence than Millennials did, and Millennials indicated a higher desire for support and feedback.[131]

A bigger issue is that colleagues of different generations tend to *presume* that they are more different from one another than they actually are.[132] To avoid what social scientists call "stereotype activation," check yourself the next time you assume that a Millennial is "lazy" if they ask for a day off, or that a Baby Boomer is "out of touch" if they don't know a particular app. Stereotypical thinking such as that can amount to ageism that divides colleagues rather than uniting them.[133]

In addition to encouraging teamwork, intergenerational relationships can offer relief from peer pressure. "We know that everyone's lives are different," says friendship expert Shasta Nelson, "but we remember this more easily with a friend who is in a different life stage. That openness can lead to less comparing, less judgment, less competition."[134]

4.4 Becoming an Effective Intercultural Communicator

Adapting to cultural differences isn't always easy. Edit Vasadi, whose story began this chapter, says that, even after 10 years in the United States, she's surprised by how fast-paced things are. "Everything is based on convenience and saving time,"

---

personal fable The sense, common in adolescence, that you are different from everybody else.

imaginary audience A heightened self-consciousness that makes it seem as if people are always observing and judging you.

she observes.[135] People order drive-through meals, coffee, and ice cream—which they devour in their cars, using paper dishes that they quickly dispose of. Even social gatherings have an air of efficiency, she says. The host is likely to have an agenda that lays out when to eat, dance, play games, and so on. By contrast, Vasadi says, "In Europe we just sit down with a bottle of wine or coffee, and just hang out and talk for hours."[136]

Scholars define **intercultural communication competence** as the ability to engage effectively with people from different cultures based on understanding of, and respect for, different perspectives.[137,138] As Vasadi's experiences illustrate, cultural nuances are so complex that it's impossible to understand everything about another culture. The concept of **cultural humility** considers effective intercultural communication to be an ongoing process in which people continually learn about others and themselves in an environment that is empowering, respectful, and adaptive.[139] The notion of cultural humility first arose in health communication, where care providers often engage with such a diverse array of people that they cannot be proficient in all the cultures they encounter—but they *can* be mindful, open-minded, humble, and self-reflective.[140] The same is true of you. Like everyone else, you're not capable of fully understanding the perspectives of everyone you encounter. By being humble and respectful, you can learn from each other in ways that benefit you both.

In this section we consider five factors that are essential to being an effective intercultural communicator: contact, tolerance for ambiguity, open-mindedness, knowledge and skills, and patience and perseverance.

> **intercultural communication competence** The ability to engage effectively with people from different cultures based on understanding of, and respect for, different perspectives.

> **cultural humility** An approach that considers effective intercultural communication to be an ongoing process in which people continually learn about others and themselves in an environment that is empowering, respectful, and adaptive.

Contact with a Diverse Array of People

More than a half century of research confirms that, under the right circumstances, spending time with people from different backgrounds leads to a host of positive outcomes: reduced prejudice, greater productivity, and better relationships.[141,142]

It's encouraging to know that increased contact with people from stigmatized groups can transform hostile attitudes. For example, door-to-door canvassers in Los Angeles were able to dramatically change attitudes of area residents by engaging in 10-minute conversations aimed at breaking down stereotypes.[143] After canvassers engaged residents in thinking more deeply about the rights of transgender people, the residents reported feeling increased empathy.

Along with face-to-face contact, technology offers a useful way to enhance contact with people from different backgrounds. Online platforms and social media make it relatively easy to connect with people you might never meet in person. The asynchronous nature of these contacts reduces the potential for stress and confusion that can easily come in face-to-face encounters.

It can feel intimidating to initiate a friendship with someone who seems different than you, but you're likely to have more in common than you think.

When was the last time you made friends with someone from a different culture than yours?

Tolerance for Ambiguity

When you encounter communicators from different cultures, the level of uncertainty is especially high. Consider the basic challenge of communicating in an unfamiliar language. Native English speaker Lauren Collins reflects on the challenges and rewards of falling in love with someone who spoke a language (French) that she didn't know.[144] As she got to know her now husband Oliver, she says, they relied heavily on nonverbal cues and sometimes misunderstood them. "I constantly thought Oliver looked irritated," Collins says, until she realized they simply had different cultural expectations about when and how much to smile. The language gap has sometimes been frustrating. "One day, Oliver told me that speaking to me in English felt like 'touching me with gloves,'" she remembers. At other times, it has offered comic relief, as when Collins inadvertently told her French mother-in-law that she "had given birth to a Nespresso machine." Over time, Collins says, she and Oliver have learned to speak each other's language pretty well. In a way she says, we are "an exaggerated version of every couple. We all have to learn how to talk" to each other.

Without tolerance for ambiguity, the number of often confusing and sometimes downright incomprehensible messages that impact intercultural interactions would be impossible to manage. Some people seem to come equipped with this sort of tolerance, whereas others have to cultivate it. One way or the other, that ability to live with uncertainty is an essential ingredient of intercultural communication competence.

Open-Mindedness

Being comfortable with ambiguity is important, but without an open-minded attitude a communicator will have trouble interacting competently with people from different backgrounds. To understand open-mindedness, it's helpful to consider three traits that are *incompatible* with it.

- **Ethnocentrism** is an attitude that one's own culture is superior to others. An ethnocentric person thinks—either privately or openly—that anyone who does not belong to their in-group is somehow strange, wrong, or even inferior.

- **Prejudice** is an unfairly biased and intolerant attitude toward others who belong to an out-group. (The root term in *prejudice* is "pre-judge.") An important element of prejudice is stereotyping—exaggerated generalizations about a group.

- **Hegemony** is the dominance of one culture over another. A common example of this is the impact of Hollywood around the world. People who are consistently exposed to American images and cultural ideas can begin to regard them as desirable.

When traveling abroad, Canadian Mariellen Ward was shocked by how inaccurate many of her assumptions were. She worried that she would encounter unhealthy conditions and archaic medical systems. Contrary to her expectations, she felt very safe. And when she sought treatment for minor illnesses, she was seen immediately and treated to world-class care. The people she encountered were friendlier than she

---

ethnocentrism The attitude that one's own culture is superior to others'.

prejudice An unfairly biased and intolerant attitude toward a group of people.

hegemony The dominance of one culture over another.

had expected, and far more diverse.[145] In short, Ward found that ethnocentric assumptions and prejudice did not reflect what she actually experienced.

Knowledge and Skill

Attitude alone doesn't guarantee success in intercultural encounters. Communicators benefit from having enough knowledge of other cultures to make a connection. The ability to shift gears and adapt one's style to the norms of another culture or coculture is an essential ingredient of communication effectiveness.

One school of thought holds that uncertainty can motivate relationship development—to a point. For example, you may be interested in a newcomer to your class because he's from another country. However, if attempting a conversation with him heightens your sense of uncertainty and discomfort, you may abandon the idea of making friends. The basic premise of anxiety uncertainty management theory is that, if uncertainty and anxiety are too low or too high, people are likely to avoid communicating.[146]

How can a communicator learn enough about other cultures to feel curious but not overwhelmed? Scholarship suggests three strategies:

- *Passive observation* involves noticing what behaviors members of a different culture use and applying these insights to communicate in ways that are most effective.

- *Active strategies* include reading, watching films, and asking experts and members of the other culture how to behave, as well as taking academic courses related to intercultural communication and diversity.

- *Self-disclosure* involves volunteering personal information to people from the other culture with whom you want to communicate. One type of self-disclosure is to confess your cultural ignorance: "This is very new to me. What's the right thing to do in this situation?" This approach is the riskiest of the three described here, because some cultures may not value candor and self-disclosure as much as others. Nevertheless, most people are pleased when strangers attempt to learn the practices of their culture, and they are usually quite willing to offer information and assistance.

Patience and Perseverance

Becoming comfortable and competent in a new culture or coculture may be ultimately rewarding, but the process isn't easy. After a "honeymoon" phase, it's typical to feel confused, disenchanted, lonesome, and homesick.[147] This stage—which typically feels like a crisis—has acquired the labels *culture shock* or *adjustment shock*.[148] See the feature "Coping with Culture Shock" for tips on navigating the adjustment to a new culture.

You wouldn't be the first person to be blindsided by culture shock. Edit Vasadi, whose story we have referenced throughout the chapter, says she sometimes still feels amazed and disoriented by cultural differences after being the United States for 10 years. All she same, she has adapted quite well. She is now married to an American. Together, they are raising two daughters in Arizona, where Vasadi, a photographer, has found her true calling "to document other multicultural couple stories and support their journey."[149]

COMMUNICATION STRATEGIES

Coping with Culture Shock

Growing up in Taiwan, Lynn Chih-Ning Chang had been taught that it was respectful for students to sit quietly and listen, so she was shocked when she moved to the United States for graduate school and found that American students speak aloud without raising their hands, interrupt one another, address the professor by first name, and eat food in the classroom. What's more, Chang's classmates answered so quickly that, by the time she was ready to say something, they were already on a new topic. The same behavior that made her "smart" and "patient" in Taiwan, she says, made her seem like a "slow learner" in the United States.[150]

Becoming comfortable and competent in a new culture may ultimately be rewarding, but the process isn't easy. Communication theorist Young Yum Kim has studied cultural adaptation extensively and offers the advice that follows.

Don't Be Too Hard on Yourself

After a "honeymoon" phase in which you feel excited to be in a new culture, it's typical to feel confused, disenchanted, lonesome, and homesick.[151] To top it off, you may feel disappointed in yourself for not adapting as easily as you expected. You aren't doing things wrong, if that happens, Kim says. It's a natural part of the process.[152]

Homesickness Is Normal

It's natural to feel a sense of push and pull between the familiar and the novel.[153] Kim encourages people acclimating to a new culture to regard stress as a good sign. It means they have the potential to adapt and grow. With patience, the sense of crisis may begin to wane, and once again, the person may feel energetic and enthusiastic to learn more.

Expect Progress and Setbacks

The transition from culture shock to adaptation and growth is usually successful, but it isn't a smooth, linear process. Instead, people tend to take two steps forward and one step back, and to repeat that pattern many times. Kim calls this a "draw back and leap" pattern.[154] Above all, she says, if people are patient and keep trying, the rewards are worth the effort.

Reach Out to Others

Communication can be a challenge while you're learning how to operate in new cultures, but it can also be a solution.[155] Blogger Benjamin Decker encourages people to open their minds to the diversity around them. "One of the most interesting individuals and someone I would never picture myself being close to is now a best friend of mine," he says. "You may never know what may come out of a hello, talking to someone you may find odd."[156]

Chang, the Taiwanese student adapting to life in America, learned this firsthand. At first, she says, she was reluctant to approach American students, and they were reluctant to approach her. Gradually, she found the courage to initiate conversations, and she discovered that her classmates were friendly and receptive. Eventually, she made friends, began to fit in, and successfully completed her degree.

MAKING THE GRADE

OBJECTIVE 4.1 **Analyze the influence of cultures and cocultures on communication, including the concepts of salience and group membership.**

- Some intercultural encounters involve people from different countries, whereas others involve communicating with people from different cocultures within a given society.
- When people come from different backgrounds, their differences may or may not be salient, depending on the situation.
- Communicating with people who seem out-group can be intimidating at first, but bridging those differences can be a learning experience that diminishes prejudice and fear.
- Although cultural characteristics are real and important, they are generalizations that do not apply equally to every member of a group.
  - > Provide two examples of intercultural communication from your own experience—one in which your affiliation with a particular group or culture was salient, and one in which it was not.
  - > Think of a cultural group with which you identify. Now describe a conversation that would be interpreted one way by members of that group and another way by out-group members.
  - > List three common generalizations about college students. How well does each of those generalizations describe you? What can you learn about generalizations from this exercise?

OBJECTIVE 4.2 **Distinguish among the following cultural norms and values, and explain how they influence communication: individualism and collectivism, high and low context, uncertainty avoidance, power distance, and talk and silence.**

- Some cultures value autonomy and individual expression, whereas others are more collectivistic.
- Some pay close attention to subtle, contextual cues (high context), whereas others pay more attention to the words people use (low context).
- Cultures vary in their acceptance of uncertainty.
- Authority figures are treated with more formality and often greater respect in cultures with high power distance than in those with low power distance.
- Depending on the cultural context, people may interpret silence as comforting and meaningful, or conversely, as an indication of awkwardness or disconnection.
  - > Consider the following two quotes by famous people: *"The most courageous act is still to think for yourself. Aloud."* (designer Coco Chanel) and *"To me, teamwork is the beauty of our sport, where you have five acting as one. You become selfless"* (Duke basketball coach Mike Krzyzewski). Which statement reflects a collectivistic perspective? Which is more individualistic? In what situations does the first quote appeal to you? Under what circumstances is the second quote more useful?
  - > Describe a time when you expected someone to know how you were feeling even though you didn't express your emotions in words. Explain how the concept of high and low context applies to that experience.
  - > If you visited a doctor with a minor health concern and the doctor said, "I'm not sure what's wrong, but I don't think it's serious," would you be reassured or frustrated? Describe how your answer relates to feelings about certainty and ambiguity.
  - > Compare and contrast how job applicants from high- and low-power-distance cultures might respond to the question "What are your qualifications?" in a job interview.
  - > Imagine that you have volunteered to greet a regional manager at the airport and drive them to the office. During the car ride, the two of you are mostly silent. Now imagine that there is an hour of silence during a road trip with your best friend. Describe how you might feel about silence in each of these encounters.

OBJECTIVE 4.3 **Evaluate how factors such as race and ethnicity, sex and gender, religion, disability, political viewpoints, and age influence communication.**

- Intersectionality theory recognizes that every person experiences life at the juncture of multiple identities that intersect in unique ways.

- People often make assumptions about others based on their apparent race or ethnicity, but appearance is an unreliable indicator of cultural and personal differences.
- When talking about race, expect strong emotions, exercise empathy, don't turn the conversation into a debate, learn from others, apologize if you offend, be an ally for others, and don't insist that people share their experiences if doing so makes them uncomfortable.
- Gender diversity goes far beyond the simplistic dichotomy of male or female.
- It can be difficult to know when or if to discuss religion, but the basic premises of effective communication apply: Judge whether the situation is appropriate and if people are interested, and be an open minded and respectful listener.
- For the most part, communicate with people who have disabilities just as you would with anyone else.
- Recognizing the emotional nature of political viewpoints, evaluate information to make sure it's credible, don't belittle others, look for common ground, and be a respectful listener.
  - > List three identities that apply to you. How does living at the intersection of those identities make your life experience different from that of a person who shares one or two of those identities, but not all three?
  - > Describe a situation in which you described your experiences (perhaps related to race, gender, age, ability, religion, or politics) to another person. How did the other person respond? How might you rewrite the conversation to make it more effective?

OBJECTIVE 4.4 Implement strategies to increase intercultural and cocultural communication competence, such as increasing contact with diverse others, and being open-minded, knowledgeable, and patient.

- Cultural competence involves knowledge of and respect for different perspectives.
- Cultural humility emphasizes the value of being mindful, open-minded, self-reflective, and committed to a process in which you continually learn about other cultures.
- An effective way to learn about cultures is to interact with people from many different backgrounds.
- Learning from different people will require that you become comfortable with uncertainty and new information.
- People tend to assume that their culture is better than others and that people from particular groups are all the same. Acknowledging those assumptions within yourself and setting them aside will help you be more open minded.
- It often takes knowledge, skill, and patience to connect with people who have diverse perspectives, but the benefits are worth the effort.
  - > List and explain three communication strategies people might use to learn about different cultures and to share their own ideas with others.
  - > Think of a time when you met someone who seemed very different from you at first but eventually became a close friend or colleague. Create a timeline that illustrates turning points in your relationship when you learned more about each other and developed rapport.
  - > How do you rate yourself in terms of intercultural sensitivity and open-mindedness? What steps might you take to keeping growing in this regard?

KEY TERMS

coculture p. 76
collectivistic culture p. 80
cultural humility p. 97
culture p. 76
ethnicity p. 88
ethnocentrism p. 98
gender p. 90
hegemony p. 98

high-context culture p. 82
imaginary audience p. 96
in-group p. 77
individualistic culture p. 80
intercultural communication competence p. 97
intersectionality p. 87
low-context culture p. 82
out-group p. 77
personal fable p. 96
power distance p. 85
prejudice p. 98
race p. 76
salience p. 77
sex p. 90
social media snarks p. 94
social media trolls p. 94
uncertainty avoidance p. 83

PUBLIC SPEAKING PRACTICE

Prepare to share your answers to the following questions: In what ways is your identity shaped by who you are as an individual? In what ways is it shaped by the groups to which you belong (e.g., your family, hometown, college, clubs, religion, and so on)? Give an example of a communication episode in which you exhibited one of the cultural patterns described in this chapter (e.g., individualism, collectivism, power distance, silence, etc.).

ACTIVITY

Examine the comments posted in response to a blog or video online. Analyze how well the people involved have followed tips in the checklist "Discussing Politics Responsibly on Social Media" in this chapter. Rewrite several of the posts to be more respectful of diverse viewpoints.

Language

CHAPTER OUTLINE

5.1 The Nature of Language 106
Language Is Symbolic
Words Have Varying Interpretations
Meaning Is Negotiated
Language Is Governed by Rules

5.2 The Power of Language 110
Gender References
Names
Accents
Assertive and Collaborative Language

5.3 Language and Misunderstandings 113
Abstract Language
Equivocal Language
Relative Words
Slang
Jargon
Euphemisms

5.4 Troublesome Language 117
Confusion About Facts, Inferences, and Opinions
Lies and Evasions
Emotive Language
Microaggressions

MAKING THE GRADE 124

KEY TERMS 126

PUBLIC SPEAKING PRACTICE 126

ACTIVITIES 126

LEARNING OBJECTIVES

5.1
Explain how symbols and linguistic rules allow people to achieve shared meaning.

5.2
Identify ways in which language shapes people's attitudes and reflects how they feel about themselves and others.

5.3
Recognize and remedy confusing language.

5.4
Distinguish among facts, inferences, and opinions, and avoid using inflammatory and microaggressive language.

Demi Lovato's social media posts used to be "all glamour shots and pictures of me looking fancy and cute,"[1] recalls the singer and actor, who felt compelled to project the role of "sexy, feminine pop star."[2] But that persona never felt genuine, says the artist. Deep down, "it feels weird to me when I get called a 'she' or a 'her,'" Lovato revealed in 2021.[3] In an effort to portray a more authentic self, Lovato has identified as nonbinary, using the pronouns *they/them*.

Pronouns are one example of how powerful language can be. Nonbinary individuals—such as Lovato and about 1 million other people in the United States[4]—do not identify exclusively with one gender or another, so it can feel confining and inaccurate for them to be called he or she. *They* has emerged as an non-gendered way to describe one person or a group of people. "I feel that [they] best represents the fluidity I feel in my gender expression and allows me to feel most authentic and true," Lovato says.[5]

Whether or not Lovato's choice resonates with you, the point is clear: Language matters. It affects people in many ways. It helps them express their identities, collaborate with others, and make sense of the world. As you will see, language is built on basic guidelines, but at the same time, it constantly evolves. This chapter explores the power of language to shape perceptions. It explains how words can either create mutual understanding or confuse or hurt others. By the end of the chapter, you should be equipped to use language more skillfully to improve your everyday interactions.

> **language** A collection of symbols governed by rules and used to convey messages between individuals.
>
> **dialect** A version of the same language that includes substantially different words and meanings.
>
> **symbol** An arbitrary sign used to represent a thing, person, idea, or event in ways that make communication possible.

5.1 The Nature of Language

When British people have a *chinwag* (chat) with someone from the United States or Canada, they might feel *knackered* (exhausted) by how hard it is to share meaning, even within the same language. Sometimes you probably feel the same way when communicating with people closer to home.

To understand what's going on, let's define some basic terms. A **language** is a collection of symbols governed by rules and used to convey messages between individuals. A **dialect** is a version of the same language that includes substantially different words and meanings.[6] English includes dozens of dialects, as do the other 7,000 or so languages of the world. Here's a closer look at four qualities of language that give it underlying structure but allow it to be highly adaptive.

Language Is Symbolic

As you read in Chapter 1, symbols represent thoughts—usually in the form of words that can be conveyed via speech, writing, or gestures. Most words are arbitrary **symbols**, meaning that they have no inherent link to what they represent. The triangle of meaning (Figure 5.1) illustrates this by showing that there's only an indirect relationship—indicated by a broken line—between a word and what it's used to represent.

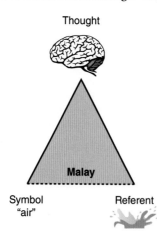

FIGURE 5.1 Triangle of Meaning

Consider the word *air*. If you grew up speaking English, it's what you breathe, but in Malay *air* means water.[7] The word is the same, but the referents are different.

Not all linguistic symbols are spoken or written words. Sign language, as "spoken" by most Deaf people, is symbolic in nature and not the pantomime it might seem to nonsigners. There are hundreds of different sign languages spoken around the world—including American, British, French, Danish, Chinese, Australian Aboriginal, and Mayan sign languages—each with its own ways of representing ideas.[8]

Words Have Varying Interpretations

Understanding some terms is fairly straightforward. You and your friends probably agree on the meaning of *chair* and *book*. However, you may disagree about other words. The term *queer* is an example. At one time, it was used in a negative way to refer to homosexuality. To people who remember that, it may still hurt to be called queer. But in the last 20 years or so, many have embraced the term as an inclusive way to describe anyone who is not cisgender or exclusively heterosexual.[9] When Twitter banned the use of *queer* on its platform in 2018, one person lamented: "Twitter thinks queer is a slur, which means basically everyone I know is going to be locked out."[10]

One reason for different interpretations is that words have meaning on two levels. One is the **denotative meaning**, the formally recognized definition of a word. Another is the **connotative meaning**, the thoughts and feelings associated with a word. Even if you memorize the dictionary, you won't know exactly how people will use and interpret the words contained in it. For example, people might feel either uplifted or insulted to be called a feminist, liberal, or conservative. In the southern United States, people realize that "bless their heart" may have multiple connotations, ranging from a sincere expression of affection or sympathy ("She's sick again, bless her heart") to an insult ("She's dumber than dirt, bless her heart").

denotative meanings Formally recognized definitions of words.

connotative meanings Informal, implied interpretations for words, often positive or negative.

Despite the potential for mix-ups using language, the situation isn't hopeless. People do, after all, communicate with one another reasonably well most of the time. Good communicators recognize that words can be taken many ways, so they engage in effortful, interactive exchanges to understand one another.

Meaning Is Negotiated

Imagine you've just been accepted for a year-long study-abroad program. Your family congratulates you and predicts this will be one of the best experiences of your life. Then you FaceTime a friend, who says, "Are you kidding?! I'd be terrified to go someplace if I didn't speak the language." These encounters and your previous experiences are likely to affect how you feel about the opportunity ahead of you.

In the film franchise starring the Minions, the little yellow creatures speak what may sound like gibberish or baby talk, but it is actually a complete language.

Can you name some syntactic rules that govern speech in your native language (for example, subject before verb)? What would it sound like to violate some of those rules?

> **coordinated management of meaning (CMM)** The perspective that people co-create meaning in the process of communicating with each other.

Coordinated management of meaning (CMM) suggests that people co-create meaning in the process of communicating with each other.[11] Is your study abroad opportunity a cause for joy or fear? The sense you make of it is influenced by an array of factors including your parents' excited reaction, your friend's comment, and your exposure to other messages throughout your life. This combination of messages has influenced your ideas about travel, independence, education, and more. From the perspective of CMM, you are influenced by others, but you also play an active role in the meaning-making process. For example, you might reply to your friend, "Not worried! I'll learn the language in no time." Your friend may accept your confident statement at face value or challenge it, perhaps by observing, "But you're a little nervous, am I right?" This exchange of messages is likely to influence the meaning that you attribute to your experience (perhaps "Actually, I *am* a little nervous") and how you interpret the communication event itself ("We're just joking around," "My friend would be more worried than I am," or "My friend is trying to scare me"). The point is that meaning is created, not just in one person's head but through an ongoing process of interaction.

Language Is Governed by Rules

The ambiguity of language is balanced by an underlying structure. Rules help people understand what words mean, as well as how to pronounce them, order them in a sentence, and use them in particular situations.

> **semantic rules** Guidelines that govern the meaning of language as opposed to its structure.

Semantic rules are guidelines about the meanings of specific words. They make it possible for us to agree that a *shoe* is for wearing on your foot and *glue* is for sticking things together. Because semantic rules differ by language, it's possible for the same word to represent different things. In Hindi and Farsi, *barf* means snow, which explains why it's the name of a popular brand of detergent in India and the Middle East.[12] However, soap by that name probably wouldn't sell well in predominantly English-speaking countries, where *barf* means vomit. Even within the same language, misunderstandings occur when words can be interpreted in more than one way.

> **phonological rules** Linguistic guidelines that govern how sounds are combined to form words.

Phonological rules govern how words are pronounced. Can you correctly say *vulnerable, comptroller, sherbet,* and *assuage*? If you pronounced them *vul-ner-uh-bull, con-troller, sher-bit,* and *ess-wage*, give yourself top marks in phonology.[13,14] The process is more complex than memorizing sounds of the alphabet because the meaning of many words depends on their pronunciation.

> **syntactic rules** Guidelines that govern how symbols can be arranged (e.g., sentence structure).

Syntactic rules govern the structure of language—the way symbols should be arranged. Correct English syntax prohibits sentences such as "Have you the cookies brought?"—which is a perfectly acceptable word order in German. Different syntactic rules apply in different situations. You probably wouldn't say aloud to someone, "That was so funny! ROTFL," but it's acceptable to text that message (which stands for rolling on the floor laughing). You're not likely to impress others if you use the casual sentence structure of a text or tweet in a college assignment, job application, or an email to your boss.

> **pragmatic rules** Guidelines that govern how people use language in everyday interaction.

Pragmatic rules help people collectively *do* things with language—such as joke around, offer comfort, and give advice.[15] Pragmatic rules help you know when it's okay to laugh, when you should be silent, how to behave at work versus at home, and so on. If you've ever wondered whether a person was joking or being serious, you know how it feels when pragmatic rules are not uniformly understood or applied. Because these rules are based on culture and situations, they change over time. For example, while your parents were growing up, they probably considered it impolite

UNDERSTANDING YOUR COMMUNICATION

How Do You Use Language?

Answer the following questions to see what orientation is suggested by the way you use language.

1. Your best friend is upset upon learning that he wasn't accepted into graduate school. What are you most likely to say?
   a. "You seem discouraged. Tell me what's going through your head."
   b. "Grad school is overrated. Tons of successful people don't have master's degrees."
   c. "There are other great schools. I can help you apply to them."
   d. "You are a great student. Don't let this get you down. The school that accepts you will be very lucky."

2. You are planning a sales pitch that could earn your company millions of dollars. What is your pitch most likely to include?
   a. A focus on the client's most deeply held values
   b. A list of the reasons your company is better than the competition
   c. Specific features that make your product highly useful and effective
   d. Jargon and other language that shows you understand the client's business

3. You hope to meet with a professor to learn more about a topic covered in class. How would you word a meeting request?
   a. "I'm excited about the ideas you shared in class. Could I meet with you to learn more?"
   b. "This topic is critical to my long-term success. Can I meet with you to learn more about it?"
   c. "I'd like to hear what steps you think I should follow to be successful at this. Can we meet?"
   d. "I like what you said in class. I know you're busy, but would it be possible to meet and talk more about it?"

4. Your family is planning a holiday celebration, but you'd like to go skiing with friends instead. How are you most likely to broach the topic with your family?
   a. "You've always been so supportive of me. I think you'll understand . . ."
   b. "Going skiing with my friends is a once-in-a-lifetime opportunity."
   c. "The ski trip is an opportunity to make new friends and maybe even some future business contacts."
   d. "I'd love to go skiing. But I won't go unless you're 100% okay with it."

INTERPRETING YOUR RESPONSES

Read the following explanations to learn more about your use of language. (More than one may apply.)

Affective

If you answered "a" to two or more questions, your language tends to focus on emotions—yours and other people's. This approach can make you a sensitive listener and a motivational speaker. Just be sure to balance this strength with an awareness of practical concerns.

Opinionated

If you answered "b" to more than one question, you tend to voice strong viewpoints. Educated opinions can be useful, but to make sure you aren't going overboard, review the units in this chapter on assertiveness and collaboration and facts versus inferences and opinions. Being open to other viewpoints can help you build teamwork and consider alternative ways of understanding the world around you.

Instrumental

If you answered "c" to two or more questions, you are inclined to adopt a goal-oriented approach to language. This strategy can be effective, but you may come off as headstrong in some situations. Make sure you don't lose sight of the emotional aspects of the issue at hand.

Affiliative

If you answered "d" to more than one question, you are disposed toward an affiliative language style. You tend to display alignment with other people and avoid actions that might place you at odds. Your thoughtfulness is no doubt appreciated. At the same time, make an effort to take a stand when it's important to do so. The discussion of assertive communication in this chapter may be helpful.

to use swear words, but these days about 1 in 4 Americans use the f-word daily, a steep increase from 10 years earlier.[16] Even book titles include expletives, with bestsellers such as *The Subtle Art of Not Giving a F—k* and *How to Get Sh-t Done*. As with all forms of communication, the challenge is to gauge what language is appropriate given the people and situation involved. Take the "How Do You Use Language?" self-assessment for insight into your approach.

5.2 The Power of Language

When you envision a beautiful scene in a faraway land, a sense of *fernweh* may wash over you. But unless you speak German, you probably don't know that word. *Fernweh* roughly translates to "farsickness"—a longing similar to homesickness, except for places you've never been.[17] Or perhaps you just found out your best friend is having a baby and you're having a hard time giving a name to your emotions. If there were such a word as *disathrilled* (simultaneously disappointed and thrilled), that might apply.

If you incorporated words such as *fernweh* and *disathrilled* into your vocabulary, would you be more sensitive to the phenomena they represent? Would you find it easier to think about them and talk about them? Your answers are at the heart of a long-running debate about language.

Philosophers and scientists have been debating whether language influences thoughts—or the other way around—for over a century. One perspective is represented by **linguistic relativism** (often called the *Sapir-Whorf hypothesis*), which proposes that the language people use influences how they experience the world.[18] For example, people who have a word for "farsickness" may recognize the emotion and talk about it more easily than those who don't. Conversely, another perspective proposes that language is more a *reflection* of people's perceptions than a guiding force in creating them. People who reject linguistic relativism propose that, when people don't have a specific word for something, they use other words in its place. They also point out that people often invent words to suit their needs (think *cyberspace*, *unfriend*, and *smartwatch*).

Although the debate about language relativity isn't likely to be resolved any time soon, it reveals a great deal about the powerful relationship between language and perception. In this unit, we consider the social implications of gender references, names, accents, and assertive and collaborative word choices.

linguistic relativism
The notion that language influences the way users experience the world.

Gender References

Linguistic elements may be especially suggestive when they describe social constructs such as gender. Let's consider a few examples.

When researchers asked Germans to describe bridges, they used words such as beautiful, fragile, pretty, and slender. Note that the German language refers to bridges as feminine. By contrast, Spanish speakers—who refer to bridges as masculine—were more likely to describe them as sturdy, strong, and towering.[19,20]

In decades past, many job titles included a masculine suffix (e.g., chairman, policeman, salesman). Over time, however, many people began to consciously change the words they used (e.g., chair, police officer, salesperson) to support a more inclusive perspective, in line with research on the effects of gendered versus gender-neutral job titles.[21,22]

If you're accustomed to a language in which people are called either he or she, you probably categorize individuals as male or female, even when you don't consciously think about it. By contrast, some languages exclusively use gender-inclusive pronouns—such as the Finnish pronoun *hän*, which refers to any individual. Finnish speakers learning English often say it requires a mental shift to categorize people by gender. (The relationship between thought and language is underscored by evidence that gender equality tends to be greater in cultures that use gender-neutral pronouns than in those that don't.[23,24])

Demi Lovato is one of about one million people in the United States who feel that binary pronouns such as he and she do not describe them.

What name and pronouns best describe you?

As these examples show, language sometimes spotlights gender and the assumptions that surround it. Around the world, about 39 percent of languages classify objects in terms of gender, and about 96 percent include gender-specific pronouns for people.[25,26] As a World Bank economist puts it, language often serves as a "psychological nudge" toward thinking in terms of male and female.[27]

Names

When Lauren learned that her parents had changed her name when she was a baby, she wondered if life would have been different with the name she'd been given at birth (Tiffany). To try it out, she changed her nickname to Tiffi on social media. As people began addressing her by the new name, she started to behave the way she thought a Tiffi would. "I said yes more—to going out at night, to dating, and to doing things that were edgy for me," she says. "All I could think was: *Lauren would never do this.*"[28]

Lauren's concept of "Tiffi" behavior may be unique, but she's onto something in recognizing that names are more than a simple means of identification. They also shape and reinforce a sense of personal identity.

Naming a baby after a family member can create a connection between the child and their namesake. Names can also make a powerful statement about cultural identity. Some names may suggest a Black identity, whereas others sound more White.[29] The same could be said for names that are Hispanic, feminine/masculine, Hindi, and so on.

Unfortunately, names can be used as the basis for discrimination. When researchers posted more than 6,000 AirBnB requests that were identical except for the users' first names, they found that would-be guests with Black-sounding names were 15 percent more likely to be declined lodging than those with White-sounding names.[30] A similar pattern is evident in employment decisions. In the United States, job applicants with names such as Mohammed and Lakisha typically receive fewer calls from employers than equally qualified candidates with names such as Thomas and Susan.[31,32] Because of this potential for discrimination, some people advocate for applications in which job candidates' names are masked during the review process.[33]

Accents

In the classic musical and film *My Fair Lady* (based on George Bernard Shaw's play *Pygmalion*), Professor Henry Higgins teaches flower-seller Eliza Doolittle an upper-crust speaking style in place of her cockney accent. Her new way of speaking has a profound impact on the way others perceive her.

An **accent** involves pronunciation perceived as different from that of other people.[34] Accents in the United States include Bostonian, Cajun, Californian, and North Carolina High Tider, among many others.[35] In a larger sense, it's important to realize that *everyone* has an accent. To your ear, you and many of the people you encounter every day might sound accent-free. But if you traveled to a different part of the world, your speech would sound decidedly accented to the people there.

An accent can either enhance or detract from a speaker's social status. Various accents may be considered pleasing to the ear, depending on the context.[36] Unfortunately, people sometimes have negative reactions to accents. Research has shown that in the United States, employers are more likely to hire candidates who sound as if English is their first language, even when the candidates speak English clearly and proficiently.[37]

accent Pronunciation perceived as different from other people's.

assertive language Wording that is clear and direct.

Assertive and Collaborative Language

Before Adam Grant was a best-selling author and organizational psychologist, he was a 25-year-old asked to speak to two groups of Air Force commanders twice his age. With the first group, he began his presentation assertively by emphasizing his expertise and credentials as a graduate of a doctoral program. It did not go well. Participants never warmed up to Grant, and most felt that the workshop was a failure. With the second group, Grant took a different approach. He began by saying, "I know that many of you in this room are thinking at this very moment, 'What can I possibly learn from a professor who's 12 years old?'" Then he gave the same four-hour presentation as the first time. After laughing at Grant's introduction, commanders in the second group were more receptive to his ideas, and they gave the workshop high marks. The experience convinced Grant of what he calls "the power of powerless communication."[38]

For our purposes, we define **assertive language** as clear and direct. It might involve speaking up in a meeting, being forthright about your accomplishments, or making a direct

Although it may seem that race car drivers win races, success actually results from the collaborative efforts of an entire team.

In what situations do you rely on others for success? What do you say or do to show the people involved that you appreciate them?

statement such as "I prefer to be called Kathryn." In some cultures and situations, assertive communication is highly effective. In the United States, employers tend to value team members who call attention to their accomplishments and say what they think.[39] Being humble in a job interview ("I still have a lot to learn") or reluctant to engage with others ("I'm not sure how I feel") might be taken as signs of insecurity or incompetence.

By comparison, **collaborative language** encourages people to think together without treating any one person's opinion as dominant.[40] Those who take this approach tend to present their opinions with qualifying phrases, such as "In my experience . . ." or "It seems to me . . . ." And they often invite other's input with tag questions such as "I think this might work, how about you?"

Tentative and self-effacing language—such as Adam Grant's opening quip about looking 12 years old—is sometimes referred to as powerless. But, as scholars have long recognized, that isn't necessarily true.[41,42] Sometimes, it's more powerful to be modest and to think *with* people than to be assertive about your own positions. Susan Cain—a former Wall Street attorney and author of *Quiet: The Power of Introverts*[43]—puts it this way:

> When people think you're trying to influence them, they put their guard up. But when they feel you're trying to help them, to muse your way to the right answer, or to be honest about your own imperfections, they open up to you. They hear what you have to say.[44]

> **collaborative language**
> A communication approach that encourages people to think together without treating any one person's opinion as dominant.

Collaborative communication has gained widespread popularity among health communication scholars and practitioners, who recognize that it's typically more effective to partner with patients—considering their reservations, questions, and preferences—than simply to give them orders or advice.[45] An interactive approach is crucial in professional settings as well. The *Harvard Business Review* has declared this an "Era of Collaboration," in which many organizations are using teamwork to generate ideas and enhance productivity and morale.[46]

Both assertive and collaborative approaches can be effective, and it's possible to combine them. You might assertively pitch an idea and then use collaborative language to encourage discussion about it. It can be tricky to know when and how much of each language to use, however. The feature "Balancing Assertive and Collaborative Language" offers tips from the experts.

5.3 Language and Misunderstandings

A patient in the hospital is distressed when a nurse tells him he "won't be needing" any clothes or personal items from home. The patient interprets the statement to mean that he's near death, but the nurse actually means that he'll be going home soon.

Even when people try hard to communicate well, misunderstandings can occur. Sometimes we laugh about them later. But in some cases, crossed wires can damage relationships and reputations. Here are six types of language that often confuse others, followed by tips for communicating more clearly.

COMMUNICATION STRATEGIES

Balancing Assertive and Collaborative Language

Imagine that you're trying to reach a decision that will affect other people. The issue might be whether to launch a new product line at work or where to go on a family vacation. On the one hand, you have researched the options and developed strong feelings about the one you think best. On the other hand, you want everyone to be happy with the decision, and you certainly don't know everything there is to know. How might you proceed? Here are some factors to keep in mind.

State Nonnegotiable Points Assertively

As a general rule, use assertive language when it's important to be clear, act quickly, or emphasize a nonnegotiable point. You're more likely to be taken seriously if you state, "It's important to meet our 5 p.m. deadline" than if you say, "We should probably finish by 5 o'clock, *okay*?"

Encourage Teamwork with Collaborative Language

If diverse ideas and team buy-in are essential to a good outcome, nothing beats collaborative language. A collaborative question such as "In your opinion, how might we make better use of social media?" is likely to stimulate more ideas than an assertive declaration such as "I think we should focus our social media on working parents."

Avoid Over-the-Top Assertiveness

Taken to extremes, assertive language can come off as presumptuous, self-centered, and bossy. You can probably think of an annoying person who always wants the last word and is determined that their ideas should win.

Don't Be a Doormat

Somewhat tentative language can be collaborative, but it's a good idea to avoid faltering, overly accommodating statements such as, "Um, well, this is a probably a stupid idea, but I sorta think, you know, that it would be really, really great if we . . . ." Your language will sound more assertive if you limit your use of hedge words (*maybe, sort of*), disfluencies (*um, uh*), and over-the-top intensifiers (*extremely wonderful; very, very, good*).

Balance Competence and Humility

Adam Grant's self-effacing opening worked because he had the expertise and confidence to follow it with a well-crafted presentation. Consider how others regard you. If your knowledge and credentials are in doubt, assertive language can help demonstrate that you know what you're talking about. At the same time, collaborative language can show people that you're approachable, have empathy, and are sincerely interested in their perspectives, which can be especially helpful if others seem put off or intimidated by you.

Be Sensitive to the Situation and Culture

Assertiveness is valued in many American workplaces, but in some other situations, a bold approach may not be well received. If the topic is sensitive or outside of your expertise, the best approach may be to listen and ask questions. Likewise, in collectivistic cultures (such as in Japan and Korea), people tend to prefer ambiguous terms and nonassertive language. In those communities, people are likely to recognize that a statement such as "I'll have to think about this a while" may be a way to say "I'm not interested" while preserving harmony and allowing the other person to save face. Similarly, in traditional Mexican culture, it's polite to add *"por favor?"* ("if you please?") to the end of requests, such as when ordering food in a restaurant. By contrast, assertive declarations, such as "I'll have the fish," are likely to seem disrespectful.

Abstract Language

It would be hard to communicate without **abstract language**—vague references to people, objects, events, and experiences. Abstract terms allow you to speak in generalities. For example, you can say your week was "busy but productive" without describing everything you did.

Abstract terms present two weaknesses, however. One is that people interpret them in many different ways. Imagine telling your partner, "We never have fun anymore." They may protest, "Of course we have fun! Just this week, we went to the movies and out to dinner." You may be using the same word, but with different interpretations. A second weakness is that abstract terms don't provide a clear picture of what you're trying to say. Consider what comes to mind if someone says, "It was the best vacation ever." Your mental image of the vacation is probably different than other people's.

When it's important to communicate clearly, use **concrete language**—words that refer to specific behaviors, objects, and events. Rather than the vague word *fun*, you could use more concrete language to tell your partner, "I miss when we used to go on road trips to the beach." In describing a great vacation, you might say, "We stayed for two weeks in a mountaintop cabin and skied in fresh snow every day."

To develop your skills using concrete language, imagine an **abstraction ladder** that begins with a general statement and gets progressively more specific with each step.[47] Consider the following statements that someone might include on a resume:

> I have fundraising experience.
>
> I have served on three fundraising committees.
>
> In the last two years, I have helped to raise $30,000 for student scholarships.

It's easy to see that job candidates who use concrete language are likely to stand out from the rest. By the same token, when it's important to convey a specific point, you'll probably be more successful using concrete terms than abstract ones.

Equivocal Language

Even when words seem specific, they can be **equivocal**, meaning that they're open to more than one interpretation. Friends of ours got a vehicle full of water after they interpreted a simple word differently. When he said, "Please leave the car open tonight," she thought he meant not to close the doors. Only after a rainy night did she realize that, by open, he meant *unlocked*.

In contrast to equivocal words, which may confuse people unintentionally, an **equivocation** is a *deliberately* vague statement that can be interpreted in more than one way.

abstract language
Wording that makes vague references to people, objects, events, and experiences.

concrete language
Wording that refers to specific people, behaviors, objects, or events.

Equivocal messages challenge the perceiver to choose from multiple or vague meanings.

What strategies do you use when you're unsure how to interpret a message?

> **abstraction ladder** A conceptualization in which a vague statement is modified to become increasingly more specific.

> **equivocal language** Words that have more than one dictionary definition.

> **equivocation** A deliberately vague statement that can be interpreted in more than one way.

> **relative words** Terms that gain their meaning by comparison.

> **slang** Language used by a group of people whose members belong to a similar coculture or other group. (Less formal and enduring than jargon).

> **jargon** Specialized vocabulary used by people with common backgrounds and experience (more formal and enduring than slang).

> **euphemism** Mild or indirect word choices used in place of more direct but less pleasant ones.

If your date asks how you like their new haircut, you might equivocate by saying, "Your hair looks so shiny," rather than admitting that you don't like the style. Equivocations can spare people the embarrassment that might come from a bluntly truthful answer. However, they can also be underhanded, as when an employee calls in sick, saying, "I'm not feeling well" when the whole truth is that she's exhausted from partying all night.

Relative Words

Relative words gain their meaning by comparison. Is the school you attend large or small? Compared with a university that has 60,000 students, it may seem small, but next to one with 2,000, it might seem large. In the same way, relative words such as *fast* and *slow*, *early* and *late*, and *short* and *long* depend on comparison for their meaning.

Using relative words without explaining them can lead to communication problems. Have you been disappointed to learn that classes you've heard were "easy" turned out to be hard, that trips you were told would be "short" were long, that "hilarious" movies were mediocre? The problem in each case came from failing to anchor the relative word to more precise measures for comparison.

Slang

Language used by a group of people whose members belong to a similar coculture or other group is called **slang**. For instance, cyclists who talk about *bonking* are referring to running out of energy. And social media enthusiasts probably recognize that *shipping* means you support the idea of a romantic relationship between particular people, as in "I ship you with the new girl." Other slang consists of *regionalisms*—terms used and understood only by people from a relatively small geographic area. Residents of the largest U.S. state know that when a fellow Alaskan says, "I'm going outside," they are leaving the state. Slang defines insiders and outsiders, creating a sense of identity and solidarity among members while outsiders are likely to be mystified or to misunderstand.[48]

Jargon

Almost everyone uses some sort of **jargon**—a specialized vocabulary used by people with common backgrounds and experiences. Whereas slang tends to be casual and changing, jargon is typically more technical and enduring. Some jargon consists of *acronyms*—initials used in place of the words they represent. In finance, P&L translates as "profit and loss," and in the military, failure to serve at one's post is known as being AWOL (absent without leave). The digital age has spawned its own jargon. For instance, "UGC" refers to user-generated content.

Jargon can be an efficient way to use language for people who understand it. The trauma team in a hospital emergency room can save time, and possibly lives, by speaking in shorthand, referring to "GSWs" (gunshot wounds), "chem 7" lab tests, and so on. But people unfamiliar with the jargon can feel left out or baffled by it. This can happen in many settings—such as classrooms, doctor's offices, and business meetings—where people may be reluctant to admit that they don't understand the jargon being used.

Euphemisms

A **euphemism** is a mild or indirect term substituted for a more direct but potentially less pleasant one. People use euphemisms when they say *restroom* instead of *toilet* or

> **COMMUNICATION STRATEGIES**
>
> ## Expressing Yourself Clearly
>
> Considering the many ways that words can be used, here are some tips to communicate clearly and minimize misunderstandings.
>
> **Use Slang and Jargon with Caution**
>
> If you say "that person needs to get woke" (educated about social issues) or "the SCOTUS is completely out of touch" (referring to the Supreme Court of the United States), make sure others understand what you mean.
>
> **Explain Your Terms**
>
> Abstract words such as *good*, *bad*, *helpful*, and *happy* mean different things to different people. Instead of saying, "Lucas is a good colleague" you might say, "Lucas always has great ideas and meets every deadline."
>
> **Be Specific**
>
> "I'll be there in 30 minutes" is clearer and more helpful than "I'll be there soon."
>
> **Focus on Behaviors**
>
> "It's important that you arrive by 9 o'clock every morning" is more likely to get results than "Be punctual."
>
> **Be Careful with Euphemisms and Equivocations**
>
> If you say, "He went to a better place," listeners may wonder if he died, got a better job, or went to a nicer restaurant for dinner.

passed away instead of *died*. Airline pilots rely on euphemisms when they tell passengers to expect *bumpy air* rather than *turbulence*.

Euphemisms often seem more polite and less anxiety-provoking than other words. However, they can be vague and misleading. Terms such as *domestic disturbance* and *battle fatigue* are easy to hear, but they downplay the harsh realities involved. In the same way, being *excessed*, *decruited*, or *made redundant* doesn't make the reality of losing one's job any easier.[49] To avoid confusion, see the feature "Expressing Yourself Clearly" for suggestions on choosing your words carefully.

5.4 Troublesome Language

Misunderstandings happen, but it's both possible and important to use language in a responsible and civil manner—one that shows respect for others. A fundamental step toward avoiding misunderstandings involves learning how to distinguish among facts, inferences, and opinions. In addition, it's essential to understand the effects of lies and evasions, emotive language, and microaggressions.

"Be honest with me Roger. By 'mid-course correction' you mean divorce, don't you."

Source: Leo Cullum The New Yorker Collection/The Cartoon Bank

Confusion About Facts, Inferences, and Opinions

When the Pew Research Center asked 5,000 adults to read 10 statements and indicate which were facts and which were opinions, the results were disappointing. Only about one-third of respondents got them all right, and nearly 1 in 4 got most of them wrong. Especially alarming was people's tendency to think that facts they disliked were opinions—and conversely, that opinions they liked were actually facts.[50]

"Chances are, you're not as open-minded as you think," concluded one writer after reviewing the research.[51] **Confirmation bias** is the tendency to reject information that's inconsistent with one's current viewpoint. It's heightened by media outlets that offer slanted versions of news stories—or opinions masquerading as facts—and by online algorithms that reinforce ideas and information people have viewed in the past.[52] As a result of all these factors, people can be closed to new ideas and relationships.

To express yourself responsibly, it's important to consider the difference between **factual statements** (which can be objectively shown to be true), **inferential statements** (which, accurate or not, are conclusions based on how someone interprets evidence), and **opinion statements** (which represent the speaker's beliefs or attitudes and may or may not be based on evidence).

> **confirmation bias** The emotional tendency to interpret new information as reinforcing of one's existing beliefs.
>
> **factual statement** An assertion that can be verified as being true or false.
>
> **inferential statement** A conclusion (accurate or not) that someone arrives at after considering evidence.
>
> **opinion statement** An assertion based on the speaker's beliefs.

Consider a few examples:

Fact: Rachel interrupted me before I finished what I was saying.
Inference: She doesn't care about what I have to say.
Opinion: She's a terrible listener.

Fact: Tom hasn't paid the rent on time in three months.
Inference: Tom doesn't manage his money well.
Opinion: Tom is immature and irresponsible.

Fact: The speaker said "um" a lot.
Inference: The speaker hasn't rehearsed very much.
Opinion: The speaker doesn't know what they're talking about.

What makes the statement "Rachel didn't like what I was saying" an *inference* and "Rachel is a terrible listener" an *opinion*? If you answered that the first statement is an interpretation based on evidence (the interruption), you are correct. Bear in mind that inferences, although they are based on evidence, are not necessarily correct. Rachel may have interrupted for any number of reasons—she wanted to express agreement, she had to rush off to class, she heard someone coming and wanted to spare the speaker embarrassment, or so on. Likewise, if someone asserts the opinion that Rachel is a terrible listener without anything to back it up, you'd be wise to judge for yourself.

One problem with inferences and opinions is that people often present them (and believe them) as if they are factual, which can lead to confusion, hurt feelings, and unfair judgments. When someone proclaims, "Liberals don't care about the budget" or "Conservatives don't care about people," they are voicing a sweeping opinion that's likely to antagonize others. Consider how much more productive it would be to focus on specific issues, as in, "Based on the candidate's proposal to . . . , I feel that they aren't paying enough attention to . . . ."

The feature "Distinguishing Between Facts and Opinions" offers guidance on interpreting different types of information. And here are some tips to help clarify your own thoughts and communicate in a respectful way:

- *Examine your own thinking.* When you form an inference or opinion about something, ask yourself "What led me to this conclusion? Is it possible that I'm overlooking something or letting emotions cloud my judgment?"
- *Focus on evidence.* Rather than believing others (or arguing with them) on the basis of opinions, ask questions such as "Why do you feel that way?" Do your own research as well.
- *Don't present opinions as facts.* For example, instead of "Online classes are best," say, "I prefer online classes."
- *Avoid emotionally inflammatory language.* Statements such as "She's a phony" imply broad or hurtful value judgments. Instead, reference a specific behavior, as in, "Her resume says she graduated, but she hasn't finished her senior year yet."
- *Don't jump to conclusions.* "I was disappointed when you missed the deadline" is better than "You don't care about the job."
- *Be polite.* Even if you disagree with someone, address them respectfully rather than resorting to insults or name calling.

Lies and Evasions

People lie more than they probably realize—on average, once or twice per day.[55] Not all lies are self-serving, though. At least some of the lies people tell are intended to be

---

COMMUNICATION STRATEGIES

Distinguishing Between Facts and Opinions

There's a place for personal opinions, but it's troublesome when they're presented as facts. Use the following guidelines to determine whether information—in a social media post, news item, or personal conversation—is fact, opinion disguised as fact, or opinion.[53,54]

If statements meet the following criteria, they are probably facts.

- The evidence presented can be objectively proven or verified.
- The information is current and relevant.
- Valid sources of information are provided.
- An effort is made to encourage additional and emerging information.

Beware of inferences or opinions masquerading as facts. The following are tell-tale signs.

- Statements seem designed mostly to stir up people's emotions.
- Claims are not supported by objective information.
- The argument is based on an isolated or unusual case.
- Assertions are overgeneralized or out of date.

Give thoughtful consideration to responsible opinions, which meet the following criteria.

- Statements are clearly acknowledged to be perceptions ("I feel that...").
- Respect is shown for other opinions.

kind or polite. Lies can do the greatest damage when the relationship is most intense, the importance of the subject is high, and there have been previous doubts about the deceiver's honesty. Of these three factors, the one most likely to cause a relational crisis is the sense that one's partner lied about something important.[56]

Experts suggest that, if you're considering deception, imagine how others would respond if they knew about it.[57] Would they accept your reasons for being untruthful, or would they be hurt by them? In light of that, we explore three types of lies here: altruistic lies, evasions, and self-serving lies.

Altruistic Lies Some lies aim to protect others' feelings or help them save face (Chapter 2). You might tell the host of a dinner party that the food was delicious even if it wasn't. Or you might compliment your significant other's new haircut or tattoo to avoid hurt feelings, even if you dislike it.

"Will the real golf pro please stand up?" Shows such as *To Tell the Truth* make a game of detecting who is lying and who isn't.

In everyday life, what signs do you look for to judge who is being honest with you?

altruistic lies Deception intended to be unmalicious, or even helpful, to the person to whom it is told.

evasion A deliberately vague statement.

Altruistic lies are defined—at least by the people who tell them—as being harmless, or even helpful, to the person to whom they are told.[58] For the most part, "white lies" such as these fall in the category of being polite, and effective communicators know how and when to use them without causing offense.

Evasions Unlike outright lies, **evasions** are vague statements that help speakers avoid telling the entire truth. Often motivated by good intentions, evasions are based on the belief that less clarity can be beneficial for the sender, the receiver, or sometimes both.[59] For instance, when your partner asks what you think of an awful outfit, you could *equivocate* by saying something truthful but vague, as in "It's really unusual—one of a kind!" Or you might *hint* when trying to escape a party saying to your host, "It's getting late," rather than, "I'm bored and want to leave now."

Self-Serving Lies Self-serving lies are attempts to manipulate the listener into believing something that is untrue—not primarily to protect the listener, but to advance the deceiver's agenda. For example, people might lie on their income tax returns or deny that they're under the influence if a cop pulls them over.

Self-serving lies may involve an *omission*—withholding information that another person deserves to know, or a *fabrication*—deliberately misleading another person for one's own benefit. For example, a romantic partner may keep a love affair secret or claim to be somewhere they weren't.

It's no surprise that self-serving lies can destroy trust and lead the deceived party to wonder what else the other person might be lying about. However, some relationships rebound from serious deceptions, particularly if the lie involves an isolated incident and the wrongdoer's apology seems sincere.[60]

Emotive Language

Particularly troublesome are opinion statements meant to incite strong emotional reactions, sometimes called **emotive language**. For example, a statement such as "Worthless bums are ruining our town" is emotionally inflammatory. A more responsible statement might sound something like this: "I worry that the rising number of unhoused people will cause families to move away from this area."

One problem with emotive language is that it tends to inspire reactions based more on emotional fervor than rational thought. This may lead people to believe an emotionally charged speaker even if the person presents no solid evidence. Or it may cause them to strike out in anger against people whose arguments are different than theirs. "Overly strong emotional language antagonizes the receiver and wipes away impulses to listen, to stay friends, or even to talk together any further," reflects psychologist Susan Heitler.[61]

Calling people ugly names is exceptionally hurtful. Calling someone lazy, ugly, or stupid is degrading, and it diverts attention away from real issues. The **fallacy of ad hominem** (translated as *to the person*) involves attacking a person's character rather than debating the issues at hand. (For more about fallacies, see Chapter 15.) If someone attacks you—perhaps by saying "You're a fool if you think that"—you might calmly remind them that you'd like to focus on the issues and evidence.

Name calling has an ugly past. Insulting language has long been used to stigmatize certain groups.[62] For example, racial and ethnic slurs suggest that certain groups of people are less human than others, implying that they don't deserve the same rights or compassion. Dehumanizing people with ugly names—or referring to them as animals (hogs, cows, pigs, monkeys)—can make it easier to accept prejudice and violence against them, observes political science professor Alexander Theodoridis.[63] And the personal effects are impossible to calculate. After Sonya Abarcar's seven-year-old daughter was called the n-word by White men, the child had nightmares and developed a fear of being attacked. When Abarcar writes about that experience and others on social media, she says, she is blasted in the comments with slurs such as the n-word, darky, animal, and ape.[64]

Social media can encourage personal attacks by creating what some theorists call "webs of hate."[65] Online, it's easy to network with people who have similar prejudices, bolstering the sense that those attitudes are more prevalent in the real world than they really are, and creating speech communities in which hateful speech is tolerated or encouraged.

Many people feel that, in today's contentious political climate, it's especially important to think clearly about information and avoid language that alienates and offends. By using language responsibly, you can encourage respectful discourse about diverse perspectives.

Microaggressions

"You speak excellent English."

People often say this to Derald Wing Sue. They aren't trying to be mean, he says. But the implication is clear: They assume that Sue, who was born and raised in Oregon, is a non-native English speaker. "What they are saying is that you are a perpetual alien in your own country, and you are not a true American," Sue says.[66] Although the speaker may mean the words as a compliment, they hurt.

Microaggressive language involves subtle, everyday messages that (intentionally or not) stereotype or demean people on the basis of sex, race, gender,

---

emotive language Opinion statements meant to stir up strong emotional reactions.

fallacy of ad hominem A statement that attacks a person's character rather than debating the issues at hand.

microaggressive language Subtle, everyday messages that (intentionally or not) stereotype or demean people on the basis of sex, race, gender, appearance, or some other factor.

In the film *Minari*, a South Korean family who has moved to rural Arkansas interacts with neighbors who clumsily try to make friends with them. In one scene, someone tells the mother she is "cute." In another, a child sputters nonsense syllables to the daughter and says, "Stop me when I say something in your language."

What qualities of microaggressions are present in these examples?

appearance, or some other factor.[67,68] Such language often takes the form of comments, questions, and even supposed compliments. Here are some more examples:

"You're really pretty for a dark-skinned girl."[69]

"I wish I was Native so I could get scholarships and stuff."[70]

"What are you ladies gossiping about?"[71]

"My neighbors are lesbians too! . . . Do you know them?"[72]

"Not to be racist, but what *are* you?"[73]

Such statements can be painful to the recipients because they are based on assumptions that feel dehumanizing and derogatory. These imply that women with dark complexions aren't usually beautiful, that Native Americans receive free handouts, that women are typically nosy and unkind, that gay women all know each other, and that a person (typically someone who appears to be nonwhite or multiracial) has an obligation to explain "what" they are so that the speaker can categorize them.

If you haven't often been the recipient of microaggressive comments, messages like these may seem innocent or trivial. But put yourself in the shoes of someone who is often burdened with a social stereotype that belies their status as a unique individual and marginalizes a group with which they identify. "The 'micro' in microaggresion doesn't mean that these acts can't have big, life-changing impacts," points out Kevin Nadal, a scholar who studies the issue.[74]

The good news is that everyone can be part of the solution by engaging in **microresistance**—everyday behaviors that call attention to hurtful language and stereotypes. The feature "Engaging in Microresistance" presents some useful strategies.

microresistance Everyday behaviors that call attention to hurtful language and stereotypes.

COMMUNICATION STRATEGIES

Engaging in Microresistance

Microaggressions present a quandary because people who initiate them aren't necessarily trying to be unkind, and it can be hard to shift an everyday conversation to talk about stereotypes and prejudice. But the stakes are high. People who are frequently the targets of microaggressive language tend to experience more stress and depression, lower self-esteem, less sense of belonging, and more worry about the future than their peers who don't.[75,76,77] Here are some tips from the experts for avoiding microaggressions yourself and for engaging in microresistance.

(Continued)

Examine Your Own Assumptions

Avoiding microaggressive language requires that everyone examine their own assumptions, educate themselves about what others may find offensive, and be willing to discuss sensitive issues respectfully. **Implicit bias** refers to prejudices and stereotypes that people harbor without consciously thinking about them.[78] For example, imagine that—deep down, without even thinking about it—you assume that overweight people are lazy or that people who don't speak English well lack intelligence. Although unconscious, these beliefs can affect the way you think about people and what you say. This may be true of many other ideas as well. You can get a sense of your implicit biases by monitoring your thoughts and by thinking before you speak. Before blurting out "You run like a girl" or "You're really outgoing for an Asian person," stop and ask yourself: *What stereotypes am I harboring and perpetuating?*

Inquire

If you are Black and someone says, "I can't believe you like country music," you might ask, "Why does that surprise you?" In considering an answer, the speaker might recognize a bias within themselves that they hadn't realized before.

Use Humor

When people tell Derald Wing Sue that he speaks excellent English, he sometimes quips in return, "Thank you, I hope so, I was born here."[79] Sometimes humor is enough, he says, to make a person stop and think.

Point Out the Underlying Assumption

Giving words to an underlying pattern can be a means of bringing it into conscious awareness. If someone says scornfully, "That's so gay," you might say, "It sounds like you have something against gay people. What's up?"

Be an Ally

An **ally** is someone from a dominant social group (e.g., White, male, cisgender, management) who actively advocates for fair treatment and social justice for others.[80] If you're on a personnel committee at work and diverse candidates are written off as "not management material," you can be an ally by saying, "I think we're overlooking a lot of great candidates because, at some level, we think everyone should be just like us. Let's agree on job-specific qualifications we can apply to everyone equally." Research shows that allies' involvement can go a long way toward changing social viewpoints and helping people who are the targets of microaggression feel less singled out and othered.[81]

Avoid Casting Individuals as "Spokespersons"

Imagine that a teacher says, "Alejandra, you're Hispanic. Tell us about the struggles of Spanish speakers in the United States." Right away, Alejandra is depicted as a "typical" member of a large group that is ostensibly "different" from everyone else. And the question presumes that she's knowledgeable about the issue and willing to discuss it. "That can be very psychologically and emotionally exhausting," points out Kevin Nadal.[82] A better option—whether you are in class or conversing with friends or colleagues—is to create an environment in which people are invited to share their experiences, but no one is put on the spot to speak for a particular group.

Choose Your Battles

There are times when it's not worth outing a microaggression. Perhaps you don't have the emotional energy to broach a sensitive topic, or you fear the other person will react aggressively toward you. Use good judgment.

Apologize If You Mess Up

If you inadvertently make a microaggressive statement, own up to it and offer a sincere apology. "We're all human beings who are prone to mistakes," says Derald Wing Sue.[83] Use the situation as an opportunity to become more aware. Inclusivity coach Pooja Kothari suggests something like this:

> I said something that I think was offensive. I have thought about it and want to apologize to you. I know my words have an impact and I am sorry for the impact they made on you. I value our friendship/relationship/camaraderie and want you to know I am aware of what I said, I take responsibility for it and am working on it.[84]

Even if your initial statement was insensitive, your sincere desire to make it right is likely to benefit the relationship and your good name.

implicit bias A prejudice and stereotype that someone harbors without consciously thinking about it.

ally Someone from a dominant social group (e.g., White, male, cisgender, management) who actively advocates for fair treatment and social justice for others.

This chapter explored the nature of language and its ability to shape and reflect reality, unite people, create misunderstandings, and help communicators think through issues. You can probably remember a time when someone's words gave you courage, comforted you, or made you feel special. Alternatively, you may recall an episode in your life when words hurt your feelings or ruined a relationship.

The words you say to yourself can be equally powerful. Based on what you have learned in this chapter, here's a tip from publisher/entrepreneur Michael Hyatt: Start using the phrase "get to" rather than "have to"—as in, "I *get to* make a presentation today" and "I *get to* learn something new." Hyatt recalls a conversation in which his friend Josh, who had just landed a book contract, said that now he "had to" write it. Josh's expression brightened when Hyatt pointed out, "No Josh, you *get to* write this book. This has been a goal of yours for as long as I have known you. You are living your dream, buddy!"[85]

MAKING THE GRADE

OBJECTIVE 5.1 Explain how symbols and linguistic rules allow people to achieve shared meaning.

- A language is a collection of symbols used to convey messages between people.
- Because of its symbolic nature, language is not a precise tool. People interpret the meaning of words differently.
- For effective communication to occur, it's necessary for people to negotiate meaning collaboratively.
- Language is shaped by semantic, phonological, syntactic, and pragmatic rules.
    > Describe a time when you and someone else interpreted the same word differently. What happened as a result?

> The denotative meaning of "green" is a shade between blue and yellow on the color spectrum. What are some connotations of the word *green*?

> In *Star Wars*, Yoda makes statements such as "Patience you must have" and "Always in motion is the future." Is this unusual use of language most relevant to phonological or syntactic rules? Explain your answer.

OBJECTIVE 5.2 Identify ways in which language shapes people's attitudes and reflects how they feel about themselves and others.

- Language affects and reflects the way people make sense of the world around them.
- Gender references in language tend to influence how people think about objects and social roles.
- A person's name shapes their sense of personal identity.

- Everyone has an accent, but stereotypes about particular accents vary.
- Assertive language is clear and direct, whereas collaborative language encourages team thinking.
  > Imagine for a moment that the words *love*, *disappointment*, and *hate* do not exist. In your opinion, would you be aware of those feelings in the same way? Could you talk about them as effectively? Why or why not?
  > Describe examples from your own experience in which language helped you connect with another person in a powerful way.
  > How might you use language to impress a prospective employer during a job interview? To show support for a friend?
  > Imagine you're on a team asked to choose between taking part in a service learning project or helping to conduct research. What might you say if you advocate assertively for your preference? Conversely, what might you say if you would like the team to take a collaborative approach to making the decision? Which approach do you prefer and why?

OBJECTIVE 5.3 Recognize and remedy confusing language.

- Vague and specialized language has the potential to create misunderstandings, either intentionally or not.
- Some words have meaning based on comparison to others (e.g., *sad* and *happy*, *difficult* and *easy*).
- Euphemisms are meant to be polite, but they can be confusing.
  > Practice moving the statement "I have studied many topics in school" up the abstraction ladder by creating several statements that are each more specific than the previous one.
  > Hoping to be excused from a meeting, Lucinda says that she "has a prior commitment." The commitment is actually going to the beach with her friends. Is this best described as a use of equivocal language or an equivocation? Explain your answer.
  > Fill in the blanks of the sentence, "Compared to making a speech, _____ is easier, but _____ is harder." Explain how comparison helps to create meaning.
  > Name several euphemisms that you consider to be clear and acceptable. Then name a few that are often confusing.

OBJECTIVE 5.4 Distinguish among facts, inferences, and opinions, and avoid using inflammatory and microaggressive language.

- Language can be disruptive when people assert their inferences and opinions as if they are facts.
- Altruistic lies fall in the category of being polite, and effective communicators know how to use them without causing offense.
- Evasions are deliberately vague and include equivocation and hinting. They are generally meant to avoid hurting people's feelings.
- Self-serving lies are attempts to manipulate the listener into believing something that is untrue. They involve omissions or fabrications.
- Emotive terms are inflammatory and tend to distract people from considering actual evidence and experiences.
- Some people think they are simply complementing someone or making an observation, but their statements may be microaggressive.
  > Identify examples of troublesome language in a movie or television show. How might you rewrite the script to reflect more effective ways of communicating?
  > Keep a tally of how many white lies and evasive comments you make in a single day. What were your reasons for making each of them?
  > Consider a self-serving lie you have communicated in a close relationship. Looking back would you do anything differently?

> Looking around you and in the media, point out examples of people discussing sensitive issues in a civil and respectful way. Point out examples of hostile and judgmental language. How would you rate your own communication skills in this regard?

> What words do people use to describe you? Which words best reflect how you feel about yourself? Do any of the words people use connote a stereotype that hurts your feelings? If so, how?

KEY TERMS

abstract language p. 115
abstraction ladder p. 115
accent p. 112
ally p. 123
altruistic lies p. 120
assertive language p. 112
collaborative language p. 113
concrete language p. 115
confirmation bias p. 118
connotative meanings p. 107
coordinated management of meaning (CMM) p. 108
denotative meanings p. 107
dialect p. 106
emotive language p. 121
equivocal language p. 115
equivocation p. 115
euphemism p. 116
evasion p. 120
factual statement p. 118
fallacy of ad hominem p. 121
implicit bias p. 123
inferential statement p. 118
jargon p. 116
language p. 106
linguistic relativism p. 110
microaggressive language p. 121
microresistance p. 122
opinion statement p. 118
phonological rules p. 108
pragmatic rules p. 108
relative words p. 116
semantic rules p. 108
slang p. 116
symbol p. 106
syntactic rules p. 108

PUBLIC SPEAKING PRACTICE

Identify an instance in the news or on social media in which someone presents an opinion as if it's a fact or uses emotive language. Prepare a brief oral presentation explaining the example and suggest more responsible ways to convey the information.

ACTIVITIES

1. Increase your ability to achieve an optimal balance between assertive and collaborative speech by rehearsing one of the following scenarios:

   - Describe your qualifications to a potential employer for a job that interests you.
   - Request an extension on a deadline from one of your professors.
   - Explain to a merchant why you want a cash refund on an unsatisfactory piece of merchandise when the store's policy is to issue credit vouchers.
   - Ask your boss for three days off so you can attend a friend's out-of-town wedding.

- Talk to a neighbor whose barking dog keeps you away at night.

2. For each of the following microaggressive statements, describe the stereotype that underlies it and provide one example of how someone may respond to it with microresistance.

- "You look good for your age."
- "You're even-tempered for a redhead."
- "What do you mean you don't like jazz? All Black people like jazz."

Listening

CHAPTER OUTLINE

6.1 The Value of Listening 130

6.2 Misconceptions About Listening 132
- Myth: Listening and Hearing Are the Same Thing
- Myth: Listening Is a Natural Process
- Myth: All Listeners Receive the Same Message
- Myth: People Have One Listening "Style"
- Myth: Women Are More Supportive Listeners Than Men
- Myth: The Majority of Listening Happens Offline

6.3 Overcoming Challenges to Effective Listening 137
- Message Overload
- Noise
- Cultural Differences

6.4 Faulty Listening Habits 140
- Pretending to Listen
- Tuning In and Out
- Missing the Underlying Point
- Dividing Attention
- Being Self-Centered
- Talking More Than Listening
- Avoiding the Issue
- Being Defensive

6.5 Listening to Connect and Support 144
- Allow Enough Time
- Be Sensitive to Personal and Situational Factors
- Ask Questions
- Listen for Unexpressed Thoughts and Feelings
- Encourage Further Comments
- Reflect Back the Speaker's Thoughts
- Consider the Pros and Cons When Analyzing
- Reserve Judgment, Except in Rare Cases
- Think Twice Before Offering Advice or Solutions
- Offer Comfort, If Appropriate

LEARNING OBJECTIVES

6.1 Summarize the benefits of being an effective listener.

6.2 Outline the most common misconceptions about listening, and assess how successfully you avoid them.

6.3 List strategies to overcome factors that make it challenging to listen well.

6.4 Identify and minimize faulty listening habits.

6.5 Describe and practice listening strategies for effectively connecting with and supporting others.

6.6 Discuss when and how to listen to accomplish a task, analyze a message, and critically evaluate remarks.

6.6 Listening to Learn, Analyze, and Critique 151
Task-Oriented Listening
Analytical Listening
Critical Listening

MAKING THE GRADE 157

KEY TERMS 158

PUBLIC SPEAKING PRACTICE 159

ACTIVITIES 159

It may seem like people are more divided than ever, especially when it comes to issues such as politics, health care, and social injustice. Listening with an open mind, even when it feels uncomfortable, is one solution.

In her TED Talk on what she calls "radical listening," Chanel Lewis describes hearing acquaintances insult someone she admired and then praise someone she considered dishonest. "I yelled, in my head," she admits. But instead of arguing or walking away, Lewis chose to engage in radical listening. She describes it as "the practice of intentionally quieting your internal voice and judgments, thereby offering your full mental space to the speaker."[1] Although Lewis has taught university-level courses and given talks on the subject, she is the first to acknowledge that radical listening is not easy. The key is to stick it out, even when conversations are difficult.

That's what Lewis did with her acquaintances, and ultimately they found an issue on which they wholeheartedly agreed. Lewis remembers thinking, "Wow! A moment I didn't expect."[2] She encourages others to listen radically, saying, "We have the opportunity to listen to and learn from people, even those that seem so different from us."[3]

Really listening involves a level of discipline and skill few people stop to consider and even fewer master. Yet the payoffs are enormous, as you will see. Masterful listening can help you make wise decisions, make a positive impression on others, and enrich your relationships. This chapter focuses on the pay-offs of being a good listener, common myths about listening, ways to overcome listening challenges and bad habits, and ways you might use listening to offer support and comfort as well as evaluate the quality of information you encounter.

6.1 The Value of Listening

You may have heard the adage, "You have two ears and one mouth for a reason." Among those who endorse the idea of listening twice as much as you talk are the legendary University of Alabama football coach Nick Saban, international business leader and bestselling author Ivan Misner, and master salesperson Tom Hopkins. Hopkins insists, "It is vital to listen more than you speak. Doing all the talking can make you come across as pushy and impersonal." Despite this advice, people tend to do the opposite: They talk more than they listen.

In his bestselling book *The 7 Habits of Highly Effective People*, Stephen Covey observes that most people only pretend to listen while they actually rehearse what they want to say themselves.[4] Rare (and highly effective) is the person who listens with the sincere desire to *understand*, observes Covey. An impressive body of evidence backs up this claim, as you will see in the following list of reasons to become a better listener:

University of Alabama football coach Nick Saban attributes much of his success to listening. Good listeners, he says, spend twice as much time listening as they do talking.

How much time do you spend listening versus talking during conversations?

- **People with good listening skills are more likely than others to be hired and promoted.**[5] Listening is the number one skill employers look for when hiring. In fact, there isn't a single occupation in which listening is not important.[6] Listening skills are also important once you get the job. Because good listeners are typically deemed appealing and trustworthy,[7] they are especially popular with employers and with customers and clients.[8,9]

- **Listening is a leadership skill.** Leaders who are good listeners have more influence and stronger relationships with team members than less attentive leaders do,[10] and they report an increased sense of personal well-being at work.[11] Workers who feel listened to by leaders tend to be more loyal and innovative, which benefits the whole organization.[12] Everyone wins when leaders listen.

- **Good listeners are not easily fooled.** People who listen carefully and weigh the merits of what they hear are more likely than others to spot what some researchers call "pseudo-profound bullshit"—statements that sound smart but are actually misleading or nonsensical.[13] Mindful listening (a topic we'll discuss more in a moment) is your best defense.

- **Listening improves your health and well-being.** You probably know from experience that talking about your problems can make you feel better. That's because engaging in what researchers call "troubles talk" improves physical and mental health.[14] When people—even strangers—listen attentively to each other speak about their everyday problems, it bolsters feelings of closeness and friendship and lowers stress.[15] Of course, the key is that people must listen actively, with genuineness and respect. You'll learn more later in the chapter about active and supportive listening.

- **Listening makes you a better friend and romantic partner.** While you're getting dressed for an evening out, make sure to clean out your ears, metaphorically speaking. Friends and partners who listen well are considered more likable[16] and more supportive than those who don't.[17] That probably doesn't surprise you, but this may: Listening well on a date can significantly increase your attractiveness rating.[18,19] The caveat is that you can't *pretend* to listen. Effective listeners are sincerely interested and engaged.

Despite the importance of listening, experience shows that much of the listening people do is not very effective. They misunderstand others and are misunderstood in return. People become bored and feign attention while their minds wander. They engage in a battle of interruptions, each person fighting to speak without hearing the other's ideas. Some of this poor listening is inevitable, even understandable. But in other cases, learning a few basic listening skills can help people be better receivers.

Do you think you're a good listener? Answer the questions in *Understanding Your Communication* for insights about your listening strengths.

6.2 Misconceptions About Listening

Stephen O'Keefe's wife often says that he "listens better than anyone she's ever met."[20] That's no accident. "I've had to work so hard to communicate,"[21] O'Keefe says, but he does the work because of the conviction that "we all pay a big price for poor listening: conflict, lack of respect, lack of understanding, [and] lack of empathy."[22]

By reading this book, you are doing important work which can help you become a better communicator. When it comes to being a better listener, part of your job is to discern listening myths from listening facts. Let's see how well you do!

Myth: Listening and Hearing Are the Same Thing

In Chapter 1, we introduced the term *receiving* to describe the process by which a message is decoded. In fact, the process of receiving a message involves multiple stages. The first stage is **hearing**, which is the physiological ability to perceive the presence of sounds in the environment. If you have that physiological ability, hearing occurs automatically when sound waves strike your eardrums and cause vibrations that are transmitted to your brain. People with physical hearing disorders lose some or all of the ability to detect sounds.

Listening occurs when the brain recognizes sounds or other input and gives them meaning. Unlike hearing, listening requires conscious effort and skill. Even when people cannot hear, they can still be attentive listeners. As Stephen O'Keefe explains, "I may be Deaf, but I figured out a way to listen better."[23] That's right! O'Keefe, whose wife says he's the best listener she knows, is Deaf. His approach to listening is simple: "Listen with your attention, listen with your eyes, and listen with your heart."[24] You'll learn more about this kind of listening in a moment.

For many people, hearing is automatic but listening is another matter. As O'Keefe jokes, "I'm amazed that people who can hear do not use their ears."[25] It's actually pretty common to hear others speak without listening to them. Sometimes people deliberately tune out speech—for example, a friend's rambling story or a boss's unwanted criticism. When you pay attention to what you hear, that's called **attending**. Your needs, wants, desires, and interests influence what you attend to as a listener.

The next step in listening is **understanding**—the process of making sense of a message. Communication researchers use the term **listening fidelity** to describe the degree of congruence between what a listener understands and what the sender was attempting to communicate.[26] Chapter 5 discussed many of the ingredients that make it possible to understand language: syntax (how words are ordered), semantics (what words mean), and pragmatic rules about using and interpreting language. Taking all of this into account, it's clear that listening isn't the passive activity you might

---

hearing The process wherein sound waves strike the eardrum and cause vibrations that are transmitted to the brain.

listening The process wherein the brain recognizes impulses as sound and gives them meaning.

attending The process of focusing on certain stimuli in the environment.

understanding The act of interpreting a message by following syntactic, semantic, and pragmatic rules.

listening fidelity The degree of congruence between what a listener understands and what the message sender was attempting to communicate.

UNDERSTANDING YOUR COMMUNICATION

What Are Your Listening Strengths?

Answer the questions below to gauge which listening approaches you tend to use most.

1. Which of the following best describes you?
   a. I'm a quick learner who can hear instructions and put them into action.
   b. I have an intuitive sense, not just of what people say, but how they are feeling.
   c. I'm a good judge of character. I can usually tell whether people are trustworthy or not.
   d. I'm a rapid thinker who is often able to jump in and finish people's sentences for them.

2. Imagine you are tutoring an elementary school student in math. What are you most likely to do?
   a. Focus on clearly articulating the steps involved in solving simple equations.
   b. Begin each tutoring session by asking about the student's day.
   c. Pay close attention to what the student says to see if they really understand.
   d. Feel frustrated if it seems the student isn't listening or isn't motivated.

3. A friend launches into a lengthy description of a problem with a coworker. What are you most likely to do?
   a. Offer some ideas for discussing the issue with the coworker.
   b. Show that you are listening by maintaining eye contact, leaning forward, and asking questions.
   c. Read between the lines to better understand what is contributing to the problem.
   d. Pretend to listen but tune out after five minutes or so.

4. If you had your way, which of the following rules would apply to team meetings?
   a. Chit-chat would be limited to five minutes so we can get to the point at hand.
   b. Everyone would get a turn to speak.
   c. People would back up their opinions with clear data and examples.
   d. There would be no meetings; they're usually a waste of time.

INTERPRETING YOUR RESPONSES

Read the explanations to below to see which listening approaches you frequently take. (More than one may apply.)

Task Oriented

If you answered "a" to more than one question, you tend to be an action-oriented listener. You value getting the job done and can become frustrated with inefficiency. Your task orientation can help teams stay on track. Just be careful that you don't overlook the importance of building strong relationships, which are essential for getting the job done. Tips for group work in Chapter 11 may be especially interesting to you.

Relational/Supportive

If you answered "b" to more than one question, you tend to be a relational and/or supportive listener. It's likely that people feel comfortable sharing their problems and secrets with you. Your strong listening skills make you a trusted friend and colleague. At work, however, this can make it difficult to get things done. Make an effort to set boundaries so people don't talk your ear off.

Analytical/Critical

If you answered "c" to more than one question, you often engage in analytical and/or critical listening. You tend to be a skeptical listener who isn't easily taken in by phony people or unsubstantiated ideas. Your ability to synthesize information and judge its merits is a strength. At the same time, guard against the temptation to reach snap judgements. Take time to consider people and ideas thoughtfully before you write them off. The tips for mindful listening can help.

Impatient

If you answered "d" to more than one question, you have a tendency to be an impatient or distracted listener. Your frustration probably shows more than you think. Review the tips throughout this chapter for ways to become more focused and active in your listening approach.

Source: © 2003 Zits Partnership Distributed by King Features Syndicate, Inc.

responding Providing observable feedback to another person's behavior or speech.

remembering The act of recalling previously introduced information. The amount of recall drops off in two phases: short term and long term.

residual message The part of a message a receiver can recall after short- and long-term memory loss.

have imagined. It relies on a sophisticated combination of effort, knowledge, skills, and physical ability.[27] Like O'Keefe said, listening is hard work.

The next listening stage involves **responding** to a message—offering observable feedback to the speaker. Feedback may include eye contact, facial expressions, questions and comments, posture, and more. Feedback serves two important functions: It helps you clarify your understanding of a speaker's message, and it shows that you are invested in what the speaker is saying.

Listeners don't always respond in obvious ways to a speaker—but research suggests that they should. When people are asked to evaluate the listening skills of others, the number-one trait they consider is whether listeners offer feedback.[28] Conversely, it's easy to see how discouraging it can be for speakers if listeners yawn, appear bored, or look at their phones. Adding responsiveness to the listening model demonstrates that communication is transactional (Chapter 1). As listeners, people are active participants in a communication transaction: While they receive messages, they also send them.

The final step in the listening process is **remembering**.[29] It has long fascinated scientists that people remember every detail of some messages but very little of others. For example, you may remember specific details about gossip you heard but forget what your roommate asked you to buy at the store. By some accounts, people tend to forget about half of what they hear *immediately after* hearing it, suggesting that they did not truly listen to and store the information.[30]

Given the amount of information you process every day—from instructors, friends, social media, TV, and other sources—it's no wonder that the **residual message** (what you remember) is a small fraction of what you hear. However, with effort, you can increase your ability to remember what is important. We'll show you ways of doing that later in the chapter.

Myth: Listening Is a Natural Process

Another common myth is that listening is like breathing—a natural activity that people usually do well. The truth is that listening is a skill much like speaking: Everybody does it, though few people do it well. In the workplace, good listeners are typically more influential than their peers because they are perceived to be more agreeable, open, and approachable than people who listen poorly.[31] However, most people are not the good listeners they think they are. In one survey, 96 percent of professionals rated themselves good listeners, but 80 percent of them admitted to multitasking while on the phone, a sure sign that they do not give callers their full attention.[32]

Sometimes it's okay to be mindless about what you hear. Paying attention to every commercial would distract you from more important matters. But problems arise when people are lazy about listening to things that really matter. For example, a college student hurt by his girlfriend's poor listening skills wrote in an online forum, "I have opened up to her about really, really personal things and then two weeks later or within the week . . . she's like, 'oh, you never mentioned it to me.' I just find this really really rude and insulting."[33]

Mindful listening involves being fully present with others—paying close attention to their gestures, manner, and silences, as well as to what they say.[34] It requires a commitment to understanding the other's perspectives without being judgmental or defensive. Chanel Lewis's approach to radical listening, which you read about at the start of this chapter, begins with mindful listening. It's what Stephen O'Keefe does when he listens with his attention, eyes, and heart. This type of listening can be difficult, especially when you are busy or when you feel vulnerable, yet the investment is worthwhile.

TING

EAR MIND EYE HEART

The Chinese word *ting* refers to deep, mindful listening. In its written form, the word combines the symbols for ears, eyes, heart, and mind.

Have you ever felt that someone listened to you with an open mind and heart, attentive to your words and your feelings? If so, how did you respond?

Myth: All Listeners Receive the Same Message

When two or more people listen to a speaker, you might assume they all hear and understand the same message. In fact, such uniform comprehension isn't the case. Recall the discussion of perception in Chapter 3, in which we pointed out the many factors that cause each of us to perceive an event differently. Perhaps you're hungry, thinking about something else, or just not interested. But you can hear things differently even when you're trying hard to understand. Sometimes your ears may trick you into hearing the wrong words. At other times, you may hear the same words but give them different meaning or significance. Your friend might find a joke funny, whereas you consider it offensive.

Misunderstandings are especially likely when remarks are interpreted out of context. When Fifth Harmony member Normani Kordei called one of her groupmates "very quirky" in an interview, she says she meant it in a good way, but some fans interpreted her remark as an insult and attacked her on social media.[35]

mindful listening
Being fully present with people—paying close attention to their gestures, manner, and silences, as well as to what they say.

Myth: People Have One Listening "Style"

Listening is a behavior, not a personality trait.[36,37] People don't have a dominant listening style. Instead, people change how they listen based on the situation and their goals. For example, you listen differently when trying to understand a new concept in class versus relaxing to your favorite album after work. You'll learn more about goal-oriented listening later in this chapter.

Myth: Women Are More Supportive Listeners Than Men

Most Americans—about 87 percent—think men and women express their emotions in substantially different ways.[38] Actually, research shows mixed results. On the one

Listening comes into play when you read and respond to social media posts and text messages.

Which of the listening stages are involved when you post online?

hand, scholars observe that feminine individuals tend to focus more on contextual and emotional cues and masculine individuals more on facts.[39,40,41] On the other hand, some researchers have found that gender only accounts for 1 percent of communication differences[42] and that culture, more than gender, shapes people's listening habits.[43]

Popular books like *Men Are from Mars, Women Are from Venus* argue that men and women speak different languages, but those arguments are often overstated, and it's important to remember that gender roles continually evolve. A number of factors interact with gender to shape how well people listen—including cultural background, personal goals, expressive style, and cognitive complexity. Overall, there is no strong evidence to suggest that women are better listeners than men.

Myth: The Majority of Listening Happens Offline

You may think that listening happens primarily in real-world contexts—during face-to-face conversations, over the phone, in meetings, and so on. The truth is that much of what you do online, over text message, and on social media involves listening too. **Social listening** is the process of attending to, observing, understanding, and responding to stimuli through mediated, electronic, and social channels.[44] It helps to think of it this way: People who post in online forums or on social media are "speaking" and the people who read and respond are "listening."[45] Some posts may have auditory components, such as spoken messages or music, while other posts are strictly visual, consisting of text or photos.[46]

Fifty-seven percent of the world's population—about 4.5 billion people—use social media,[47] which means social listening is something most people do regularly. In Chapter 2, you read about social listening in the context of masspersonal communication and viral posts—how Nathan Apodaca's skateboarding video landed him a spot in a music video with Snoop Dogg[48] and Brianna Ramirez's mac and cheese TikTok got her fired from Panera Bread.[49] Their experiences underscore an important lesson: You never know who is listening to you online.

People aren't the only social listeners though. Organizations do it too. Social listening (sometimes called social monitoring) helps organizations and marketers tap into customers' likes and interests.[50,51] Moreover, customers who use social media channels to ask questions, request assistance, and so on expect that companies will listen to them and respond.[52] Carter Wilkerson, however, wasn't sure he'd hear back when he tweeted fast-food restaurant Wendy's. He asked how many retweets it would take for him to get free chicken nuggets for a year. Wendy's replied, "18 million." #NuggsForCarter was born! Carter managed to get

social listening The process of attending to, observing, understanding, and responding to stimuli through mediated, electronic, and social channels.

more retweets than anyone else in the history of Twitter—3.4 million.[53] Although he fell short of the goal, Wendy's gave him the nuggs anyway!

6.3 Overcoming Challenges to Effective Listening

With the number of people and devices clamoring for your attention, listening can seem harder than ever before.[54] Consider these examples:

- Your phone vibrates during a meeting at work. You sneak a peek at the screen to see what's up.
- You just sat down to dinner. A neighbor drops by to warn you about some car break-ins nearby. You know the issue is important, but you're hungry and irritated by the interruption.
- Over coffee, a friend complains about having a bad day. You want to be supportive, but you're preoccupied with problems of your own, and you need to get to class soon.

Distractions such as these fall into several categories.

Message Overload

It's impossible to listen carefully to all the information that bombards you daily. Along with face-to-face interactions, you probably deal with phone calls, emails, DMs, texts, and social media notifications. You're likely engaging with mass media at some point too—streaming a movie, listening to a podcast, or watching TV, etc. This deluge of communication makes the challenge of attending to messages tougher than at any time in human history.[55,56]

Social media seems to have exacerbated the problem. Although it helps people stay in touch, frequent updates and notifications are driving some folks nuts. At least that's what participants told a team of researchers studying social media's effect on family communication. One 40-year-old woman complained about the near-constant, senseless messages she receives, saying, "I don't care that you just went to the grocery store and found bacon on sale."[57] She explained that her son isn't a fan either, "He can't stand when there's a group text going around about what color shirt you're wearing today."[58]

People who consider themselves good multitaskers may think they are immune to message overload, but the data say otherwise—multitasking takes a toll on performance and attention,[59,60] as well as on short-term memory.[61] To combat this, experts suggest turning off communication technologies while you work on complex tasks, sending clear and brief emails with specific subject lines, and thinking twice before sharing trivial information with everyone you know.[62]

Noise

Effective listening often requires overcoming various forms of noise.

Psychological Noise People are often wrapped up in personal concerns that seem more important to them than the messages others are sending. It's hard to pay

Source: DILBERT © 2009 Scott Adams. Used by permission of UNIVERSAL UCLICK. All rights reserved.

attention to someone else when you're worried about an upcoming test, thinking about how much fun you had last night with friends, or planning what you're going to say when it's your turn to talk. To avoid such distractions, CEO Mark Fuller pretends he's doing improvisational comedy, because improv is all about responding.[63] As Fuller explains, "If we're on stage, I don't know what goofball thing you're going to say, so I can't be planning anything. I have to really be listening to you so I can make an intelligent—humorous or not—response."[64] Try Fuller's strategy the next time you find yourself wrestling with internal distractions.

Physical Noise Sometimes external distractions make it hard to pay attention to others. The sound of traffic, construction work, music, phones going off, other people's conversations, and the like interfere with your ability to hear well, let alone listen. You can listen better by eliminating sources of noise whenever possible, by turning off the television, silencing your phone, closing the window, and so on. In some cases, you may need to find a more hospitable place to have a conversation.

Physiological Noise Another listening distraction may arise from the way your body functions. You've probably noticed that it's harder to listen if you are tired, hungry, too hot, or too cold. If physiological noise is too distracting, it may be better to have an important conversation another time.

Cultural Differences

The behaviors that define a good listener vary by culture. Americans are most impressed by listeners who ask questions and make supportive statements.[65] By contrast, Iranians tend to judge people's listening skills based on more subtle indicators, such as their posture and eye contact. That's probably because the Iranian culture relies more on context.[66] (As you may remember from Chapter 4, members of high-context cultures are particularly attentive to nonverbal cues.) Meanwhile "dozing listening" is common in Japan. Occasionally nodding off in a lecture or meeting is tolerated, whereas picking up a phone or reading a book would be considered rude.[67] Unless you're in Japan, we don't recommend you try dozing listening!

Expectations vary by generation as well. For instance, researchers found that members of Generation X (those born between 1965 and 1980) and Generation Z (those born after 1996) have different views on what it means to listen well.[68] For example, many Gen Zers prioritize evaluative listening and associated behaviors such as asking questions and critiquing information. Gen Xers, on the other hand, value relational listening and gestures like leaning in and eye contact. These generational differences are especially important because they are a factor in many parent–child, manager–employee, and professor–student relationships. The likelihood of miscommunicating or even offending one another is high. If a Gen Z employee checks their phone in a meeting or focuses their gaze on the floor during a presentation, a Gen X manager may assume they are not listening given the lack of eye contact.

One lesson is that whereas people in some cultures and age groups may overlook silence, nodding off, or not looking at the speaker, others may interpret such behavior as rudely inattentive. This applies whether you are listening in person or online. See the feature "Listening in a Virtual Space" for more information.

Cultural and generational differences shape perceptions of what it means to be a good listener. For instance, older adults value eye contact more than younger adults.

Has a parent or boss falsely accused you of not listening to them? What were the consequences?

COMMUNICATION STRATEGIES

Listening in a Virtual Space

Chances are pretty good that, like most people, you hadn't heard of the teleconferencing company Zoom before the coronavirus pandemic. Now, Zoom is a part of everyday life for many people.[69] Around 300 million Zoom meetings were held every day in 2020 alone.[70] And it's likely that Zoom is here to stay, at least to some extent.[71] In one study, 97 percent of people said they wanted to continue working remotely after COVID restrictions were lifted.[72] More than three-quarters of managers and employees say remote work will soon be the new normal,[73,74] and a third of college students, and more than 40 percent of graduate students, say they prefer online learning to in person classes.[75] Listening well in these virtual spaces is crucial if you want to succeed.

You may already consider yourself a good listener, but virtual listening is different from real-world listening. For example, while attending an in-person meeting you might nod and say things like "*Uh huh*" and "*Yeah*" to let the speaker know you're listening. In a Zoom call, nodding is fine, but to make audible responses such as "*Yeah*," audience members would have to keep their microphones on, which can lead to overlapping talk, automatically putting your image center screen, and broadcasting distracting background noises.[76,77] Instead of encouraging the speaker, your audible feedback may inadvertently steal the show.[78] To help you avoid such mistakes and hone your virtual listening skills, we've compiled the following tips.

(Continued)

Eliminate Distractions Beforehand

Reading email, checking your phone, and scrolling social media during an online meeting are big no-no's. Even though you would not (hopefully) do these things in an in-person meeting, you might be tempted to try them on a Zoom call. After all, who would be the wiser? The answer is everyone! Alison DeNisco Rayome, a technology and software writer for CNET, offers a warning: "Just because you're at home doesn't mean everyone can't see you staring at your phone instead of paying attention.... Looking away from the camera at other content on your laptop or monitor is also pretty obvious."[79] To avoid temptations that will interfere with your ability to listen, silence your phone and put it to the side before you join a meeting. Minimize tabs and apps on your computer as well, so you can focus your full attention on the meeting window.

Pay Attention to What Your Camera Sees

Another potential distraction is what appears behind you. Each Zoom square offers a look into the private lives of participants—how they decorate their homes, which books they keep on their shelves, and so on.[80, 81] You may even glimpse their pets, children, or partners. It's hard for people to listen carefully when so many things can catch their eye. To create a better listening environment, take a few moments before a virtual meeting to tidy your space and minimize distracting elements or try using the blur feature (available in most teleconferencing platforms) to obscure your background. Also, be sure the camera angle doesn't add to the visual noise—make certain it's straight on, at eye level, and not pointing at the ceiling or up your nose.

Check Audio and Video Settings

A good Zoomer knows to log in early and make sure audio and video are working. It's especially important to check the video settings, lest you end up like attorney Rod Ponton. He appeared in a virtual court hearing—with a kitten filter turned on![82] Luckily the judge had a sense of humor, but you may not fare as well if you turn up as a cat or potato in your next virtual meeting. It's hard for others to listen to you well when they can't see or hear the real you clearly.

Practice Good Teleconferencing Etiquette

Once the meeting is underway, there a few rules you should follow. Mute your microphone when you aren't speaking to cut down on noise and avoid disrupting the speaker. When the camera is on, make a point to smile, nod, and look interested. Interject, politely. If you have something to add, raise your hand to get the facilitator's attention. Once you have the floor, it helps to acknowledge what others have said before changing the topic (people are more inclined to listen to you if they feel they've been heard by you).[83] Stay focused. It's easy to be a distracted listener when a lot is happening on the screen. Experts have two suggestions to help. First, turn off the self-view feature, this way "you're not preoccupied with looking at yourself."[84] (Just remember that others can still see you.) Second, consider switching to speaker view. With this feature enabled, you'll only see the person talking.

Write It Down

It helps to take notes in meetings. For starters, it can keep you focused on what the speaker is saying. Sarah Gershman, a communication scholar and writer for *Harvard Business Review*, says jotting things down can help in another way: "Writing down wandering thoughts allows you to put the thought 'somewhere' so that you can return to it later, after the meeting has ended."[85] The sooner you set aside the distracting thought, the sooner you get back to listening.

Virtual listening is hard. *Really* hard. It's exhausting and frustrating at times too,[86] but it has the power to shape how you live, learn, and work for decades to come. In time, it will get easier. Follow these suggestions and you'll be well on your way to mastering virtual listening.

6.4 Faulty Listening Habits

Kaleigh makes a point of really listening when friends talk about their problems or brag about their accomplishments. "I always want people to feel heard and valued when they're talking to me," she says.[87] But, as Kaleigh laments, "I rarely get that in

Calvin and Hobbes   by Bill Watterson

Source: CALVIN and HOBBES © 1995 Watterson. Reprinted by permission of UNIVERSAL UCLICK. All rights reserved.

return. . . . I feel very much like the sidekick, or the 'best friend in the movie,' whose storyline isn't that important. It's discouraging."[88] If you've ever been in Kaleigh's position, you know how upsetting it can be when others don't listen to you. Perhaps, like Kaleigh's friends, you're guilting of not listening as attentively as you could sometimes.

Here are eight bad habits to kick if you want to be a better listener who is fully present with the people around you.

Pretending to Listen

When people **pseudolisten**, they give the appearance of being attentive, but they aren't really. They may look people in the eye, nod, and smile at the right times, and may even answer occasionally. That appearance of interest, however, is a polite facade. When it happens at work, inauthentic listening leaves people feeling stressed, unhappy, and burned out.[89]

pseudolistening An imitation of true listening.

Tuning In and Out

When people respond only to the parts of a speaker's remarks that interest them, they are engaging in **selective listening**. Everyone is a selective listener from time to time, but it's a habit that can lead to confusion, misunderstandings, and hurt feelings.

selective listening A listening style in which the receiver responds only to messages that interest them.

Missing the Underlying Point

Rather than looking below the surface, **insensitive listeners** tend to take remarks at face value. An insensitive listener might miss the warble in a friend's voice that suggests they are more upset than their words let on. Or when a partner complains, "I always take out the trash," an insensitive listener might miss that what's wanted is a thank-you.

insensitive listening The failure to recognize the thoughts or feelings that are not directly expressed by a speaker, and instead accepting the speaker's words at face value.

Dividing Attention

A lot of people try balancing "real-world listening" and social listening. The truth is that most folks aren't very good at it. In one experiment, college students who attempted to listen to a lecture while checking Instagram scored a full letter grade lower on tests than students who abstained from social media during class.[90] Juggling online and in-person interactions and experiences also strains relationships. Reflecting on a family vacation, blogger Aisha Taylor wrote, "I felt a tug of war inside me between living

Trying to balance real-world listening and social listening is harder than most people think. The truth is that very few do it well.

What are some drawbacks of attending to social media while present with others?

for social media vs. being present in the moment."[91] Aisha's time on Instagram "turned into a point of frustration" for her family. [92] They may have felt neglected or hurt because Aisha spoke with and listened to virtual others instead of giving them her full attention.

Being Self-Centered

People who focus on themselves and their interests instead of listening to and encouraging others are **conversational narcissists**.[93,94] There's a scientific explanation behind conversational narcissism—it feels good! Talking about yourself tends to activate a part of your brain that releases pleasure hormones (similar to the hormones released during sex and while eating).[95] It's no wonder you may be tempted to go on and on about yourself. Unfortunately, the experience probably won't be as enjoyable for the people listening to you.

conversational narcissists People who focus on themselves and their interests instead of listening to and encouraging others.

Talking More Than Listening

Comedian Paula Poundstone once quipped, "It's not that I'm not interested in what other people have to say, it's that I can't hear them over the sound of my own voice." If you can relate, the acronym WAIT may help; it stands for "Why Am I Talking?" Your answer reveals your motivation for speaking, explains *New York Times* writer Adam Bryant. "Is it about the other person—to show them that you understand what they're saying, because maybe you've had a similar experience? Or is there subtext of needing to brag a bit?"[96] A mutually satisfying conversation is more like a game of catch—you should be throwing and catching the ball in equal measure.[97] There's no calculating the esteem and wisdom people earn by listening well rather than talking all the time.

Avoiding the Issue

insulated listening A style in which the receiver ignores or is oblivious to undesirable information.

People who avoid difficult subjects are **insulated listeners**. It's understandable that someone might tune out when the topic is touchy. Maybe you really don't want to hear your boss go over the reasons you didn't get a promotion. Even so, there are advantages to listening. You might earn the boss's respect and learn some valuable tips if you are a receptive listener. When the tables are turned and you're the one who has to deliver touchy information, you can help people listen to you by being upfront about the sensitive nature of what you are about to disclose. One blogger offers this example: "I have something to tell you. I'm not proud of what I've done, and you might be mad. But I know I need to tell you. Can you hear me out?"[98]

defensive listening A response style in which the receiver perceives a speaker's comments as an attack.

Being Defensive

People who perceive that they're being attacked—even when they aren't—can become **defensive listeners**, more interested in justifying themselves than in understanding

6.4 Faulty Listening Habits 143

the other person's point of view. Guilt or insecurity are often to blame, with the effect that casual remarks are interpreted as threats ("Why do you care if I enjoyed lunch? It's none of your business if I cheat on my diet!"). A common defensive response is counterattacking the perceived critic, but that's not likely to win you many friends. Instead of proving your point, you will probably make the other person feel defensive and angry. By contrast, nondefensive listeners—like Chanel Lewis, who you read about in the chapter opener—are sincerely interested in understanding the other person's perspective, even when the topic makes them uncomfortable.

You may recognize the egotism behind many of these bad habits but still feel tempted to engage in them sometimes. As we said, listening is hard. It's worth the effort, however. The feature "Active Listening" offers suggestions for becoming a more focused and engaged listener.

Nodding and maintaining eye contact show the speaker that you are paying attention.
How else might you show someone that you are listening?

active listening Listening to understand contextual and emotional components of a message while empathizing with the speaker.

COMMUNICATION STRATEGIES

Active Listening

Even before reading this chapter, you likely encountered the phrase "active listening" at some point. It tops the list of highly sought-after job skills employers look for when hiring.[99] Most college students know they need to be good at it if they want to be successful.[100] **Active listening** is associated with conversational competence, an assortment of social skills, and supportive communication.[101,102] In terms of strengthening your interpersonal relationships, it's been called the "biggest gift you can offer to someone else."[103]

But what exactly *is* active listening? The simple answer is: listening to gain understanding while accurately perceiving the speaker's state of mind.[104] In practice, however, it's not so simple. The listener's aim is three-fold: Recognize what someone is saying and how they feel; empathize with the speaker; and respond in a way that conveys your understanding and acknowledges the speaker's feelings. With time and effort, you can excel at active listening. The following suggestions will help.

Pay Attention

Before you can attempt processing a speaker's message, you need to attend to it. That means focusing your full attention on the speaker. Start by minimizing disruptions, including extraneous thoughts and worries. Set aside or silence electronic devices whenever possible and look at the speaker. Listen for underlying messages as well as surface meanings. Pay attention to cues about how the speaker feels.

(Continued)

Be Mindful of Body Language

Take note if the speaker is sitting upright, slouching, grinning, pursing their lips, or so on. Nonverbal cues such as these can tell you a lot about the speaker's state of mind, whether they are confident, agitated, happy, and so on. Be aware of your own body language too. Nodding, smiling, maintaining eye contact, and leaning forward can signal your understanding and encourage the speaker to continue.[105,106] Here are a few things to keep in mind. Nodding doesn't necessarily mean you agree with a message, rather it says, "I'm listening. Go on." Smiling, on the other hand, tells the speaker you approve and/or are happy about what they have to say. A smile can also help diffuse any tension or discomfort the speaker may feel. While maintaining eye contact, keep your gaze relaxed and natural. Don't stare, which may come off as aggressive, angry, or intimidating. Avoid distracting movements as well. Don't fidget, sigh, cross your arms, or the like. Keep your body oriented toward the speaker. (You'll learn more about nonverbal communication in Chapter 7.)

Encourage the Speaker

Verbal affirmations also let the speaker know you're listening. Short phrases like "Uh huh," "I see," and "I understand" work well. Paraphrasing (repeating back what the speaker said in your own words) and asking questions also show that you're interested, as long as you don't interrupt the speaker or cut them off. Open-ended and probing questions encourage elaboration ("Which aspect of the situation is most concerning to you?") and help clarify your understanding ("If I have this right, you're upset about what happened? Is that a good reflection of how you feel?"). There are a few temptations to avoid when asking questions. As a general rule, let the speaker decide what details are most important from their perspective. Satisfying your curiosity with questions like "What did they say then?" might distract from the real issue. Skipping ahead ("So, what are you going to do about it?") may imply that you're not interested in hearing the whole story. Judgments disguised as questions ("You didn't really think that was a good idea, or did you?") could shut down the conversation.

Reflect on Content *and* Emotion

It's important to communicate that you grasp the meaning of the message and recognize the speaker's emotions. For instance, if a speaker tells you they missed work because their mother was in a car accident, you may respond, "Your mom was in a wreck?" Your question reflects content (you heard and understood the facts). But if your next comment is along the lines of "Whose fault was it?" or "I was in a wreck once," then you're missing the underlying emotion. You're also missing an opportunity to connect with the speaker on a deeper level. If, instead, you say something akin to "Oh my gosh. I'm sorry. I can understand why you seem upset.," then you're reflecting on the speaker's emotions and showing compassion. This invites further exploration of the speaker's thoughts and feelings, and may ultimately lead to more satisfying conversations, enhanced wellbeing, and stronger relationships.[107]

At work and elsewhere, you'll find that active listening can help you connect with and offer support to others.[108] It's easy to appreciate why active listening is considered one of the most essential job skills[109] and the foundation for effective leadership.[110,111] While active listening takes effort, it pays off in countless ways!

6.5 Listening to Connect and Support

Listening Ears is a London-based nonprofit organization with one purpose: improving people's lives through supportive listening.[112] And there's no doubt that supportive listening can make a difference. Research shows that it can reduce loneliness and stress and build self-esteem.[113] Volunteers with Listening Ears provide companionship and support for people living alone and/or experiencing social isolation. They do this by listening at "Feel-Good Centers," where people gather for conversation, snacks, and fun activities. For those who cannot attend in-person events, "Community Angels" make house calls. But listening volunteers aren't unique to London. For instance, there are listening cafés in Japan where anyone can go to enjoy conversations with dedicated listening volunteers.[114] The benefits go both ways. People who provide social support often feel an enhanced sense of wellbeing themselves.[115]

In this section, we explore the role of **relational listening**, which involves emotionally connecting with others; and **supportive listening**, which goes a step further, with the goal of helping a speaker deal with personal dilemmas, whether they be minor stressors or life-changing situations. There are some overlaps with active listening, as you will see.

Despite the benefits of relational and supportive listening, there can be drawbacks. It's easy to become overly involved with others' feelings. When that happens, listeners may feel overwhelmed and unable to offer an objective perspective.[116] And, in some cases, listeners may experience what experts call secondary traumatic stress when upsetting events are shared with them.[117] While helping others is commendable, you should not attempt it if doing so puts your own health and wellbeing at risk. It's okay to suggest that someone seek professional help (from a therapist or physician) if their problems are too much for you to handle.

Even if you feel confident that you can help others, you may come off as opinionated, intrusive, or even bossy. Talking about problems in certain ways can actually be dysfunctional, ultimately making matters worse.[118] To avoid some common listening mistakes, here are 10 strategies when the goal is connecting with and/or helping someone through a difficult time.

The goal of supportive listening is to help the speaker with a personal dilemma.

When you need help with a problem, who is there for you? What makes that person a supportive listener?

relational listening
A listening style that is driven primarily by the desire to build emotional closeness with the speaker.

supportive listening
Listening with the goal of helping a speaker deal with personal dilemmas.

Allow Enough Time

Connecting with and supporting others can take time. If you're in a hurry, it may be best to reschedule important conversations for a better time. You might say, "I want to give you my undivided attention. Can we meet later today?"

In other situations, brief interactions can be meaningful. A quick "Hello" or "Thank you" might be all it takes. "I work at a bakery where the customers tend to be rude,"[119] explains a young cashier. "One day, a customer looked at my name tag and said, 'Thank you, (my name)' and it was the nicest feeling to be acknowledged."[120]

The gift of attention often speaks for itself, even when you don't know what to say. Medical studies show that, even when doctors cannot offer a cure, patients' coping skills are positively linked to the amount of time their doctors spend listening to them.[121]

Be Sensitive to Personal and Situational Factors

Before committing yourself to helping another person—even someone in obvious distress—make sure your support is welcome. Even then, there is no single best way to provide support. There is enormous variability in terms of what will work with a given person in a given situation. This explains why communicators who are able to use a wide variety of helping styles are usually more effective than those who rely on just one or two styles.[122]

You can boost the odds of choosing the best helping style in each situation by considering three factors.

- *Personal preference:* Some people prefer to handle difficult situations on their own. They may benefit from the opportunity simply to voice their thoughts. By contrast, other people welcome advice and analysis.[123]

- *The situation:* Timing is everything. Sometimes the most supportive thing you can do is listen quietly. At other times, listening well may help you realize that running an errand or offering to assist with a task may be the best way to help.[124]

- *Your own strengths and weaknesses:* You may be best at listening quietly, offering a prompt from time to time. Or perhaps you are especially insightful and can offer useful analysis of problems. These can be listening strengths. Just be careful that you don't use an approach that is comfortable for you when a different one might be more effective. In some situations, speakers may assume that quiet listeners aren't paying attention, or that "helpful" listeners are overly judgmental or eager to tell them what to do.

In most cases, the best way to help is to use a combination of responses in a way that meets the needs of the other person and suits your personal communication style.[125]

Ask Questions

Asking questions can help a conversational partner define vague ideas more precisely. You might respond to a friend by asking, "You said Greg has been acting 'differently' toward you lately. What has he been doing?" or "You told your roommates that you wanted them to help keep the place clean. What would you like them to do?" Questions can also encourage people to keep talking, which is particularly helpful when you are dealing with someone who is quiet or fearful of being judged.

Listen for Unexpressed Thoughts and Feelings

People often don't say what's on their minds or in their hearts, perhaps because they're confused, fearful of being judged, or trying to be polite. However, these unstated messages can be as important as the spoken ones. It can be valuable to listen for unexpressed messages. Consider a few examples:

STATEMENT	POSSIBLE UNEXPRESSED MESSAGE
"Don't apologize. It's not a big deal."	"I'm angry (or hurt, disappointed) by what you did."
"You're going to a concert tonight? That sounds like fun!"	"I'd like to come along."
"Check out this news story. That's my little sister!"	"I'm proud of what she did."
"That was quite a party you [neighbors] had last night. You were going strong at 2 a.m."	"The noise bothered me."

There are several ways to explore unexpressed messages. You can *paraphrase* by restating the speaker's thoughts and feelings in different words, as in "It sounds like that really surprised you" or "So you aren't sure what to do next, right?" You can prompt the speaker to volunteer more information with questions such as "Really?" and "Is that right?" Or you can ask questions such as "What are the pros and cons, as you see them?"

Encourage Further Comments

Sometimes you can strengthen relationships and support others simply by encouraging them to say more. While questioning requires a great deal of input from the respondent, another approach is more passive. **Prompting** involves using silences and brief statements of encouragement to draw others out, and in so doing to help them solve their own problems. Consider this example:

> **Pablo:** Julie's dad is selling his old MacBook and printer for $900! If I want it then I have to buy it today because somebody else is interested. It's a great deal. But it would wipe out my savings.
> **Tim:** That's a dilemma.
> **Pablo:** I wouldn't have the cash to upgrade my phone for a while. . . but I'd have my own printer. I wouldn't have to go to the library to print stuff out anymore. That would save me a lot of time.
> **Tim:** That's for sure.
> **Pablo:** Do you think I should buy it?
> **Tim:** I don't know. What do you think?
> **Pablo:** I can't decide.
> **Tim:** *(silence)*
> **Pablo:** I'm going to do it. I probably won't get a deal like this again.

> **prompting** Using silence and brief statements to encourage a speaker to continue talking.

Prompting works best when it's done sincerely. Your nonverbal behaviors—eye contact, posture, facial expression, tone of voice—must show that you are concerned with the other person's problem but not advocating for one outcome over another. Great teachers harness this power regularly. They know that students often learn more when asked questions and encouraged to work through problems instead of being given the answers up front.[126]

When encouraging others to talk, be careful to not redirect the conversation back to yourself. Avoid responses like these:

> **Abel:** "I don't know whether to quit or stay in a job I hate."
> **Brianna:** "You think your job is bad? Let me tell you about the job I had last summer . . ."

> **Carlo:** "My grandma is having health problems and I want to go visit her, but midterms are coming up and I'd hate to miss them."
> **Danielle:** "Family always comes first. When my grandfather had an accident . . ."

Reflect Back the Speaker's Thoughts

Both active and supportive listeners often **reflect** aloud about the thoughts and feelings they have heard a speaker express. This is akin to paraphrasing, but the goal of reflecting isn't so much to clarify your understanding as to help the other person reflect on their thoughts. The following conversation between two friends shows how reflecting can offer support and help a person find the answer to their own problem:

> **reflecting** Listening that helps the person speaking hear and think about the words they have just spoken.

> **Jill:** I've had the strangest feeling about my boss lately.
> **Mark:** What's that? *(A simple question invites Jill to go on.)*

You can show that you understand a speaker and help them clarify their thoughts by listening carefully and paraphrasing what you hear.

Describe an instance in which you used these techniques.

Jill: I'm thinking maybe he has this thing about women—or maybe it's just about me.

Mark: You mean he's coming on to you? *(Mark paraphrases what he thinks Jill has said.)*

Jill: Oh no, not at all! But it seems like he doesn't take women—or at least me—seriously. *(Jill corrects Mark's misunderstanding and explains herself.)*

Mark: What do you mean? *(Mark asks another simple question to get more information.)*

Jill: Well, whenever we're in meetings, he never asks any women what they think. But I do know he counts on some women in the office.

Mark: Now you sound confused. *(Reflects her apparent feeling.)*

Jill: I am confused. I don't think it's just my imagination.

Mark: Maybe you should . . . *(Starts to offer advice but catches himself and decides to ask a sincere question instead.)* So, what do you think you should do?

Jill: Well, I could ask him if he's aware that he never asks women's opinions. But that might sound aggressive. Maybe I could just tell him I'm confused about what is going on. But what if it's nothing?

Mark: *(Mark thinks Jill should confront her boss, but he isn't positive that this is the best approach, so he paraphrases what Jill seems to be saying.)* And that might make you look bad.

Jill: It might. Maybe I could talk it over with somebody else and get their opinion . . .

Mark: . . . see what they think . . .

Jill: Yeah. Maybe I could ask Brenda. She's easy to talk to, and I respect her judgment. Maybe she could give me some advice.

Mark: Sounds like you're comfortable talking to Brenda first.

Jill: *(Warming to the idea.)* Yes! Then if it's nothing, I can calm down. But if I do need to talk to the boss, I'll know I'm doing the right thing.

Reflecting a speaker's ideas and feelings can be surprisingly helpful. First, reflecting helps the other person sort out the problem. In the dialogue you just read, Mark's paraphrasing helped Jill consider carefully what bothered her about her boss's behavior. The clarity that comes from this sort of perspective can lead to solutions that weren't apparent before. Reflecting also helps the person unload more of the concerns they have been carrying around, often leading to the relief that comes from catharsis. Finally, listeners who reflect the speaker's thoughts and feelings (instead of judging or analyzing, for example) show their involvement and concern.

Consider the Pros and Cons When Analyzing

There are many reasons to analyze what you hear—to interpret it and consider it from multiple perspectives. In a few minutes, you'll learn about the importance of analytical listening when assessing whether information is trustworthy or not. But when the goal is supportive listening, analysis serves another function—to offer an interpretation of a speaker's message that may help them achieve more clarity. Analysis statements meant to be supportive are probably familiar to you:

"I think what's really bothering you is . . ."

"She's doing it because . . ."

"I don't think you really meant that."

"Maybe the problem started when he . . ."

Interpretations sometimes help people with problems consider alternative meanings. Under the right circumstances, an outside perspective can make a confusing problem suddenly clear by suggesting a solution or providing an understanding of what is occurring.

There can be two problems with analyzing, however. First, your interpretation may be wrong, and it may confuse the speaker even more upon hearing it. Second, even if your interpretation is correct, saying it out loud might not be useful. There's a chance it may arouse defensiveness in the speaker because analysis can imply superiority and judgment. Even if it doesn't, the person may not be willing or able to understand your view of the problem. Use active listening skills to help you decide if, or when, to share your analysis of a speaker's situation.

Reserve Judgment, Except in Rare Cases

Judgments can be helpful, but for the most part, they are a risky way to respond to someone in distress. A **judging response** evaluates the sender's thoughts or behaviors in some way. The evaluation may be favorable ("That's a good idea" or "You're on the right track now") or unfavorable ("An attitude like that won't get you anywhere"). But in either case, it implies that the listener is qualified to pass judgment on the speaker, which can cause hurt feelings.

Sometimes negative judgments are purely critical. How many times have you heard responses such as "Well, you asked for it!" or "I told you so!" or "You're just feeling sorry for yourself"? Statements like these can sometimes serve as a verbal wake-up call, but they often make matters worse. At other times, negative judgments involve constructive criticism, which is less critical and more intended to help a person improve in the future. Friends may offer constructive criticism about everything from what you wear to where you work or attend school, and teachers or bosses may evaluate your work to help you master concepts and skills. Whether or not it's justified, even constructive criticism can make people feel defensive.

> **judging response**
> A response that evaluates the sender's thoughts or behaviors in some way. The evaluation may be favorable or unfavorable.

Think Twice Before Offering Advice or Solutions

A Reddit user observed that when people talk about their concerns, the listener's "first inclination is to fix the problem."[127] But offering a solution, what scholars call an **advising response**, might not be the way to go. Although giving advice is sometimes valuable, often it isn't as helpful as you might think. For one thing, it can be hard to tell when someone actually wants your advice. Statements such as "What do

> **advising response**
> A response in which the receiver offers suggestions about how the speaker should deal with a problem.

you think of the new employee?," "Would that be an example of harassment?," and "I'm really confused" may be intended more to solicit information than to get advice.

Sometimes people just want to be heard. Venting is a way to let off steam, and some scholars believe it also reinforces trust in relationships.[128] The trick is letting the speaker vent without interjecting, even when the answer to their problem seems painfully obvious to you. There's a popular video on YouTube, *It's Not About the Nail*, that illustrates this point perfectly: A young woman complains about feeling a lot of pressure and having headaches to her befuddled husband—as he stares curiously at a nail protruding from her forehead! When he suggests removing the nail, thereby solving the problem, she yells, "You always try to fix things when what I really need is for you to just listen!"[129] Versions of that line are probably spoken every day.

As linguist Deborah Tannen points out, people's listening approaches and preferences are influenced by how they were socialized to deal with emotions. For instance, because men have traditionally been socialized to avoid expressing intense emotions, they may consider it supportive to offer solutions or distractions such as "Here's what you should do . . ." or "Don't worry about it."[130] People accustomed to acknowledging their feelings may see solutions and distractions as brushing off their concerns or belittling their problems. To avoid misunderstandings, a blogger suggests being upfront about the type of support you want: "Lead with something along the lines of 'Can I vent for a minute?'—or anything that signals to us that this is just a time for active listening rather than a problem-solving session."[131] If you're the listener, it doesn't hurt to ask the other person if they're blowing off steam or seeking input.

Even when people do ask for advice outright (as in "What do you think I should do?"), offering it may not be helpful for several reasons. For one, what's right for one person may not be right for another. If your suggestion is not the best course to follow, you could inadvertently cause harm or get blamed for a bad outcome. Another downside is that advice can seem like a put-down in that it casts the advice giver as wiser or more experienced than the recipient. And finally, advice may discourage others from making their own decisions or feeling accountable for them. For the most part, advice is most welcome when it has been clearly requested and when the advisor respects the face needs of the recipient.[132]

Offer Comfort, If Appropriate

Sometimes a listener's goal is to offer **comfort** by reassuring, supporting, encouraging, or distracting the person seeking help. Comforting responses can take several forms:

Agreement	"You're right—the landlord is being unfair."
Offer of help	"I'm here if you need me."
Praise	"I don't care what the boss said, I think you did a great job!"
Reassurance	"I know you'll do a great job."
Diversion	"Let's catch a movie and get your mind off this."
Acknowledgment	"I can see that really hurts."

Even if you mean well, some responses may fail to help. Telling someone who is obviously upset that everything is all right, or joking about a serious matter, can leave the other person feeling worse, not better. They might see your comments as a put-down or an attempt to trivialize their feelings. It's usually hurtful to be

comforting A response style in which a listener reassures, supports, encourages, or distracts the person seeking help.

judgmental as well—"There are people worse off than you are" *or* "No one ever said life was fair."[133] And it may also be frustrating to hear "I understand how you feel" from someone who can't really know what the other person is going through.[134] An American Red Cross grief counselor explains that simply being present can be more helpful than anything else when people who are distressed or grief-stricken:

> Listen. Don't say anything. Saying "it'll be okay," or "I know how you feel" can backfire. . . . Be there, be present, listen. The clergy refer to it as a "ministry of presence." You don't need to do anything, just be there or have them know you're available.[135]

This section has considered listening as a means of connecting with and helping others. Although you may envision these aims playing out in face-to-face interactions or over the phone, don't forget that supportive listening also happens online and on social media. People can share interests and concerns and potentially gain support from one another, even when they don't know each other in real life.

6.6 Listening to Learn, Analyze, and Critique

If you have ever been a patient in an emergency room, chances are the doctor prompted you "to get to the crux of the story quickly."[136] Busy ER doctors need to hear the facts *fast*: chief complaint, symptom onset, allergies, and so on. The same doctor out to dinner with family or friends would likely listen differently, perhaps with more patience and attention. That's because how people listen is determined largely by the situation and their goals. Maybe the goal is understanding a new concept in class or evaluating a salesperson's pitch for accuracy and consistency. In this section, we consider the value and techniques of task-oriented, analytical, and critical listening.

task-oriented listening A listening style in which the goal is to secure information necessary to get a job done.

Task-Oriented Listening

The purpose of **task-oriented listening** is to secure information necessary to complete the job at hand. This might involve following your bosses' instructions at work, a friend's tips for figuring out a new game, a teacher's suggestions—the list goes on.

Task-oriented listeners are often concerned with efficiency. They may view time as a scarce and valuable commodity and grow impatient when they think others are wasting it. A task orientation can be an asset when

The situation and your goals inform your listening behaviors.

How might listening to friends differ from listening in a classroom or at work?

deadlines and other pressures demand fast action (such as working in an emergency room). These listeners keep a focus on the job at hand and encourage others to be organized and concise.

Despite its advantages, a task orientation can be off-putting to others when it seems to disregard their feelings. Even when the situation calls for task-oriented listening, it's important that you don't overlook others' emotional issues and concerns. Also, an excessive focus on getting things done quickly can hamper the kind of thoughtful deliberation that some jobs require.

You can become more effective as a task-oriented listener by approaching others with a constructive attitude and by using some simple but effective skills. The following guidelines should help you be more effective.

Listen for Key Ideas It's easy to lose patience with long-winded speakers, but good task-oriented listeners stayed tuned in. They are able to extract the main points, even from a complicated message. If you can't figure out what the speaker is driving at, you can always ask in a tactful way by using the skills of questioning and paraphrasing.

Ask Questions Questioning involves asking for additional information to clarify your understanding of the sender's message. If you are heading to a friend's apartment for the first time, typical questions might be, "How is the traffic between here and there?" or "Where should I park when I arrive?" One key element of these types of questions is that they ask the speaker to elaborate. See the feature "When Is a Question Not a Question?" for information on sincere and counterfeit questions.

Paraphrase Another type of feedback can also help you confirm your understanding. **Paraphrasing** involves restating in your own words the message you think the speaker has just sent. For example, you might say to a direction giver, "You're telling me to drive down to the traffic light by the high school and turn toward the mountains, is that it?" or ask a professor, "When you said, 'Don't worry about the low score on the quiz,' did you mean it won't count against my grade?"

In other cases, a paraphrase will reflect your understanding of the speaker's feelings. You might say to your boss, "You said not to worry about the customer's complaint, but I get the feeling it's a problem. Am I mistaken? Is this cause for concern?"

In each case, the key to success is to restate the other person's comments in your own words as a way of cross-checking the information. If you simply repeat the speaker's comments verbatim, you will sound foolish—and you still might misunderstand what has been said and why. Notice the difference between simply parroting (repeating without understanding) a statement and really paraphrasing:

> **Speaker:** I'd like to go to the retreat, but I can't afford it.
> **Parroting:** You'd like to go, but you can't afford it.
> **Paraphrasing:** So if we could find a way to pay for you, you'd be willing to come. Is that right?

As these examples suggest, effective paraphrasing is a skill that takes time to develop. It can be worth the effort, however, because it offers two very real advantages. First, it boosts the odds that you'll accurately and fully understand what others are saying. You've already seen that using one-way listening or even asking questions may lead you to think that you've understood a speaker when, in fact, you haven't. Second, paraphrasing guides you toward sincerely trying to understand another person instead of using faulty listening styles such as tuning in and out, selective

paraphrasing
Feedback in which the receiver rewords the speaker's thoughts and feelings to confirm understanding and/or express empathy.

COMMUNICATION STRATEGIES

When Is a Question Not a Question?

Have you ever been asked a question along the lines of "You didn't like the poem I wrote, did you?" or "Don't you think my cooking is good?" Questions like that can put you in a difficult position as a listener. If you, in fact, did not like the poem, do you say so? If you think your friend is a horrible cook, do you tell them? Before answering such questions, you might want to consider underlying meanings. Stephen O'Keefe once found himself in a similar predicament. As he tells it, "My wife bought a hideous bag. It looked like roadkill. She goes, 'Stephen! What do you think of my bag?'"[137] O'Keefe contemplated his options, and then a realization struck him: "She's not asking me for my opinion. She's fishing for a compliment!"[138]

Not all questions are **sincere**, or genuine requests for information. **Counterfeit questions** are attempts at disguising the real meaning of a message, thereby masking a speaker's true agenda. A speaker's motive may be innocuous (to get a compliment), but it could just as easily be to antagonize or insult the listener. While you probably can't stop a speaker from posing counterfeit questions, you can certainly stop yourself from asking them. Here are a few tips.

Avoid Leading Questions

From a listener's perspective, leading questions feel like a trap because they imply that there is a "correct" answer. For example, if a controlling boss asks, "Don't you think this is great?," savvy listeners know that a brutally honest answer may not be well received. Sometimes leading questions are veiled requests for validation or praise, and may stem from a speaker's insecurity ("I'm doing a fantastic job on this project, don't you agree?" or "Do you think I'm an idiot because I don't understand this?"). Rather than put listeners on the spot, you could try being frank about what you really want or need: "I think I'm doing a good job, but I could really use some encouragement."

Don't Disguise Assertions as Questions

When someone says, "Are you going to stand up and give him what he deserves?" they are clearly voicing an opinion rather than seeking a yes or no answer. If you want to share your view on a topic with a listener, phrase it as a declaration. Lead with "I believe," "I think that," or "In my opinion."

Don't Use Questions to Check Your Assumptions

It's one thing to ask sincere questions as a listener, but be careful not to use counterfeit questions to present your own assumptions. For example, if someone asks, "Why aren't you listening to me?," they are advancing the unchecked assumption that the other person isn't paying attention. Likewise, "What's the matter?" assumes that something is wrong. As Chapter 3 explains, perception checking is a much better strategy: "When you kept looking out the window during our meeting, I thought you weren't listening to my idea, but maybe you were considering the implications of what I suggested. What was on your mind?"

Steer Clear of Hidden Agendas

If someone asks, "Why don't we reorganize our workspace?" and what they really want is the corner office for themselves, the question is merely a vehicle for advancing a personal goal. As a speaker, it's better to be upfront about your objective in a tactful manner so you don't come across as manipulative, especially if the listener is on to you! As a listener, you might ask questions such as "What would you most like to achieve?" Hidden agendas aren't always bad, but it's hard to achieve understanding if they remain unspoken.

As a speaker, resist the urge to ask counterfeit questions. As a listener, consider how to respond to them with communication competence (Chapter 1) and emotional intelligence (Chapter 3).

counterfeit question
A question that is not truly a request for new information.

sincere question
A question posed with the genuine desire to learn from another person.

analytical listening
A listening style in which the primary goal is to understand a message.

Feynman Technique
A process proposed by physicist Richard Feynman to make sense of complicated information.

listening, and so on. Listeners who paraphrase to check their understanding are judged to be more socially adept than listeners who do not.

Take Notes Understanding others is crucial, of course, but it doesn't guarantee that you will remember everything you need to know. As you read earlier in this chapter, listeners usually forget about half of what they hear immediately afterward.

Sometimes recall isn't especially important. You don't need to retain many details of the vacation adventures recounted by a neighbor or the childhood stories told by a relative. At other times, though, remembering a message—even minute details—is important. The lectures you hear in class are an obvious example. Likewise, it can be important to remember the details of plans that involve you: the time of a future appointment, instructions on taking your medications, or the orders given by your boss at work. At times like these it's smart to take notes instead of relying on your memory.

Analytical Listening

The goal of **analytical listening** is to fully understand a message. Your own experience will prove that full understanding is rare. Just think about the times when others fail to understand *you*. You also know from experience how hard it can be to understand complicated ideas—in your studies, at work, and about the world at large.

Nobel Prize–winning physicist Richard Feynman knew a thing or two about grasping complicated information. The four-part process he shared with the world can be useful whether you are trying to understand a political candidate's position, a chemistry lecture, a new policy at work, or anything else that is both complicated and important.[139] To use the **Feynman Technique**, follow these four steps:

- Listen carefully to information about the new concept and then describe it as best you can on a sheet of paper. Feel free to use words, images, arrows, or anything else. (It's okay if your understanding is not perfect at this stage.)
- Explain the concept as if you were talking to a child or a new student, using words and graphics as if you are a teacher.
- Consider which aspects of the concept seem clear to you and which are still a little foggy.
- Review the original information (using your notes, follow-up questions, a recording, or printed material) to better understand details you haven't mastered yet.

Repeat the process, if necessary, until you can explain the concept in simple language. This gradual approach should help you analyze even highly complex information.

Sometimes the challenge isn't the complexity of a message, but the emotional reaction you feel as a listener. When you encounter messages that stir up strong feelings, here are some tips from the experts to avoid tuning out or jumping to conclusions.

Listen for Information Before Evaluating Although it's tempting to avoid, unfriend, and unfollow people whose beliefs are different than your own, "it's worth listening to people you disagree with," urges Zachary R. Wood, who wrote the book *Uncensored* about difficult conversations surrounding race, free speech, and dissenting viewpoints.[140] Wood has made it a practice to bring people with diverse beliefs together simply to listen to one another.

The principle of listening to information with an open mind seems almost too obvious to mention, yet most people are guilty of judging a speaker's ideas before completely understanding them. The tendency to make premature judgments is especially strong when the idea you are hearing conflicts with your own beliefs. As one writer put it:

> The right to speak is meaningless if no one will listen.... It is simply not enough that we reject censorship ... we have an affirmative responsibility to hear the argument before we disagree with it.[141]

You can avoid the tendency to judge before understanding by following the simple rule of paraphrasing a speaker's ideas before responding to them. The effort required to translate the other person's ideas into your own words will keep you from arguing, and if your interpretation is mistaken, you'll know immediately.

Source: Courtesy of Ted Goff

Separate the Message from the Speaker The first recorded cases of blaming the messenger occurred in ancient Greece. When messengers reported losses in battles, their generals sometimes responded to the bad news by having the messengers put to death. This sort of irrational reaction is still common today (though fortunately less violent). Consider a few situations in which there is a tendency to get angry with someone bearing unpleasant news: an instructor tries to explain why you did poorly on a major paper; a friend explains what you did to make a fool of yourself at the party last night; the boss points out how you could do your job better. At times like this, becoming irritated with the messenger may not only cause you to miss important information but can also harm your relationship with the other person.

There's a second way that confusing the message and the messenger can prevent you from understanding important ideas. At times you may mistakenly discount the value of a message because of the person who is presenting it. Even the most boring instructors, the most idiotic relatives, and the most demanding bosses occasionally make good points. If you write off everything a person says before you consider it, you may be cheating yourself out of valuable information.

Search for Value Even if you listen with an open mind, sooner or later you will end up hearing information that is either so unimportant or so badly delivered that you're tempted to tune out. Although you may want to escape from such tedious situations, there are times when you can profit from paying close attention to seemingly worthless communication. This is especially true in situations where the only alternatives to attentiveness are pseudolistening or downright rudeness.

Once you try, you probably can find some value in even the worst conditions. Consider how you might listen opportunistically when you find yourself locked in a boring conversation with someone whose ideas you think are worthless. Rather than torture yourself until escape is possible, you could keep yourself amused—and perhaps learn something useful—by listening carefully until you can answer the following (unspoken) questions:

"Is there anything useful in what this person is saying?"

"What led the speaker to come up with ideas like these?"

You may be familiar with the saying "Don't believe everything you hear." It's good advice. Critical listeners know when and how to regard conversations with skepticism.

What types of situations merit critical listening?

critical listening
Listening in which the goal is to evaluate the quality or accuracy of the speaker's remarks.

"What lessons can I learn from this person that will keep me from sounding the same way in other situations?"

Listening with a constructive attitude is important, but not all information or all speakers are trustworthy. We turn next to a listening approach that involves evaluating the merit of what you hear.

Critical Listening

The goal of **critical listening** is to go beyond trying to understand the topic at hand, and instead, to assess its quality. At their best, critical listeners apply the techniques described here to consider whether an idea holds up under careful scrutiny.

Examine the Speaker's Evidence and Reasoning Trustworthy speakers usually offer support to back up their statements. A car dealer who argues that domestic cars are just as reliable as imports might cite statistics from *Consumer Reports* or refer you to online reviews written by satisfied customers. A professor arguing that students are more community-oriented than they used to be might recount stories about then and now.

Chapter 13 describes several types of supporting material that can be used to prove a point: definitions, descriptions, analogies, statistics, and so on. For now, here are a few things to consider when evaluating a speaker's message.

Evaluate the Speaker's Credibility The acceptability of an idea often depends on its source. If your longtime family friend, a self-made millionaire, says you should invest your life savings in jojoba fruit futures, you might be grateful for the tip. If your deadbeat neighbor makes the same offer, you would probably laugh off the suggestion.

Chapter 15 discusses credibility in detail, but two questions provide a quick guideline for deciding whether or not to accept a speaker as an authority:

- **Is the speaker competent?** Does the speaker have the experience or the expertise to qualify as an authority on this subject? Note that someone who is knowledgeable in one area may not be well qualified to comment in another area. For instance, your friend who can answer any question about writing code might be a terrible advisor when the subject turns to romance.

- **Is the speaker impartial?** Knowledge alone isn't enough to certify a speaker's ideas as acceptable. People who have a personal stake in the outcome of a topic are more likely than others to be biased. The unqualified praise that a commission-earning salesperson lavishes on a product may be more suspect than the mixed review you get from an actual user. This doesn't mean you should disregard all comments you hear from an involved party—only that you should consider the possibility of intentional or unintentional bias.

Examine Emotional Appeals Sometimes emotion alone may be enough reason to persuade you. In other cases, it's a mistake to let yourself be swayed by emotion when the logic of a point isn't sound. For example, the excitement an advertisement promises probably isn't a good enough reason to buy a product you can't afford. Again, the fallacies described in Chapter 15 will help you recognize flaws in emotional appeals.

As you read about various approaches to listening throughout this chapter, you may have noted that you habitually use some more than others. But you can control the way you listen and choose approaches that best suit the situation at hand. "Listening is not easy or simple at all. It's a constant practice,"[142] concedes Chanel Lewis, whose approach to radical listening you read about at the start of the chapter. It's a worthwhile endeavor because, as Lewis explains, "Being listened to and understood are ways in which we affirm each other's existence."[143]

MAKING THE GRADE

OBJECTIVE 6.1 Summarize the benefits of being an effective listener.

- Listening—the process of giving meaning to an oral message—is a vitally important part of the communication process.
- There are many advantages to being a good listener. People with good listening skills are more likely than others to be hired, promoted, and respected as leaders. In addition, listening improves relationships.
- Identify people from your own experience whose listening ability illustrates the advantages outlined in this chapter.
  > Listen, *really* listen, to someone important in your life. Note how that person responds and how you feel about the experience.
  > How could better listening benefit your life?

OBJECTIVE 6.2 Outline the most common misconceptions about listening, and assess how successfully you avoid them.

- Listening and hearing are not the same thing. Hearing is only the first step in the process of listening. Beyond that, listening involves attending, understanding, responding, and remembering.

- Listening is not a natural process. It takes both time and effort. Mindful listening requires a commitment to understand others' perspectives without being judgmental or defensive.
- It's a mistake to assume that all receivers hear and understand messages identically. Recognizing the potential for multiple interpretations and misunderstanding can prevent problems.
- People do not have just one listening style. Situation and goals influence how people listen.
- Women are not better supportive listeners than men. Several factors interact with gender to shape how well people listen—including cultural background, personal goals, expressive style, and cognitive complexity.
- Listening happens in several contexts, including online. Social listening is the process of attending to, observing, understanding, and responding to messages through mediated, electronic, and social media channels.
  > Give examples of how common misconceptions about listening create problems.
  > How can you use the information in this section to improve your listening skills?

OBJECTIVE 6.3 **List strategies to overcome factors that make it challenging to listen well.**

- A variety of factors contribute to ineffective listening, including message overload, psychological noise, physical noise, physiological noise, and cultural differences.
  > Which factors described in this chapter interfere with your ability to listen well?
  > Develop an action plan to minimize the impact of these listening challenges.

OBJECTIVE 6.4 **Identify and minimize faulty listening habits.**

- Faulty listening habits include pseudolistening, selective listening, insensitivity, dividing your attention, conversational narcissism, assuming that talking is more impressive than listening, avoiding difficult issues, and defensiveness.
  > From recent experience, identify examples of each type of ineffective listening described in this section.
  > Focus on the bad listening habit you are most guilty of and brainstorm ways to overcome it.

OBJECTIVE 6.5 **Describe and practice listening strategies for effectively connecting with and supporting others.**

- Relational and supportive listening involve the willingness to spend time with people to better understand their feelings and perspectives. The goal is to help the speaker, not the receiver.
- These approaches are most effective when the listener is patient, considers the people and situation involved, asks questions, is sensitive to underlying meanings, encourages the speaker to continue, reflects back the speaker's thoughts, and is careful to offer analysis, judgment, advice, and comfort when they are likely to help the listener.
- Listeners can be most helpful when they use a variety of styles, focus on the emotional dimensions of a message, and avoid being judgmental.
  > What listening behaviors do you find most helpful when you want to connect with another person or share a problem with them? Which do you find least helpful?
  > Think of someone close to you. What listening behaviors do you think that person most appreciates? Which do they find least helpful? How might you increase your use of listening behaviors they find most helpful?

OBJECTIVE 6.6 **Discuss when and how to listen to accomplish a task, analyze a message, and critically evaluate remarks.**

- Task-oriented listening helps people accomplish mutual goals. It involves an active approach in which people often identify key ideas, ask questions, paraphrase, and take notes.
- Analytic listening involves the willingness to suspend judgment and consider a variety of perspectives to achieve a clear understanding. This type of listening requires people to discern between what is true and what is not.
- Critical listening is appropriate when the goal is to judge the quality of an idea. A critical analysis is most successful when the listener ensures correct understanding of a message before passing judgment, when the speaker's credibility is taken into account, when the quality of supporting evidence is examined, and when the logic of the speaker's arguments is examined carefully.
  > Identify situations in which each type of listening described in this section is most appropriate.
  > On your own or with feedback from others who know you well, assess your ability to listen for each of the following goals: to accomplish a task, analyze ideas, and critically evaluate messages. Which types of listening are your strongest and weakest?

KEY TERMS

active listening p. 143
advising response p. 149
analytical listening p. 154
attending p. 132
comforting p. 150
conversational narcissists p. 142
counterfeit question p. 153
critical listening p. 156
defensive listening p. 142
Feynman Technique p. 154
hearing p. 132
insensitive listening p. 141

insulated listening p. 142
judging response p. 149
listening p. 132
listening fidelity p. 132
mindful listening p. 135
paraphrasing p. 152
prompting p. 147
pseudolistening p. 141
reflecting p. 147
relational listening p. 145
remembering p. 134
residual message p. 134
responding p. 134
selective listening p. 141
sincere question p. 153
social listening p. 136
supportive listening p. 145
task-oriented listening p. 151
understanding p. 132

PUBLIC SPEAKING PRACTICE

Describe a situation in which listening poorly caused a misunderstanding. Explain what happened and what you learned from the experience.

ACTIVITIES

1. You can see how listening misconceptions affect your life by identifying important situations in which you have fallen for each of the following assumptions. In each case, describe the consequences of believing these erroneous assumptions.

   - Thinking that because you were hearing a message you were listening to it
   - Believing that listening effectively is natural and effortless
   - Assuming that other listeners were understanding a message in the same way as you
   - Thinking people have unique listening styles, like fixed personality traits
   - Believing women are better supportive listeners than men
   - Not recognizing that social listening is practiced by both people and organizations online and on social media

2. Imagine that a friend says, "I'm so overwhelmed by work and school demands that I just don't know what to do."

   - Experiment with listening responses by writing a separate response for each of the following that illustrates what it might sound like.
     > Advising
     > Judging
     > Analyzing
     > Questioning
     > Comforting
     > Prompting
     > Reflecting
   - Do the same for each of the following.
     > At a party, a guest you have just met for the first time says, "Everybody seems like they've been friends for years. I don't know anybody here. How about you?"
     > Your best friend has been quiet lately. When you ask if anything is wrong, they snap, "No!" in an irritated tone of voice.
     > A fellow worker says, "The boss keeps making sexual remarks to me. I think it's a come-on, and I don't know what to do."
     > It's registration time at college. One of your friends asks if you think they should enroll in the communication class you've taken.
     > Someone with whom you live remarks, "It seems like this place is always a mess. We get it cleaned up, and then an hour later it's trashed."
   - Discuss the pros and cons of using each response style considering the situation and who is involved.

Nonverbal Communication

CHAPTER OUTLINE

7.1 Characteristics of Nonverbal Communication 163
Nonverbal Communication Is Unavoidable
Nonverbal Communication Is Ambiguous
Nonverbal Cues Convey Emotion
Nonverbal Cues Help Manage Identities
Nonverbal Cues Affect Relationships

7.2 Functions of Nonverbal Communication 169
Repeating
Substituting
Complementing
Accenting
Regulating
Contradicting
Deceiving

7.3 Types of Nonverbal Communication 175
Body Movements
Voice
Appearance
Touch
Space
Time

7.4 Influences on Nonverbal Communication 186
Culture
Gender

MAKING THE GRADE 189

KEY TERMS 190

PUBLIC SPEAKING PRACTICE 190

ACTIVITIES 191

LEARNING OBJECTIVES

7.1 Explain the characteristics of nonverbal communication and the social goals it serves.

7.2 Describe key functions served by nonverbal communication.

7.3 List the types of nonverbal communication, and explain how each operates in everyday interaction.

7.4 Explain the ways in which nonverbal communication reflects culture and gender differences.

If you are on TikTok, chances are you are familiar with Khaby Lame. Even if you don't recognize the name, you have almost certainly seen his face. After all, it's the second most famous face on TikTok—Lame has over 116 million followers. Even more impressive is the fact that he amassed all those followers without saying a word! He silently pokes fun at complicated, often ridiculous life hack videos circulating on social media by completing the same tasks in a simple, straightforward manner. He peppers his videos with humorous, often exaggerated expressions and ends them with a dramatic sweep of the hand. It's what CNN and *The New York Times* call his signature "duh" punchline.[1,2] The secret to Lame's success is twofold. First, he has a "universal exasperated everyman quality" that many can relate to.[3] Second, he uses nonverbal communication—facial expressions and hand gestures—to "transcend language barriers and make connections across cultures."[4] Lame's silence highlights an important point, according to marketing executive Christina Ferraz, "You don't need to speak to be seen or understood."[5]

Nonverbal communication is important in other ways as well. Nonverbal encoding and decoding skills can be strong predictors of your overall well-being and how attractive others find you.[6] In general, people with good nonverbal communication skills are more persuasive than those who are less skilled, and they have a greater chance of success in settings ranging from careers to poker to romance. Nonverbal sensitivity is a major part of emotional intelligence (Chapter 3), and researchers have come to recognize that it's impossible to study spoken language without paying attention to its nonverbal dimensions.[7]

Although people commonly assume that nonverbal communication is silent, that's not always the case. By definition, **nonverbal communication** is the process of conveying messages without using words. Nonverbal cues include facial expressions, touch, use of space, clothing, gestures (if they don't convey specific words), and other cues you can't hear. But nonverbal communication also includes audible cues that aren't linguistic—such as humming, sobs, how loudly or quickly someone speaks, and so on. To test your understanding of what qualifies, answer the following questions:

- Is American Sign Language mostly verbal or nonverbal?
- What about an email?
- How about laughter?

If you answered "verbal," "verbal," and "nonverbal," respectively, you are correct. American Sign Language doesn't require sound, but it is word-based, thus verbal.[8] Emails are usually verbal for the same reason. (If you imagined an email filled with nothing but emojis, that would indeed be nonverbal.) Laughter involves vocal

nonverbal communication
Messages expressed without words, as through body movements, facial expressions, eye contact, tone of voice, and so on.

Khaby Lame knows a thing or two about nonverbal communication. His facial expressions and hand gestures helped him amass more than 116 million followers on TikTok.

How attentive are you to your own nonverbal cues? What nonverbal cues displayed by others catch your attention most? What do you tend to overlook?

chords, but doesn't rely on words, so it's nonverbal. These distinctions only begin to convey the richness of nonverbal messages.

This chapter will help you become more mindful about the nonverbal cues you display to others and become more sensitive to wordless messages other people convey. In the following pages, you will explore the defining characteristics of nonverbal communication, the functions served by nonverbal cues, types of nonverbal communication, and how culture and social ideas about gender affect nonverbal communication. Reading about these topics won't transform you into a mind reader, but it should make you a far more accurate observer of others—and yourself.

7.1 Characteristics of Nonverbal Communication

You have probably noticed that there's often a gap between what people say and how they actually feel. An acquaintance says, "I'd like to get together again" in a way that leaves you suspecting the opposite. A speaker tries to appear confident but acts in a way that almost screams out, "I'm nervous!" You ask a friend what's wrong, and the "Nothing" you get in response rings hollow. Then there are other times when a message comes through even though there are no words at all. A look of irritation, a smile, or a sigh can say it all. Situations like these have one thing in common: Messages were sent nonverbally. Although you have certainly recognized nonverbal messages before, this chapter should introduce you to a richness of information you have never noticed. Here are a few qualities that reveal the nature and richness of nonverbal communication.

Nonverbal Communication Is Unavoidable

Even if you try not to send nonverbal cues—perhaps by closing your eyes or leaving the room—those behaviors transmit messages to others. It's virtually impossible not to communicate nonverbally.[9] Think about the last time you blushed, stammered, or cried even when you didn't want to. The chances are that people around you assumed your behavior "said" something about how you were feeling.

Some theorists argue that while unintentional behavior might provide information, it shouldn't count as communication. Others draw the boundaries of nonverbal communication more broadly, suggesting that because

Signing is a language because it uses gestures to express particular words. But, like proficient speakers, a good signer uses nonverbal expression to add depth to a message.

How expressive are you as a communicator? How do you think others would respond if you were more expressive? Less expressive?

unintentional behavior often conveys messages it is communication and thus worth studying.[10] This book takes the broad view here because, whether or not nonverbal behavior is intentional, people use it to form impressions about one another.

You may wonder why people can decide on the words they use, but the nonverbal cues they send are sometimes beyond their control. Scientists think it's because of the way the brain works. One part of the brain (the cerebrum) governs speech, but a different part (the limbic system) processes emotional reactions,[11] responding to situations automatically. For instance, you probably jump when something pops out of the bushes, get a burst of adrenaline in stressful situations, and cry or laugh when something moves you. These responses occur without consciously thinking about them, and can persist even if you don't want them to. (Think about a shaky voice in a job interview or when giving a speech.)

Of course, you can control *some* aspects of your nonverbal communication, such as eye contact and posture. But even when you send nonverbal cues deliberately, it's hard to be aware of everything you are doing. In one study, participants were asked to show nonverbally that they either liked or disliked another person.[12] Immediately afterward, less than a quarter of the participants could describe all the nonverbal behaviors they used to send those messages. That's probably because people send messages simultaneously in many ways (eyes, face, body, space, touch, and so on). The feature "Managing Nonverbal Communication at Work" takes a closer look at some of the ways nonverbal messages may be helping or hindering your career prospects.

The fact that you and everyone around you constantly send nonverbal clues is important because it means that you have a constant source of information available about yourself and others. If you tune in to these signals, you will be more aware of how those around you are feeling and thinking, and you will be better able to respond to them. But nonverbal behaviors do not always have clear or obvious meanings.

Nonverbal Communication Is Ambiguous

You may have encountered some of the countless books, podcasts, or videos that promise to reveal the hidden meanings behind nonverbal behavior. Nose twitching? *They're lying.* Eyes looking up? *He's skeptical.* Arms crossed? *She's angry.* Former FBI agent and body language specialist Joe Navarro says such simplistic interpretations are "crap."[29] Maybe you cross your arms because you're angry, or maybe you're just cold. Or maybe, Navarro suggests, "it's a self-hug."[30] The reality is that nonverbal cues are difficult to interpret accurately because cues can have multiple meanings. Consider an example: You text someone you met recently, saying "Let's get together soon." Their reply: 😬. What's that supposed to mean? Could it be *"Let me figure out when I'm free,"* *"Thinking about whether I want to see you,"* or *"Do I know you?"*? In situations like this, there's no way to be certain.[31] Your best guess about the meaning of this nonverbal message, like all others, will depend on several factors:

- The *context* in which the nonverbal behavior occurs. (Have you been in touch since the first meeting, or is this your first contact since then?)
- The *history and tone* of your relationship with the sender. (Did your first conversation take place at church? In a dive bar? Did the other person express a desire to see you again then?)

COMMUNICATION STRATEGIES

Managing Nonverbal Communication at Work

There is no doubt that nonverbal cues can affect your success at work. A research team asked MBA candidates to prepare a two-minute pitch to give to a prospective employer.[13] The researchers recorded the pitches and asked study participants to evaluate them. Some of the evaluators watched videotapes of the pitches, others listened to just the audio, and a third group read the pitches in transcript form. Candidates in the video and audio recordings were rated as more intelligent compared with candidates whose pitches were read. One implication is that, whenever possible, it's a good idea to meet with prospective employers or clients in person so that they can see and hear you. Facial expressions, body language, and tone of voice can help sell an idea better than words on a page. Resumes, cover letters, and written proposals, although indispensable, don't have the same impact of nonverbally rich, face-to-face interactions.

Of course, nonverbal cues can also wreck your chances of success. You may be sending the wrong message, inadvertently suggesting to others that you are angry, bored, incompetent, and so on. Here are a few tips to help you harness the power of nonverbal communication and reduce the likelihood of your cues being misinterpreted.

Consider Your Expressions

Without knowing it, your facial expressions could undermine your efforts at work. That's what happened to Kim Korte's niece, Amber. Although Amber is a kind and considerate person, her coworkers frequently complained to the boss about her negative attitude. Some said they were afraid to approach her because she looked mean. As Korte explained, "Amber felt uncomfortable, not knowing who complained or how many people felt this way about her."[14] Without her meaning to, Amber's expression gave the impression that she was angry. Although it may seem unfair, people judge (and are judged by) others based on their nonverbal communication all the time, even when the communication is unintentional.

Korte, an employment coach, acknowledges that, like Amber, you cannot always guard against others' assumptions, but you can check your expressions. If there's a disconnect between what you are feeling (or doing) and what your face is saying, take note. For example, do you look stern when you're deep in thought? Angry when concentrating? Aloof when tired? If so, it might be time for a smile break. Not only will smiling boost your mood and productivity, but studies find that people who smile regularly are considered more confident, successful, and approachable than those who don't.[15] Plus, smilers are more likely to be promoted![16]

Make Eye Contact, but Not Too Much

Sometimes, locking eyes with someone can come off as aggressive. On the other hand, lack of eye contact can make you seem unconfident or disinterested. In most situations, people find moderate eye contact to be pleasant and appealing.[17] It signals "You have my attention" and "I'm interested in what you have to say." To strike a balance, experts recommend making eye contact with conversational partners but glancing away briefly every 7 to 10 seconds.[18]

Mind Your Posture

You know from Chapter 6 that your body language shows others you are actively listening. Standing up straight sends additional messages too, namely that you are confident and capable.[19] It also influences how likable you appear to others. People with good posture are typically rated as more likable than those who slouch.[20]

Dress Appropriately

Although good posture can work wonders, it can't offset bad clothing choices. In a series of experiments, researchers found that people who demonstrated good posture but wore casual clothes (vs. business attire) were rated lower on professionalism and competency measures.[21] The takeaway is that you should stand up straight *and* wear professional-looking clothes.

(Continued)

Monitor Your Tone of Voice

Vocal cues "show that we are alive inside—thoughtful, active," says researcher Nicholas Epley.[22] Without vocal cues, Epley observes, it's hard to judge a speaker's personality or intent. Tone has four basic dimensions: funny vs. serious, formal vs. casual, respectful vs. irreverent, and enthusiastic vs. matter of fact.[23] Even a simple remark such as "That's not what I expected" can be interpreted many ways based on the tone. Thinking about these options can help you choose a tone that supports the message you hope to convey.

Use Haptics to Your Advantage

The old phrase "keeping in touch" takes on new meaning once you understand the relationship between **haptics**, the study of touch, and career effectiveness. Some of the most pronounced benefits of touching occur in the health and helping professions. For example, patients are more likely to take their medicine as prescribed when physicians give a gentle touch while prescribing it.[24] In counseling, touch often increases patients' self-disclosures.[25]

Touch can also enhance success in sales and marketing. When salespersons touch customers in a way that doesn't make them feel rushed or crowded, customers tend to shop longer, evaluate the store more highly, and buy more.[26] When an offer to sample a product is accompanied by touch, customers are more likely to try the sample and buy the product.[27] While appropriate contact can enhance your success, it isn't welcome in all cultures or all situations. Too much contact can be bothersome, annoying, or even creepy.

Attending to nonverbal cues at work can feel like a full-time job all on its own. There is a lot to keep track of. Bringing awareness to your nonverbal cues is the first step. To help with this, you might ask a friend to record you giving a speech or interacting with others and then study the video. Most people have blind spots when it comes to their own communication.[28] You may overestimate how well you hide your anxiety, boredom, or eagerness from others. As you watch the video, consider: *How does your voice sound? How closely does your appearance match what you've imagined?* and *What messages are suggested by your posture, gestures, and face?*

---

haptics The study of touch.

expectancy violation theory The proposition that nonverbal cues cause physical and/or emotional arousal, especially if they deviate from what is considered normal.

- The *sender's mood* at the time. (This is difficult to determine with no face-to-face information but easier if you are asking in person.)
- Your *own feelings*. (If you're feeling insecure, almost anything can seem like a threat.)

No matter the meaning you think you detect, you should consider nonverbal behaviors, not as facts, but as clues to be explored further.

Nonverbal communication can be ambiguous even in seemingly innocuous situations. Years ago, employees of the Safeway supermarket chain discovered this fact firsthand when they tried to follow the company's new "superior customer service" policy that required them to smile and make eye contact with customers. Twelve employees filed grievances over the policy, reporting that several customers had propositioned them, misinterpreting their actions as come-ons.[32]

Expectancy violation theory helps explain how people manage ambiguity when it comes to nonverbal communication.[33,34] The theory is comprised of three main propositions:

- People experience a degree of physical and/or psychological arousal when processing nonverbal cues, especially if those cues violate what they normally expect from others. For example, you might have a strong reaction if a Safeway cashier makes steady eye contact while smiling at you or if a stranger stares at you, touches you, or gets very close to you.

- A violation may be perceived either positively or negatively. If the person is attractive, their smile or prolonged gaze may create positive feelings. But in other situations, you may feel afraid or uncomfortable if someone grins or stares at you longer than usual.
- People either *accommodate* or *compensate* based on how they feel about the perceived violation. If someone you like invades your personal space, you might not mind, and you might even lean closer toward them. That's *accommodating*. On the other hand, if the violation is unwelcome, you might *compensate* by moving away, avoiding eye contact, or ignoring the person.

When is a smile good customer service and when is it a come-on? It might be hard to discern the difference because nonverbal communication can be ambiguous.

Have you ever made mistaken assumptions about others' nonverbal behaviors? Have others misinterpreted yours? When should you be more cautious about jumping to conclusions?

Expectancy violation theory illustrates how nonverbal communication can create a sense of closeness or distance between people. It presents several implications. One is that nonverbal communication is highly potent. It can stimulate strong reactions in others, even when that isn't the sender's intention. Another implication is that a violation can be perceived many ways. One person may consider a gentle touch on the arm to be sweet and thoughtful, whereas someone else may interpret the same gesture as aggressive or overly familiar. By the same token, a friend may not mind if you have your phone out during dinner (a nonverbal cue), but a date may consider it insulting, as if they aren't important enough to have your undivided attention.[35] The best advice is to think about the implications of your nonverbal cues, proceed slowly, and pay close attention to people's reactions.

Nonverbal Cues Convey Emotion

A friend who knows you well might recognize that you are shocked, happy, stressed, or sad, even when you're trying to hide those feelings or when you haven't fully acknowledged them within yourself.

One reason nonverbal cues are so powerful is that they can convey certain meanings better and more concisely than words can. You can prove this for yourself by imagining how you could express each item on the following list nonverbally:

- You're bored.
- You are opposed to capital punishment.
- You are attracted to another person in the group.
- You want to know if you will be tested on this material.

Your face can communicate your emotions without you having to say a word.

Have your facial expressions ever revealed your feelings, even though you were trying to keep them hidden?

affect displays Facial expressions, body movements, and vocal traits that reveal emotional states.

The second and fourth items in this list would be difficult, if not impossible, to convey without using words because they involve ideas more than emotions. The first and third items, however, involve feelings. You can imagine how boredom and attraction might be expressed nonverbally through what social scientists call **affect displays**—facial expressions, body movements, and vocal traits that convey emotion. (*Affect* is another word for emotion.) A sustained gaze, forward lean, head tilt, and sincere smile can say "I'm into you" in less time than it takes to deliver your best pick-up line. Hopefully, the response isn't a furrowed brow and wrinkled nose!

Nonverbal Cues Help Manage Identities

In Chapter 3 you explored the notion that people strive to create images of themselves as they want others to view them. A great deal of this occurs nonverbally. According to social psychologist Amy Cuddy, your body language not only influences the way others think about you; it also affects how you feel about *yourself*.[36] Consider what happens when you attend a party where you are likely to meet people you may want to get to know better. Instead of conveying your desired image verbally ("*Hi! I'm confident, friendly, and easygoing*"), you behave in ways that support the identity you want to project. You might smile a lot and adopt an open, relaxed posture. These behaviors will probably help you feel more confident too, Cuddy predicts. It's also likely that you will dress carefully—even if you are trying to create the illusion that you haven't given a lot of attention to your appearance.

Nonverbal Cues Affect Relationships

Psychologist Paul Ekman is an expert when it comes to facial expressions (he's been studying them for decades) and he says that "emotional expressions are crucial to the development and regulation of interpersonal relationships."[37] People with facial paralysis, because of congenital conditions like Mobius syndrome or acquired through injury or illness (Lyme disease, Bell's palsy, and stroke, for example), have difficulty forming and maintaining relationships.[38] They also experience depression and anxiety more than people who have a full range of facial expressions available to them.[39] This makes sense to Kathleen Bogart, a psychologist specializing in facial paralysis. As she explains, "At first glance, a person with a paralyzed face may look unfriendly, bored, unintelligent, or even depressed. And indeed, people with facial paralysis are often mistakenly ascribed these characteristics."[40] The result, according to Bogart, is people have less desire to form friendships with them.[41,42] People who *choose* to paralyze muscles in their faces with elective cosmetic procedures can also experience communication challenges as a result. Botox injections, for instance, temporarily eliminate wrinkles by freezing certain muscle groups (typically around the eyes and forehead) and thus people have a much harder time showing emotions such as anger and surprise.[43,44] The inability to express certain emotions can diminish the ability to effectively communicate with and be understood by others.[45] That's why actress Julia Roberts refuses to get Botox. She once joked, "I want my kids to know when I'm pissed, when I'm happy, and when I'm confounded."[46]

Botox injections are becoming increasingly popular, especially among younger adults. Actress Julia Roberts, however, is having none of it. She says she prefers moving her face, even if it means having wrinkles.

If unable to move your face, how would you show surprise, shock, happiness, or anger? How might this affect your ability to communicate with others?

When it comes to nonverbal cues, sometimes the problem isn't making them—it's reading them. For example, people with autism spectrum disorder and those born with a syndrome called nonverbal learning disorder (NVLD) have trouble understanding nonverbal cues.[47,48] They often misread signs of boredom and confusion during social interactions, and take humorous or sarcastic messages literally, because cues for interpreting such messages are based heavily on nonverbal signals (tone of voice, for instance). Consequently, many people with autism and NVLD have trouble making friends.[49] Even for neurotypical people, the nuances of nonverbal behavior can be confusing.

7.2 Functions of Nonverbal Communication

Although verbal and nonverbal messages differ in many ways, these two forms of communication operate together on most occasions. The following discussion explains the many functions of nonverbal communication and shows how nonverbal messages relate to verbal ones.

Repeating

If someone asks you for directions to the nearest drugstore, you could say "North of here about two blocks" and then point north with your index finger. This sort of repetition isn't just decorative. People remember comments accompanied by gestures more than those made with words alone.[50]

Substituting

When a friend asks you what's new, you might shrug your shoulders instead of answering in words. In this way, your nonverbal behavior is a substitute for a verbal response. In other situations, you might raise your eyebrows after a colleague makes an off-color comment or hug someone for longer than usual to show that you missed them. Sometimes a nonverbal cue says it all.

Some gestures substitute for specific words. **Emblems** are deliberate nonverbal behaviors that have precise meanings known to everyone within a cultural group. For example, most Americans consider that a head nod means "yes," a head shake means "no," a wave means "hello" or "good-bye," and a hand to the ear means "I can't hear you." Keep in mind that the meaning of many emblems varies by culture. (Take the "Understanding Your Communication" quiz for insights about your nonverbal strengths.)

Complementing

Sometimes nonverbal behaviors reinforce the content of a verbal message. Consider, for example, a friend who apologizes for forgetting an appointment with you. You will be most likely to believe your friend if they use a sincere-sounding tone of voice and show an apologetic facial expression. We often recognize the significance of complementary nonverbal behavior when it is missing. If your friend's apology is delivered with a shrug, a smirk, and a light tone of voice, you will probably doubt its sincerity, no matter how profuse the verbal explanation is.

Much complementing behavior consists of **illustrators**—nonverbal behaviors that accompany and support spoken words. Scratching your head when searching for an idea and snapping your fingers when it occurs are examples of illustrators that complement verbal messages.

Accenting

Just as we use *italics* to emphasize an idea in print, we use nonverbal devices to emphasize oral messages. Pointing an accusing finger adds emphasis to criticism (and probably creates defensiveness in the receiver). Stressing certain words with the voice ("It was *your* idea!") is another way to add nonverbal accents.

Regulating

Nonverbal behaviors can control the flow of verbal communication. For example, conversational partners send and receive turn-taking cues.[54] The speaker may use nonverbal fillers, such as "um" or an audible intake of breath,

> **emblems** Deliberate gestures with precise meanings, known to virtually all members of a cultural group.

> **illustrators** Nonverbal behaviors that accompany and support verbal messages.

Since the early days of email and text messages, people have devised emoticons, using keystrokes to create sad expressions :-(, surprised looks :-0, and more. Even now, when graphic emojis are readily available, people crave more nonverbal cues in cyberspace. Answering the call, Facebook added a series of new graphics to expand on the familiar but overused "Like" button.

UNDERSTANDING YOUR COMMUNICATION

How Worldly Are Your Nonverbal Communication Skills?

Answer the following questions to test your knowledge about nonverbal communication in different cultures.

1. In the United States, touching your index finger to your thumb while your other fingers point upward means "OK." But what does it mean elsewhere? (Two of the following are correct.)

    a. It signifies money in Japan.
    b. People in Greece and Turkey interpret it to mean 30.
    c. In France, it means "you're worth zero."
    d. It's a compliment in Russia, implying that "you and I are close friends."

2. People around the world recognize the sign for "peace" or "victory"—two fingers up, thumb holding down the other fingers, palm facing out. But in many places, the same gesture means something different if you show the back of your hand instead. What does that gesture mean to people in England, New Zealand, and Australia?

    a. "May I have seconds?"
    b. "I'll be right back."
    c. "Goodbye."
    d. "Up yours!"

3. In the United States, people convey "come closer" by alternately extending and curling their index finger in someone's direction. Where might the same gesture be considered a serious insult?

    a. Egypt
    b. The Philippines
    c. Spain
    d. Saudi Arabia

4. The "thumbs up" sign that means "yes" or "job well done" in the United States means something else in other cultures. Two of the following are true. Which ones?

    a. It means "it's my turn to talk" in Tahiti and neighboring South Pacific Islands.
    b. It means "up yours" in Australia, Greece, and the Middle East.
    c. It stands for the number 5 in Japan, or 1 in Germany and Hungary.
    d. In Myanmar, it's a symbol of mourning, meaning "someone has died."

INTERPRETING YOUR RESPONSES

Read the following explanations to see how many answers you got right.

Question 1

The gesture that means "OK" in the United States means "money" in Japan. But it has a darker meaning in France, where it conveys "you're worth zero." It's risky to use this gesture in other parts of the world as well. In Brazil, Germany, and Russia, it depicts a private bodily orifice; and in Turkey and Greece it's taken as a vulgar sexual invitation. The correct answers are "a" and "c."

Question 2

The palm-forward V sign is popular around the world, especially in Japan, where it's customary to flash a peace sign while being photographed.[51] But a slight variation makes a big difference. Winston Churchill occasionally shocked audiences during World War II by "flipping them off" (knuckles forward) when he really meant to flash a victory sign (palm forward).[52] Years later, U.S. president Richard Nixon made the same mistake in Australia, essentially conveying "f--- you" to an Australian crowd when he got the gesture wrong.[53] The correct answer is "d."

Question 3

The gesture Americans use to mean "come closer" is offensive in many places, including the Philippines, Slovakia, China, and Malaysia. People in those cultures summon a dog that way, so it's a put down to use it with a person. Answer "b" is correct.

Question 4

"Thumbs up" has a positive connotation in the United States. Meanwhile, people in Germany and Hungary interpret the gesture to mean number 1, and people in Japan use it for the number 5. However, the gesture is taken as an insult (akin to "up yours!") in the Middle East. Both "b" and "c" are correct.

Source: DILBERT © 2006 Scott Adams. Used by permission of UNIVERSAL UCLICK. All rights reserved.

to signal that they would like to keep talking.[55] Or they may hold up a finger to suggest that the listener wait to speak. Conversely, long pauses are often taken as opportunities for others to speak. Nonverbal regulators also include signs that you would like to wrap up the conversation, such as an extended *okaaaay* or a glance at your phone.

When you are ready to yield the floor, the unstated rule is as follows: Create a rising vocal intonation pattern, then use a falling intonation pattern, or draw out the final syllable of the clause at the end of your statement. Finally, stop speaking. If you want to maintain your turn when another speaker seems ready to cut you off, you can often suppress the attempt by taking an audible breath, using a sustained intonation pattern (because rising and falling patterns suggest the end of a statement), and avoiding any long pauses in your speech.

Contradicting

It's not unusual for people to say one thing but display nonverbal cues that suggest the opposite. A classic example is when someone with a red face and bulging veins yells, "I'm not angry!" As you have no doubt experienced, when verbal and nonverbal messages are at odds, people are more likely to believe the nonverbal cues than the words.[56]

Many contradictions between verbal and nonverbal messages are unintentional. You may say you understand something, not realizing that your puzzled expression suggests otherwise. But people sometimes send mixed messages on purpose. For example, if you become bored with a conversation while your companion keeps rambling on, you probably wouldn't bluntly say, "I want you to stop talking." Instead, you might nod politely and murmur "uh-huh" and "no kidding?" at the appropriate times but subtly signal your desire to leave by looking around the room, turning slightly away from the speaker, or picking up your bag or backpack. These cues may be enough to end the conversation without the awkwardness of expressing what you are feeling in words.

Deceiving

Lie to me* was a popular TV show about "human lie detectors," psychologists who helped solve crimes by analyzing suspects' micro-expressions and body language.

The show was loosely based on scientific research about deception detection, but the main characters' abilities to decode nonverbal cues were grossly exaggerated for dramatic effect. In reality, most people—even trained law enforcement experts—accurately detect deception only about half of the time (54 percent, according to several decades' worth of studies[57,58]). The truth is that "there is no universal indicator of deception," says former FBI agent Jim Clemente.[59] However much you may have been led by popular media to think someone clearing their throat, covering their mouth, or touching their nose is a sure sign of deception, that just isn't the case. Joe Navarro, a retired FBI agent, emphatically states, "Scientifically and empirically, there's no Pinocchio effect. Humans are lousy at detecting deception."[60]

Even when it comes to people you know well, it can be hard to tell if they are lying or telling the truth. For the most part, close friends and romantic partners are little better at spotting each other's lies than strangers are.[61,62] This may be partly because people *want* to believe those they know well and are reluctant to believe strangers. If you have a **truth bias** regarding someone you know, you probably assume they're being honest unless you have a compelling reason to suspect otherwise.[63] Conversely, you may harbor a **deception bias**, assuming that people you don't care for are likely to be dishonest. A deception bias might arise from knowing about someone's past lies or bad behavior, your own feelings of insecurity, previous experiences with other liars, or a history of telling lies yourself.[64]

Despite the evidence, misconceptions about deception cues persist.[65] Unfortunately, misreading someone's body language can lead to real-world consequences, especially when the person misreading it is an authority figure. For example, judges and jurors may assume that body language and facial expressions are an accurate indication of someone's innocence or guilt.[66] Airport security is another problem area. Members of the Transportation Security Administration developed a list of 94 suspicious behaviors, which they believed could be used to identify potential terrorists (for example, avoiding eye contact, emitting a strong body odor, covering the mouth when speaking, and having a pale face indicative of a recent beard shaving).[67] The program has been criticized for unfairly targeting immigrants, especially Middle Eastern and Muslim travelers.[68]

By now, you probably realize that "common sense" notions about lying are faulty. Liars might avoid eye contact, stutter, fidget, sweat, and so on. But some of them don't. And many people who display these nonverbal cues are telling the truth. With this awareness in mind, it *is* slightly easier to catch someone in a lie if you know what to look *and* listen for. See the feature "Deception Detection Hacks" for more information.

Jennifer Beals starred in *Lie to me**, a television show about fictional human lie detectors. Experts agree that most people only correctly detect deception about half of the time.

Do you think you are good at lying? Invite a friend to guess whether you are lying when you describe something you have (or pretend to have) in your bag or backpack. Was your friend's conclusion correct?

truth bias The tendency to assume that others are being honest.

deception bias The tendency to assume that others are lying.

COMMUNICATION STRATEGIES

Deception Detection Hacks

Face it, most people are liars. Estimates vary, but research suggests that 95 percent of folks fib at least once a week.[69] Most people stretch the truth more often than that though: One study found that 60 percent of people tell an average of three lies during every 10-minute conversation they have.[70] That's a whole lot of lying! But according to sociologists and psychologists, including author Robert Feldman, being equivocal or telling white lies can be essential in interpersonal relationships.[71] For example, if your grandmother's cookies aren't very good, saying so may hurt her feelings. The best approach for saving face and preserving your relationship may be a simple "Thank you for the cookies" rather than a brutally honest assessment ("These cookies are like hockey pucks"). So long as the intent is benevolent, most white lies are generally okay. Real lies, on the other hand, are self-serving and may have negative consequences, including hurting others. This is the type of lie you want to be on the lookout for (and avoid telling).

Catching someone in a lie isn't easy (research predicts you'll fail half of the time). But there are a few things you can zero in on when trying to discern fact from fiction. Here are some tips that may help.

Consider the Motive

People are more inclined to lie in certain situations than in others. For example, a child is likely to lie to avoid punishment. Men are likely to exaggerate their assets and attributes when on dates.[72,73] And it seems as if everyone lies in job interviews! Data suggests 80 percent of people embellish their experience and skills to some degree, while up to 30 percent claim to have degrees they did not earn.[74] Interviewers are not completely honest either: They may overstate benefits packages and opportunities for advancement.[75] Depending on the stakes and your viewpoint on the matter at hand, the lie may be trivial (avoid scolding), somewhat egregious (land a second date or get a job), or reprehensible (entice a job applicant with false promises).

If you suspect you're being lied to, ask yourself, "What does this person have to gain by lying to me?" Then ask, "What are the conceivable consequences?" If you decide the potential deception is inconsequential, doing nothing may be the best course of action. If the outcome is potentially negative, digging deeper may be the way to go.

Ask Questions

Lying is taxing. It requires a lot of mental energy to keep the details straight. That is why most people keep their lies simple.[76,77,78] Compared with lies, truthful statements have much more contextual detail.[79,80,81] If you are confronted with an explanation that seems too simple or a story that's scant on specifics, former FBI agent Jim Clemente's advice is "drill deeper."[82] You can do this by asking probing questions. Ask the person to repeat their story, but have them start with the ending and work their way back to the beginning. Request exact details about the way something smelled, how things might have tasted, whether it was hot or cold outside, where people stood in relation to one another, and so on (interestingly, most liars focus on what they claim to have seen and heard and neglect their other senses).[83] Listen carefully for contradictions. Is the person getting their "facts" mixed up? Searching for answers? Growing agitated? If so, you may be dealing with deception.

Listen

When it comes to catching someone in a lie, Clemente says active listening is essential (Chapter 6). In fact, studies show that audio cues are far better indicators of deception than visual cues.[84] There are a few key indicators you should listen for. In addition to telling more detailed stories, truth tellers typically use more first-person pronouns.[85] Liars, on the other hand, make more references to "they" and "them."[86] They also use more qualifiers, words and phrases such as "like," "almost," "maybe," and "sort of."[87] Clemente cautions that what follows a qualifier is usually "something the teller doesn't have confidence in."[88] And sometimes it is not what's said, but how it's said. Nonlinguistic verbal cues, such as vocal tone and pace, can tell you a lot.[89,90] For instance, when some people lie, they talk in a higher-than-normal pitch.[91]

Departures from a person's usual way of speaking (from fast to slow, or soft to loud) can be a red flag.

Watch for Changes
When a deceiver behaves the way they think you would expect an honest person to act, they are engaged in what Clemente calls perception management. For example, if they think you will associate fidgeting with lying, they may sit very still. Liars often make more deliberate eye contact than truth tellers, for the simple reason that people assume liars avert their gaze.[92,93] Of course, some people do sit still or make eye contact, as a matter of habit, while others might tap their feet, rock side to side, look down, and so on. The trick is to watch for altered patterns and sudden changes. If someone has been slouching, for instance, but sits up straight when you ask a pointed question, Clemente says you should take notice.[94] It doesn't necessarily mean the person is lying, but when considered along with their general demeanor and verbal cues (what they say and how they say it), abrupt shifts often point to deception.[95]

Look for Coherence
When a person's verbal and nonverbal communication match, it's called **communicative coherence**. Honest statements usually have a higher degree of communicative coherence compared with falsehoods.[96] Observe the person's nonverbal cues. Do their behaviors and gestures complement or contradict what they are saying? The less coherent the message, the more questionable it becomes.

Following these tips will not make you a human lie detector (those only exist on TV), but they may help you become a bit better at detecting deception. Keep in mind that some people are very skilled deceivers, and you may never catch them in a lie. The more motivated they are to conceal the truth and the longer they have to plan the deception, the less likely you are to uncover it.[97]

7.3 Types of Nonverbal Communication

Now that you understand how nonverbal messages operate as a form of communication, it's time to look at the various types of nonverbal behavior. The following pages explain how people send messages with their bodies, artifacts (such as clothing), environments, and the way they use time.

Body Movements

Stop reading for a moment and notice how you are sitting and what movements and facial expressions you have displayed over the last few minutes. What do these say about how you feel? Are there other people near you now? What messages do you get from their posture and movements? Watch a randomly selected video on YouTube without sound to see what messages are suggested by the movements of people on the screen. These simple experiments illustrate the communicative power of **kinesics**, the study of how people use their bodies and faces to communicate with others. This section explores three types of kinetic nonverbal behaviors: posture, fidgeting, and facial expressions.

Posture   There's a reason parents often tell their children to stand up straight: Good posture sends a message to others that you are confident and capable.[98] It can also influence how likable others find you (experimenters have observed a link between good posture, outward-focused gestures, and likability).[99] Standing tall may make you *feel* more confident too.[100] Some research suggests that standing up straight can

communicative coherence Consistency of a person's verbal and nonverbal communication.

kinesics The study of body movement, facial expression, gesture, and posture.

also help you better perceive others' nonverbal cues thereby improving your chances of detecting deception.[101]

Sometimes postural messages are obvious. If you see a person drag themselves through the door or slump over while sitting in a chair, it's apparent that something is going on. But most postural cues are subtler. For instance, mirroring the posture of another person can have positive consequences. One experiment showed that counselors who used "posture echoes" while communicating with clients were considered more empathic than those who did not.[102] Researchers have also found that partners in romantic relationships tend to mirror each other's behaviors.[103]

Because posture sends messages about how alert and vulnerable someone is, criminals may use it to identify people they think will be easy targets. Rapists sometimes use postural clues to select victims they believe will be easy to intimidate.[104] Walking slowly and tentatively, staring at the ground, and moving your arms and legs in short, jerky motions can suggest to others that you lack confidence and aren't paying attention to your surroundings.

Fidgeting For a lot of people, fidgeting helps "burn off" nervous energy, soothe anxiety, and reduce stress. It's likely that you fidget (at least occasionally). You might tap your fingers, wring your hands, play with a pen, fuss with your clothing, and so on. Social scientists call these behaviors **manipulators**, in the sense that they involve manipulating or fiddling with things.[105] Greater than normal use of manipulators is often a sign of discomfort. It can also suggest waning attention. In one experiment, researchers found the longer students sat through a lecture the more they fidgeted (and the less they remembered from the lecture).[106] Sustained attention (or what researchers call "time on task") can take its toll, especially for people prone to daydreaming or with attention deficit disorders.

Some data suggest that "focused fidgeting" (playing with a fidget spinner, fidget ring, or Pop It!, for example) boosts attention and improves listening.[107] This is because fidgeting increases physiological arousal, helping people feel more alert (at least temporarily).[108] If you have a tendency to fidget at school or work, you might experiment with fidget devices to see if they help. Fidget rings make a good choice because they are subtle and less likely to draw attention.[109] Be advised—sometimes fidget devices can have the opposite effect, as was the case with students in one experiment who had lower test scores when they were allowed to use them in class.[110]

Of course, fidget toys are probably out of the question if you are giving a speech (a Pop It! would be pretty distracting for your audience and may detract from your credibility). If you are nervous and worried you might fidget while speaking, follow these tips: Wear clothing that makes you feel confident and comfortable, don't hold anything in your hands, and try to relax your body and use gestures naturally.

Eye Contact How important is eye contact? A visit to the grocery store will suggest an answer. Check out the Quaker Oats man, the Trix rabbit, the Sun-Maid girl, and Chef Boy-Ar-Dee. The odds are that all of them will be looking back at you. Researchers at Cornell University found that people were more likely to choose Trix over other cereals if the rabbit was looking at them rather than away.[111] "Making eye contact, even with a character on a cereal box, inspires powerful feelings of connection," said Brian Wansink, one of the researchers.[112] By the same token, panhandlers, salespeople, and petitioners may try to catch your eye because after they've managed to establish contact with a glance it becomes harder for you to disengage.

People's need for eye contact begins at birth. Newborns instinctively lock eyes with their caregivers.[113] And eye contact is important for establishing friendships and

> **manipulators**
> Movements in which one part of the body (usually the hands) massages, rubs, holds, pinches, picks, or otherwise involves an object or body part.

romantic relationships. People for whom eye contact is difficult or uncomfortable, such as those with autism spectrum disorder, sometimes have trouble forming and maintaining interpersonal relationships.[114]

The meaning people give eye contact throughout their lives varies by culture. In Euro-American culture, meeting someone's glance with your eyes is usually a sign of involvement or interest, whereas looking away signals a desire to avoid contact. However, in some cultures—such as traditional Asian, Latin American, and Native American—it may be considered aggressive or disrespectful to make eye contact with a stranger or an authority figure.[115] It's easy to imagine the misunderstandings that occur when one person's "friendly gaze" feels rude to another, and conversely, how "politely" looking away might be interpreted as indifference.

Facial Expressions As you read at the start of this chapter, Khaby Lame's TikTok stardom is possible because his facial expressions transcend language and culture. In other words, his expressions are universal. According to psychologist Paul Ekman, there are at least six basic human emotions that are universally recognized: fear, joy, anger, disgust, sadness, and surprise.[116,117] There is more than one unique expression for each emotion, but people can get a pretty good idea about what others are feeling by looking at their faces.[118] For example, exposed to something funny, people everywhere display an expression that can usually be identified as a smile, even if they don't speak the same language or share the same culture. The same is true for expressions of disgust, sadness, and so on. Of course, there are some exceptions: Members of certain remote tribes in Africa, for example, have trouble distinguishing between Westerners' expressions of surprise and joy.[119]

Despite the universality of some emotional expressions, it doesn't mean that facial expressions are always easy to understand. For one thing, people can produce a large number of expressions and change them very quickly. A split-second frown can be replaced by a look of affection or surprise. As mentioned earlier, the growing prevalence of Botox may further confound the issue. A failure to raise one's eyebrows upon hearing shocking news might come across as indifference. Additionally, people can "say" one thing with part of their face and something different with another. A smile might be accompanied by a scrunched-up nose that suggests disapproval along with mirth.

In the movie *Inside Out*, a young girl experiences a range of emotions as she adjusts to life in a new city. Her feelings are depicted as animated characters (from left to right) anger, joy, fear, sadness, and disgust. Nonverbal cues for these emotions are the same in virtually every culture.

Watch a video in a language you don't understand. What cues give you the impression of anger, joy, fear, sadness, and disgust?

affect blend The combination of two or more expressions, each showing a different emotion.

Affect blends are simultaneous expressions that show two or more emotions, such as fearful surprise or angry disgust.

Culture also makes it complicated to interpret facial expressions (and other forms of nonverbal communication) with certainty. For one thing, cultural display rules govern when and how emotions are expressed.[120] Many Japanese individuals, for instance, tend to suppress emotional displays in public settings.[121] Second, some emotions appear to be unique to certain cultures. *Schadenfreude* is the German experience of deriving joy or amusement from a rival's misfortunes, but there isn't an English word for this phenomenon. English speakers, therefore, may have a hard time recognizing and describing the emotion.[122] Finally, different cultures may ascribe different values to expressions. For example, in much of the United States and Europe, people who smile are regarded as friendlier, less aggressive, and more confident than those who don't.[123,124] But people in some cultures are put off by smiling, especially when whole-face grins are involved. In Russia, wearing a serious expression can be a status booster, since it suggests that one is serious, reliable, and powerful.[125] As with all nonverbal communication, it pays to be mindful of the situations, cultures, and relationships involved.

Facial expressions not only reflect what people feel, they can also affect *how* people feel. Numerous experiments dating back to the 1970s found that when people held a pen between their teeth, thereby inducing a smile, they rated cartoons funnier and reported feeling happier than people who either did not smile or who frowned.[126,127] Smiling also appears to influence perception and memory: Smiling people see and remember happy faces more than sad faces.[128] The next time you're down in the dumps, try smiling. Chances are good it will make you feel better!

Voice

"That kind of joke can get you in trouble" says the boss to your coworker with a chuckle. You and your colleague may be left wondering if the boss was serious or only kidding. Social scientists use the term **paralanguage** to describe nonverbal cues that are audible. These include tone, speed, pitch, volume, and laughter, to name just a few. Paralanguage also describes what you do when you emphasize particular words with your voice, or pepper your speech with **disfluencies** (such as stammering and use of "uh," "um," and "er").

paralanguage Nonlinguistic means of vocal expression: rate, pitch, tone, and so on.

disfluencies Vocal interruptions such as stammering and use of "uh," "um," and "er."

Source: Alex Gregory The New Yorker Collection/The Cartoon Bank

The impact of paralinguistic cues is powerful. When asked to determine a speaker's attitude, listeners typically pay more attention to vocal nuances than to the speaker's words.[129] And when vocal cues contradict a verbal message, listeners usually trust the cues more than the words.[130] To experiment with this, say each of these statements as if you really mean them:

- Thanks for waking me up.
- I really had a wonderful time on my blind date.
- There's nothing I like better than waking up before sunrise.

Now say them again in a sarcastic way. You'll probably find that paralanguage can change the meaning of the statements, even suggesting that they mean the opposite of what they say on the surface. That makes it important to match a compliment, apology, command, or any other statement with vocal cues that support the message you mean to convey.

Some vocal factors influence the way a speaker is perceived by others. For example, communicators who speak loudly and without hesitations are often viewed as more confident than those who pause and speak quietly.[131] Lower-pitched voices are consistently rated as more competent and authoritative.[132,133] They may also be considered more desirable. Researchers find that, when a person is romantically interested in someone, they tend to lower their vocal intonations.[134] And it seems to work—members of all genders tend to gravitate toward would-be partners who have slightly lower-pitched voices than normal.

When it comes to paralanguage and gender, there are some distinct differences. The voices of most cisgender people fit their gender identity. For instance, feminine voices are usually marked by greater variability in pitch, frequent upward tonal shifts, and a breathy quality.[135] In comparison, masculine voices tend to be louder, more monotone, lower-pitched, and rougher sounding.[136] Assumptions about paralanguage can be hurtful for transgender persons because people may misgender them based on how their voices sound.[137]

Appearance

How people appear can be just as revealing as how they sound and move. This section explores the communicative power of physical attractiveness, clothing, and body art.

Physical Attractiveness Some observers have coined the phrase "the Tinder trap" (after the popular dating app) to describe the tendency, especially online, to judge people primarily or even solely on how they look. As experience has no doubt taught you, looks are not the most important factor in judging relationship potential, but they contribute significantly to first impressions.

People who are considered physically appealing are more likely than others to be satisfied with life.[138] And no wonder. Attractive people are more likely than their peers to win elections,[139] influence their peers,[140] and succeed in business. They are less likely to be convicted in court,[141,142] and when they are found guilty, they typically receive shorter sentences.[143] Attractive people also get preferential treatment both in hiring decisions and on the job.[144] One study even found that people with attractive avatars enjoy an advantage in online job interviews.[145]

What some researchers call the *beauty premium* involves a reciprocal cause-and-effect pattern. People tend to assume that attractive people are more confident and capable than others, so they treat them as if they are—which can lead attractive people to actually be more confident (although not necessarily more talented) than others.[146] And because people tend to smile at attractive people, they tend to smile back, which can boost their attractiveness rating even more.[147] One lesson is that presenting yourself in a confident and friendly way can improve others' opinion of your appearance, even if you don't consider yourself the best looking person in the room.

All this being said, if you aren't extraordinarily gorgeous or handsome by society's standards, don't despair. Evidence suggests that, as people get to know one another, they perceive that likable people are more attractive, and unpleasant people less attractive, than they originally thought.[148] Moreover, posture, gestures, facial expressions, and other behaviors can increase the attractiveness of an otherwise unremarkable person. And occasionally, physical attractiveness has a downside. Employers sometimes turn down especially good-looking candidates because they perceive them to be threats.[149] All in all, while attractiveness generally gets rewarded, over-the-top good looks can be intimidating.[150]

Clothing As a means of nonverbal communication, clothing can be used to convey economic status, educational level, group membership, athletic ability, interests, and more. It also exerts a powerful influence on how people think and feel about themselves and, to a large extent, how they act.

You may have heard the expression "Clothes make the person." That sentiment dates to the ancient Greeks, possibly earlier. The gist is that what you wear shapes who you are, largely owing to the way people treat you based on your clothing. People who don smart clothes, such as tailored suits and business attire, are often taken more seriously and thought to be more competent, confident, and successful than those who dress more casually.[151]

enclothed cognition
The phenomenon of adopting traits or enacting behaviors associated with articles of clothing.

What you may not know is that your clothes may also affect how well you do on tests, how likely you are to attend class regularly, your charitable intentions, and even your hormone levels. Here's a brief rundown of the research:

- *Clothes can make you feel and act smarter.* When participants in one study wore a physician's laboratory coat, they performed better on cognitive tasks.[152] Because physicians are regarded as smart, participants believed wearing a physician's coat would increase their intelligence. When they were told the same coat was a painter's smock, their scores deteriorated. Researchers coined the term **enclothed cognition** to explain the phenomenon of adopting traits associated with articles of clothing.

Clothes can affect how you feel about yourself and, to some extent, how you act. They might even make you smarter! Participants in an experiment performed better on cognitive tests when they wore a physician's lab coat.

How do your clothes make you feel? How do your clothing choices inform the way others perceive you?

- *School uniforms boost attendance and test scores.* Psychologists explain that uniforms encourage people—not just school children—to pay more attention to their work.[153] However, authority uniforms (such as those worn by police and the military) tend to increase wearers' tolerance for risk and may lead them to make mistakes because they are overconfident.[154]
- *High-status clothes can make you less charitable.* Women given designer handbags and accessories in a series of experiments were considerably less likely to engage in prosocial, charitable behaviors compared with women given unbranded items.[155,156] The status boost that luxury goods bestow appears to encourage resource hoarding.
- *Casual clothing can make you feel less assertive.* Researchers have found that people wearing casual clothes tend to have lower testosterone levels and inhibited negotiating skills. Suit wearers, on the other hand, tend to exhibit more dominant behaviors, which seem to help them make better deals.[157]
- *The colors you wear may influence how others perceive you.* In one study, experimenters asked participants to rate the attractiveness of individuals shown in photos. The photos were identical except that the experimenters had digitally changed the color of the models' shirts. Participants consistently rated both male and female models to be more attractive when they were shown wearing red or black shirts compared to yellow, blue, or green.[158]

Of course, not everyone who wears Prada is stingy, and not all suit wearers fare well in business, but these studies showcase how powerful clothes can be.

While both men and women are critiqued for their clothing choices, women often face additional challenges and criticism. Some of these include the following:

- What's considered fashionable for women changes every year, whereas men's designs change less frequently.
- The latest fashions might not look good on all women or be designed to fit them.[159]
- Clothing and shoes that society expects women to wear can be uncomfortable and impractical.[160]
- Women receive mixed messages about how they should appear. Media images tend to depict and celebrate women in revealing clothing, but dressing that way can lead others to take women less seriously. In one study, participants judged women in photos to be more intelligent, powerful, organized, professional, and efficient when they were dressed in business clothes rather than in form-fitting or revealing clothing.[161] An extra blouse button undone can make the difference between a woman being deemed intelligent or incompetent at work.[162]
- Gender stereotypes mean women encounter more problems managing their appearance at work.[163]

Body Art There's no denying that tattoos are having a moment. Across much of Europe and in the United States, close to half of all adults sport at least one tattoo.[164,165] A decade ago, only about 20 percent of Americans were inked.[166] According to experts who study body art, the rise in popularity is attributed to two main factors. First is the ever-growing number of celebrities and athletes who have tattoos and the second

Most clothing and body art are conscious choices, crafted to support a desired identity.
What does your physical appearance say about you?

is Instagram, where millions of tattoo-related posts are helping normalize body art.[167,168] For a long time, tattoos were associated with rebels and criminals,[169,170] but Millennials (41 percent of whom are tattooed[171]) are changing people's ideas about that. For example, consider Dr. Katherine Palmisano, a Millennial and internal medicine physician who has tattoos covering both of her arms. She says they help her connect with her patients, many of whom have tattoos as well. "The idea that hair color or tattoos implies that someone is unprofessional is short-sighted," maintains Palmisano.[172]

That being said, when it comes to ink, beauty is in the eye of the beholder *and* the identity of the design wearer. Age and gender have a lot to do with it. People tend to consider tattoos more attractive on younger bodies than on older ones.[173] Tattooed women are ascribed more negative qualities than tattooed men.[174,175] For example, women with visible tattoos are often considered promiscuous and rank low on credibility, attractiveness, motivation, honesty, generosity, and intelligence.[176,177] Tattooed men, on the other hand, enjoy more social acceptance and popularity. They may even be more desirable than ink-free men. When researchers showed people images of the same men with or without tattoos, the respondents tended to rate the tattooed versions as more masculine and aggressive.[178] That can be both good and bad. Women in the study found the tattooed men appealing, but not necessarily good husband or father material.

While things are changing, a tattoo taboo still exists in many workplaces. In a survey of business managers, half reported that tattooed job candidates are taken less seriously.[179] Overall, people with visible tattoos have more difficulty getting hired.[180]

If you are considering a tattoo, Amanda Mull with *The Atlantic* has some advice: "The try-before-you-buy model of body art is not a bad idea."[181] A temporary or semi-permanent tattoo lets you experiment without commitment. (Inkbox offers over 4,000 designs that last up to two weeks, while Ephemeral sells longer-lasting options).

Touch

A supportive pat on the back, high five, handshake, or fist bump can often be more powerful than words. Haptics, the study of touch, has revealed that physical contact is even more powerful than you might think.

Experts propose that one reason actions speak louder than words is because touch is the first language infants learn.[182] In many ways, touch is essential. Mortality rates in orphanages decreased two thirds in the early 20th century, when people recognized that babies need to be touched and held in order to thrive.[183,184] As children grow, appropriate touch seems to boost their mental functioning as well as their physical health. Children given plenty of physical stimulation tend to develop significantly higher IQs and less communication apprehension than those who have less contact.[185]

When it comes to health and well-being, touch is important for adults as well.[186] Physicians and nurses often soothe patients, establish rapport, and show empathy via compassionate touch, such as a hand placed on the shoulder or a hug.[187,188] It seems to be appreciated. Patients report being more satisfied with healthcare providers who demonstrate compassion through touch.[189] But you don't have to be a physician or nurse for your touch to make a difference. Studies show that affectionate touch from a loved one can reduce people's perceptions of pain and lower their blood pressure.[190] It can also make them happier. For example, romantic partners who frequently touch each other are typically more satisfied with their relationships than other couples are.[191,192]

Touch can also be persuasive. When people were asked to sign a petition, researchers found that they were more likely to cooperate when the person asking touched them lightly on the arm. In one variation of the study, 70 percent of those who were touched complied, whereas only 40 percent of the untouched people agreed to sign.[193]

Touch is welcomed in other scenarios as well. Studies consistently find that fleeting touches on the hand and shoulder often result in larger tips for restaurant servers.[194,195] Even athletes reflect the power of touch. One study of the National Basketball Association revealed that the touchiest teams had the most successful records, while the lowest scoring teams touched each other the least.[196]

Although touch may be appreciated in some situations, it can be annoying or frightening in others, especially when the person touching you is someone you consider an out-group member or a stranger.[197] The #MeToo movement brought attention to unwelcomed touching and sexual advances. Hopefully, the result is that more people will take time to consider if their touch is appropriate and desired by the recipient.[198]

Space

There are two ways that the use of space can create nonverbal messages: the distance people put between themselves and the territory they consider theirs. This section looks at each of these dimensions.

Distance   The study of the way people use space is called **proxemics**. Preferred spaces are largely a matter of cultural norms. For example, people living in hyperdense Hong Kong manage to live in crowded residential quarters that most North Americans would find intolerable. Anthropologist Edward T. Hall defined four distances used in mainstream North American culture.[199] He observed that people choose a particular distance depending on how they feel toward others at a given time, the context of the conversation, and their personal goals.

Intimate distance begins with skin contact and ranges to about 18 inches. The most obvious context for intimate distance involves interaction between people who are emotionally close—and then mostly in private situations. Intimate distance between individuals also occurs in less intimate circumstances such as visiting the doctor or getting your hair cut. Allowing someone to move into the intimate zone usually is a sign of trust.

Personal distance ranges from 18 inches at its closest point to 4 feet at its farthest. The closer range is the distance at which most relational partners stand in

proxemics   The study of how people and animals use space.

intimate distance   One of four distance zones, ranging from skin contact to 18 inches.

personal distance   One of four distance zones, ranging from 18 inches to 4 feet.

public. Many people are often uncomfortable if someone else "moves in" to this area without invitation. The far range of personal distance runs from about 2.5 to 4 feet. This is the zone just beyond the other person's reach—the distance at which someone can be kept "at arm's length." This term suggests the type of communication that goes on at this range: Interaction is still reasonably personal, but less so than communication that occurs a foot or so closer.

Social distance ranges from 4 to about 12 feet. Within it are the kinds of communication that usually occur in business situations. Salespeople, customers, and coworkers usually maintain a distance of 4 to 7 feet, whereas people engaging in formal and impersonal interactions tend to stay 7 to 12 feet apart. The next time you're in a conversation with your boss or best friend, consider how far apart you are.

social distance One of four distance zones, ranging from 4 to 12 feet.

Public distance is Hall's term for the farthest zone, running outward from 12 feet. Teachers tend to stand about 12 feet or a little more from students. When people are 25 feet apart or more, two-way communication is difficult. In some cases, it's necessary for speakers to use public distance owing to the size of their audience, but we can assume that anyone who voluntarily chooses to use it when they could be closer is not interested in having a dialogue.

public distance One of four distance zones, ranging from 12 feet or more.

Choosing the optimal distance can have a powerful effect on how others respond to you. For example, students are more satisfied with teachers who reduce the normal distance between themselves and the class. They also are more satisfied with the course itself and more likely to follow the teacher's instructions.[200] Likewise, medical patients are often more satisfied with physicians who don't "keep their distance."[201]

Allowing the right amount of personal space is so important that scientists who create interactive robots take proxemics into account. They have found that people tend to be creeped out by robots who get too close, especially if the robot makes consistent "eye contact" with them. However, people don't mind as much if robots whose nonverbal behaviors seem friendly move into their personal space.[202]

Territoriality Scholars use the term **territoriality** to describe people's tendency to claim places and spaces they consider to be more or less their own. These spaces might include your bedroom, house, or the chair you usually occupy in class or at your favorite coffee shop. People tend to use nonverbal markers to declare which territory is theirs. You might erect a fence around your yard or spread a blanket on the beach to mark the area as yours, at least temporarily. You may feel annoyed or disrespected if someone encroaches on your territory without permission. Indeed, honoring boundaries is one way of showing respect. People with higher status generally are granted more personal territory and privacy. You probably knock before entering the boss's office if the door is closed, whereas a boss can usually walk into your work area without hesitating. In the military, greater space and privacy usually come with rank: Privates typically sleep 40 to a barracks, sergeants have their own private rooms, and high-ranking officers have government-provided houses.

territoriality The tendency to claim spaces or things as one's own, at least temporarily.

Environment Google, Microsoft, and some other employers have begun to design communication-friendly work environments in which the furniture can be moved at will and there is plenty of space for conversations and group interaction. They know what social scientists have long found to be true: People's physical environments shape the communication that occurs within them. Especially appealing are flexible work spaces that feature both areas for team collaboration and private spaces in which people can work without interruptions.[203]

The sweet spot, with respect to employee happiness and productivity, lies somewhere between an open office plan with shared work areas and one with walled-off, private spaces. Open plans were popular for many years because it was widely believed such spaces fostered face-to-face communication and collaboration. In reality, open plans can actually decrease interaction by as much as 70 percent.[204] Researchers observe that people who need uninterrupted time to focus tend to sidestep conversations by avoiding eye contact and using headphones—clear signals they don't want to be disturbed.[205] Open offices can be noisy as well, and there's evidence showing noise is distracting, contributes to stress, and lowers performance[206] (this is true both at work and at school[207]).

In many situations, the best environment is one with natural light, flexible seating configurations, and views of nature. Such spaces have been shown to enhance productivity at work and learning in the classroom.[208] If you can, arrange to work or study in a place that meets these conditions.

Time

Chronemics is the study of how people use and structure time.[209] The way people handle time can send both intentional and unintentional messages.[210] If you have ever been "left on read" after sending a text message, you know response time can say a lot. Sometimes, a delayed response or no response at all is a message in and of itself.

Social psychologist Robert Levine describes several ways that time can communicate.[211] In a culture that values time highly, such as the United States, waiting can be an indicator of status. "Important" people (whose time is supposedly more valuable than that of others) may be seen by appointment only, whereas it is acceptable to intrude without notice on "less important" people. To see how this rule operates, consider how natural it is for a boss to drop by a subordinate's office unannounced, whereas most employees would never intrude into the boss's office without an appointment. A related rule is that low-status people must not make more important people wait. It would be a serious mistake to show up late for a job interview, although the interviewer might keep you waiting in the lobby. Important people are often whisked to the head of a restaurant or airport line, whereas the presumably less exalted are forced to wait their turn.

The use of time depends greatly on culture.[212] Some cultures (e.g., North American, German, and Swiss) tend to be **monochronic**, emphasizing punctuality, schedules, and completing one task at a time. Other cultures (e.g., South American, Mediterranean, and Arab) are more **polychronic**, with flexible schedules in which people tackle multiple tasks simultaneously. One psychologist discovered the difference between North and South American attitudes when teaching at a university in Brazil.[213] He found that some students arrived halfway through a two-hour class and that most of them stayed put and kept asking questions when the class was scheduled to end. A half hour after the official end of the class, the professor finally closed off discussion, because there was no indication that the students intended to leave. This flexibility of time is quite different from what is common in most North American colleges.

> **chronemics** The study of how humans use and structure time.

> **monochronic** The use of time that emphasizes punctuality, schedules, and completing one task at a time.

> **polychronic** The use of time that emphasizes flexible schedules in which multiple tasks are pursued at the same time.

How people handle time sends a message.

Do you typically respond to messages in a timely manner? What do you think a delayed response, or no response, communicates to others?

Even within a culture, rules of time vary. Sometimes the differences are geographic. In New York City, the party invitation may say "9 P.M.," but nobody would think of showing up before 9:30. In Salt Lake City, guests may be expected to show up on time, or perhaps even a bit early.[214] Even within the same geographic area, different groups establish their own rules about the use of time. Consider your own experience. In school, some instructors begin and end class punctually, whereas others are more casual. With some people you feel comfortable talking for hours in person or on the phone, whereas with others time seems to be precious and not meant to be "wasted."

7.4 Influences on Nonverbal Communication

Sometimes nonverbal cues are easy to interpret no matter who is involved. As mentioned before, facial expressions that reflect emotions such as happiness and sadness are similar around the world.[215] Other nonverbal cues can be culture specific or influenced by gender identity. This section explores those influences. First, you'll learn some of the ways culture shapes nonverbal communication.

Culture

Use of Space   Edward Hall pointed out that, whereas Americans are comfortable conducting business at a distance of roughly 4 feet, people from the Middle East stand much closer.[216] It's easy to visualize the awkward advance-and-retreat pattern that might occur when two diplomats or businesspeople from these cultures meet. The Middle Easterner would probably keep moving forward to close a gap that feels wide to them, whereas the American would probably continually back away. Both would feel uncomfortable, although they may not know why.

Eye Contact   Like distance, patterns of eye contact vary around the world.[217] A direct gaze is considered appropriate for speakers in Latin America, the Arab world, and southern Europe. On the other hand, Asians, Indians, Pakistanis, and northern Europeans typically gaze at a listener peripherally or not at all. In either case, deviations from the norm are likely to make a listener uncomfortable.

Source: DILBERT © 1991 Scott Adams. Used by permission of UNIVERSAL UCLICK. All rights reserved.

Nonverbal Focus Culture also affects how nonverbal cues are monitored. In Japan, for instance, people tend to look to the eyes for emotional cues, whereas Americans and Europeans focus on the mouth.[218] These differences can often be seen in the text-based emoticons commonly used in these cultures. (Search for "Western and Eastern emoticons" in your browser for examples.)

Paralanguage Even within a culture, various groups can have different nonverbal rules. For instance, younger Americans often use "uptalk" (statements ending with a rise in pitch) and "vocal fry" (words ending with a low guttural rumble). Celebrities such as Kim Kardashian and Zooey Deschanel popularized these vocalic styles. It's therefore not surprising that females use them more than males do,[219] although age of the speaker (i.e., Millennial or younger) has more of an impact than gender. There is some debate whether vocal fry diminishes one's credibility[220] or enhances it.[221] Either way, vocal mannerisms are one way that speakers affiliate with their communities.

Affect Displays Although some nonverbal expressions are more or less universal, the way they are used varies widely around the world. In some cultures, display rules discourage the overt demonstration of feelings such as happiness or anger. In other cultures, displaying those feelings is perfectly appropriate. Thus, a person from Japan may appear much more controlled and placid than an Arab, when in fact their feelings may be nearly identical.

The same principle operates among cocultures. For example, observational studies have shown that, in general, Black women in all-Black groups are more nonverbally expressive than White women in all-White groups.[222] This doesn't mean that Black women feel more intensely than their White counterparts. A more likely explanation is that the two groups follow different cultural rules. Researchers found that in racially mixed groups, both Black and White women moved closer to the others' style (i.e., Black women dialed it back while White women amped up their expressions and gestures). This nonverbal convergence shows that skilled communicators can adapt their behavior when interacting with members of other cultures or cocultures to make the exchange smoother and more effective.

All in all, communicators are likely to be more tolerant of others once they understand that the nonverbal behaviors they consider unusual may be the result of cultural differences. In one study, American adults were presented with videotaped scenes of speakers from the United States, France, and Germany.[223] When the sound was off, viewers judged foreigners on their body language alone and rated them more negatively than their fellow citizens. But when the speakers' voices were added (allowing viewers to recognize that they were from a different country), participants were less critical of the foreign speakers.

Gender

Although differences by gender are often smaller than people think, women in general are more nonverbally expressive than men, and women are typically better at recognizing others' nonverbal behavior.[224] More specifically, research shows that, compared with men, women tend to smile more, use more facial expressions and gestures, touch others more, stand closer to others, be more vocally expressive, and make more eye contact. These patterns are observed across numerous cultures.[225,226,227] Most communication scholars agree that these differences are influenced more by social conditions than by biological differences.

When it comes to evaluating people's nonverbal cues, socially constructed ideas about gender mean that men—especially gender-conforming heterosexual men—often get preferential treatment. For example, women are frequently judged more harshly than men based on what they wear to work. Women's wardrobe choices can lead others to gauge them as less professional, credible, intelligent, and competent than they are.[228] Women who eschew more feminine clothes in favor of traditionally masculine-style suits may do better in the workplace—to a point. Research has suggested that if their clothes are *too* masculine, women may receive lower competency ratings.[229] Gender nonconforming people—those whose identities and/or appearances do not align with masculine and feminine gender norms—face similar judgments.

Makeup is another example of how people may evaluate others' nonverbal cues in gendered ways. Some employers, particularly those in the service industry, expect or even require women to wear a certain amount of makeup on the job, but not men. Makeup is not only costly but requires time to apply.[230] The average woman in America wears $8 worth of cosmetics a day and spends an hour putting on beauty products. Over a lifetime, that adds up to about $300,000 and a couple of years spent on makeup.[231] According to U.S.-based employment lawyers, gender-based discrimination related to personal appearance is legal.[232] But the real kicker is that women who wear too much makeup are deemed less competent, less moral, and less human.[233] A series of experiments to test what social scientists call the *cosmetics dehumanization hypothesis* revealed that heavy eye makeup and bright lipstick contributed to women being objectified and sexualized.[234] Wearing too little or no makeup may cost a woman her job, but wearing too much may objectify her in others' eyes.

Of course, wearing makeup is not limited to women. It has become more common over the past decade for men to wear makeup regularly outside of drag events and the music industry. Male world leaders (including Emmanuel Macron and Donald Trump) commonly wear makeup to mask dark circles and skin imperfections.[235] The recent mainstreaming of men's cosmetics has been helped along by Millennials, who have embraced brands such as Tom Ford for Men and Boy de Chanel.[236] But society's acceptance extends mainly to cisgender men whose makeup is subtle.[237] A stigma persists against men wearing eye shadow, eyeliner, mascara, or lipstick in most contexts.

Gender informs many people's understanding of another nonverbal phenomenon having to do with facial expressions. It's often called *resting bitch face*, although scholars prefer the term *resting blank face* (RBF).[238]

Eminem and Kristen Stewart have the type of neutral expression neuroscientists call *resting blank face* or RBF. People often interpret those with RBF as unfriendly.

How might your neutral expression inform the way others perceive you? Have you ever misread someone's neutral expression? What were the consequences?

Behavioral neuroscientists explain that, even when you relax your face, others can pick up on faint traces of emotion.[239] Neutral expressions may come across as contempt. (One study found neutral expressions on women, but not men, resulted in significantly lower likability scores.[240]) A neuroscientist speaking to *The Washington Post* noted, "RBF isn't necessarily something that occurs more in women, but we're more attuned to notice it in women because women have more pressure on them to be happy and smiley and to get along with others."[241]

Despite these differences, men's and women's nonverbal communication patterns have a good deal in common.[242] You can prove this by imagining what it would be like to use radically different nonverbal rules. Standing only an inch away from others, sniffing strangers, or tapping people's foreheads to get their attention would mark you as bizarre no matter your gender.

At the start of this chapter, you read how Khaby Lame's nonverbal communication—namely, his expressions and gestures—propelled him to TikTok stardom. Nonverbal communication is much more than facial expressions and gestures, however. It encompasses a wide range of behaviors and choices that affect your interpersonal relationships, health, and career success. Hopefully, this chapter has helped you become more attuned to the nonverbal messages you send and receive.

MAKING THE GRADE

OBJECTIVE 7.1 Explain the characteristics of nonverbal communication and the social goals it serves.

- Nonverbal communication helps people manage their identities, define their relationships, and convey emotions.
- It is impossible to avoid communicating nonverbally. Humans constantly send messages about themselves that are available for others to receive.
- Nonverbal communication is ambiguous. There are many possible interpretations for any behavior. This ambiguity makes it important for the receiver to verify any interpretation before jumping to conclusions about the meaning of a nonverbal message.
- Nonverbal communication is different from verbal communication in complexity, flow, clarity, impact, and intentionality.
  > Describe three messages that qualify as nonverbal and one message that is verbal. Explain the difference between these two types of communication.
  > Considering that people cannot *not* communicate, what messages do you think you send nonverbally to strangers who observe you in public?
  > If a friend were preparing for a job interview and asked your advice about appearing and feeling confident, what advice would you give about managing their nonverbal communication?

OBJECTIVE 7.2 Describe key functions served by nonverbal communication.

- Nonverbal communication serves many functions: repeating, substituting, complementing, accenting, regulating, and contradicting verbal behavior, as well as deceiving.
- Detecting deception based on nonverbal cues is much harder than most people think because people tend to have truth or deception biases and because nonverbal behavior is highly ambiguous.
  > Give an example of each of the following nonverbal behaviors: repeating, substituting for, complementing, accenting, regulating, and contradicting verbal messages.

> Try interacting with people for an hour without using words, then reflect on the experience. What were you able to convey easily through nonverbal means? What was most difficult?

OBJECTIVE 7.3 **List the types of nonverbal communication, and explain how each operates in everyday interaction.**

- We communicate nonverbally in many ways: through posture, fidgeting, eye contact, facial expressions, voice, physical attractiveness, clothing, body art, makeup, touch, distance and territoriality, environment, and time.
- Members of some cultures tend to be monochronic, whereas others are more polychronic.
  > Describe the difference between a monochronic time orientation and a polychronic orientation.
  > Pause to look at your surroundings. How conducive are they to a positive state of mind? To social interaction or contemplation? How does your environment influence the way you feel right now?
  > Keep a tally of how many disfluencies you utter in one day. Consider whether you are happy with the results. If not, what would you like to change?
  > Notice the nonverbal cues of someone around you right now. Write down two or three interpretations of how that person is feeling. Ask if any of them are accurate.

OBJECTIVE 7.4 **Explain the ways in which nonverbal communication reflects culture and gender differences.**

- Based on their culture, people may focus on different nonverbal cues and have different expectations about how to use space, eye contact, paralanguage, and affect displays.
- In some cultures, eye contact and physical closeness are interpreted as signs of attentiveness, in others as challenges or indications of disrespect.
- In general, women are socialized to be more attentive to nonverbal cues than are men.
- Socially constructed ideas about gender mean people's nonverbal behaviors are judged differently. Cisgender heterosexual men often benefit, whereas women and gender nonconforming persons face harsher criticism for their choices, especially when it comes to clothing and makeup.

> Name three rules of appropriateness for making or avoiding eye contact that are familiar to you, but may be unfamiliar to someone from a different culture.
> What advice would you offer someone who is packing clothing for a trip during which they will encounter people from many different cultures?
> How would you explain the reasons that men and women may use and interpret nonverbal cues differently?

KEY TERMS

affect blend p. 178
affect displays p. 168
chronemics p. 185
communicative coherence p. 175
deception bias p. 173
disfluencies p. 178
emblems p. 170
enclothed cognition p. 180
expectancy violation theory p. 166
haptics p. 166
illustrators p. 170
intimate distance p. 183
kinesics p. 175
manipulators p. 176
monochronic p. 185
nonverbal communication p. 162
paralanguage p. 178
personal distance p. 183
polychronic p. 185
proxemics p. 183
public distance p. 184
social distance p. 184
territoriality p. 184
truth bias p. 173

PUBLIC SPEAKING PRACTICE

Think of a time from your personal experience or the plot of a movie or tv show when someone's nonverbal behavior was baffling or offensive to someone else. Create a brief oral presentation in which you describe what happened and speculate on why the people involved interpreted the nonverbal cues differently.

ACTIVITIES

1. You can become more adept at both conveying and interpreting vocal messages by following these directions.

    a. Join with a partner and designate one person A and the other B.

    b. Partner A should choose a passage of 25 to 50 words from a newspaper or magazine, using their voice to convey one of the following attitudes:
        > Egotism
        > Friendliness
        > Insecurity
        > Irritation
        > Confidence

    c. Partner B should try to detect the emotion being conveyed.

    d. Switch roles and repeat the process. Continue alternating roles until each of you has both conveyed and tried to interpret at least four emotions.

    e. After completing the preceding steps, discuss the following questions:
        > What vocal cues did you use to make your guesses?
        > Were some emotions easier to guess than others?
        > Given the accuracy of your guesses, how would you assess your ability to interpret vocal cues?
        > How can you use your increased sensitivity to vocal cues to improve your everyday communication competence?

2. To consider the influence of culture on nonverbal communication, do the following:

    a. Identify at least three significant differences between nonverbal practices in two cultures or cocultures (e.g., ethnic, age, or socioeconomic groups) within your own society.

    b. Describe the potential difficulties that could arise out of the differing nonverbal practices when members from the cultural groups interact. Are there any ways of avoiding these difficulties?

    c. Now describe the advantages that might come from differing cultural nonverbal practices. How might people from diverse backgrounds profit by encountering one another's customs and norms?

Understanding Interpersonal Communication

CHAPTER OUTLINE

8.1 Characteristics of Interpersonal Communication 194
What Makes Communication Interpersonal?
Content and Relational Messages

8.2 Interpersonal Relationship Building 197
How People Choose Relational Partners
Metacommunication
Self-Disclosure
Interpersonal Communication Online

8.3 Communicating with Friends and Family 205
Friendships Have Unique Qualities
Friendships Develop with Communication
Friendships Can Build Bridges
Family Relationships

8.4 Communicating with Romantic Partners 213
Stages of Romantic Relationships
Love Languages

8.5 Relational Dialectics 220
Connection Versus Autonomy
Openness Versus Privacy
Predictability Versus Novelty

MAKING THE GRADE 223

KEY TERMS 225

PUBLIC SPEAKING PRACTICE 225

ACTIVITIES 225

LEARNING OBJECTIVES

8.1
Explain what makes some communication interpersonal and how content and relational messages differ.

8.2
Describe how people choose relational partners and the role of metacommunication, self-disclosure, and online communication in developing those relationships.

8.3
Identify common communication patterns in friendships, parent–child dynamics, and sibling relationships.

8.4
Describe stages of romantic relationships and options for conveying intimate messages.

8.5
Compare dialectical continua and strategies for managing them.

"Will you accept this rose?" This line from *The Bachelor* and *The Bachelorette* invites a contestant to be part of another round in the reality television game of love. To date, viewers have watched 33 couples say "I will" in marriage proposals at the shows' conclusions. However, only 5 of those couples have actually said "I do."[1] What happens after the show to turn "I love you forever" into "Maybe not"?

This chapter explores the role of communication in important relationships. As you have no doubt experienced, close relationships—whether with romantic partners, friends, or family members—sometimes seem effortless and rewarding. At other times, they're challenging and downright confusing.

Reality shows barely scratch the surface, of course. Sean Lowe, one of the few who found lasting love on *The Bachelor*, describes his wake-up moment: "You leave the show, then you get into the real world and find out like, 'Oh crap! Being in a relationship isn't always easy and it actually takes work.'"[2] Communicating effectively when powerful emotions are involved requires know-how and awareness. This chapter will explore communication patterns that help people develop close relationships; the different ways people communicate with friends, family members, and romantic partners; and how relational partners manage the give-and-take of competing relational and personal needs.

8.1 Characteristics of Interpersonal Communication

> **interpersonal communication** Two-way interactions between people who share emotional closeness and treat each other as unique individuals.

This section defines **interpersonal communication** and explores two levels of communication—one that focuses on the topic at hand and another that simultaneously conveys how people feel about one another.

What Makes Communication Interpersonal?

To clarify what communication is and isn't interpersonal, consider a few of the people you're likely to encounter on a typical day. Which of the following encounters do you think qualify as interpersonal communication?

- In your morning class, you strike up a conversation with a classmate you've just met.
- Later, you take part in a group project meeting that's so lively and productive you can't remember who said what.
- At midday, you pick up lunch and chat briefly about the weather with the cashier.
- After work, you enjoy a conversation with your housemate, who will understand what a crazy day you've had and will share some personal experiences and feelings in return.
- Before bed, you exchange texts with someone you met online but have never seen in person.

If you picked the scenario with your housemate, you're right. If you also selected the online relationship, score that one as a maybe for now. The other encounters are important in their own ways, but they probably don't qualify as interpersonal.

To understand why some encounters qualify and others don't, think of *inter*personal communication as the opposite of *im*personal communication. Chatting with strangers and colleagues may be enjoyable, but not necessarily personal. As this book uses the term, **interpersonal communication** involves interaction between people who share emotional closeness and treat each other as unique individuals. Consider the implications of that definition by returning to the previous examples.

You might eventually develop a close relationship with the classmate you just met, but for now, you don't know each other well enough to appreciate one another's unique qualities. Interpersonal relationships require a sense of closeness that doesn't occur instantly. Rather, it evolves over time.

Interpersonal communication involves more than chatting occasionally or greeting people in passing. Emotional closeness makes these relationships irreplaceable.

What factors make the most important relationships in your life special?

The project meeting is a good example of effective task-related communication. But since there's no evidence that your colleagues have a close personal attachment, you may assume that it's not interpersonal.

Although interactions with strangers and casual acquaintances serve an important role, your exchange with the cashier isn't interpersonal because there's no exchange of personal information, and you're not likely to care if a different cashier helps you next time. By contrast, if a different person showed up tomorrow playing the role of your roommate, mother, or brother, you would probably be disoriented and upset.

The conversation with your housemate is interpersonal because you're invested in listening to and sharing personal information with each other as unique individuals.

The relationship with your online friend may or may not be interpersonal. It depends—you guessed it—on how much personal information you each share and whether you treat one another as unique individuals.

Of course, it isn't always so easy to categorize relationships. On a continuum between impersonal and interpersonal, some relationships fall in the middle. The main point here is that interpersonal relationships are marked by emotional closeness and a sense of uniqueness.

"She's texting me, but I think she's also subtexting me."

Source: Leo Cullum The New Yorker Collection/The Cartoon Bank

Content and Relational Messages

When Lucien asked his friend Haris to go to lunch, Haris said, "Sorry, I'm busy." Lucien wondered if his friend was actually unable to go or whether something was bothering him.

Lucien's uncertainty in this example involves the difference between content and relational meaning. Virtually every verbal statement contains both a **content message**, which focuses explicitly on the subject being discussed (in this case "Sorry, I'm busy"), and a **relational message** that suggests how the parties feel toward one another. If Haris declined the lunch invitation with an apologetic tone of voice, Lucien might interpret it differently than if he sounded angry or frustrated. Because relational messages are often implied rather than stated outright, they can be difficult to interpret accurately.

Affinity The degree to which one person likes another is called **affinity**. Sometimes people indicate feelings of affinity explicitly (as in "You're a great friend" or "I love you"), but often the clues are nonverbal, such as a pat on the back or a friendly smile. If you ask someone to lunch and they bluntly tell you no, you might assume (rightly or wrongly) that they don't like you.

Immediacy Whereas affinity involves attraction, **immediacy** reflects the level of engagement between people in a relationship. If two people interact openly and frequently, immediacy is high. If one or both of them is detached and distant, it's low. Affinity and intimacy can interact in different ways. Perhaps you have two friends you really like (high affinity), but you engage with one of them regularly (high immediacy) and the other one very rarely (low immediacy).

Respect The degree to which you admire another person and hold them in high esteem is known as **respect**. While respect and affinity might seem similar, they're actually different dimensions of a relationship.[3] If you have a charming coworker who goofs off on the job, you may like them but not respect them. Likewise, you might respect a boss or teacher's talents without liking them.[4] In our example, Lucien may have been disappointed that Haris declined lunch, but respected him as a busy adult with the right to spend time as he chooses.

Control In every conversation and every relationship, there is some distribution of **control**—the amount of influence exercised by the individuals involved. Control can be shared evenly among relational partners, or one person can have more or less than the other. An uneven distribution of control in some ways isn't necessarily problematic if it balances out in others, but relationships suffer if control is lopsided, as when one partner orders the other around or always gets their way.[5] In our example, perhaps Lucien invites Haris to lunch nearly every day and Haris sidestepped the invitation today to have more control over his own schedule.

In the end, we can speculate on Lucien and Haris's feelings, but we are only guessing. Keep this in mind when you evaluate the relational messages of people around you. The impatient tone of voice you take as a sign of anger might be due to fatigue, and the interruption you consider belittling might arise from enthusiasm about your idea. Ultimately, relational meanings are ambiguous and situational. It's a good idea to check your understanding before making assumptions about them.

---

content message A message that communicates information about the subject being discussed.

relational message A message that expresses the social relationship between two or more individuals.

affinity The degree to which people like or appreciate one another, whether they display that outwardly or not.

immediacy Expression of interest and attraction communicated verbally and/or nonverbally.

respect The degree to which a person holds another in esteem, whether or not they like them.

control The amount of influence one has over others.

8.2 Interpersonal Relationship Building

There are probably those who make you smile and, conversely, others who frustrate you on a regular basis. In this section, you will explore how people evaluate relationship potential and how they can develop closeness by talking about the underlying dynamics of their relationship and sharing personal information.

How People Choose Relational Partners

Considering the number of people with whom you communicate, truly interpersonal interaction is rather scarce. That isn't necessarily unfortunate. Most people don't have the time or energy to create personal relationships with everyone they encounter—or even to act in a personal way all the time with the people they know and love best. In fact, the scarcity of interpersonal communication contributes to its value. Like precious jewels and one-of-a-kind artwork, interpersonal relationships are special because they are rare.

Sometimes people don't have a choice about their relationships. Children can't select their parents, and most workers aren't able to choose their bosses or colleagues. In many other cases, though, you are likely to seek out some people and actively avoid others. Following are eight common reasons why you might establish a close relationship with someone.

You Have a Lot in Common People are typically drawn to those who remind them of themselves. Coworkers in one study were most likely to pursue friendships with colleagues if they felt they had a lot in common.[6] A quick survey of your friends probably shows that you share many common interests and perspectives. (Despite this tendency, you will see in a moment that there are also good reasons to make friends with people who seem dissimilar to you on the surface.)

You Balance Each Other Out The folk wisdom that "opposites attract" seems to contradict the similarity principle just described. In truth, both are valid. Differences strengthen a relationship when they are *complementary*—that is, when each partner's characteristics satisfy the other's needs. For example, when introverts and extroverts pair up as friends, they typically report that the quieter person serves as a steady anchor for the friendship, and the more gregarious partner propels the other to take part in activities they might otherwise avoid.[7]

You Like and Appreciate Each Other Of course, you aren't drawn to everyone who likes you, but to a great

A fist bump or pat on the back can convey "I like you" and "I enjoy being around you."

How do you show your friends that you like and appreciate them?

extent, you probably like people who like you and shy away from those who don't. It's no mystery why this is so. Approval tends to bolster people's self-esteem. It is rewarding in its own right, and it can confirm the part of your self-concept that says, "I'm a likable person."

You Admire Each Other It's natural to admire people who are highly competent in something you care about. Forming relationships with talented and accomplished people can inspire you and offer the validating knowledge that someone you admire admires you back.

You Open Up to Each Other People who reveal important information about themselves often seem especially likable, provided that what they share is appropriate to the situation and the stage of the relationship.[8] Self-disclosure is appealing partly because it's validating to know you aren't alone in your feelings ("I broke off an engagement, myself" or "I feel nervous with strangers, too"). And when people share private information, it suggests that they respect and trust each other. (You'll read more about self-disclosure later in the chapter.)

You Interact Frequently In many cases, proximity leads to liking.[9] For example, you're more likely to develop relationships with students who take the same classes as you than with those in different fields. Proximity allows you to learn more about each other and to engage in relationship-building behaviors. Plus, people in close proximity may be more similar than those who live, work, and play in different places. At the same time, social media allows you to experience "virtual proximity" with people online even if they are not physically nearby.

The Relationship Is Rewarding Relationships involve give and take, and the balance must be right. **Social exchange theory** proposes that people stay in relationships if the rewards are greater than or equal to the costs.[10] The rewards might include fun, favors, and feeling valued and admired. The costs might involve a sense of obligation, putting up with annoying habits, emotional pain, and so on. According to social exchange theorists, people use this formula (usually unconsciously) to decide whether dealing with another person is a "good deal" or "not worth the effort."

social exchange theory The idea that relationships are worth maintaining if the rewards are greater than or equal to the costs involved.

There's a Good Balance At this point, it may seem that individuals who are practically perfect have the best chance of building strong relationships. Actually, perfection isn't required or desirable (or even possible). Read the following before you fall into the trap of thinking you must be supermodel stunning, Mensa smart, or Olympic-level talented for people to find you appealing.

- *First impressions can mislead.* While it's true that people gravitate to those whose interests and attitudes seem similar to their own, they tend to overestimate how similar they are to their friends and underestimate how similar they are to people they don't know well.[11,12] In reality, there's strong evidence that superficial similarities such as appearance do not predict long-term happiness with a relationship, and when you're willing to communicate with a range of diverse people, you're likely to find that your differences are not as great as you thought.[13,14]

- *Perfection can be a turn-off.* People tend to like individuals who are attractive and talented, but be uncomfortable around those who seem *too* perfect. Let's face it, no one wants to look bad by comparison.

- *It's not all about communication, but it's a lot about communication.* The online dating service eHarmony matches people based on "29 dimensions of compatibility," and other online dating sites make similar promises. It's easy to imagine mathematical formulas for finding friends as well. However, the long-term success of people matched by computer algorithms is no greater than that of people who meet on their own.[15] That's because long-term compatibility relies less on superficial similarities and more on how people interact with each other once they start a relationship and encounter stressful issues.

A few lessons emerge from these observations. One is to break free of your comfort zone and give new people a chance. Another is that, when you are the person who seems different from others, you can help reduce the stranger barrier by being friendly and approachable and letting people get to know the real you. Finally, don't be discouraged if you aren't "perfect" by society's standards. Being perfect is overrated. Being nice matters more.

Metacommunication

Try to guess what the following statements have in common: "You seem upset." "I was only joking." "I appreciate your honesty." "I was confused by your text." The answer? They all involve metacommunication.

The term **metacommunication** describes messages that refer to other messages.[16] In other words, metacommunication is communication about communication. You are engaging in metacommunication when you say "I appreciate your honesty" or you text "jk" (just kidding). Here are some key things to know about metacommunication.

> **metacommunication**
> Messages that refer to other messages; communication about communication.

Metacommunication Can Bring Issues to the Surface With metacommunication, you can call attention to underlying meanings. You might say to a friend, "I think I offended you during our conversation the other day" or ask your partner "Are you quiet because I was late for our date?" In this way, you may be able to show that you care and are paying attention, and you can create opportunities to discuss how you both feel.

Metacommunication Can Be Risky As you have no doubt experienced, talking about underlying issues can make relational partners feel vulnerable. If one of you is offended or angry, you're both likely to be sensitive about it. It may help to realize that important issues in a relationship often lie beneath the surface, and ignoring them can make things worse. Every relationship involves some degree of conflict and misunderstanding. Successful partnerships stand out because the people involved are willing to do the repair work needed to address them, heal, and move on.[17]

The authors of *Difficult Conversations* offer a number of tips that may help:

- Take time to think about how you are feeling.
- Share your feelings honestly without blaming the other person.
- Listen rather than debating who is right or wrong.[18]

The good news is that, as risky as it feels in the moment, metacommunication can help clear up misunderstandings and keep resentment from festering.

Metacommunication Isn't Just for Problem Solving Metacommunication doesn't always involve heavy conversations. It's also a way to reinforce the good aspects of a

"There's something you need to know about me, Donna. I don't like people knowing things about me."

Source: Leo Cullum The New Yorker Collection/The Cartoon Bank

self-disclosure The process of deliberately revealing information about oneself that is significant and that would not normally be known by others.

relationship, as in "Thank you for praising my work in front of the boss." Comments such as this let others know that you value their behavior, and they boost the odds that the other person will continue that behavior in the future.

Self-Disclosure

"I don't have any secrets," some people claim. If that's true, they engage in a great deal of **self-disclosure**, which is the process of deliberately revealing significant information about oneself that would not normally be known by others.

Under the right conditions, self-disclosure is rewarding.[19] Talking about your feelings and experiences can help you understand yourself better and connect with others at a deeper level. It can be validating to know that others know and like the real you and are willing to share something about themselves in return. At the same time, however, self-disclosure involves an element of vulnerability. What you say might be used against you. It might also make others uncomfortable. You've probably experienced moments of TMI (too much information) when someone shared intimate details that made you feel overwhelmed or embarrassed. Here are two models of self-disclosure that offer insights about relationship development and self-disclosure.

Social Penetration Model Social psychologists Irwin Altman and Dalmas Taylor describe two ways in which communication can be more or less disclosive.[20] Their **social penetration model** (Figure 8.1) proposes that communication occurs within two dimensions: (a) *breadth*, which represents the range of subjects being discussed and (b) *depth*, which reflects how personal the information is. For example, as you start to reveal information about your personal life to coworkers—perhaps what you did over the weekend or stories about your family—the breadth of your disclosure is likely to increase. The depth may also expand if you shift from relatively nonrevealing messages ("I went out with friends.") to more personal ones ("I went on this awful date set up by my mom's friend. . . ."). Depth is also reflected in the difference between a relatively casual statement such as "I love my partner" and an emotionally vulnerable statement such as "I don't think my partner loves me back."

Each of your personal relationships has a different combination of breadth and depth. As relationships become more intimate, disclosure increases (usually gradually) in terms of both dimensions. Even so, it can be difficult to

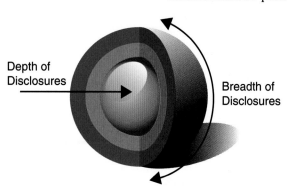

FIGURE 8.1 Social Penetration Model

know how much to share and how soon. As a rule of thumb, theorists recommend the "Goldilocks principle:" Pay attention to the other person's reaction to gauge if you are offering "too much" or "too little" and aim instead for "just right."[21]

The Johari Window   Another model that represents self-disclosure is the **Johari Window**, which describes information based on what individuals know and share about themselves.[22] Imagine a frame that contains everything there is to know about you: your likes and dislikes, your goals, your secrets, your needs—everything.

Of course, you aren't aware of everything about yourself. Like most people, you're probably discovering new things all the time. The Johari Window makes the most sense if you consider it in the context of particular relationships, since you share different information with different people. The Johari Window depicts what you know and think about yourself in quadrants on the left (Figure 8.2) and the parts you don't know about in quadrants on the right. You can also divide this frame in another way. The top row contains the things about you that the other person knows, and the bottom row things about you that you keep to yourself.

Altogether, the Johari Window presents everything about you divided into four parts. It's worth considering each block in the model separately.

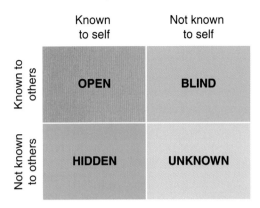

FIGURE 8.2   The Johari Window

social penetration model   A theory that describes how intimacy can be achieved via the breadth and depth of self-disclosure.

Johari Window   A model that describes the relationship between self-disclosure and self-awareness.

- The *open area* represents information about you that both you and your relational partner are aware of. For example, you may both know that you aspire to be a CEO one today.
- The *blind area* includes information you are unaware of but the other person knows. Perhaps you display talents that aren't obvious to you but are appreciated by the other person. You learn about information in the blind area primarily through feedback.
- The *hidden area* is made up of information that you know but aren't willing to reveal to others. Do you secretly have romantic feelings for the other person? Do you have more social anxiety than you let on? Information in this area may eventually become known to the other person if you self-disclose it.
- The *unknown area* represents information about you that neither you nor the other person knows. For example, even if you're afraid of public speaking now, you may surprise yourself and others by emerging as a confident speaker in the future. Or, a crisis or challenge may bring out aspects of your personality that you never knew were there.

The Johari Window presents three main implications for communication.

- Individuals keep some aspects of themselves hidden from others.
- Some secrets are okay, but interpersonal relationships of any depth require that individuals have some open area. You've probably found yourself in situations in which you felt frustrated trying to get to know someone who was very reserved.
- Disclosing too much too soon can make things uncomfortable.
- Through open communication, individuals are likely to learn things about themselves and their relational partners.

If you are a *Bachelor* or *Bachelorette* fan, you know the cringeworthy sensation of self-disclosure pushed to its limits. Reality TV manufactures such moments by pressuring strangers to develop intimate relationships very quickly with almost no privacy. But you needn't fall prey to the same traps in real life. As you can imagine, sharing every detail of your personal life with people you barely know usually isn't effective, but refusing to self-disclose can also hurt your relationships. The feature "Questions to Ask Yourself Before Self-Disclosing" presents eight considerations that can help you determine when and how self-disclosing may be beneficial to you and your relationships.

Interpersonal Communication Online

Maya and Jad live in different countries and know each other only online. Although they've never met in person, Maya considers Jad to be one of her closest friends. The two have shared personal experiences—including some they haven't shared with their RL (real life) friends. "I trust him enough to ask things I can't ask face-to-face," Maya says, "I like to think I have found a safe little corner where I can talk and joke with someone who I would not have met otherwise."[26]

Is the communication between Maya and Jad interpersonal? Scholars' early definitions specified that interpersonal communication could only be achieved in person.[27] Times have changed, however. Today, virtually everyone agrees that interpersonal communication can occur via technology—as through texts, emails, video chats, and social media.[28] Based on Maya's description, her online relationship with Jad does seem to be interpersonal. In this section, you will read about five of the most rewarding aspects of online interpersonal communication and two of its less appealing qualities.

Online Communication Helps People Stay Connected When their in-person routines don't match up well, technology can help people stay in touch. Evidence suggests that close relationships are often enhanced through regular use of instant messaging,[29] and email and video chats can help older adults who can't get around very well maintain relationships and receive social support.[30] Low mobility was a factor for nearly everyone during the COVID-19 pandemic, when people were asked mostly to stay at home and minimize personal contact. During that time, researchers found that people of all ages who used electronic means to stay in touch with others tended to be less lonely, anxious, and depressed than those who didn't.[31]

There's More Diversity Online Most face-to-face communication networks are limited to people in a relatively small geographic region. But the number of people you can befriend online is virtually endless. "I grew up in a fairly small town, so being a sci-fi and comics nerd who loved makeup, '80s

There are often advantages to communicating online with special people in your life.

Under what circumstances, if any, do you consider online interpersonal communication to be as good or better than face-to-face interactions?

and '90s pop music, fancy cake, and sushi pretty much made me a peer group of one," reflects Rachel. Online, however, she has access to "an entire world's worth of people," including many who share her interests and passions.[32]

Online Communication Can Feel Nonthreatening   When a researcher in Turkey interviewed college students from four different parts of the world, they said they would

---

COMMUNICATION STRATEGIES

Questions to Ask Yourself Before Self-Disclosing

No single style of self-disclosure is appropriate for every situation. However, there are some questions you can ask to determine when and how self-disclosure may be a good idea.[23]

Is the Other Person Important to You?

Disclosure may be the path toward developing a more personal relationship with someone. However, it can be a mistake to share personal information with people you don't know very well.

Is the Disclosure Appropriate?

This can be tricky to answer because appropriateness relies on personal preferences and culture. North Americans, with their individualistic orientation, are often comfortable disclosing personal information sooner than are people from more collectivistic cultures, such as Japan's.[24] As a result, North Americans may come off as overly personal, and they may assume that people who don't disclose as quickly are standoffish or uninterested.

Is the Risk of Disclosing Reasonable?

You're asking for trouble when you open up to someone who's likely to share your secrets with others or make fun of you. On the other hand, knowing that your relational partner is trustworthy and supportive makes it more reasonable to speak up.

Are the Amount and Type of Disclosure Appropriate?

Telling others about yourself isn't an all-or-nothing decision. Before sharing very important information with someone, test their reaction by disclosing something less personal first.

Is the Disclosure Relevant to the Situation at Hand?

A study of classroom communication revealed that it was ineffective for people to share all their feelings and be completely honest. A better alternative was to establish a relatively honest climate in which pleasant but less-than-fully disclosive relationships were the norm.[25] Even in close personal relationships, constant disclosure isn't a useful goal. Instead, the level of sharing in successful relationships rises and falls in cycles.

Is the Disclosure Reciprocated?

There's nothing quite like sharing vulnerable information about yourself only to discover that the other person is unwilling to do the same. Unequal self-disclosure creates an unbalanced relationship.

Will the Disclosure Be Constructive?

Self-disclosure can be a vicious tool if not used carefully. Disclosures that are needlessly hurtful toward others may damage your relationships. Comments such as "I've never thought you were very smart" may be devastating—to the listener and to the relationship.

Is the Self-Disclosure Clear and Fair?

Express yourself clearly and avoid assumptions. It's better to say "When you don't text me back I feel..." than to complain vaguely "When you avoid me..." It's also vital to express your thoughts and feelings clearly. "I feel like you no longer want to spend time with me" is more understandable than "I don't like the way things have been going."

like to have more international friends, but they felt anxious about approaching people from different cultures in person.[33] They were afraid they would say the wrong thing or not know what to say. The students were less anxious about communicating *online* with people from different cultures. After the chance to strike up cross-cultural friendships in cyberspace, most students in the study said they would like to meet their new international friends in person, providing evidence that mediated relationships can be a gateway to face-to-face interaction.

Online Communication Can Be Validating   One appealing quality of online technology is the opportunity it presents to receive social approval. Posting news of the A+ you earned in English is likely to be rewarded almost instantly with "likes" and congratulations. Considering that, have you ever pondered which has more impact online—"likes" or comments? Researchers who wondered the same thing asked more than 300 Facebook users to describe recent posts they had made based on memory alone. Participants were then asked to log onto Facebook and record how many likes and comments (positive, neutral, or negative) those posts garnered. It turns out that "likes" and supportive comments were both appreciated, but in different ways. Participants tended to gauge how interested and caring others were based on their *comments*. But they had a better memory for posts that received a high number of *likes*, and they considered those posts to be especially rewarding. One conclusion is that the personal nature of comments makes people feel cared about, whereas a high quantity of likes makes them feel important.

Online Communication Has a Pause Option . . . Sometimes   Many forms of online communication are asynchronous, meaning that they allow you to think about messages and then reply when you are ready. This is an advantage because you can catch mistakes or avoid blurting out something you might regret later. However, by the same token, asynchronous electronic communication can feel less spontaneous and more calculated. "Arguing over text messages is cheating," asserts one person, who feels that relational partners more easily mask their true feelings when technology is in the middle.

Online Communication Can Be Distracting   On the downside, social media use can interfere with in-person relationships. Paying more attention to your devices than to the people around you is known as **phubbing** (a combination of phoning and snubbing).[34] Researchers in one study found that the mere presence of a mobile device—even if it's sitting unused on the table—can have a negative effect on closeness, connection, and conversation quality while people discuss personal topics.[35]

Perceived Infidelity Can Be an Issue   Tara was furious when she found out that her boyfriend Michael had been looking up past girlfriends online. She saw it as cheating on their relationship, whereas Michael said he was just curious and had no intention of connecting with any of his old flames.[36]

There's no hard and fast rule about what qualifies as digital infidelity. The best bet is for romantic partners to discuss their expectations. Here are some questions to consider together:

- Does looking up previous romantic partners online constitute cheating to you? Why or why not?
- If someone in a committed relationship engages in romantic talk online with someone they will never meet in person, do you think that is cheating? Why or why not?
- How about posting sexually provocative comments or photos to no one in particular?

> **phubbing**   A mixture of the words *phoning* and *snubbing*, used to describe episodes in which people pay attention to their devices rather than to the people around them.

Too Much Online Communication Can Be Problematic In moderation, social media can boost feelings of connection and identity. However, it's possible to have thousands of superficial online "friendships" but few you can count on during hard times. As a consequence, communicators who spend excessive time online tend to be lonelier than their peers.[37] Considering the pros and cons of online communication, most experts agree that moderation is key and the best relationships often include both.

8.3 Communicating with Friends and Family

"You are more than just my best friend—you are the sister I never knew I needed— you are family." With these words, poet Marisa Donnelly acknowledges the powerful influence of both friends and family.[38] This section explores factors that influence those relationships.

From the outset, it's important to note that some people are both friend and family. Sometimes, a parent, uncle, or sibling is also a friend. At other times, as poet Marisa Donnelly reflects, "people with whom we don't share DNA or a roof over our heads" become so close that they are family.[39]

Friendships Have Unique Qualities

Everyone has a friend who has said or done something that made a powerful difference to them. Friendships are unique for a number of reasons:

- Friends typically treat each other as equals, unlike parent–child, teacher–student, or doctor–patient relationships, in which one partner has more authority or higher status than the other.[40]
- People can have as many friends as they want or have time for, in contrast to family and romantic relationships, which are limited in number.
- Friends are relatively free to design relationships that suit their needs. You may have close friends you talk to every day and others you see only once in a while.

Good friends help keep people healthy, boost their self-esteem, and make them feel loved and supported.[41,42] Friends also help one another adjust to new challenges and uncertainty.[43] It's not surprising then that people with strong and lasting friendships are typically happier than those without them.[44,45]

Most characters in the TV show *Friends* lived with or across the hall from each other.

How do you create opportunities to be with people whose interests are similar to yours?

Friendships Develop with Communication

What's a recipe for a good friendship? The simple answer is time, talk, and shared activities. In one study, a researcher tracked how people developed friendships after relocating to a new city and/or beginning college.[46] In both circumstances, it took an average of 50 hours of time with someone to create a casual friendship, 90 hours before they became real friends, and about 200 hours to become close friends.

How and where you spend time is another element of developing friendships. You might spend hours each day with acquaintances at work or school, but you wouldn't consider them friends. One key to developing a friendship is time spent together in different settings—like going out for coffee, seeing a concert together, or visiting each other's homes. Shared leisure activities are often a sign that a friendship is deepening.

Communication also plays an important role. Close friendships are characterized by self-disclosure and discussing personal issues. But everyday talk is also important. Catching up, checking in, and joking around are important ways to keep a friendship strong.

Think of several friends in your life—perhaps a new friend, a longstanding one, and a colleague at work. Then consider how they compare on the dimensions described here.

Short-Term Versus Long-Term Short-term friends tend to change as your life does. You may say goodbye because you move, graduate, or switch jobs. Or perhaps you spend less time at parties or on the ball field than you used to. It's natural for your social networks to change as a result. However, long-term friends are with you even when they aren't. Particularly today, with so many ways to stay in touch, people report that—as long as trust and a sense of connection are there—they feel as close to some of their long-term friends who live far away as to those who are nearby. For example, Natasha and Katie became friends in college and then launched careers in separate countries. But their friendship didn't end. "We might be in different jobs, different relationships and, quite literally, a world apart," Natasha says, "but ever since I've known [Katie] she's taught me what friendship really is. Timezones and distance are just numbers."[47]

Low Disclosure Versus High Disclosure Some of your friends know more about you than other friends do. Self-disclosure is associated with greater levels of intimacy such that only a few confidants are likely to know your deepest secrets. One interesting exception occurs among people who are highly self-disclosive online. They might announce personal news to hundreds of friends and acquaintances with a single post or tweet. This isn't necessarily bad. However, it's easy to cross the line and go public with information you might later wish you had kept private. As one blogger points out, you have several hundred online "friends," but not all of them need or want to hear that you were cheated on last night.[48]

Doing-Oriented Versus Being-Oriented Some friends experience closeness *in the doing*. That is, they enjoy performing tasks or attending events together and feel closer because of those shared experiences. In these cases, different friends are likely to be tied to particular interests—a golfing buddy or shopping partner, for example. Other friendships are *being-oriented*. For these friends, the main focus is on simply spending time together, and they meet up just to talk or hang out. Even long-distance friends may send texts or photos to "be" together when they are miles apart.

COMMUNICATION STRATEGIES

How to Be a Good Friend

Experts suggest the following communication strategies to keep your friendships strong.

Be a Good Listener

To show how much you care, put aside distractions and pay close attention to your friend's words and nonverbal cues. Chapter 6 offers many tips on how to be a good listener.

Give Advice Sparingly

Despite your good intentions, offering advice, especially when it's not requested, can come off as insensitive and condescending. A better option is to listen attentively and, if appropriate, ask your friend what options they can think of and what they consider the pros and cons of each.

Share Your Feelings Appropriately

When you feel good about your friendship, say so. And don't be reluctant to let your friend know when you feel hurt or frustrated. Making snide remarks, or saying "It's nothing" when you feel upset, is likely to damage your friendship.[50] Share your feelings without attacking the other person.

Be Validating and Appreciative

Find ways to let your friend know they matter to you. Depending on the relationship, this might involve a thoughtful phone call, spending time together, or remembering a birthday. The best validations come from knowing what your friend will appreciate most.

Apologize and Forgive

If you slip up—perhaps by forgetting an important date or saying something that embarrasses your friend—admit the mistake, apologize sincerely, and promise to do better in the future. Remembering your goof may inspire you to offer forgiveness, rather than harboring a grudge, the next time your friend makes a mistake.[51]

Keep Confidences

Two of the most dreaded violations of trust are revealing private information to others and saying unkind things about friends behind their back. Your friendships will grow and deepen when you demonstrate your loyalty by keeping confidences.

Give and Take Equally

The best friendships are characterized by equal give and take. One payoff of being a giver is the sense that you make a difference in someone's life, not just that someone makes a difference in yours.[52]

Stay Loyal in Hard Times

People who believe they can count on their friends typically experience less everyday stress and more physical and emotional resilience than others.[53] Saying "I'll always be there for you" and backing that up with attentive behaviors can make a world of difference.

Low Obligation Versus High Obligation There are probably some friends for whom you would do just about anything—no request is too big. For others, you may feel a lower sense of obligation. Culture may play a pivotal role. Friends raised in low-context cultures such as the United States' are more likely than those raised in high-context cultures such as China's to express their appreciation for a friend out loud (see Chapter 4). Friends who are Chinese are more likely to express themselves indirectly—mostly often by doing favors for friends and by showing gratitude and reciprocity when friends do favors for them.[49] It's easy to imagine the misunderstandings that might occur when one friend puts a high value on words and the other on actions.

UNDERSTANDING YOUR COMMUNICATION

What Kind of Friendship Do You Have?

Think of a particular friend and select the answers that most accurately describe your relationship.

1. Which of the following best describes the time you spend with this friend?
   a. We see each other a lot. I'd really miss our time together if something prevented that.
   b. Sometimes we spend time together and sometimes not. It's not a big deal either way.
   c. We don't see each other very often, but when we do, we're as in sync as if no time has passed.
   d. We haven't spent much time together yet, but I hope we will.

2. If you were on a long car ride together, what would you most likely talk about?
   a. Whatever is on my mind. I can tell this friend anything.
   b. Current events or what we've been up to at school or work.
   c. Funny memories. We've had many adventures together through the years.
   d. Where we were raised, what we're studying in school, and other topics to get to know each other better.

3. If your friend were in bed with the flu for several days, what you would be most likely to do?
   a. Stop by to cheer them up and help out.
   b. Send a "get well soon" text.
   c. Call to say I wish I could be there in person.
   d. I probably wouldn't know about it until later.

4. If this friend said something that hurt your feelings, what would you probably do?
   a. Talk about it together and repair the rift.
   b. Avoid them for a while.
   c. Let it go. It's nothing compared to all we've been through together.
   d. Rethink my desire to be friends.

INTERPRETING YOUR RESPONSES

Read the following explanations to reflect on the qualities of your friendship. More than one may apply.

Loyal

If you answered "a" more than once, this is close a friend you can count on. You are likely to disclose a great deal to each other and to back each other up, even when things are difficult. This is likely to be a long-term relationship with the rewards that come from knowing that someone knows you well and is there for you no matter what.

Independent

If you answered "b" two or more times, this is a friendship that doesn't require a great deal of commitment. You are able to get together when you like without feeling a strong sense of obligation to do so. Although not as close as some friendships, this one may be valuable, particularly if other obligations claim a lot of your time right now. Not every one has to be a best friend. Just be careful not to let your desire for independence keep you from forming strong bonds with one or two friends you can always count on.

Far Yet Close

If you answered "c" more than once, this seems to be an enduring, long-term friendship that remains strong even though you're not able to be together in person as much as you would like. You are likely to enjoy the benefits of being emotionally connected without much obligation to do things together or for each other. It can be a great feeling to know that distance and time cannot dim the memories you have shared together. At the same time, be sure not to take this friendship for granted. A thoughtful text or call may help you feel close even when you're not together physically.

Evolving

If you answered "d" two or more times, it's likely that your friendship is still developing. Your sense of obligation is likely to be low at this stage, as you venture to disclose more about yourselves to each other. It remains to be seen if this will be a short- or long-term relationship, but the benefits of having strong friendships suggest that it may be worth the effort to find out.

Frequent Contact Versus Occasional Contact You probably keep in close touch with some friends. Perhaps you work out, travel, socialize, or FaceTime daily with them. But you might connect with other friends only at reunions or via occasional phone calls or text messages. Some friends go years without seeing each other and then reconnect as if they were never apart.

Regardless of the type of friendship you have, the feature "How to Be a Good Friend" offers ways you can nurture it. And the self-assessment quiz "What Kind of Friendship Do You Have?" can help you appreciate the different types of friends in your life.

Friendships Can Build Bridges

You learned earlier in the chapter that people gravitate to those who seem similar to them. That may be a natural inclination, but there's a lot to be gained by broadening your notion of who might make a good friend.

"We went to school where I was in the one percent who were black; Amy was in the one percent Jewish," says Malaika Adero, describing a lifelong friendship that began in the fourth grade.[54] As they have learned, it can be mind-opening to be friends with people from different backgrounds.

The **intergroup contact hypothesis** reflects evidence that prejudice tends to diminish when people communicate with individuals they might otherwise stereotype.[55] For example, a study in the United States showed that college students who have Muslim friends usually have a higher opinion of Muslims than those who don't.[56] Positive feelings usually increase after even a small amount of interaction and continue to grow as friends spent more time together. A similar pattern has been noted between people of different races,[57] abilities,[58] ages,[59] nationalities,[60] and genders.[61] Intergroup friendships reduce prejudice in three main ways, each of which involves communication.

> **intergroup contact hypothesis** A proposition based on evidence that prejudice tends to diminish when people have personal contact with those they might otherwise stereotype.

Stereotypes Fade As you get to know strangers, preconceived notions about them tend to fall away. Chiarra, an exchange student from Italy, says she was surprised to find that college students in the United States don't all "wear high-fashion clothes and live in huge houses with swimming pools." As she made friends with Americans, Chiarra realized that they have many different personalities and lifestyles. "When I get back to Italy I'll object when people say things about Americans based on TV shows. I know better now," Chiarra says.

Trust Grows Fear and distrust tend to diminish as you realize how much you have in common with people who seem different from you on the surface.[62] For this reason, some people say the best antidote to racism is intergroup friendship.[63,64] Musician and author Daryl Davis, who is Black, spent 30 years befriending members of the

Stereotypes usually dissolve when people get to know each other. *How diverse is your social network?*

Can you think of at least one person who seems different from you on the surface whom you'd like to know better? How might you accomplish that?

Ku Klux Klan, a racist hate group. Many of the Klan members whom Davis met had never talked to a Black person before. He reasoned that, if he could lessen people's fear of the unknown by communicating with them, their hate would fade as well. An estimated 200 KKK members left the group after getting to know Davis.[65] Few people may go as far as Davis did, but the power of friendship in dispelling fear and distrust cannot be overstated.

Understanding Blooms Learning about another person can lead to greater appreciation for them and a better understanding of the challenges they face.[66] Since becoming close friends with Keith, who is paralyzed, Leah says she knows firsthand how much pain he endures yet how optimistic and energetic he is.[67] She reflects that interacting with Keith has changed the way she approaches other people with disabilities.

The implications of intergroup friendships can be far reaching. People who let go of their stereotypes about one group are often less prejudiced toward other groups as well.[68] As a result, they're more likely to select qualified members of nondominant groups for jobs, promotions, and education opportunities.[69,70] As Jeffrey Tucker puts it: "The more diverse friendship networks we cultivate, the less we think in categories of 'us vs them.'"[71]

Even knowing the importance of intergroup friendships, you might feel shy about initiating them. It's natural to worry that you might say the wrong thing. The feature "How to Make Friends with a Wide Range of People" offers some tips for overcoming your fears.

COMMUNICATION STRATEGIES

How to Make Friends with a Wide Range of People

Jen is 19 years older than her good friend Melissa, but they both love hiking, spa days, and sharing adventures together. Jen says that Melissa keeps her young, and Melissa says that Jen is always there for her.[72] Friendships with a diverse array of people can be rewarding, but it may seem hard to bridge your differences at first. Here are some tips to help.

"Accept the Awkward"

That's the advice of experts who acknowledge that it's natural to feel self-conscious when approaching new people—but who encourage you to do it anyway. "Guess what? You can't have friends without getting vulnerable," says one advisor.[73] A good strategy is to introduce yourself and ask a few nonthreatening questions such as "What are you studying in school?" or "Are you a fan of this artist? . . . Me too!"

Don't Fixate on Differences

Focus on what you have in common, not just on what makes you dissimilar. For example, if you meet someone who is substantially older or younger than you, don't let that distract you from appreciating them as a unique person. "We're all just people, and we all have our favorite things, our fears, and our shared humanity," says writer Alice Zhang, who encourages others, "Over time, you'll come to see them as a 'friend,' not your 'older friend' or 'younger friend.'"[74]

Be Approachable

When you are the one who stands out as different, you can help reduce the stranger barrier by smiling, being friendly, and letting people get to know the real you. Keep in mind that other people probably feel as intimidated as you do, but almost everyone likes to make a new friend.

Family Relationships

Defining what makes a family isn't a simple matter. This book adopts the position of theorist Martha Minnow, who proposes that people who share affection and resources as a family and who think of themselves and present themselves as a family *are* a **family**.[75] Your own experiences probably tell you that this concept of a family might encompass (or exclude) bloodline relatives, adopted family members, stepparents, honorary aunts and uncles, and blended families in which the siblings were born to different parents. This makes it easy to understand why people can be hurt by questions such as "Is he your natural son?" and "Is she your real mother?" Calling some family members "natural" or "real" implies that others are fake or that they don't belong.[76]

family People who share affection and resources and who think of themselves and present themselves as a family, regardless of genetics.

Parenting Relationships   Power and influence play a role in any relationship, but especially in the communication dynamic between parents and children. Imagine establishing the curfew for teenage members of a family as you focus on the following parenting styles.[77]

- **Authoritarian** parents are strict and expect unquestioning obedience. You might characterize this as a "do it because I said so" style consistent with a *conformity* approach. In the curfew example, teens would be expected to follow their parents' rules, beliefs, and values without challenging them.[78]

authoritarian An approach in which parents (or other types of leaders) are strict and expect unquestioning obedience.

- **Authoritative** parents are also firm, clear, and strict, but they encourage children to communicate openly with them. These parents have high expectations, but they are willing to discuss them and to listen to children's input. This family communication pattern emphasizes *conversation*. Teens and their parents would probably negotiate the curfew by talking openly about it and listening to each other.

authoritative An approach in which parents (or other types of leaders) are firm, clear, and strict, while encouraging open communication.

- **Permissive** parents do not require children to follow many rules. Based on this approach, parents and children may communicate about other topics, but they probably don't spend a lot of time setting firm guidelines such as curfews.

Most evidence suggests that children who grow up with an authoritarian/conformity pattern don't get much experience sharing their emotions and negotiating with authority figures, so it may take work to develop these skills as adults.[79]

permissive An approach in which parents do not require children to follow many rules.

Children who grow up with authoritative parents are typically more adept at expressing their emotions confidently and effectively as they grow older.[80] As two researchers in this field put it, "authoritative parents provide the dual benefits of structure and compassion—they are warm, responsive, assertive without being overly intrusive or restrictive."[81]

As you might imagine, children who don't engage in much give-and-take communication about rules as they grow up are usually less comfortable negotiating expectations in other relationships.

Siblings   In the midst of what one theorist calls the "playing and arguing, joking and bickering, caring and fighting"[82] of sibling life, people learn a great deal about themselves and how to relate to others.

Just as people who think of themselves as family *are* family, a sibling relationship isn't limited to people who share biological parents. It's more about shared life experiences. As one person who grew up in a blended family puts it, "You feel

In the 2022 movie *Cheaper by the Dozen*, 10 children in a blended family negotiate when to assert their independence, when to compete, and when to cooperate.

If there are people in your life whom you regard as siblings, are you mostly supportive, longing, competitive, apathetic, or hostile toward one another? Do you spend enough time together, or do you long for more?

supportive siblings Those who talk regularly and are accessible and emotionally close to one another.

longing siblings Those who admire and respect one another but interact less frequently and with less depth than they would like.

competitive siblings Those who consider themselves to be rivals vying for scarce resources such as their parents' time and respect.

beyond annoyed when explaining your family structure and someone says, 'Oh, so you're only half sisters.'" It's tempting, she says, to reply, "Only? ONLY? Well, you're my half-friend now."[83]

Whatever the origin, sibling relationships involve an interwoven, and often paradoxical, collection of emotions. Children are likely to feel both intense loyalty and fierce competition with their siblings and to be both loving and antagonistic toward them. Here are five types of sibling relationships people might settle into as they become adults.[84]

- **Supportive siblings** talk regularly and consider themselves to be accessible and emotionally close to one another. Mutually supportive relationships are most common among siblings who are close in age.[85]
- **Longing siblings** typically admire and respect one another. However, they interact less frequently and with less depth than they would like. This can be especially difficult for younger siblings who watch older ones move out. One teen lamented when his brother left home: "[I] look at his empty desk, the table where we would sit and talk, and start bawling... I know I'll see him again, but nothing will be the same."[86]
- **Competitive siblings** behave as rivals, vying for scarce resources such as their parents' time and respect.[87] It's not uncommon for siblings to feel competitive as they grow up. That feeling sometimes extends into adulthood, especially if they perceive that their parents continue to play favorites.[88,89]

> **COMMUNICATION STRATEGIES**
>
> ## Strengthening Family Ties
>
> Communicating with family members can be a joy and a challenge. Following are some strategies for successful communication.
>
> **Share Family Stories**
>
> Family stories contribute to a shared sense of identity. They also convey that adversity is an inevitable part of life, and they can suggest strategies for overcoming it.[91]
>
> **Listen to Each Other**
>
> People who are involved in reflection and conversation learn how to manage and express their feelings better than people who don't. They tend to have better relationships as a result.[92]
>
> **Negotiate Privacy Rules**
>
> Privacy violations among family members can have serious consequences.[93] At the same time, too much privacy can lead family members to overlook dangerous behavior and avoid distressing but important topics. Experts suggest that families talk about and agree on privacy expectations and rules.
>
> **Coach Conflict Management**
>
> Effective conflict management doesn't just happen spontaneously. It's a sophisticated process that often goes against people's fight-or-flight instincts. Families can help by creating safe environments for discussing issues and striving for mutually agreeable solutions.
>
> **Go Heavy on Confirming Messages**
>
> Supportive messages from family members can give individuals the confidence to believe in themselves. Compliments such as "You're a very thoughtful person" and "I know you will do a great job" tend to be self-fulfilling. For example, teens and young adults whose parents compliment and encourage them are less likely than others to experience mental illness.[94]
>
> **Have Fun Together**
>
> Happy families make it a point to minimize distractions and spend time together on a regular basis. They establish togetherness rituals that suit their busy lives, such as sharing dessert even when they can't eat dinner together,[95] and they share in adventures, both large and small.

- **Apathetic siblings** are relatively indifferent toward one another. They communicate with one another only on special occasions, such as holidays or weddings.
- **Hostile siblings** harbor animosity toward each other and often stop communicating.[90] Unlike apathetic siblings, who may drift apart without hard feelings, hostile siblings usually feel a sense of jealousy, resentment, and anger toward one another.

See the feature "Strengthening Family Ties" for tips on communicating well with the people you call family.

apathetic siblings Those who are relatively indifferent toward one another and seldom or never communicate.

hostile siblings Those who harbor animosity toward each other and often stop communicating.

8.4 Communicating with Romantic Partners

Romantic love is the stuff of songs, fairytales, and happy endings. So it might surprise you that the butterflies-in-your-belly sense of romantic bliss isn't a great predictor of happiness. A much better indicator is the effort that couples put into their communication. Factors such as trust, agreeableness, and emotional expressiveness are primarily responsible for long-term relationship success.[96,97] This section explores the role of communication in forming and sustaining romantic relationships.

Stages of Romantic Relationships

Some romances ignite quickly, whereas others grow gradually. Either way, couples are likely to progress through a series of stages as they define what they mean to each other and what they expect in terms of shared activities, exclusivity, commitment, and their public identity.

One of the best-known explanations of how communication operates in different phases of a relationship was developed by communication scholar Mark Knapp. His **developmental model** depicts five stages of intimacy development (coming together) and five stages in which people distance themselves from each other (coming apart).[98] Other researchers have suggested that the middle phases of the model can also be understood as relational maintenance—keeping stable relationships operating smoothly and satisfactorily.[99] Figure 8.3 shows how Knapp's 10 stages fit into this three-part view of communication in relationships. Consider how well these stages reflect communication in the close relationships you have experienced.

developmental model (of relational maintenance) Theoretical framework based on the idea that communication patterns are different in various stages of interpersonal relationships.

Initiating The initiating stage occurs when people first encounter one another. In person, this might involve simple conversation openers such as "It's nice to meet you" and "How's it going?" People may also send electronic messages to initiate contact. For example, after noticing Spencer at the gym, Cate sent a private message to him via Instagram to say hi and introduce herself.[100] In this stage, people form first impressions and have the opportunity to present themselves in an appealing manner.

Experimenting People enter the experimenting stage when they begin to get acquainted by asking questions such as "Where are you from?" and "What do you do?" Viewing a person's social media profiles can also offer insight. Although small talk and information in online profiles may seem relatively superficial, they provide information that helps people decide who they would like to get to know better.

In the experimenting stage, face-to-face interactions tend to have an edge over online communication. When researchers asked college students to meet a new person either in person or via technology, those in the face-to-face group more often said they liked the new person, felt they had a lot in common, and wanted to know them better.[101]

Communication in the initiating and experimenting stage doesn't qualify as interpersonal since the participants have not yet formed an emotional connection, but it may lay the foundation for a closer relationship in the future.

Intensifying In the intensifying stage, truly interpersonal relationships develop as people begin to express how they feel about each other. Romantic intimacy usually requires that partners express themselves through some combination of physical contact, shared experiences, intellectual sharing, and emotional disclosures.[102] If couples are in sync, they may experience heightened intimacy at this stage. But if they aren't, they may feel either pressured or rejected by the other person.

Dating couples often navigate the uncertainty of the intensifying stage by flirting, hinting around, asking hypothetical questions, giving compliments, and being more affectionate than before. They may become bolder and more direct if their partners seem receptive to these gestures.[103]

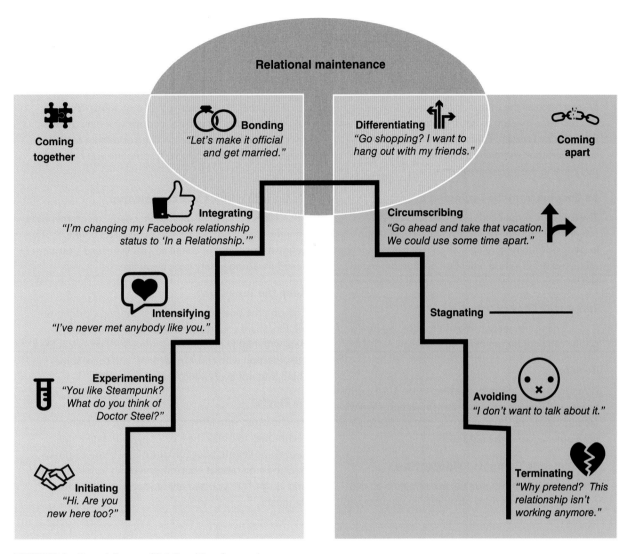

FIGURE 8.3 Knapp's Stages of Relational Development

Even though online relationships may be slow to start, they typically progress just as quickly as other relationships once the partners make contact via video chats or in-person meetings.[104] Switching from cyberspace to face-to-face can be intimidating, however. For experts' suggestions, see the feature "Meeting an Online Date for the First Time."

Integrating In the integration stage, couples begin to take on an identity as a social unit. And as it becomes a given that they will share resources and assistance, partners become comfortable making straightforward requests of each other. They're likely to be pictured together in social media and to meet each other's families. Couples begin to share possessions and memories—our apartment, our dog, our song. As one observer puts it, you know a relationship is on stable ground when you can have an argument but know that you'll still be together afterward.[112]

COMMUNICATION STRATEGIES

Meeting an Online Date for the First Time

Researchers call the transition from online-only chatting to a real-time meetup *modality switching*. Daters often just call it *awkward*.[105] Following are some tips from communication researchers and dating experts to help put you and the other person at ease.

Be Genuine from the Beginning

The computer-enhanced selfie that looks great on screen may cause an online admirer to be disappointed in person. Ultimately, relationships fare best when your mediated self is a close reflection of your in-person self.[106]

Talk on the Phone First

If your online connection shows promise, see how a phone call goes before meeting in person.

Be Safe

Arrange to meet in a public setting rather than a private home, and provide your own transportation. (Avoid your favorite hangouts just in case you'd rather not run into each other again in the future.)[107]

Put Romantic Thoughts Aside for Now

It may sound counterintuitive, but as dating advice columnist Jonathan Aslay observes, "most successful long-term relationships are built on a solid friendship."[108] Approaching the encounter as a new friendship can be less anxiety provoking and more realistic than expecting fireworks with someone you barely know.

Begin with a Quick and Easy Meetup

"Don't meet for a meal on your first date," recommends Jennifer Flaa, who met her future husband online.[109] She suggests meeting for coffee or a drink instead. You can schedule a longer date later, if you like.

Keep the Tone Light

This isn't the time to make an over-the-top clothing selection, share your deepest secrets, or interrogate your date with highly personal questions. Instead, encourage conversation with casual questions, and share something (but not everything) about yourself.

Be Playful

Psychologist Seth Meyers suggests the following techniques to break the ice:[110] pay your date a sincere compliment, pose fun questions, share a lighthearted story about yourself, and don't forget to smile and laugh. "Let yourself enjoy your first dates," he suggests.[111]

Bonding The bonding stage is likely to involve a wedding, a commitment ceremony, or some other public means of communicating to the world that this is a relationship meant to last. Bonding generates social support for the relationship and demonstrates a strong sense of commitment and exclusivity.

Not all relationships last forever, however. And even when the bonds between partners are strong and enduring, it's sometimes desirable to create some distance. The following stages accomplish that.

Differentiating In the differentiating stage, the emphasis shifts from "how we are alike" to "how we are different." For example, a couple who moves in together may find that they have different expectations about doing chores, sleeping late, what to watch on TV, and so on. This doesn't necessarily mean the relationship is doomed. Differences remind partners that they are distinct individuals. To maintain this balance, partners may claim different areas of the home for their private use and reduce their use of nicknames, gestures, and words that distinguish the relationship as intimate and unique.[113]

Circumscribing In this stage, communication decreases significantly in quantity and quality. Rather than discuss a disagreement, which requires some degree of energy on both people's part, partners may withdraw mentally by using silence, daydreaming, or fantasizing. They may also withdraw physically by spending less time together, indicating reduced interest and commitment.

Stagnating If circumscribing continues, the relationship may begin to stagnate. In this stage, partners behave toward each other in familiar ways without much feeling. Like workers who have lost interest in their jobs yet continue to go through the motions, some couples unenthusiastically repeat the same conversations, see the same people, and follow the same routines without any sense of joy or novelty.

Avoiding When stagnation becomes too unpleasant, partners distance themselves in more overt ways. They might block or unfollow each other on social media, use excuses such as "I've been busy lately," or make direct requests, such as "Please don't come over. I don't want to talk to you."

At this stage, a relationship is in trouble, but it isn't necessarily over. Partners may recognize the warning signs in time to reverse the trend.

Terminating Characteristics of this final stage include conversations about where the relationship went wrong and the desire to break up. The relationship may end in many ways—with a cordial dinner, a text, a note left on the table, or a legal document stating the dissolution.

Using email and instant messaging to end intimate relationships is popular because technology makes it easy—some say too easy—to call the whole thing off without an emotional face-to-face encounter. That can be a wise choice if the other person may become abusive, but in most instances, it's considered more respectful to talk in person so that you both have a chance to express yourself and ask questions. Breaking up in person also makes it more likely that you can be friends (or even get back together) in the future. By contrast, says one advice columnist, the "rude jolt" and "bitter after taste" of a break-up text may create resentment that never goes away.[114]

Not every breakup is forever. One key difference between couples who get together again after a breakup and those who go their separate ways is how well they communicate about their dissatisfaction and negotiate for a mutually appealing fresh start. Unsuccessful couples deal with their problems by avoidance, indirectness, and reduced involvement with each other. By contrast, couples who repair their relationships more often air their concerns and spend time and effort negotiating solutions to their problems.

You may assume that you can predict what will happen next if your relationship shows signs of coming together or coming apart. But couples differ in terms of how quickly they move through the stages, and just because you've drifted apart lately, the relationship isn't necessarily hopeless. Communication plays a significant role in the process. A number of practical lessons emerge from the developmental perspective:

- *Each stage requires different types of communication.* If you don't find yourself sharing secrets on your first date, don't worry. Partners may find that talking about highly personal issues deepens their bond in the intensifying stage but can be overwhelming in earlier stages. Likewise, the polite behavior of the first two stages may seem cool and distant as intimacy increases. Every relationship, and every stage of involvement, has its own pace and rhythm.

- *Partners can change the direction a relationship is headed.* They may recognize the early signs of coming apart in time to reverse the trend. For example, if they realize they are differentiating or stagnating, they might refresh their relationship by doing more of the things they did while coming together, such as going on dates, sharing feelings, and pursuing new experiences together.
- *Relational development involves risk and vulnerability.* At any stage—even those associated with coming together—the relationship may falter. Intimacy only evolves if people are willing to take the chance of becoming gradually more self-disclosive.[115] Your knowledge of relational stages can help you understand whether the relationship is trending more toward the positive or the negative.

Love Languages

Relationship counselor Gary Chapman observes that people typically orient differently to five love languages.[116]

Affirming Words This language includes compliments, thanks, and statements that express love and commitment. Even when you know someone loves and values you, it's nice to hear it in words. The happiest couples continue to flirt with each other, even after they have been together for many years.[117]

Quality Time Some people show love by completing tasks together, talking, or engaging in some other mutually enjoyable activity. The good news is that, even when people can't be together physically, talking about quality time can be an important means of expressing love. For example, partners separated by military deployments often say they feel closer to each other just talking about everyday activities and future plans.[118]

Sometimes a love language that resonates with one partner isn't meaningful to the other, which can lead to hurt feelings.

What's your main love language? How well does it match with the preferences of special people in your life? How can you communicate in ways that are meaningful to people you care about?

Acts of Service People may show love by performing favors, such as caring for each other when they are sick, doing the dishes, or making meals. Committed couples report that sharing daily tasks is one of the most important ways that they show love and commitment.[119] Although each person need not contribute in exactly the same ways, an overall sense that they are putting forth equal effort is essential to long-term happiness.[120]

Gifts It's no coincidence that many people buy gifts for loved ones on holidays, birthdays, anniversaries, and other occasions. For some people, receiving a gift—even an inexpensive or free one such as a flower from the garden or a handmade card—adds to their sense of being loved and valued.[121]

Physical Touch Loving touch may involve a hug, a kiss, a pat on the back, or sex. Touch is potent even in long-term

UNDERSTANDING YOUR COMMUNICATION

What Is Your Love Language?

Answer these questions to learn more about the love languages you prefer. If you're in a romantic relationship, consider inviting your partner to answer the same questions and then compare your responses. You may be surprised to find they aren't identical.

1. You have had a stressful time working on a team project. The best thing your romantic partner can do for you is:
   a. Set aside distractions to spend some time with you
   b. Do your chores so you can relax
   c. Give you a big hug
   d. Pamper you with a dessert you love
   e. Tell you the team is lucky to have someone as talented as you

2. What is your favorite way to show that you care?
   a. Go somewhere special together
   b. Do a favor without being asked
   c. Hold hands and sit close together
   d. Surprise your romantic partner with a little treat
   e. Tell your loved one how you feel in writing

3. With which of the following do you most agree?
   a. The most lovable thing someone can do is give you their undivided attention.
   b. Actions speak louder than words.
   c. A loving touch says more than words can express.
   d. Your dearest possessions are things loved ones have given you.
   e. People don't say "I love you" nearly enough.

4. Your anniversary is coming up. Which of the following appeals to you most?
   a. An afternoon together, just the two of you
   b. A romantic, home-cooked dinner (you don't have to lift a finger)
   c. A relaxing massage by candlelight
   d. A photo album of good times you have shared
   e. A homemade card that lists the qualities your romantic partner loves about you

INTERPRETING YOUR RESPONSES

For insight about your primary love languages, see which of the following best describes your answers.

Quality Time

If you answered "a" to one or more questions, you feel loved when people set aside life's distractions to spend time with you. Keep in mind that everyone defines quality time a bit differently. It may mean a thoughtful phone call during a busy day, a picnic in the park, or a few minutes every evening to share news about the day.

Acts of Service

Answering "b" means you feel loved when people do thoughtful things for you, such as washing your car, helping you with a repair job, bringing you breakfast in bed, or bathing the children so you can put your feet up. Even small gestures say "I love you" to people whose love language involves acts of service.

Physical Touch

Options labeled "c" are associated with the comfort and pleasure you get from physical affection. If your partner texts to say, "Wish we were snuggled up together!" they are speaking the language of touch.

Gifts

If you chose "d," you treasure thoughtful gifts from loved ones. Your prized possessions are likely to include items that look inconsequential to others but have sentimental value to you because of who gave them to you.

Words of Affirmation

Options labeled "e" refer to words that make us feel loved and valued, perhaps in a card, a song, or a text. To people who speak this love language, hearing that they are loved (and why) is the sweetest message imaginable.

relationships. Researchers in one study asked married couples to increase the number of times they cuddled for four weeks.[122] On average, the couples reported greater satisfaction with their relationships by the end of the study.

A good deal of research supports the importance of love languages in promoting harmony.[123,124,125] Most people value all the languages to some degree, but they give some greater weight than others. Good intentions may lead you astray if you assume that your partner feels the same way you do. The golden rule—do unto others as you would have them do unto you—isn't very useful when your partner's love language preferences differ from yours.[126]

You can learn more about your preferences by completing the self-assessment "What Is Your Love Language?"

8.5 Relational Dialectics

Relationships can feel like a balancing act. You want connection, but also independence. You want to share your thoughts and feelings, but also to have some privacy. You want the relationship to be fresh, but still have predictability you can count on.

The model of **relational dialectics** suggests that partners in every close relationship constantly seek a balance between opposing forces such as togetherness versus independence, sharing versus privacy, and comfortable routines versus new adventures.[127] As you read about each set of opposing needs, consider how they operate in your life.

> **relational dialectics**
> The perspective that partners in interpersonal relationships deal with simultaneous and opposing forces of connection versus autonomy, predictability versus novelty, and openness versus privacy.

Connection Versus Autonomy

The conflicting desires for togetherness and independence are embodied in the *connection–autonomy dialectic*. One of the most common reasons for breaking up is that one partner doesn't satisfy the other's need for connection:[128]

> "We barely spent any time together."
> "My partner wasn't committed to the friendship."
> "We had different needs."

But relationships can also split up for the opposite reason. One, or perhaps both partners may feel stifled by what seem like excessive demands for staying connected.[129] In this case, complaints may sound like this:

> "I felt trapped."
> "I needed freedom."
> "My partner was too needy."

Individuals are faced with this dilemma even within themselves. You may desire intimacy, but also feel the need for some time to yourself.

At different stages, the desire for connection or autonomy can change. In the classic book *Intimate Behavior*, Desmond Morris suggests that people in close relationships repeatedly go through three stages:[130]

> "Hold me tight."
> "Put me down."
> "Leave me alone."

In marriages and other committed relationships, for example, the "Hold me tight" bonds of the first year are often followed by a desire for independence. This need for autonomy can manifest in many ways, such as making friends or engaging in activities that don't include one's partner or making a career move that might disrupt the relationship. Movement toward autonomy may lead to a breakup, but it can also be part of a cycle that redefines the relationship and allows people to recapture or even surpass the closeness they had previously. For example, you might find that spending some time apart makes you miss and appreciate your partner more than ever.

Openness Versus Privacy

Self-disclosure is one characteristic of interpersonal relationships. Yet it's also important to maintain some emotional space between yourself and others. These sometimes-conflicting drives create the *openness–privacy dialectic*. When the drive for openness is strong, you might crave the closeness that comes from sharing personal information. The following statements are clues that one person (or both) desires openness in the relationship:

> "What's on your mind?"
> "How are you really feeling?"
> "Tell me more."

When couples desire the same amount of openness, they may share information with pleasure. But if one partner craves more privacy than the other, they might be frustrated into thinking or saying:

> "Why are you interrogating me?"
> "Stop pushing so hard!"
> "Don't try to read my mind!"

Predictability Versus Novelty

Sharing new experiences can keep a relationship fresh and exciting. At the same time, shared routines can create a sense of security. These opposing needs represent different ends of the *predictability–novelty dialectic*. Too much predictability can sap the excitement out of a relationship, whereas too much novelty can lead to uncertainty and insecurity. "I don't know who you are anymore," you might say or hear.

Source: ©2006 Zits Partnership Distributed by King Features Syndicate Inc.

People differ in their desire for stability and surprises—even from one time to another. The classic example is becoming engaged just before graduation or military deployment, when life may seem particularly novel and uncertain. Commitment may counterbalance some of the uncertainty people feel in those situations. However, things may feel *too* predictable once life settles into a new routine.

All in all, dialectical tensions are a fact of life in close relationships. In this sense, "happily ever" never happens. Here's the reality:

- *Relationships involve continual change and negotiation.* Relational partners who understand dialectical tensions can give up the unrealistic notion that they will always be in sync or that negotiating relationship options should be effortless.
- *Partners can be in sync in some ways, but not in others.* Recognizing different dialectical tensions may help you identify what's going on when you feel the tension of opposing drives.

COMMUNICATION STRATEGIES

Managing Dialectical Tensions

As you have probably experienced, some strategies for managing dialectical tensions are more productive than others. Consider which of these you might avoid and which you'd like to use more.[131]

Denial

One of the least functional responses to dialectical tensions is to deny that they exist. People in denial insist that everything is fine even if it isn't. They may refuse to deal with conflict, ignoring problems or pretending that they agree about everything.

Selection

When partners employ the strategy of selection, they respond to one end of the dialectical spectrum and ignore the other. For example, a couple caught between the conflicting desires for stability and novelty may decide that predictability is the "right" or "responsible" choice and put aside their longing for excitement.

Moderation

The moderation strategy is characterized by compromises in which couples back off from expressing either end of the dialectical spectrum. A couple might decide that taking separate vacations is too extreme for them, but they will make room for some alone time while they are traveling together.

Alternation

Communicators sometimes alternate between one end of the dialectical spectrum and the other. For example, partners may spend time apart during the week but reserve weekends for couple time.

Polarization

In some cases, couples find a balance of sorts by each staking a claim at opposite ends of a dialectical continuum. For example, one partner might give up nearly all personal interests in the name of togetherness, while the other maintains an equally extreme commitment to being independent. In the classic *demand–withdraw pattern*, the more one partner insists on closeness the more the other feels suffocated and craves distance.[132]

Reframing

Communicators can also respond to dialectical challenges by reframing them in ways that redefine the situation so that the apparent contradiction disappears. Consider partners who regard the inevitable challenges of managing dialectical tensions as "exciting opportunities to grow" instead of as "relational problems."

- *Some approaches are more conducive than others to relational satisfaction.* The feature "Managing Dialectical Tensions" outlines some of the ways people (deliberately or unconsciously) negotiate a place on dialectic continua.

After reviewing why and how people develop interpersonal relationships, you may be in a better position to talk about your impressions, share personal information effectively, and manage the give-and-take that comes with being close.

Although you may not end up on a reality TV show, perhaps you can learn a few lessons from them about relationships: (1) Very often, it's not what you say but how you say it; (2) Sharing either too much too soon or nothing about yourself can derail a relationship; and (3) Having the courage to work through difficult issues is critical to long-term happiness. Former *Bachelorette* star Jillian Harris adds one more to the list: "It sounds cliché, but be yourself."[133]

MAKING THE GRADE

OBJECTIVE 8.1 Explain what makes some communication interpersonal and how content and relational messages differ.

- Interpersonal communication involves emotional closeness between people who treat each other as unique individuals.
- Interpersonal communication consists of both content (literal) messages and relational (usually implied) messages that suggest how people feel about each other in terms of affinity, respect, and control.
  > List a few of your close friends and family members. Now list people who are present in your life but not emotionally close to you (a bank teller, casual acquaintance, and so on). Which of these qualify as interpersonal relationships? Are there some people in the gray zone between interpersonal and impersonal—perhaps a professor or neighbor you would miss but don't know well on a personal level?
  > Think of a recent conversation with someone who is important to you. What about the conversation felt satisfying (or not) to you? If you were a relationship coach using this conversation as a teaching tool, what lessons might you share in terms of what worked well and what could have been more effective?

OBJECTIVE 8.2 Describe how people choose relational partners and the role of metacommunication, self-disclosure, and online communication in developing those relationships.

- People typically gravitate to those with whom they have a good deal in common, whose characteristics complement their own, who like them back, who open up to them, and who offer rewards that are worth the costs required to maintain the relationship.
- Metacommunication involves interpersonal exchanges in which the parties talk about the nature of their interaction, as in "Are you being serious?"
- The social penetration model describes how intimacy can be achieved via the breadth and depth of self-disclosure.
- The Johari Window describes the relationship between self-disclosure and self-awareness in terms of what you know (or don't know) about yourself and what a relational partner knows (and doesn't know) about you.
- Online communication can facilitate connections and social support that might otherwise be difficult or intimidating. However, people who overuse technology may find that it detracts from their in-person relationships and can lead them to feel lonely and isolated.

- Have you ever been attracted to someone based on physical similarities and then found out you were incompatible in other ways? If so, what qualities might have made you more compatible?
- Transcribe a recent conversation, including as much detail as you can about what people said and how they said it. See if you can identify examples of metacommunication and self-disclosure.
- When is the last time you told someone a deeply personal secret about yourself? How was it received?
- How might you maximize the benefits of online communication in relationships that are important to you? How might you minimize the potential drawbacks?
- Do you feel that social media use ever interferes with your ability to get things done and to be fully present with the people around you? If so, what might you do to cut down a bit?

OBJECTIVE 8.3 **Identify common communication patterns in friendships, parent–child dynamics, and sibling relationships.**

- Friendships vary in terms of how long they last, how much the friends share with each other, what they do together, how obligated they feel toward one another, and how they communicate.
- Parents have an influence on whether children grow up to value *conversation* or *conformity* as a means of solving problems.
- Sibling relationships often involve a complex mixture of camaraderie and competition.
  - Think of your closest friend. Are you mostly similar to each other, or are your characteristics complementary? What is most rewarding about the friendship? How do your differences influence the way you communicate together?
  - Think of a close friend you have known a long time. Do you communicate differently with them than with friends you have made recently? If so, how?
  - While you were growing up, were decisions such as curfews decided mostly through conversation or through conformity with rules set by your parents or guardians? Or were there few rules? How do you think the communication patterns you experienced as a child affect the way you communicate now?
  - If there is anyone in your life you consider to be a sibling, which of the styles described in this chapter best represents your relationship? How does that style influence the way you communicate with each other?

OBJECTIVE 8.4 **Describe stages of romantic relationships and options for conveying intimate messages.**

- Romantic relationships typically pass through stages of coming together (initiating, experimenting, intensifying), sustaining the relationship (integrating, bonding, differentiating, and circumscribing), and sometimes, of coming apart (stagnating, avoiding, and terminating).
- People typically value some love languages more than others. While all are important, it's a mistake to assume that everyone's preferences are the same as yours.
  - Create a timeline of a relationship (a friendship or romance), including key turning points. What did you say and do to get better acquainted? How did the way you communicated early on affect what happened between you? Does the relationship reflect any of the stages in the developmental model presented in this chapter?
  - The intensifying stage can be both exciting and unsettling. Is there a time when you felt vulnerable and wondered if you were setting yourself up to get hurt? Conversely, have you ever been so cautious that you lost the chance to get to know someone better? What did you learn from these experiences?
  - Which of the love languages (affirming words, quality time, acts of service, gifts, or physical touch) are most meaningful to you? Which are most meaningful to the significant people in your life?
  - Have you ever experienced a sense of drifting apart in a relationship? If so, what communication strategies did you use to either increase or decrease the emotional distance? Will you do anything differently if you find yourself in a similar situation again?

OBJECTIVE 8.5 **Compare dialectical continua and strategies for managing them.**

- The relational dialectic perspective calls attention to the way relational partners negotiate a balance between opposing desires.
    > Have you ever grappled with wanting to be with someone but also craving time for yourself? If so, how did you handle those competing needs?
    > What strategies from the "Managing Dialectical Tensions" feature do you typically use? How successful are those strategies? What others might you use?

KEY TERMS

affinity p. 196
apathetic siblings p. 213
authoritarian p. 211
authoritative p. 211
competitive siblings p. 212
content message p. 196
control p. 196
developmental model p. 214
family p. 211
hostile siblings p. 213
immediacy p. 196
intergroup contact hypothesis p. 209
interpersonal communication p. 194
Johari Window p. 201
longing siblings p. 212
metacommunication p. 199
permissive p. 211
phubbing p. 204
relational dialectics p. 220
relational message p. 196
respect p. 196
self-disclosure p. 200
social exchange theory p. 198
social penetration model p. 200
supportive siblings p. 212

PUBLIC SPEAKING PRACTICE

Describe the qualities that first attracted you to a special person in your life. Did the importance you placed on those characteristics change as the relationship developed? If so, how? Prepare to share your experiences in a brief oral presentation.

ACTIVITIES

1. Draw two columns on a sheet of paper.
    - In one column, list the people who are closest to you.
    - In the other column, list people you are acquainted with but whom you don't know well.
    - List several things you might say to people in one column that you probably wouldn't say to people in the other.

2. Answer the following questions as you think about a relationship with a person who is very important to you.
    - Is the relationship more in a getting-to-know-you stage or more in an intensifying or sustaining stage? How does the stage you are in influence your communication with each other?
    - What love languages do you use most? Are you satisfied with these, or could you do a better job showing how much you care?

3. Draw three horizontal lines on a sheet of paper.
    - On one line, write "autonomy" at one end and "connection" at the other end.
    - On another line, write "openness" at "privacy" at opposite ends.
    - On the third line, label one end "predictability" and the other end "novelty."
    - Think of a close relationship, and draw a star on each continuum representing your personal preference (e.g., mostly prefer autonomy, closer to the middle, or mostly prefer connection). Then draw a circle on each continuum where you think your relational partner usually falls. In what ways are you similar? How are you different?

Managing Conflict

CHAPTER OUTLINE

9.1 Understanding Interpersonal Conflict 228
- Expressed Struggle
- Interdependence
- Perceived Incompatible Goals
- Perceived Scarce Resources

9.2 Communication Climates 231
- Confirming and Disconfirming Messages
- How Communication Climates Develop

9.3 Conflict Communication Styles 237
- Nonassertiveness
- Indirect Communication
- Passive Aggression
- Direct Aggression
- Assertiveness

9.4 Negotiation Strategies 245
- Win–Lose
- Lose–Lose
- Compromise
- Win–Win

9.5 Cultural Approaches to Conflict Communication 251
- Individualism and Collectivism
- High and Low Context
- Emotional Expressiveness

MAKING THE GRADE 253

KEY TERMS 254

PUBLIC SPEAKING PRACTICE 255

ACTIVITIES 255

LEARNING OBJECTIVES

9.1 Explain the key facets of interpersonal conflict, including expressed struggle, interdependence, and the perception of incompatible goals and scarce resources.

9.2 Describe the role of communication climate and relational spirals, and practice communication strategies for keeping relationships healthy.

9.3 Identify characteristics of nonassertive, indirect, passive aggressive, directly aggressive, and assertive communication, and explain how these conflict approaches vary.

9.4 Explain the differences between win–lose, lose–lose, compromising, and win–win negotiation strategies, and apply the steps involved in achieving win–win solutions.

9.5 Compare and contrast conflict management approaches that differ by culture.

CHAPTER 9   Managing Conflict

It was a magical moment—a couple who met as lifeguards 25 years before, now happily married with children, sharing a nostalgic swim in a beautiful lake. As the couple paused to tread water, "our eyes met," remembers the wife. "I let my sentiments roam freely, tenderly telling Steve, 'I'm so glad we decided to do this together.'" She luxuriated in the moment, expecting "an equally gushing response." Instead, Steve said, "Yeah. Water's good," and started paddling again.[1]

As quickly and unexpectedly as that, the seeds of conflict can emerge. It's no one's fault, necessarily. Goals and expectations differ. When they do, hurt feelings and frustration can quickly escalate into resentment or arguments.

The woman sharing a nostalgic swim with her husband was Brené Brown, a social work scholar and author of numerous books about embracing one's imperfections and daring to be vulnerable. That doesn't make her immune to hurt feelings, of course. "Didn't he hear me?" she remembers thinking, as her husband swam away. She reflects, "My emotional reaction was embarrassment, with shame rising."[2]

You've probably found yourself at odds with someone who is important to you. Conflict management is one of the biggest challenges people face—whether with romantic partners, friends, coworkers, or family members.

Although you may wish that every conflict you experience could be resolved or disappear, for a variety of reasons that doesn't always happen. Sometimes, that's a good thing. Communicating about a conflict may stimulate creative thinking and deeper understanding. At the very least, it gives you and others the chance to put your thoughts into words. As you will see here, those goals can be even more valuable than finding a neat and tidy solution.

The bottom line is that you don't control other people's behavior, so you can't guarantee particular outcomes. But you can approach conflict in a constructive and collaborative way. The main point of this chapter is that managing conflict skillfully can often lead to healthier, stronger, and more satisfying relationships. As famed problem-solver Bernard Meltzer once said, "If you have learned how to disagree without being disagreeable, then you have discovered the secret of getting along—whether it be business, family relations, or life itself."[3]

9.1 Understanding Interpersonal Conflict

Hurt but not defeated, Brené Brown decided to try again when she and her husband reached the opposite shore of the lake. "I flashed a smile in hopes of softening him up and doubled down on my bid for connection," she recalls. She again looked Steve in the eyes, and this time said, "This is so great. I love that we're doing this. I feel so close to you." He replied, "Yep. Good swim," and swam back toward the other shore. After being twice disappointed, Brené remembers thinking indignantly, "This is total horseshit."[4]

You might like to think that such an experience would never happen in your relationships. But regardless of what people may wish for or dream about, a conflict-free world just doesn't exist. Even the best communicators, the luckiest people, are bound to find themselves in situations in which their needs don't match the needs of others. Money, time, power, humor, and aesthetic taste, as well as a thousand other issues, arise and keep people from living in a state of perpetual agreement.

9.1 Understanding Interpersonal Conflict

Whatever form it may take, every interpersonal **conflict** involves an expressed struggle between at least two interdependent parties who perceive incompatible goals, scarce resources, and/or interference from one another in achieving their goals.[5] A closer look at four parts of this definition helps illustrate the conditions that give rise to interpersonal conflict.

Expressed Struggle

At times, you may fume to yourself rather than expressing your frustration. For example, you may be upset for months because a neighbor's loud music keeps you from sleeping. That's most accurately described as internal conflict. Actual interpersonal conflict requires that both parties know a disagreement exists, such as when you let the neighbor know that you don't appreciate the decibel level. You might say this in words, or you might use nonverbal cues—as in giving the neighbor a mean look, avoiding them, or slamming your windows shut. One way or another, once both parties know that a problem exists, it's an interpersonal conflict. In Brené Brown's swimming story, the conflict has yet to be expressed, but it will be.

No matter how satisfying your relationships, some degree of conflict is inevitable.
When you do find yourself at odds with the people who matter the most to you? How do you handle conflicts when they arise?

conflict An expressed struggle between at least two interdependent parties who perceive incompatible goals, scarce rewards, and/or interference from the other party in achieving their goals.

Interdependence

People in a conflict are usually dependent on each other. The welfare and satisfaction of one depend on the actions of another. After all, if they didn't need each other to solve the problem, they could solve it themselves or go their separate ways. Although this seems obvious from a distance, many people don't realize it in the midst of a disagreement. One of the first steps toward resolving a conflict is to take the attitude that "we're in this together."

Perceived Incompatible Goals

Conflicts often look as if one party's gain will be another's loss. If your neighbor turns down their loud music, they lose the enjoyment of hearing it the way they want, but if they keep the volume up, then you're awake and unhappy. It helps to realize that goals often are not as oppositional as they seem. Solutions may exist that allow both parties to get what they want. For instance, you might achieve peace and quiet by closing your windows and getting the neighbor to do the same. You might use earplugs. Or perhaps the neighbor could get a set of headphones and listen to the music at full volume without bothering anyone. If any of these solutions proves workable, then the conflict disappears.

Unfortunately, people often fail to see mutually satisfying answers to their problems. And as long as they perceive their goals to be mutually exclusive, they may create a self-fulfilling prophecy in which the conflict is very real.

COMMUNICATION STRATEGIES

Managing Conflict in Online Classes and Teams

Online interactions are here to stay. About 6 in 10 workers and 7 in 10 college students say they like the option to engage in both virtual and in-person experiences.[6,7] The challenge, most say, is that—although virtual interactions can be convenient and efficient—they often can make it difficult to communicate openly and build close relationships.[8] These factors are especially important when conflict arises. Here are tips from the experts that may help.

Build Trust

Classmates and team members who know each other well are more likely to work through conflicts than those who don't. To build trusting relationships remotely, encourage conversations via online breakout rooms, one-on-one phone conversations, and video chats.

Ask Rather Than Assuming

Mediated communication makes it's hard to read between the lines. If you suggest something via email or text and receive a reply that simply says "Okay," should you assume that the sender is pleased, annoyed, distracted? It's hard to know. Summon the courage to ask. You might say, "I'd love your feedback" or ask, "Do you have any reservations about this idea?"

Pause to Cool Off

Even people who wouldn't be hostile in person sometimes lash out via technology. "Everyone has done it at some point," says career advisor Dana Brownlee. "You're so ticked off at someone or something that you fire off an email in the heat of anger."[9] A better option, Brownlee advises, is to vent on paper or in a Word document and then come back to it later. That way, your thoughts won't accidentally end up in someone's inbox before you decide if it's a good idea to send them.

Remember That Written Messages Are Permanent

Because emails, texts, and posts are in written form, there's an everlasting "transcript" that doesn't exist when communicators deal with conflict face to face. Written communication can help a team share ideas and important information. But if you aren't careful, you may wish you could take back some messages that are now part of the record forever.

Don't Harbor Hurt Feelings

"The nature of remote communication means that people can wallow up and let concerns fester over time," observes a writer for The Virtual Hub.[10] Rather than harboring grudges, let people know when something bothers you and encourage others to do the same. You might say, "My feelings were hurt when someone said, 'Men are jerks' and everyone laughed," or "I wanted to follow up on your comments about my project. Are you disappointed in my performance?"

Use Your Resources

Online communication presents a wide range of options than can be useful. If your team is trying to make a difficult decision, you might hold a real-time video meeting to brainstorm ideas, follow up with an online discussion board, and post a collaborative document that team members can review, comment on, and update. Apps such as IdeaBoardz, GroupMap, and Stormz are designed to help virtual groups brainstorm and make decisions.

Even if you mostly work and attend school in person, virtual communication skills are a must today. "Everyone can increase their virtual intelligence," says a productivity consultant, and "communication is key" to accomplishing that.[11]

Perceived Scarce Resources

In a conflict, people often believe that there isn't enough of the desired resource to go around. That's one reason conflict so often involves money. If a person asks for a pay raise and the boss would rather keep the money or use it to expand the business, then the two parties are in conflict.

Time is another scarce commodity. Should you spend the afternoon doing school work? Take an extra shift at work? Go out with friends? Spend time with your family? Enjoy the luxury of being alone? With only 24 hours in a day, you're bound to end up in conflicts with classmates, colleagues, friends, and loved ones—all of whom may want more of your time than you have available to give.

Having laid out the ingredients for conflict and acknowledged that it's a fact of life, it's time to focus on ways you can manage conflict effectively and even use it to strength your relationships. See the feature "Managing Conflict in Online Classes and Teams" and then read on for information about creating positive communication environments.

9.2 Communication Climates

As Brené and Steve swam back across the lake, she envisioned the day unfolding in a pattern they had enacted many times before when they were frustrated with each other. She predicted that Steve would say, "What's for breakfast, babe?" and she would roll her eyes and say, "Gee, Steve. I forgot how vacation works. I forgot that I'm in charge of breakfast. And lunch. And dinner. And laundry. And packing and goggles. And . . ."[12]

You get the point. Every relationship has a **communication climate**—an emotional tone. It's a lot like the weather. Some communication climates are fair and warm, whereas others are stormy and cold. Some are polluted and others healthy. Some relationships have stable climates, whereas others change dramatically—calm one moment and turbulent the next. Although the sun was shining, Brené predicted that a metaphorical dark cloud was brewing for her and her husband.

communication climate The emotional tone of a relationship as it is expressed in the messages that the partners send and receive.

A communication climate doesn't involve specific activities as much as the way people feel about one another as they carry out those activities. Consider two communication classes, for example. Both meet for the same length of time and follow the same syllabus. It's easy to imagine how one of these classes might be a friendly, comfortable place to learn, whereas the other might be cold and tense—even hostile. The same principle holds for families, coworkers, and other relationships. Communication climates are a function more of the way people feel about one another than of the tasks they perform.

Communication climate influences how people respond when conflict emerges in a relationship. As you will see in the following section, some relationships involve trust and respect, whereas others are steeped in criticism and defensiveness.

Confirming and Disconfirming Messages

What makes some relationship climates positive and others negative? A short but accurate answer is that the communication climate is determined by the degree to which people see themselves as valued. When you believe that the other person views you as important, you're likely to feel good about the relationship. By contrast, the relational climate suffers when you think others don't appreciate or care about you. This section describes two types of messages that shape relational climates.

Disconfirming Messages A message is **disconfirming** if it denies the value of another person.[13] Disagreeing with someone can be disconfirming, and it can be hurtful to point out something that bothers you about another person. That's not to say you will (or should) always agree with other people or find their behavior 100 percent

disconfirming messages Actions and words that imply a lack of agreement or respect for another person.

Source: Ted Goff, North America Syndicate, 1994

COMMUNICATION STRATEGIES

Rules for Fighting Fair

Here are eight ways to engage in relationship-friendly conflict management, based on the work of Jack Gibb.[14]

Avoid Judgment Statements

Don't make "you" statements, such as "You don't know what you're talking about" or "You smoke too much," which are likely to cause defensiveness and escalate conflict.

Use "I" Language

Use statements such as "I get frustrated when you interrupt me" that focus on a specific behavior and your thoughts and feelings about it.

Be Honest

Think about what you want to say, and plan the wording of your message carefully so that you can express yourself clearly.

Show Empathy

Empathic messages show that you accept another person's feelings and can put yourself in their place. You might say, "I can understand why you thought I was ignoring you at the party."

Treat Others as Your Equal

People who convey an attitude of equality communicate that others have just as much worth as they do. Demonstrate that you are willing to listen to others and consider their needs and goals, not just your own.

Avoid Attempts to Control or Manipulate Others

Be careful not to impose your preferences without regard for other people's needs or interests. For example, avoid guilt-provoking proclamations such as "If you cared about me, you would . . ." Instead, share your feelings and invite the other person to do the same.

Don't Be a Know-It-All

Comments such as "Only an idiot would vote for him" can come off as arrogant and dismissive. A better alternative is to acknowledge that you don't have a lock on the truth. You might say, "My impression is that the candidate has very little experience. What do you know about him?"

Focus on Mutually Beneficial Problem Solving

The bottom line is that solutions have to work for *both* of you for a relationship to thrive. You can help build a healthy relational climate by being respectful, being a good listener, and seeking solutions that satisfy both your needs and the other person's.

acceptable. The point is more to handle those inevitable conflicts fairly and respectfully. The feature "Rules for Fighting Fair" provides some ways to accomplish that.

Unfortunately, people sometimes handle conflict in ways that erode their relationships. After four decades of studying how partners communicate, psychologist John Gottman can predict with a rate of accuracy approaching 90 percent whether or not a couple is headed toward a breakup.[15] Gottman calls the most hurtful conflict tactics the Four Horsemen of the Apocalypse because, when they are present on a regular basis, a relationship is usually is serious trouble and unlikely to survive.[16,17] Although Gottman studied married couples, the same types of messages can damage all types of relationships. As you read about the Four Horsemen, consider if you are ever guilty of any of them with partners, friends, roommates, family members, or anyone else you know.

- *Partners criticize each other.* Whereas it can be healthy for relational partners tactfully to point out specific behaviors that cause problems ("I wish you would let me know when you're running late"), **criticism** goes beyond that to deliver a personal, all-encompassing accusation such as "You're lazy" or "The only person you think about is yourself."

- *Partners show contempt.* **Contempt** takes criticism to an even more hurtful level by mocking, belittling, or ridiculing the other person. Expressions of contempt can be explicit ("People laugh at you behind your back" or "You disgust me") or nonverbal (sneering, eye rolling, or a condescending tone of voice). Whereas criticism implies "You are flawed," contempt implies "I hate you."[18] Gottman has found that contempt is the single best predictor of divorce.[19] Experience probably shows you that it often spells doom in other relationships as well.

- *Partners are defensive.* When faced with criticism and contempt, it's not surprising that partners often react with **defensiveness**—protecting their self-worth by counterattacking ("You're calling me a careless driver? You're the one who got a speeding ticket last month"). Once an attack-and-defend pattern develops, conflict often escalates or partners start to avoid each other.

- *One or both partners engage in stonewalling.* **Stonewalling** is a form of avoidance in which one person refuses to engage with the other. Walking away or giving one's partner the silent treatment sends the message "You aren't even worth my attention." Disengagement may seem like a better option than arguing, but it robs partners of the chance to understand each other better.

These are the big offenders on Gottman's list, but people may engage in a number of other disconfirming messages as well. Table 9.1 lists a variety of behaviors that create distance between people in a relationship. It's easy to see how each of them is inherently disconfirming.

It's important to note that disconfirming messages, like virtually every other kind of communication, are a matter of perception. That's why it can be a good idea to engage in perception checking before jumping to conclusions: "Were you laughing at my joke because you think I look stupid, or was it something else?" You might find that a message you thought was disconfirming was actually delivered with good intentions.

Confirming Messages Consider times when someone made you feel good with a compliment, a smile, or encouraging words. **Confirming messages** convey that you are valued by implying that "you are important."[20] Brené was trying to engage Steve

---

criticism A message that is personal, all-encompassing, and accusatory.

contempt Verbal and nonverbal messages that ridicule or belittle another person.

defensiveness Protecting oneself by counterattacking the other person or justifying one's own behavior.

stonewalling Refusing to engage with the other person.

confirming messages Actions and words that express respect for another person.

TABLE 9.1
Distancing Behavior

TACTIC	DESCRIPTION
Avoidance	Evading the other person
Deception	Lying to or misleading the other person
Disrespect	Treating the other person in a degrading way
Detachment	Acting emotionally uninterested in the other person
Discounting	Disregarding or minimizing the importance of what the other person says
Humoring	Not taking the other person seriously
Impersonal demeanor	Treating the other person like a stranger; interacting with that person as a role rather than a unique individual
Inattention	Not paying attention to the other person
Nonimmediacy	Displaying verbal or nonverbal clues that minimize interest, closeness, or availability
Reserve	Being unusually quiet and uncommunicative
Restraint	Curtailing normal social behaviors
Restriction of topics	Limiting conversation to less personal topics
Shortening of interaction	Ending conversations as quickly as possible

Source: Adapted from Hess, J. A. (2002). Distance regulation in personal relationships: The development of a conceptual model and a test of representational validity. *Journal of Social and Personal Relationships, 19,* 663–683.

in a confirming exchange when she told him she was glad to be there with him. She remembers how she felt when she didn't receive the validation she had expected in return: "I thought *What's going on? I don't know if I'm supposed to feel humiliated or hostile. I wanted to cry and I wanted to scream.*"[21]

On the other side, put yourself in Steve's shoes. He might have been at a loss about how to respond. You can learn some valuable tools from scholars who have identified three main categories of confirming communication.[22] Here are those categories, in order from the most basic to the most powerful.

- *Show recognition.* Recognition seems easy and obvious, and yet there are many times when people don't respond to others on this basic level. Brené remembers that when Steve tossed off his "Yep. Good swim" response, "he seemed to be looking through me rather than at me."[23] Your friends may feel a similar sense of being invisible or ignored if you don't return phone messages or if you fail to say hi when you encounter each other at a party or on the street. Of course, this lack of recognition may simply be an oversight. You might not notice your friend, or the pressures of work and school might prevent you from staying in touch. Nonetheless, if the other person *perceives* you as avoiding contact, the message has the effect of being disconfirming.

- *Acknowledge the other person's thoughts and feelings.* Acknowledging the ideas and emotions of others is an even stronger form of confirmation than simply recognizing them. Attentive listening is probably the most common form of acknowledgment. Not surprisingly, leaders who are supportive of others and their ideas tend to be more successful than those who are more concerned with promoting their own image and ideas.[24,25]

- *Show that you agree.* Whereas acknowledgment means you are interested in other people's ideas, endorsement means that you agree with them. People tend to be attracted to those who agree with them.[26] The message is: We have a lot in common and are in sync. You can probably find something in a message to endorse even if you don't agree with it entirely. You might say, "I can see why you were so angry," to a friend, even if you don't approve of their outburst. Of course, outright praise is a strong form of endorsement and one you can use surprisingly often if you look for opportunities to compliment others.

It's hard to overstate the importance of confirming messages. People who offer confirmation generously are considered to be more appealing candidates for romantic relationships than their less appreciative peers.[27] Confirming messages are just as important in other relationships. Family members are most satisfied when they regularly encourage each other, joke around, and share news about their day.[28] And in the classroom, students are more likely to be actively engaged when instructors show an active interest in them and acknowledge their efforts.[29]

How Communication Climates Develop

As Brené and Steve continued their swim across the lake, she was feeling hurt and was already imagining the bickering that might lay ahead for them. One challenge of conflict management is that people tend to feel defensive and angry when their expectations are thwarted or they don't agree with their relational partners. One comment can escalate into hours of snide comments or tense silence.

A **relational spiral** is a communication pattern in which one person's behavior is followed by a similar or even more intense response by another person, which tends to inspire an even greater reaction in the first person, and so on.[30] There is a natural tendency to strike back when one's feelings are hurt, as captured in the old saying "what goes around comes around." But acting defensively can make a difficult situation even worse. The good news is that relational spirals aren't always negative.

Escalatory Conflict Spirals In an **escalatory conflict spiral**, one perceived slight leads to another until the communication escalates into a full-fledged dispute.[31] Perhaps you feel that a friend said something unkind about you, so you say insulting things about your friend, and the cycle continues until you are both even more hurt and furious.

Avoidance Spirals Not communicating can also be destructive. In **avoidance spirals**, rather than fighting, individuals slowly lessen their dependence on one another, withdraw, and become less invested in the relationship.[32] If you have ever said, "I'm not calling them. If they want to talk, they can call me," you have been part of an avoidance spiral. Even the best relationships can go through periods of conflict

relational spiral A reciprocal communication pattern in which each person's message reinforces the other's emotional tone.

escalatory conflict spiral A reciprocal pattern of communication in which messages between two or more communicators reinforce one another in an increasingly negative manner.

avoidance spiral Occurs when relational partners reduce their involvement with one another, withdraw, and become less invested in the relationship.

positive spiral Occurs when one person's confirming message leads to a similar or more confirming response from the other person.

and withdrawal. However, too much negativity may lead to a point of no return from which the relationship cannot be saved.

Positive Spirals Fortunately, spirals can escalate in beneficial ways as well. In **positive spirals**, one person's confirming message leads to a similar response from the other person, and so on. Offering a sincere compliment, apology, invitation, or simply one's undivided attention can inspire more of the same from your relational partner.

UNDERSTANDING YOUR COMMUNICATION

What's the Forecast for Your Communication Climate?

Think of an important person in your life—perhaps a friend, a roommate, a family member, or a romantic partner. Choose the option in each of the following groups that best describes how you communicate with each other.

1. When I am upset about something, my relational partner is most likely to:
   a. Listen to me and provide emotional support
   b. Say I should have tried harder to fix or avoid the problem
   c. Ignore how I feel

2. When we are planning a weekend activity and I want to do something my partner doesn't want to do, I tend to:
   a. Suggest another option we will both enjoy
   b. Beg until I get my way
   c. Cancel our plans and engage in the activity with someone else

3. When my partner and I disagree about a controversial subject, we usually:
   a. Ask questions and listen to the other person's viewpoint
   b. Accuse the other person of using poor judgment or ignoring the facts
   c. Avoid the subject

4. If I didn't hear from my partner for a while, I would probably:
   a. Call or text to make sure everything is okay
   b. Feel angry about being ignored
   c. Not notice

5. The statement we are mostly likely to make during a typical conversation sounds something like this:
   a. "I appreciate the way you . . ."
   b. "You always forget to . . ."
   c. "Were you saying something?"

INTERPRETING YOUR RESPONSES

Read the following explanations for a climate report about your relationship.

Warm and Sunny

If the majority of your answers are "a," your relational climate is warm and sunny, with a high probability of confirming messages. You seem to be experiencing a positive spiral. Use suggestions throughout this chapter to strengthen and nurture your relationship even more.

Stormy

If you answered mostly "b," your relationship tends to be turbulent, with outbreaks of controlling or defensive behavior. Storm warning: You seem to be in a downward escalatory conflict spiral that can damage your relationship. That's not to say it's hopeless, but you may want to consider underlying feelings—yours and the other person's. Guidance on self-disclosure may be helpful.

Chill in the Air

If most of your answers are "c," beware of falling temperatures. It's natural for people to drift apart sometimes, but your relationship shows signs of chilly indifference and avoidance spiraling. Consider whether you are guilty of the damaging patterns described in this chapter. You may be able to change the weather by engaging in more supportive communication.

It often feels that relational spirals have a life of their own. People may be inclined, even without thinking about it, to mirror and escalate their partners' behaviors, even if doing so is harmful to the relationship. The best communicators recognize this tendency and make mindful choices instead. They may switch from negative to positive messages without discussing the matter. Or they may engage in metacommunication (Chapter 8). "Hold on," one might say, "This is getting us nowhere."

Take the "What's the Forecast for Your Relational Climate?" quiz and then follow along as the chapter takes a closer look at useful approaches and techniques for managing conflict.

9.3 Conflict Communication Styles

When Steve and Brené reached the dock where their swim had started, she decided to talk about her problem. Rather than blaming Steve for her hurt feelings, which was likely to escalate the conflict, she said to him instead, "I've been trying to connect with you and you keep blowing me off. I don't get it."[33]

Consider what you might have done in a similar situation. Are you prone to avoiding sensitive issues? Do you tend to hint around when something upsets you, or are you likely to say outright how you feel?

This section describes five conflict communication styles, which are common patterns of behavior. As you will see, each style varies on two dimensions: concern for self and concern for others. As you read, ask yourself which styles you use most often and how these styles affect the quality of your close relationships.

Nonassertiveness

Anticipating their 10-year wedding anniversary, Alex asked his wife Danielle what she would like. She said, "Just surprise me!" So Alex planned a surprise get-away at a bed and breakfast Danielle likes. She didn't seem thrilled to be there, however. "I asked her what was wrong, and she kept saying, 'Nothing,'" Alex remembers. "Finally, the last day there, she told me she really had her heart set on a new wedding ring. How was I supposed to know that?"[34]

The inability or unwillingness to express one's thoughts or feelings in a conflict is known as **nonassertion**. A nonassertive person may insist that "nothing is wrong" even when it is. Sometimes nonassertion comes from a lack of confidence. At other times, people lack the awareness or skills to use a more direct means of expression. It's easy to imagine that Danielle felt it would be unseemly to ask outright for a new ring. She might have dropped subtle hints or hoped that Alex would surprise her with a ring because he wanted to give her one, not because she asked him to.

In conflict situations, nonassertion can take a variety of forms. One is *avoidance*—either steering clear of the other person or avoiding the topic, perhaps by talking about something else, joking, or denying that a problem exists. People who avoid conflicts usually believe it's easier to put up with the status quo than to face the problem head-on and try to solve it. *Accommodation* is another type of nonassertive response, which involves dealing with conflict by giving in, thus putting others' needs ahead of your own. If you agree to see a movie that's unappealing to you because others in your group want to see it, you are accommodating.

> **nonassertion** The inability or unwillingness to express one's thoughts or feelings.

While nonassertion won't solve a difficult or long-term problem, here are a few situations in which accommodating or avoiding is a sensible approach.

- *The conflict is minor or short-lived.* Avoidance may be the best choice if the matter isn't serious. You might let a colleague's occasional grumpiness pass without saying anything, knowing they are under a lot of stress. Or you might not mind seeing a movie someone else prefers, knowing you'll get first choice next time.
- *The relationship is new or sensitive.* You might reasonably choose to avoid conflict with some people in your life. For example, you might not object if a new friend or your grandfather eats a snack you had put aside for yourself.
- *The risks are great.* You might choose to keep quiet if speaking up would put you at risk, as when it might get you fired from a job you can't afford to lose or provoke someone who might do you physical harm.

Nonassertion displays low concern for self and high concern for others. That can be a virtue. But in some cases, being selfless can damage your relationships. For one, you might begin to resent that people don't "listen to you" or honor your preferences, even though you may not be voicing your thoughts in a way others can understand. For another, relational partners may feel that they don't know the real you, and they may be frustrated that you don't give them honest feedback about what you like and don't like. As one analyst puts it, "trying to make everyone happy can make you [and others] miserable."[35]

Indirect Communication

Whereas a nonassertive person resists dealing with conflict at all, someone using an indirect style addresses the conflict but in subtle ways. **Indirect communication** conveys a message in a roundabout manner. The goal is to get what you want without causing hard feelings. If your partner keeps forgetting to turn off the lights, you might say, "That bulb is so dim, it's easy to forget that it's on, isn't it?" Or, if your neighbor tends to play loud music, one indirect approach would be to strike up a friendly conversation with them and ask if anything you are doing is too noisy for them, hoping they will get the hint.

Because indirect communication saves face for the other party, it's often kinder than blunt honesty. If your guests are staying too long, it's probably more polite to yawn and hint about your big day tomorrow than to bluntly ask them to leave. Likewise, if you're not interested in going out with someone who has asked you for a date, it may be more compassionate to claim that you're busy than to say, "I'm not interested in seeing you."

At other times, you might communicate indirectly to protect yourself. For example, you might joke around about being underpaid rather than directly asking the boss for a raise. At times like these, a subtle approach may get the message across while softening the blow of a negative response. The risk, of course, is that the other person will misunderstand you or will fail to get the message at all. There are times when an idea is so important that hinting lacks the necessary punch.

Indirect communication involves a moderate concern for self and others. The next section covers two reactions to conflict that demonstrate low concern for others, and one approach that balances concern for self with concern for others.

Even when communication is indirect, it can have powerful consequences. Sexual harassment in the workplace is one example. Suggestive looks, "accidental"

indirect communication
Hinting at a message instead of expressing thoughts and feelings directly.

COMMUNICATION STRATEGIES

Dealing with Sexual Harassment

Sexual harassment takes many forms. It can be a blatant sexual overture or a verbal or nonverbal behavior that creates a hostile work environment. The harasser can be a supervisor, peer, subordinate, or even someone outside the organization. Here are several options to consider if you or someone you care about experiences harassment:

Consider Dismissing the Incident

This nonassertive approach is only appropriate if you truly believe that the remark or behavior is trivial. Dismissing incidents that you believe are important can result in self-blame and diminished self-esteem, and the offender may assume it's permissible to keep acting that way.

Tell the Harasser to Stop

Assertively tell the harasser that the behavior is unwelcome, and insist that it stop immediately. Your statement should be firm, but it doesn't have to be angry. Remember that many words or deeds that make you uncomfortable may not be deliberately hostile.

Write a Personal Letter to the Harasser

A written statement may help the harasser understand what behavior you find offensive. Just as important, it can show that you take the problem seriously. Detail specifics about what happened, what behavior you want stopped, and how you felt. You may want to include a copy of your organization's sexual harassment policy. Keep a record of when you delivered your message.

Ask a Trusted Third Party to Intervene

This indirect approach can sometimes persuade the harasser to stop. The person you choose should be someone whom you trust and whom understands your discomfort. It's a plus if the harasser respects and trusts this person as well.

Use Formal Channels

Report the situation to your supervisor, personnel office, or a committee that has been set up to consider harassment complaints.

File a Legal Complaint

If all else fails or the incident is egregious, you may file a complaint with the federal Equal Employment Opportunity Commission or with your state agency. You have the right to obtain the services of an attorney regarding your legal options.

---

touches, and other behaviors can be frightening or demoralizing, even if they seem subtle on the surface. See the feature "Dealing with Sexual Harassment" for some response strategies if you are ever in that situation.

Passive Aggression

If you know someone who responds to conflict with unkind humor or snide comments and then acts like they didn't intend to hurt your feelings, you have experienced **passive aggression**, which occurs when a communicator expresses hostility in an ambiguous way. Scholar George Bach describes five types of passive aggressive people:[36]

- *Pseudoaccommodators* only pretend to agree with you. A passively aggressive person might commit to something ("I'll be on time from now on") but not actually do it.
- *Guiltmakers* try to make you feel bad. A guiltmaker will agree to something and then make you feel responsible for the hardship it causes them ("I really should be studying, but I'll give you a ride").

> **passive aggression** An indirect expression of aggression, delivered in a way that allows the sender to maintain a facade of innocence.

- *Jokers* use humor as a weapon. They might say unkind things and then insist they were "just kidding," insinuating that you are being too sensitive ("Where's your sense of humor?").
- *Trivial tyrannizers* do small things to drive you crazy. Rather than express their feelings outright, they might "forget" to clean the kitchen or to put gas in the car just to annoy you.
- *Withholders* keep back something valuable. A withholder punishes others by refusing to provide thoughtful gestures such as courtesy, affection, or humor.

Direct Aggression

Patricia was so furious with members of her project group that she sent out an angry group text calling them "lazy and irresponsible." Later, she wished she had handled the situation differently.

A **directly aggressive message** confronts another person in a way that attacks their character, intelligence, or dignity, rather than debating the issues. Aggressive people often use intimidation and insults to get their way. Many directly aggressive messages are easy to spot:

"You don't know what you're talking about."
"That was a stupid thing to do."
"What's the matter with you?"

Direct aggression can also involve nonverbal cues, such as sneering, shouting, or using an intimidating tone of voice. It's easy to imagine a hostile way of expressing statements such as:

"What is it now?"
"I need some peace and quiet."

You may get what you want in the short run using verbal aggressiveness. Yelling "Shut up!" might stop the other person from talking, and saying "Why don't you just get it *yourself*?" may save you from some exertion. But the relational damage of such an approach probably isn't worth the cost. Direct aggression can be hurtful, and the consequences for the relationship can be long lasting.[37]

In some cases, aggressive behavior is downright abusive. See "Protecting Yourself from an Abusive Partner" for experts' advice on how to stay safe.

Assertiveness

Winston Churchill is said to have proclaimed, "Courage is what it takes to stand up and speak. Courage is also what it takes to sit down and listen."[41] Assertiveness, which represents a balance between high self-interest and high concern for others, involves a good deal of both.

An **assertive communicator** handles conflicts by expressing their needs, thoughts, and feelings clearly and inviting others to do the same. They communicate directly but without judging others or dictating to them. Assertive communicators believe that it's usually possible to resolve problems to everyone's satisfaction.

---

direct aggressive message Actions and words that attack the position and perhaps the dignity of another person.

assertive communication A style of communicating that directly expresses the sender's needs, thoughts, or feelings, delivered in a way that does not attack the receiver.

COMMUNICATION STRATEGIES

Protecting Yourself from an Abusive Partner

There are no magic communication formulas to prevent or stop the behavior of an abusive person, but there are steps you can take to protect yourself.

Don't Keep Abuse a Secret

Abusers often isolate their partners from friends and loved ones because it's easier to control them if they don't have a strong network of people who know what's going on.[38] Avoid this trap by keeping close contact and open communication with people you trust. At the very least, tell someone what's happening and ask that person to assist you in getting help.

Protect Sensitive Information

Avoid sharing passwords that will allow the abuser to access your communication with others or learn your whereabouts.

Use Secret Codes, If Necessary, to Alert Others

If explicitly calling for help might enrage an abuser, try making use of code words or gestures instead. Let trusted people know in advance that, if you mention certain words (e.g., pelican, garage, or palm tree) in a conversation, they should come right away or call the police. In the United States, you may also be able to send a cryptic cry for help via 911. There is no universal code word, and operators may not pick up on your intentions at first, but some callers have alerted authorities by pretending to talk to a friend or order pizza, providing their address, and answering the operator's yes or no questions. Even if you cannot say anything, calling 911 and leaving the phone line open may help authorities realize that you are in danger.[39] And if someone can see you in person or via video, you can discretely signal that you need help by talking normally but displaying your palm with your thumb sideways (the way you would signify the number 4) and then closing your fingers over your thumb.

Don't Blame Yourself

Abused people often believe they are at fault and that they "had it coming." Remember—*no one deserves abuse*. Abusive people make the choice to be abusive. No one makes them behave that way or prevents them from making that choice.[40] Many resources are available to help you. One source of information and assistance is www.healthyplace.com/abuse.

When Özlem Cekic became the first female Muslim in the Danish Parliament, she was the target of racist letters, emails, and social media posts. At first, she deleted and ignored them. Then she decided to respond assertively to the aggressive messages. "I started answering those emails and suggesting to the writer that we might meet for coffee and a chat," Cekic says.[42] Since then, she has engaged in hundreds of #coffeedialogues with a wide range of people. In the process, she has found that fear underlies a great deal of what feels like hate. "The people I visit are just as afraid of people they don't know as I was afraid of them before I started inviting myself for coffee," she reflects.[43] But once people get to know one another, she says, they often discover that they have a lot in common and that their fears were baseless. Cekic shares her experiences in a TED talk[44] and in the book *Overcoming Hate Through Dialogue: Confronting Prejudice, Racism, and Bigotry with Understanding—and Coffee*.[45]

You may not go so far as having coffee with people who behave aggressively toward you on social media, but you might dialogue with them via technology. And you can be assertive closer to home as well. Imagine that Monday morning arguments tend to erupt in your household. You might engage with your loved ones by saying something like this:

> I've noticed that we're often impatient with each other on Monday mornings. I think I'm especially tense because I dread the weekly staff meeting. I'd like to spend some time Sunday preparing for that meeting. I think that will make me less stressed, and maybe that will help us start the week together on a positive note. Is there something we can to do make Monday mornings less stressful for you?

As this scenario suggests, being assertive usually means talking about an issue when you have a cool head rather than in the heat of the moment. Assertive individuals avoid accusations and assumptions. Their motto might be: *We are good people with good intentions who can work this out together.* Being assertive requires self-awareness, patience, and good listening skills. It's not always easy. People who manage conflicts assertively may experience feelings of discomfort while they are working through problems. However, they usually feel better about themselves and one another afterward. For example, romantic partners who approach conflict in a patient and caring way often feel closer to each other as a result.[46,47]

Here are five topics you might address when you want to approach a conflict assertively:

Describe the Behavior in Question   An assertive description is specific without being evaluative or judgmental.

> *Behavioral description:* "You asked me to tell you what I thought of your new car, and when I told you, you said I was too critical."

> *Evaluative judgment:* "Don't be so *touchy*! It's hypocritical to ask for my opinion and then get mad when I give it to you."

Judgmental words such as *touchy* and *hypocritical* invite a defensive reaction. The target of your accusation can reply, "I'm not touchy or hypocritical!" It's harder to argue with the facts stated in an objective, behavioral description. Furthermore, neutral language reduces the chances of a defensive reaction.

Share Your Interpretation of the Other Person's Behavior   This is where you can use the perception-checking process outlined in Chapter 3. Remember that, after referencing a specific behavior, a complete perception check includes two possible interpretations of the behavior and an invitation for the other person to respond:

> When it took me two days to call you back [*behavior*], perhaps you thought I didn't care [*one interpretation*]. Or you might have assumed our plans were off [*another interpretation*]. Did I hurt your feelings [*invitation to respond*]?

The key is to label your hunches as such instead of suggesting that you are positive about what the other person's behavior means.

Describe Your Feelings   Expressing your feelings adds a new dimension to a message. For example, consider the difference between these two responses:

> "When you call me in the middle of the day [*behavior*], I think you miss me and care about me [*interpretation*], and I feel special [*feeling*]."

"When you call me in the middle of the day [*behavior*], I think something must be wrong [*interpretation*], and I feel stressed [*feeling*]."

Applied to our previous example, an assertive message that conveys feeling might sound something like this:

"When you said I was too critical after you asked my opinion of your car [*behavior*], it seemed to me that you were disappointed [*interpretation*], and I felt bad for being so blunt [*feeling*]."

Describe the Consequences A consequence statement explains what happens (or might happen) as a result of the behavior you have described. There are three kinds of consequences:

- *What happens to you, the speaker*: "When you tease me, I'm tempted to avoid you."
- *What happens to the target of the message*: "When you drink too much, you start to drive dangerously."
- *What happens to others*: "When you play the radio so loud, it wakes up the baby."

It's important that a consequence statement not sound like a threat or an ultimatum. It isn't meant to manipulate the other person. Instead, the goal is to explain what impact someone's behavior has, at least from your perspective.

State Your Intentions Intention statements can communicate three kinds of messages:

- *Where you stand on an issue*: "I wanted you to know how hurt I felt" or "I wanted you to know how much I appreciate your support."
- *Requests of others*: "I'd like to know whether you are angry" or "I hope you'll come again."
- *Descriptions of how you plan to act in the future*: "I've decided to stop lending you money."

In our ongoing example, adding an intention statement would complete the assertive message:

"When you said I was too critical after you asked my opinion of your car [*behavior*], it seemed to me that you were disappointed [*interpretation*]. That made me feel bad for being so blunt [*feeling*]. Now I realize that it hurt your feelings when I called your new car a gas guzzler [*consequence*]. I'm going to be more supportive and less critical in the future [*intention*]."

It's good to know these steps in being assertive, but keep in mind that they are only a general guide. Depending on the situation, you may use a different order, combine steps, or return to some steps to make sure you both understand each other. In communication, as in many other activities, patience and persistence are essential. Take the "How Assertive Are You?" quiz to consider what approach you usually take.

UNDERSTANDING YOUR COMMUNICATION

How Assertive Are You?

Choose the answer to each question that best describes you.

1. You feel you deserve the corner office that has just become available. What would you do?
   a. Stay quiet and hope the boss realizes that you deserve the office.
   b. Hint around that you have outgrown your cubicle.
   c. Tell your coworkers, "You deserve it more than I do," but secretly ask the boss if you can have it.
   d. Meet with your supervisor and lay out the reasons you think you deserve the office.
   e. Threaten to quit if you aren't assigned to the office.

2. Your best friend just called to cancel your weekend trip together at the last minute. This isn't the first time your friend has done this, and you are very disappointed. What do you do?
   a. Reassure your friend that it's okay and there are no hard feelings.
   b. Declare, "But I've already packed," hoping your friend will take the hint and decide to go after all.
   c. Resolve to cancel the next trip yourself to teach your friend a lesson.
   d. Say, "I feel disappointed, because I enjoy my time with you and because we have made nonrefundable deposits. Can we work this out?"
   e. Announce that the friendship is over. That's no way to treat someone you care about.

3. During a classroom discussion, a fellow student makes a comment that you find offensive. What do you do?
   a. Ignore it.
   b. Tell the instructor after class that the comment made you uncomfortable.
   c. Say nothing, but tell other people how much you dislike that person.
   d. Join the discussion, say that you see the issue differently, and invite your classmate to explain why they feel as they do.
   e. Announce that the statement is the most stupid thing you have ever heard.

4. You are on a first date when the other person suggests seeing a movie you are sure you will hate. What do you do?
   a. Say, "Sure!" How bad can it be?
   b. Lie and say you've already seen it.
   c. Say, "Okaaay" and raise your eyebrows in a way that suggests your date must be either stupid or kidding.
   d. Suggest that you engage in another activity instead.
   e. Proclaim that you'd rather stay home and watch old reruns than see that movie.

EVALUATING YOUR RESPONSES

Based on your answers, see which of the following describes your usual conflict management style. (More than one may apply.)

Nonassertive

If you chose "a" multiple times, you tend to rank low on the assertiveness scale. The people around you may be unable to guess when you have a preference or hurt feelings. It may seem that "going with the flow" is the way to go, but research suggests that relationships flounder when people don't share their likes and dislikes with one another. Try voicing your feelings more clearly. People may like you more for it.

Indirect

If more than one of your answers is "b," you often know what you want, but you rely on subtlety to convey your preferences. This can be a strength because you aren't likely to offend people. However, don't be surprised if people sometimes fail to notice when you are upset. Research suggests that indirect communication works well for small concerns, but not for big ones. When the issue is important to you, step up to say so.

Passively Aggressive

Answering "c" multiple times is an indication that you are sometimes passively aggressive. Rather than taking the bull by the horns, you are more likely to seek revenge, complain to people around you, or use snide humor to make your point. These techniques can make the people around you feel belittled and frustrated. Plus, you are more likely to alienate people than to get your way in the long run. Try to break this habit by saying what you feel in a clear, calm way.

Assertive

If you chose "d" more than once, you tend to hit the bulls-eye in terms of healthy assertiveness. You say what you feel without infringing on other people's right to do the same. Your combination of respectfulness and self-confidence is likely to serve you well in relationships.

Aggressive

If "e" answers best describe your approach, you tend to overshoot assertive and land in the aggressive category instead. Although you may mean well, your comments are likely to offend and intimidate others. Try toning it down by stating your opinions (gently) and encouraging others to do the same. If you refrain from name calling and accusations, people are likely to take what you say more seriously.

9.4 Negotiation Strategies

You may think of negotiating as a formal process—something people only do when buying a car or establishing the salary for a new job. But when you consider **negotiation** as a process designed to help people reach agreement when one person wants something from another,[48] you probably realize that you negotiate more than you thought. You may negotiate to determine who will do specific household chores, what days you have off at work, where you go on a family vacation, what each group member will contribute to a shared project, and much more. This unit describes four negotiation strategies. As you read about them, consider which ones you use now, and whether others might serve you better.

negotiation An interactive process meant to help people reach agreement that best suits their goals.

win–lose problem solving An approach to conflict resolution in which one party strives to achieve their goal at the expense of the other.

Win–Lose

Win–lose problem solving occurs when one party achieves their goal at the expense of someone else. People resort to this method of resolving disputes when they perceive a situation as being either–or, as in "Either I get what I want, or you get your way." The most clear-cut examples of win–lose situations are games such as baseball or poker, in which the rules require a winner and a loser. Some interpersonal issues seem to fit into this win–lose framework, such as when two coworkers seek a promotion to the same job, or when a couple disagrees on how to spend their limited money.

Power is a distinguishing characteristic in win–lose problem solving, because it's necessary to defeat an opponent to get what you want. The most obvious kind of power is physical. Some parents threaten their children with warnings such as "Stop misbehaving, or I'll send you to your room." Power can also involve rewards or punishments. In most jobs, supervisors have the power to decide who does what, when they will work, who is promoted, and even who is fired.

"It's not enough that we succeed. Cats must also fail."

Source: Leo Cullum The New Yorker Collection/The Cartoon Bank

Even the usually admired democratic principle of majority rule is a win–lose method of resolving conflicts. However fair it may be, it results in one group getting its way and another group being unsatisfied.

There are some circumstances when win–lose problem solving may be necessary. For instance, if two people want to marry the same person, they can't both succeed at the same time. And it's often true that only one applicant can be hired for a job. Sometimes people adopt a win–lose approach with the intention of intimidating others. See the feature "Negotiating with a Bully" for tips on how you might respond if the other person is playing dirty.

All the same, don't be quick to assume that your conflicts are necessarily win–lose. As you will soon read, many situations that seem to require a loser can actually be resolved to everyone's satisfaction.

Lose–Lose

In **lose–lose problem solving**, neither side is satisfied with the outcome. Although it's hard to imagine that anyone would willingly use this approach, in truth, lose–lose is a fairly common way to handle conflicts. In many instances both parties strive to be winners, but as a result of the struggle, both end up losers. On the international scene, many wars illustrate this sad point. A nation that gains military victory at the cost of thousands of lives, large amounts of resources, and a damaged national consciousness hasn't truly won much. On a personal level the same principle holds true. You have probably seen battles of pride in which both parties strike out and both suffer. Perhaps your roommate is angry because his sister hasn't called in a while, so he resolves not to call her. The distance between them grows and they both lose.

> **lose–lose problem solving** An approach to conflict resolution in which neither party achieves their goals.

COMMUNICATION STRATEGIES

Negotiating with a Bully

"Bullying doesn't end when you grow out of your playground days," observes negotiation specialist Alexandra Dickinson.[49] As an adult, how might you respond if someone with whom you are trying to negotiate raises their voice, makes inflammatory or insulting statements, threatens you, or refuses to listen to your perspective? Experts suggest the following strategies:

Remain Calm
Don't try to "out-bully the bully" by raising your own voice or making threats in return.[50]

Take a Break to Let Tempers Cool
Try to identify the person's interests even if they aren't expressing them well. "I make a point of listening and taking notes . . . to reinforce that I'm paying close attention," says one professional.[51]

Show Sympathy and Respect
Sometimes bullies just want to be heard and treated as if they matter. Meeting those needs might reduce their inclination to rant and rave.

Withdraw
If nothing else works, consider leaving the room or discontinuing the conversation.

9.4 Negotiation Strategies

Compromise

Unlike lose–lose outcomes, a **compromise** gives both parties at least some of what they wanted, although both sacrifice part of their goals. For example, imagine that one partner says, "You're either at work or school every night Monday through Thursday. I hardly ever get to see you." After talking about it, the couple might compromise by agreeing to devote two nights a week to work and two nights a week to each other.

Although a compromise may be better than nothing, it's often not the cure-all that people make it out to be. Conventional wisdom, such as "The key to a good marriage is compromise," overlooks the reality that better solutions may be available. In our example, the two-nights-a-week compromise isn't likely to be successful if one partner resents having to delay graduation by taking fewer night classes. It might be more gratifying to find a solution that allows the partners to meet their obligations *and* spend quality time together. If that sounds difficult to accomplish, you may be right. Negotiation isn't always easy. But win–win problem solving is often possible once people let go of the notion that problem solving always means that at least one person must lose or make concessions.

The three-sequence hand gesture shown here can silently let others know that you're in a dangerous situation and need help. If you see someone use this gesture, call the authorities immediately. In 2021, a teen who had been abducted used the gesture to alert a nearby motorist, who called 911. As a result, the girl was rescued by the police.

Under what circumstances might it be important to silently ask for help?

Win–Win

The goal in **win–win problem solving** is a solution that satisfies both people's needs. Neither tries to win at the other's expense. Instead, both parties believe that, by working together, it's possible to find a solution that reaches all their goals. This is typically the most satisfying and relationship-friendly means of negotiating.

Roger Fisher and William Ury of the Harvard Negotiation Project are well-known proponents of win–win problem solving. They recommend that participants begin by focusing on their *interests* (*why* they want something) rather than their *positions* (what *solution* they think is best).[52] For example, suppose you'd like a quiet evening at home but your partner wants to go to a party. Those are your positions. Based on them, clearly one of you will win and the other will lose. However, after listening to each other and sharing your interests, you may realize that you can both get your way. You have an interest in spending a quiet evening rather than getting dressed up and talking to a room full of people. Your partner isn't crazy about going out tonight either but has an interest in connecting with two old friends who are going to be at the party. Once you understand both parties' underlying goals, a solution presents itself: Invite those two friends over for a casual dinner at your place before they head off to the party. This way, neither you nor your partner compromises on what you want to achieve. Indeed, the evening may be more enjoyable than either of you expected.

compromise An agreement that gives both parties at least some of what they wanted, although both sacrifice part of their goals.

win–win problem solving An approach to conflict resolution in which the parties work together in an attempt to satisfy all their goals.

Following are the steps involved in win–win problem solving, which is consistent with being assertive. Here too, you'll probably find that, although it's good to know all the recommended steps, you may choose to focus on some more than others, depending on the situation.

Identify Your Problem and Unmet Needs Before you speak up, it's important to realize that if something bothers you, the problem is *yours*. Perhaps you're frustrated by your partner's tendency to yell at other drivers. Because *you* are the person who is dissatisfied, you are the one with a problem. Realizing this may make a big difference when the time comes to approach your partner. Instead of feeling and acting in a way that blames the other person, you'll be more likely to describe your feelings, which will reduce the chance of a defensive reaction.

Before you voice your problem to the other person, pause to think about why their behavior bothers you. Perhaps you're afraid they will offend someone you know. Maybe you're worried that their anger will lead to unsafe driving, or you would like to use the car ride as an opportunity for conversation.

If you feel vulnerable making your needs known, you aren't alone. Brené Brown points out that conflict can stir up deep-seated fears. "We don't know what to do with the discomfort and vulnerability," Brown says, adding that "emotion can feel terrible, even physically overwhelming. We can feel exposed, at risk, and uncertain."[53] Considering this, it's no wonder that many people avoid conflict, accommodate other's wishes, or disguise their vulnerability with aggression.

The irony, points out Brown, is that avoiding conflict and handling it badly usually make people feel worse and more disconnected from others, when what they usually want is to be understood and accepted. The good news, Brown says, is that you don't have to be an expert at understanding emotions—yours or other people's. You need only to be curious about them in an open and nonjudgmental way. This might involve saying, "I'm having an emotional reaction to what's happened and I want to understand."[54]

Agree on a Time to Talk Unconstructive fights often start because the initiator confronts someone who isn't ready. A person may not be in the right frame of mind to face a conflict if they are tired, busy with something else, or not feeling well. At times like these, it's unfair to insist on having a difficult discussion without notice and expect to get the other person's full attention. Instead, you might say, "Something's been bothering me. Can we talk about it?" If the answer is "yes," then you're ready to go further. If it isn't the right time for a serious discussion, find a time that's agreeable to both of you.

Describe Your Problem and Needs Other people can't meet your needs without knowing why you're upset and what you want. It's up to you to describe your problem as specifically as possible without judging the other person. Include a need statement, empathy, and a specific reference to the behavior in question. That might sound something like this:

> I look forward to riding home from work together because I like the chance to hear about your day and make plans for later [*need/desire*]. I know you get frustrated with city traffic [*empathy*], but I feel disappointed when you yell at other drivers instead of talking with me [*problem*].

When Brené and her husband reached the dock, she told him her feelings had been hurt by his brief responses, and she explained why this way:

> I feel like you're blowing me off, and the story I'm making up is either that you looked over at me while I was swimming and thought, *Man, she's getting old.* She

can't even swim freestyle anymore. Or you saw me and thought *She sure as hell doesn't rock a Speedo like she did twenty-five years ago.*[55]

With this statement, Brené showed the self-awareness and courage to say outright why Steve's half-hearted responses were so painful to her. She recommends the phrase "the story I'm making up" as a way to express yourself without blaming the other person.

Check the Other Person's Understanding   After you have shared your problem and described what you need, make sure the other person has understood what you've said. As you may remember from the discussion of listening in Chapter 6, there's a good chance of your words being misinterpreted, especially in a stressful situation. If your partner says, "You're telling me I'm a bad driver," you can take the opportunity to say something like, "I'm not judging your driving. I know it's stressful. I'm just saying I'd love to have some quality time with you on the ride home."

Ask About the Other Person's Needs   After you've made your position clear, it's time to find out what would make the other person feel satisfied about the issue. There are two reasons why it's important to discover their needs. First, it's fair. They have as much right to feel satisfied as you do to. Second, it's good for the relationship. You are most likely to continue interacting if you both feel that your needs are being met.

You might learn about the other person's needs simply by asking about them: "Now that you know what I want and why, tell me what you need from me." After they begin to talk, your job is to use the listening skills discussed in Chapter 6 to make sure you understand.

Back at the lake, Brené was surprised to learn that her husband was dealing with his own fears. He said he had suffered a vivid nightmare the previous night in which he had tried desperately to save all five of their children when a boat had come at them suddenly in the water. He told her:

> I don't know what you were saying to me today. I have no idea. I was fighting off a total panic attack during that entire swim. I was just trying to stay focused by counting my strokes.[56]

Brené understood. As a lifelong swimmer, she and Steve were aware of the dangers posed by sharing the waterway with motorboats. Such a nightmare would have unnerved her, too.

Check Your Understanding of the Other Person's Needs   Paraphrase or ask questions about your partner's needs until you're certain you understand them. You might say, "It sounds like traffic is extra frustrating on the way home because you're tired and hungry. Is that how it feels?"

In her Netflix special *The Call to Courage*, best-selling author Brené Brown talks about the rewards available to people who dare to be vulnerable.

When you feel that others are criticizing or disrespecting you, how do you usually react? Are you willing to be yourself even when others don't approve of your actions? Why or why not?

Negotiate a Solution   Now that you understand each other's needs, the goal is to find a way to meet them. First, partner in thinking of as many potential solutions as you can. If your interest is quality time on the way home and your partner's interest is to avoid a stressful driving situation, brainstorm as many options as you can that will allow you to meet both of those needs.

The key word here is *quantity*. Write down every thought that comes up, no matter how unworkable it seems at first. Next, evaluate the solutions. This is the time to talk about which solutions are likely to meet both of your interests (win–win) and which probably wouldn't.

After evaluating the options, pick the one that looks best to both of you. Your decision doesn't have to be final, but it should seem potentially successful. It's important to be sure that you both understand and support the solution.

To go back to the driving example, perhaps you decide to meet for dinner after work and then drive home once the rush-hour traffic has subsided. Or you might alternate who drives each day or take the train together instead. These solutions might satisfy your interest in spending time together and your partner's interest in avoiding a stressful commute. The beauty of win–win problem solving is that you will probably *both* benefit from more quality time and less stress, which underscores the value of effective conflict management.

Follow Up on the Solution   After you have tried your solution for a while, talk about how things are going. You may find that you need to make some changes or even choose a different option. The idea is to keep on top of the problem and keep using creativity to solve it.

All of this being said, win–win solutions aren't always possible. There will be times when even the best-intentioned people simply won't be able to find a way of meeting all their needs. When that happens, compromising may be the most sensible approach. You will even encounter instances when pushing for your own solution is reasonable, and times when it makes sense to willingly accept the loser's role. Table 9.2 describes some factors to consider. But even when win–win problem

TABLE 9.2
Choosing the Most Appropriate Method of Conflict Resolution

1. Consider deferring to the other person:
   - When you discover you are wrong
   - When the issue is more important to the other person than it is to you
   - To let others learn by making their own mistakes
   - When the long-term cost of winning may not be worth the short-term gains
2. Consider compromising when:
   - There is not enough time to seek a win–win outcome
   - The issue is not important enough to negotiate at length
   - The other person is not willing to seek a win–win outcome
3. Consider competing when:
   - The issue is important and the other person will take advantage of your noncompetitive approach
4. Consider striving for a win-win solution when:
   - The issue is too important for a compromise
   - A long-term relationship between you and the other person is important
   - The other person is willing to cooperate

solving isn't a perfect success, the steps we've discussed haven't been wasted. A genuine desire to learn what the other person wants and to try to satisfy those desires will build a climate of goodwill that can help you improve your relationship.

9.5 Cultural Approaches to Conflict Communication

"If you're planning to travel or live in Italy," says Italian native Ewa Niemiec, "be prepared for a land of strong feelings, loud voices, and even bigger hand gestures."[57] That's not necessarily bad, she says, unless people mistake the passion for conflict or aggression.

Misunderstandings may arise when people from emotionally reserved cultures observe or interact with people from more emotionally expressive ones. Americans visiting Italy and Greece, for example, often think they're witnessing an argument when actually they are overhearing a friendly conversation.

So far this chapter has addressed the dual goals of conflict management (concern for self and concern for others) as if they are of equal merit. But cultural expectations sometimes privilege one of these interests over the other. This section considers the impact of two cultural variables discussed in Chapter 4—individualism and collectivism, and high and low context—and a third variable—emotional expressiveness.

Individualism and Collectivism

In individualistic cultures—like many of those in the United States—the goals, rights, and needs of each person are considered important, and most people would agree that it's an individual's right to stand up for themselves. People in such cultures typically value direct communication in which people say outright if something is bothering them.[58]

By contrast, people in collectivist cultures (common in Latin America and Asia) usually consider the concerns of the group to be more important than those of any individual. Preserving and honoring the face of the other person are prime goals, and communicators go to great lengths to avoid communication that might embarrass a conversational partner. The kind of assertive behavior that might seem perfectly appropriate to an American or Canadian would seem rude and insensitive in these cultures.

Depending on the culture and context, this person may seem either friendly or threatening.

How emotionally expressive are you? What role does culture play in what feels natural to you?

High and Low Context

As you might imagine, people in low-context cultures—like many in the United States—often place a premium on being direct and literal. By contrast, people in high-context cultures, like that of Japan, more often value self-restraint and avoid confrontation. Communicators in high-context cultures derive meaning from a variety of unspoken cues, such as the situation, social conventions, and hints. For this reason, what seems like "beating around the bush" to an American might seem polite to an Asian. In Japan, for example, even a simple request like "close the door" may seem too straightforward. A more indirect statement such as "it's somewhat cold today" would be more appropriate. Or a Japanese person may glance at the door or tell a story about someone who got sick in a drafty room. They may also be reluctant to simply say "no" to a request. A more likely answer would be, "Let me think about it for a while," which anyone familiar with Japanese culture would recognize as a refusal, but others may take to mean "It's possible. I'll let you know my decision soon."

Emotional Expressiveness

From the examples so far, you might expect the United States to top the charts in terms of directness when it comes to conflict management. However, a mediating factor is at play—emotional expressiveness. In this regard, many people in the United States have a great deal in common with Asian cultures, namely, a preference for calm communication rather than heated displays of emotion.[59] From this perspective, it may seem rude, frightening, or incompetent to show intense emotion during conflict. Indeed, people who become passionate are warned about the danger of saying things they don't mean, and highly expressive people may labeled "loose cannons" or "hot heads."

By contrast, in cultures that value emotional expressiveness, people who do *not* show passion are regarded as hiding their true feelings. Latin Americans, African Americans, Arabs, Greeks, Italians, Cubans, and Russians are typically considered highly expressive.[60,61] To them, behaving calmly in a conflict episode may be a sign that a person is unconcerned, insincere, or untrustworthy.

It's easy to imagine how people from different cultural backgrounds might have trouble finding a conflict style that's comfortable for them both. Conflict management styles can differ even within one culture or family. Sensitive communication can often help bridges those gaps, however.

Brené Brown reflects that the conversation in the lake, which might have ended in bickering and withheld affection, instead resulted in a renewed sense of love and commitment. Expressing their fears and listening to each other brought her and Steve closer. Afterward, as the couple walked back up to the lake house, he popped her playfully with his wet towel and said, "Just so you know: You still rock a Speedo."

MAKING THE GRADE

OBJECTIVE 9.1 Explain the key facets of interpersonal conflict, including expressed struggle, interdependence, and the perception of incompatible goals and scarce resources.

- Interpersonal conflict is an acknowledged struggle between at least two interdependent people who perceive that they have incompatible goals, scarce resources, and interference from one another in achieving their goals.
- Many people think that the existence of conflict means that there's little chance for happy relationships with others. Effective communicators know differently, however. They realize that although it's impossible to *eliminate* conflict, there are ways to *manage* it effectively.
- Managing conflict skillfully can lead to healthier, stronger, and more satisfying relationships.
  > What distinguishes interpersonal conflict from the frustration you may feel with the behavior of a stranger you will never see again?
  > Think of a time in which you and a relational partner experienced conflict. What goals and resources were involved? Were you able to express your feelings to each other and reach a mutually satisfying conclusion? Why or why not?
  > Imagine that someone you care about begins to say, "You're right. I'm wrong," any time conflict between the two of you emerges. Do you think this is a good or a bad sign for your relationship? Why?

OBJECTIVE 9.2 Describe the role of communication climate and relational spirals in interpersonal relationships, and practice communication strategies for keeping relationships healthy.

- Communication climate refers to the emotional tone of a relationship.
- Confirming communication occurs on three increasingly positive levels: recognition, acknowledgment, and endorsement.
- Disconfirming messages deny the value of others and show a lack of respect. Four particularly damaging forms of disconfirming messages are criticism, contempt, defensiveness, and stonewalling.
- Relational spirals are reciprocal communication patterns that escalate in positive or negative ways.
- Communication strategies that enhance relational climates include using "I" language, striving for mutually satisfying options, being honest, showing empathy, and respecting other people's viewpoints.
  > Recall a recent verbal or nonverbal message that made you feel good about yourself. Now think of one that made you feel frustrated or unappreciated. What was different about these episodes?
  > Describe a confirming message you have sent to someone else. Then describe a disconfirming message you have sent, even if you didn't mean to. How did the other person react in each episode?
  > How might you react if your partner is upset with you and you don't think you have done anything wrong? Do your answers suggest that you are ever guilty of criticism, contempt, defensiveness, or stonewalling? If so, how might you behave differently to avoid damaging your relationship?
  > Describe a time when you and a relational partner were involved in an increasingly negative spiral. What did you do (or what might you have done) to help stop the downward spiral?
  > Identify several disconfirming messages from your own experience, and rewrite them as confirming ones using the tips for creating positive communication climates in this chapter.
  > Is the emotional tone of your most important relationships warm and welcoming, stagnant, or chilly and unsatisfying? How so?

OBJECTIVE 9.3 Identify characteristics of nonassertive, indirect, passive-aggressive, directly aggressive, and assertive communication, and explain how conflict approaches vary.

- Nonassertive behavior reflects a person's inability or unwillingness to express thoughts or feelings in a conflict, as when one engages in avoidance or accommodation.

- Indirect communication involves hinting about a conflict rather than discussing it directly.
- Passive aggressive behavior is somewhat indirect, but it has a hostile tone meant to make the recipient feel bad.
- Directly aggressive communicators seek to intimidate or belittle people rather than striving for production solutions collaboratively.
- Assertive communicators share their feelings and goals and encourage others to do the same.
  > Describe the pros and cons of each of the following conflict styles: nonassertive, indirect, passive aggressive, directly aggressive, and assertive.
  > Think of a behavior that bothers you. Write out what you might say to the person involved, including the components of assertive messages described in this chapter.
  > Scan several Twitter feeds or the comments sections on a blog or YouTube video. Identify examples in which people said things they might not have in person.

OBJECTIVE 9.4 Explain the differences among win–lose, lose–lose, compromising, and win–win negotiation strategies, and apply the steps involved in achieving win–win solutions.

- Although win–lose and lose–lose conflicts do not sound appealing, in rare occasions they are the best option.
- Compromise is often heralded as effective conflict management, but it may not always be the best option, considering that it involves less-than-optimal results for everyone involved.
- Win–win outcomes involve goal fulfillment for everyone. This is often possible if the parties involved have the proper attitudes and skills.
  > What happened the last time you openly disagreed with someone? Was your relationship with that person better or worse afterward?
  > Think of a time when you compromised to resolve a conflict. Were you satisfied with the result? Can you think of a way in which both of you might have met your goals completely?
  > List and describe the eight steps for win–win negotiating described in this chapter.
  > Imagine that you and a friend have just signed the lease on an apartment with one regular bedroom and one master suite. Using the principles of win–win negotiating, how might you work together to decide who gets which bedroom?

OBJECTIVE 9.5 Compare and contrast conflict management approaches that differ by culture.

- In different cultures, it may be expected that people will approach conflict indirectly or that they will be expressive and direct about it.
  > Describe conflict management approaches that feel most comfortable to you. What has influenced you to respond in those ways?
  > Describe how you might adapt your conflict management approach to suit different people and situations, such as with your best friend, at work, with older adults in your family, and with your boss or professor.

KEY TERMS

assertive communication p. 240
avoidance spiral p. 235
communication climate p. 231
compromise p. 247
confirming messages p. 233
conflict p. 229
contempt p. 233
criticism p. 233
defensiveness p. 233
direct aggressive message p. 240
disconfirming messages p. 231
escalatory conflict spiral p. 235
indirect communication p. 238
lose–lose problem solving p. 246
negotiation p. 245
nonassertion p. 237
passive aggression p. 239
positive spiral p. 236
relational spiral p. 235
stonewalling p. 233
win–lose problem solving p. 245
win–win problem solving p. 247

PUBLIC SPEAKING PRACTICE

Think of a conflict in your life. What were your needs and fears regarding the conflict? What were the other person's? Were you able to meet both of your needs? If so, how? If not, what might you have done differently? Create a brief oral presentation in which you share the lessons you learned about conflict management from this experience.

ACTIVITIES

1. Deepen your understanding of how confirming and disconfirming messages create communication spirals by doing the following:

    - Think of an interpersonal relationship. Describe several confirming or disconfirming messages that have helped create and maintain the relational climate. Be sure to identify both verbal and nonverbal messages.

    - Show how the messages you have identified have created either negative or positive relational spirals.

2. Composing assertive responses for each of the following situations:

    - A neighbor's barking dog is keeping you awake at night.

    - A friend hasn't repaid the $20 she borrowed two weeks ago.

    - Your boss made what sounded like a sarcastic remark about the way you put school before work.

Now develop two assertive messages you could send to a real person in your life. Discuss how you could express these messages in a way that is appropriate for the situation and that fits your personal style.

Communicating for Career Success

CHAPTER OUTLINE

10.1 Communication Skills Are Essential 258

10.2 Setting the Stage for Career Success 259
   Developing a Good Reputation
   Managing Your Online Identity
   Cultivating a Professional Network

10.3 Preparing Job Search Materials 263
   Create a Portfolio of Your Work
   Write a Confidence-Inspiring Cover Letter
   Construct a High-Quality Resume
   Follow Application Instructions
   Keep Organized Records of Your Interactions

10.4 Taking Part in a Job Interview 269
   Preparing for an Interview
   Participating in a Job Interview

10.5 Adapting to a New Work Environment 277
   Culture in the Workplace
   Patterns of Interaction
   Communication and Workplace Etiquette
   Working Remotely

   MAKING THE GRADE 285
   KEY TERMS 286
   PUBLIC SPEAKING PRACTICE 287
   ACTIVITIES 287

LEARNING OBJECTIVES

10.1 Describe the reasons why good communicators are likely to excel in the workplace.

10.2 Engage in communication behaviors conducive to becoming an appealing job candidate.

10.3 Create application materials that make it easy for prospective employers to see how your skills and aspirations match their needs.

10.4 Practice communication strategies to make a good impression during employment interviews.

10.5 Demonstrate skills to communicate effectively as a new employee in person and remotely.

While she was growing up, Nathália Rodrigues watched her parents struggle financially.[1] As a college student, she experienced the stress of high-interest credit card debt herself.[2] Determined to make a difference, Rodrigues launched Nath Finanças, a social media crusade to provide everyday people with easy-to-understand information about personal budgets, saving money, interest rates, and more. Her catchphrase "ARE YOU SURE YOU NEED THIS?" has inspired many followers to avoid unnecessary spending, and when they slip up, to post the lament "Nath, I have failed you."[3]

Rodrigues's posts are now viewed by more than half a million people in her native Brazil and around the world. At age 22, she was named one of the "World's 50 Greatest Leaders" by *Fortune* magazine.[4] Even if you're not an aspiring influencer and entrepreneur like Rodrigues, her experience underscores many lessons that can boost your career success in any field:

- Build great communication skills.
- Network with others.
- Develop a professional identity.
- Present yourself knowledgeably and confidently.

This chapter explores these strategies and others to help you succeed in the job market and in the career of your choice.

10.1 Communication Skills Are Essential

Employers across the board agree that communication is vital to getting a good job and excelling in the workplace. Verbal and written communication skills are among the top five qualities that employers look for in job candidates, along with teamwork and problem-solving skills that rely heavily on communication. Communication ability even outranks technical knowledge on employers' wish lists.[5] They're right to place such a high premium on it. Here are four reasons that outstanding communicators flourish in the professional world:

Good Communicators Work Well in Teams "Life is a team sport," observes human resource specialist Robert Half.[6] This is especially true in the workplace, where effective teamwork is linked to successful outcomes, high morale, efficiency, problem solving, satisfaction, and loyalty to the organization.[7]

Good Communicators Build Public Awareness Along with promoting the organization during one-on-one interactions, team members can serve as brand ambassadors to larger audiences. They might make sales pitches, inform audiences about the value of a product or service, or advocate for change or public policies. (Chapters 12 through 15 will help you refine your public speaking skills.)

Good Communicators Enhance Customer Satisfaction A resounding 9 in 10 people say that customer service is a deciding factor in who they do business with, and the number-one factor in customer service is—you guessed it—effective communication.[8,9] Great service keeps people coming back and can enhance an organization's reputation among potential customers. For example, when a toddler became fussy at a Trader Joe's store in Florida, employees began dancing and singing to entertain him. After the child's mother videotaped and shared the incident on social media, it went viral and was broadcast by *Good Morning America*, ABC News, and other news outlets.[10] Ultimately, the employees' over-the-top effort to help a customer was viewed more than 500,000 times.

Good Communicators Make Good Leaders Leaders spend most of their time communicating—mostly about organizational problems that boil down to poor communication, observes business analyst Matt Myatt.[11] When leaders don't communicate effectively, team members are likely to experience added stress, low morale, and costly delays and failures.[12] On the other hand, employee productivity, satisfaction, and loyalty tend to be higher when leaders are good listeners with a genuine interest in understanding the people around them.[13,14]

"No matter what industry you work in and no matter what your role is, effective communication moves a business forward," says business communication expert Mary Shores. That doesn't mean it's easy. "Let's be honest," Shores says, "becoming a master communicator takes serious work."[15]

10.2 Setting the Stage for Career Success

Launching a successful career begins long before you apply for a job. "Ideally, college students should take steps to lay the foundation for an effective job search as early as the second semester of their freshman year," advises Mike Profita, a career development specialist with more than 20 years' experience.[16] Forward-thinking students realize that college is an audition for the work they will do later. They also know it's never too soon to invest in career-related networking. Here are some communication strategies you can use now to enhance your future success.

Developing a Good Reputation

Put yourself in the shoes of a professor, classmate, or current coworker. If you frequently show up late, they can only assume you will do the same in the future. If your work or appearance are sloppy, they probably can't imagine anything different from you. The opposite is also true. People who go above and beyond now will probably be first in line for glowing recommendations and opportunities later.

Jim Kellam, a biology professor at Saint Vincent College, devoted personal time over winter break to write an enthusiastic letter of recommendation for a student who had excelled in his classes. Would he do the same for a poor or mediocre student? No. There's too much at risk, Kellam says. For one thing, insincere praise won't help a graduate succeed, even if they get the job. For another, praising an underperformer may lead employers to believe that the program's "excellent" students aren't actually very good. Finally, Kellam says, "my reputation will be harmed because I lied or exaggerated how good the student would be for the job."[17]

Managing Your Online Identity

Think of your digital presence as "online marketing—but for yourself," recommends one social media leader.[18] What you post now can work for or against you later.

The greater your online presence, the more vigilant you need to be to ensure that potential employers won't find fault with how you present yourself.

What would a potential boss think after reviewing your online presence? Can you boost your employability by cleaning up your identity?

About 7 in 10 employers look at applicants' social media posts. On the bright side, about 43 percent say a favorable online presence has inspired them to hire a candidate. However, 57 percent have ruled out candidates based on what they learned about them online.[19] Here are experts' tips for creating an online presence that works in your favor.

Showcase Your Strengths and Goals Creating an online identity isn't about fooling anyone—it's about portraying yourself in an authentic and favorable way. Take stock of your interests, talents, and goals. Make sure that people who encounter you online have a clear sense of who you are.

Build a Professional Identity Your online photos and information should create a sense that you're ready for the career of your choice. For example, hoping to land a job in Washington, DC, Joseph Cadman used LinkedIn to post a profile photo of himself in a suit and tie in front of the Capitol Building.[20] While you're at it, make sure your email address and profile names are dignified. An unprofessional identifier (such as BeachBum, Juicy, or MuscleMan) can land you in the "no" pile, say job recruiters.[21]

Avoid Embarrassing Posts When employers review candidates' online presence, the most common deal-breakers are evidence of drinking or partying, posts that are disrespectful toward others, and critical comments about a current or former employer.[22] Even if an off-color meme seems like a harmless joke, do you want it to represent you? Think carefully about everything you post. (See Chapter 2 for more about using social media effectively.)

Monitor Your Online Presence Even information you think is private may be accessible online. Search for yourself on a range of digital platforms (Google, Yahoo!, MSN Search, MetaCrawler, Dogpile, Ask.com) and see what comes up. Also double-check the privacy settings on your social media accounts (although you shouldn't consider them foolproof) and sign up for Google Alerts to receive a notification when your name pops up online. While you're at it, create Google Alerts for potential employers so you can stay current about them as well.

digital dirt Unflattering information (whether true or not) that has been posted about a person online.

Engage in Damage Control The phrase **digital dirt** refers to unflattering information posted about a person online, whether it's true or not. If possible, remove incorrect, unfair, and potentially damaging information about yourself. If you can't remove it, consider a service such as ReputationDefender.com that will monitor your online identity and ask the managers of offending websites to remove the information.

Beware of Mistaken Identities You might find that unfavorable information pops up about someone with the same name as you. One job seeker googled herself out of curiosity, only to find that the first hit was the Facebook page of a person with the same name whose personal profile was loaded with immature comments. To minimize the chance of being mistaken for someone else, you might distinguish yourself by including your middle name or middle initial on your resume and online.

Don't Be Scared Off With all of these warnings, you may be tempted to avoid having a digital footprint at all. That's probably unwise. Nearly half of employers say that job candidates with no online presence aren't likely to make the short list. They reason that candidates who have established an impressive online presence are likely to do the same in person.[23]

Remain Careful When You Get Hired Your social media conduct remains important once you're on the job. Never post profanity or information that disparages your employer or clients, reveals confidential information, or makes you (hence your employer) look bad. And if you see a post that disparages your employer, don't post a confrontational response, advises a human resource specialist. Instead, notify your supervisor or public relations team and let them handle it.[24]

Cultivating a Professional Network

Here are a few real-life examples that show the power of networking:

- Aspiring to launch a career in marketing, Andrei Petrik requested a meeting with the CEO of a marketing firm to get his advice. The CEO was so impressed by Petrik that he hired him.[25]
- Hoping to move into management, Oyinkan Akinmade shared her career goals with her supervisor. When a leadership position became available, the supervisor made sure Akinmade knew about it. She applied and got the promotion.[26]
- When Ed Han learned that a high school friend knew a hiring manager in publishing, he asked his friend to pass along his resume. The manager felt that Han had promise and gave him a job.[27]

The list of people who have landed jobs by **networking** (meeting people and maintaining relationships) is nearly endless. By some accounts, 8 in 10 people in the workplace today got their jobs by networking.[28] That's because most positions are never advertised,[29] and even when they are, employers are (naturally) influenced by the recommendations of people they trust. As a consequence, jobs go to those who form relationships with people in the industry. It's never too soon to start.

> **networking** The process of creating and maintaining relationships that further one's goals.

Networking will only work, of course, if you are the kind of person others recognize as worth endorsing. If you are willing to work hard and you have the necessary skills to do a job (or are willing to learn those skills), there are several steps you can take to create and benefit from a personal network.

Look for Networking Prospects Besides the people you see every day, you probably have access to a wealth of other contacts. These might include (current and past) co-workers, classmates, teammates, social acquaintances, and more. The feature "Building a Career-Enhancing Network" provides some tips for being in the right place to meet people who may be able to help with your career.

Engage in Online Networking Numerous websites offer professional networking opportunities. A few of the most popular are LinkedIn, LetsLunch, Opportunity, and Shapr. Most offer basic memberships for free. It's not too early to join a few of these and

The saying "It isn't just what you know, it's who you know" illustrates the value of cultivating a network of people who can offer you career guidance and support.

Networking events are one way to meet people. What are some other ways you might expand your network?

set up a personal profile—which might include career-relevant projects you've been part of, volunteer work, awards, accomplishments, and interests. Above all, consider how your information will look to people who don't know you well. You may be proud of your membership in the National Rifle Association, Planned Parenthood, or a religious or political group, but a prospective employer might not find your affiliations so admirable. (You'll read more about managing your online identity later in the chapter.)

Seek Referrals Each contact in your immediate network has connections to other people who might be able to help you. Social scientists have verified that the "six degrees of separation" hypothesis is true. The average number of links separating any two people in the world is indeed only half a dozen.[30]

You can take advantage of this by seeking out people who are removed from your personal network by just one degree. If you ask 10 people for referrals, and each of them knows 3 others who might be able to help, you have the potential of support from 30 people.

Conduct Informational Interviews Andrei Petrik, whose story begins this section, had the courage to reach out and ask for an **informational interview**—a structured meeting in which the interviewer seeks knowledge from someone. Interviews of this sort are usually conducted in person, but if necessary, you might also use the phone or an online meeting format. A good informational interview can help you achieve the following goals:

- Learn more about a job, organization, or field.
- Make a positive impression on the person you are interviewing.
- Gain referrals to other people who might be able to help you.

Unless you know the person well, it's usually best to send your request for a meeting in a letter or email. (A text message is usually too casual, and DMs are easily overlooked.) Keep your message brief. Introduce yourself, explain your reason for wanting the interview, and emphasize that you're seeking information, not asking for a job. Request a specific amount of time for a meeting. (Shorter requests are more likely to be granted.)

After initial pleasantries, be sure to use your limited time to focus on career-related information, such as "What are the three fastest-growing companies in this field?" and "What do you think about the risks of working for a start-up company?"

Often, the best way to get information is to ask straightforward questions. But there are times when it's more gracious to be indirect. For instance, instead of asking

> **informational interview**
> A structured meeting in which a person seeks answers from someone whose knowledge might help them succeed.

> **COMMUNICATION STRATEGIES**
>
> ## Building a Career-Enhancing Network
>
> **Take Part in Volunteerism and Service Learning**
> In addition to gaining experience you may use in a career, you're likely to come in contact with civic and business leaders while working in the community.
>
> **Attend Lectures, Forums, and Networking Events**
> Make it your goal to speak personally with several people whose interests are similar to yours. Although it's natural to feel a little nervous and out of place at first, networking coach Darrah Brustein says, "Be yourself.... The people you connect with when you are authentic are the ones you'll want to stay in touch with."[31] After the event, send a personal note thanking the presenters and others you met there.
>
> **Keep Up with Local News**
> There's no better way to know who is involved and what the latest issues are. This can help you identify potential mentors and speak knowledgeably about current issues when you meet them.
>
> **Join Career-Related Organizations**
> Many professional groups offer discounted membership fees for students. Take advantage of the opportunity to meet people in the career field you hope to join. And don't just sit in the back of the room. Volunteer to serve on committees or hold an office. Your hard work will be noticed.
>
> **Use Online Resources**
> Search LinkedIn and similar sites for people who have something in common with you—perhaps those who attended the same school as you or majored in the same discipline. Common interests can be great conversation starters.

"What's your salary?" you might say, "What kind of salary might I expect if I ever held a position like yours?" Listen closely and ask follow-up questions. If the primary question is "Who are the best people to ask about careers in the financial planning field?," a secondary question might be "What do you think is the best way for me to go about meeting them?"

Show Appreciation Don't forget to show gratitude to the people in your network. Always send a thank-you message after meeting with them. Beyond that, take the time to maintain relationships and let your contacts know when their help has made a difference in your career advancement. In addition to being the right thing to do, your thoughtfulness will distinguish you as the kind of person worth hiring or helping again in the future.

10.3 Preparing Job Search Materials

First impressions aren't reserved for face-to-face meetings. In fact, most potential employers will form an impression of you before they even meet you, which makes it especially important to represent yourself well on paper and online. Here are some plan-ahead strategies for presenting yourself in the best (most accurate) light.

Create a Portfolio of Your Work

One of the best ways to show employers what you *can* do is to share with them what you *already* have done. A professional **portfolio** is a collection of your best work, along with awards, letters of commendation, and other materials that provide evidence of your credentials. An **eportfolio** (also called a digital portfolio or efolio) is essentially a personal website on which you present the same information. It used to be that only artists and designers created portfolios, but that has changed. "Above all, employers [of all types] want to see proof of your qualifications," advises one recruiter.[32] You'll benefit from developing both a printed portfolio and an online efolio. Here are some tips for creating them:

> **portfolio** A collection of a person's best work along with awards, letters of commendation, and other materials that provide evidence of their credentials.

> **eportfolio** A personal website on which a person presents portfolio information, such as work samples, awards, and letters of commendation.

- *Acquaint yourself with eportoflio options.* Search online for eportfolios in your discipline or career field to see what others have done.
- *Begin with the end in mind.* Review job postings in the career of your choice. Note what employers are looking for, and then create a blueprint for a portfolio that showcases those qualifications. For example, if hiring managers list teamwork, communication skills, and fundraising experience, those should be the main focus of your portfolio.
- *Start building now.* Even if you don't have professional experience yet, your academic achievements, group projects, and volunteer work are excellent opportunities to create portfolio elements. Use instructors' feedback to help you polish up writing samples before you post them. If relevant, include press releases, news items, videos, and other work you have created. Describe a relevant group project, what you contributed, what the group accomplished, and how the experience has helped you develop professional skills. If someone you worked with wrote a letter praising your efforts, include that too.
- *Categorize information.* Organize information into clear categories, such as Resume, Leadership Experience, Social Media Design Work, Writing Samples, Design Awards, or whatever is relevant to your career aspirations. Categories will make your materials easy to navigate and will call attention to your most important qualifications.
- *Keep the format simple.* Most people will skim your portfolio contents quickly. They probably won't spend time wrestling with a complicated format or sorting through extraneous information. Your portfolio design should be clear, professional, and attractive. Templates for eportfolios are provided by platforms such as Wix, Squarespace, WordPress, FolioSpaces, Google Sites, Weebly, and PortfolioGen.
- *Make your eportfolio searchable.* The online templates just mentioned make it easy to present your information in HTML, which is conducive to online searches. Include key words people might search for, and consider creating a webpage meta description to make it easier for people to find and preview your work. (Search "create a meta description" online for instructions.)
- *Share your work.* Your portfolio won't count for much if people don't see it. Bring a printed copy to job interviews, and include links to your eportfolio on your resume, your LinkedIn profile, and other networking sites.

Write a Confidence-Inspiring Cover Letter

As one expert puts it, a cover letter is "an introduction, a sales pitch, and a proposal for further action all in one."[33] Write a letter that provides a brief summary of your interests and experience, and include the contents listed here:

- If possible, direct your letter to a specific person. (Be sure to get the spelling and title correct.)
- Indicate the position you're applying for and introduce yourself.
- Briefly describe your accomplishments as they apply to qualities and duties listed in the job posting.
- Demonstrate your knowledge of the company and your interest in the job.
- State the next step you hope to take—usually a request for an interview.
- Conclude by expressing appreciation to the reader for considering you.

See Figure 10.1 for an example of a cover letter that includes all of these.

Templates are available to create an electronic portfolio of your work that you can easily share with prospective employers.

Considering a job you would like to have one day, what skills and accomplishments might you include in a portfolio about yourself?

Construct a High-Quality Resume

Whereas a cover letter is an introduction meant to pique an employer's interest, your resume should provide details to back up the claim that you're a qualified candidate. A good resume provides specific information about your professional strengths and achievements without embellishing. "I've received thousands of resumes," says Gary Burnison, the CEO of a global consulting firm. His favorite resume of all time "had no gimmicks, no Fortune 500 company listed and wasn't folded into an origami airplane," he says. Instead, it stood out because it was easy to read, clear, truthful, and specific.[34] Here are some tips from Burnison and other recruiters to help you create a top-notch resume.

Begin with Your Name and Contact Information   "Even if you're the most qualified person in the world, it's not going to matter much if the hiring manager can't contact you," points out one recruiter.[35] Include your full name, phone number (including area code), mailing address, and email address. Make sure everything is correct and that your name appears exactly as it does on social media and professional correspondence. (If you will soon change your last name or you have done so recently, indicate both.) Give your personal phone number rather than your current work number. And, since some employers keep applications on file for the future, provide an email address you will continue to monitor.

Include URL Links to Your Work   If your resume will be viewed electronically (as most are), include URLs to your eportfolio, LinkedIn profile, or other electronic means of showcasing your work.

Rose L. Magnon
[mailing address]
[permanent phone number]
[permanent email address]

[Date]

Renée Robinson, Executive Director
International Society for the Advancement of Children
2525 West 37th Avenue
Landersville, MD 55555

Dear Ms. Robinson:

I am interested in the public relations assistant position at the International Society for the Advancement of Children, as advertised in *The Philanthropy Newsletter*.

As the attached resume shows, I currently serve as a Communication Coordinator at the University of East Florida, in which position I have gained experience that would allow me to make significant contributions to your organization and its mission.

I believe I would be effective in the public relations assistant position at ISAC for three main reasons.

- I am a skilled communicator who crafts messages carefully and adapts well to different audiences and formats. I currently author three newsletters, each designed to reach a specific stakeholder group. I also curate content on numerous social media platforms including Facebook, Instagram, and Twitter. I believe my success in these endeavors would allow me to recruit sponsors for ISAC, coordinate and publicize ISAC events such as your annual Children's Festival, and share your success stories.

- I have demonstrated success in fundraising. As a volunteer workplace campaign manager, I coordinated fundraising efforts involving 100 volunteers. The result was a total contribution of more than $800,000. I would enjoy the chance to help coordinate and promote your semiannual Education for All event and other fundraisers.

- I am an accomplished speaker who enjoys interacting with audiences of all sizes. In the last few years, I have won a regional Toastmasters Competition and ranked in the Top 3 at state collegiate debate and forensics tournaments. I would like to use this skill at conferences and civic events to share stories about the good work that ISAC does.

I became interested in the International Society for the Advancement of Children while working on an international project with United Way. I particularly admire your efforts to provide education to children in impoverished areas of the world. My career goal is to coordinate humanitarian efforts, and I would be honored by the chance to help with the good work that you do.

I have attached my resume, which includes links to my online portfolio and my LinkedIn profile, where you can see samples of my work. I hope you will consider me for the assistant public relations position. I am available for an interview at your convenience. Thanks very much for your consideration.

Sincerely,

Rose L. Magnon

FIGURE 10.1 Sample Cover Letter

Weigh the Benefits of a Summary or Objective Consider whether to include a career objective (a one-sentence explanation of what type of position you hope to attain), a resume summary (a brief explanation of how your qualifications apply to the job), or neither. A career objective that's a perfect fit for the position can work in your favor, but if it doesn't exactly match, it may knock you out of the running. If a summary effectively calls attention to your qualifications, you might include one. But if your qualifications are already straightforward, and including a summary would take up space you can use more effectively in another way, you might skip the objective and summary altogether.[36]

Present Your Most Important Qualifications First This usually means listing your academic accomplishments early on and describing your work experiences from the most recent to the least recent.[37]

List Accomplishments, Not Just Job Duties Instead of a vague statement, such as "Volunteer and fundraising experience," be specific about what you have achieved. For example, you might write "Directed a five-student team that raised $12,000 for college scholarships."

Use Powerful Verbs You should use full sentences in your cover letter, but you probably won't in your resume. That means that most of your descriptions will begin with verbs. Make sure they are powerful and specific. Verbs such as *created, implemented, achieved, exceeded, increased*, and *completed* convey energy and specific accomplishments,[38,39] as in "Achieved a 97% satisfaction rating over a 24-month period as a customer care representative."[40] Also make sure verb tense is correct. Prior experiences should be described in past tense (e.g., "Supervised . . .") and ongoing experiences in present tense ("Supervise . . .").

Be Honest "Make no mistake: Employers *will* do a reference check—and if they find out that you lied about something, it's game over," says Burnison. His advice is twofold: "Tell the truth" and include links to an online portfolio so employers can see evidence of your achievements.[41] (Even if someone does get a job with a less-than-honest resume, they're likely to be fired if the deception comes to light.)

Edit Thoroughly Many employers immediately put cover letters and resumes that include typos or grammatical errors in the "no" stack. If possible, have a staff member at your school's career center critique your materials to be sure they are free of errors.

Make Your Materials Easily Searchable Many companies use software to electronically scan cover letters and resumes to help identify the most appealing candidates. This software is most likely to spot your qualifications if your materials include words (spelled correctly) that are relevant to job qualifications and if the format you use is easily scannable. Resumes with complicated or unusual design elements may throw off electronic scanners and knock you out of the running. Unless the employer specifically asks for a text-based format such as Word, upload the file as a PDF, which is most compatible with current search software.[42]

Use Two Pages If Necessary Most people have abandoned the old advice that a resume should never be longer than one page. Use two pages, if necessary, to include your relevant qualifications and keep the format from looking cramped or hard to read.[43] But don't pad your resume to make it longer. Always keep the information you present clear and concise.

You might use the sample resume in Figure 10.2 as a guide, but keep in mind that there are different resume formats for different purposes. Consult with your school's

Rose L. Magnon

[mailing address]
[permanent email address]
[permanent phone number]
[URLs for online portfolio, LinkedIn page, or other online materials]

SUMMARY OF QUALIFICATIONS

Experienced in using social media to promote mutually beneficial collaborations in the community and with members of other cultures. Dedicated to a career in the nonprofit sector coordinating humanitarian efforts and securing funding for international partnerships.

EDUCATION

University of East Florida, Oceanview, FL
B.A. in Communication/Public Relations, 3.86 GPA on a 4-point scale

EMPLOYMENT EXPERIENCE

Communication Coordinator, University College of Arts, Social Sciences, and Humanities
2021 to present

- Write, design, and distribute a semiannual dean's report to reflect the college's goals and achievements
- Create and publish approximately 300 social media posts per year on Twitter, Instagram, and Facebook
- Organize and publicize a university/community lecture series attended by an average of 850 people per year
- Analyze two-way communication with stakeholders as the basis for an annual strategic communication plan

Director of Student Recruitment (temporary contract) January to August 2020

- Led nationwide marketing effort to recruit university accounting majors to take part in a new online employment platform. My efforts helped to attract 50 new students in 8 months.

ACTIVITIES & VOLUNTEER EXPERIENCE

Co-Chair of Cultural Team for United Way Global Resident Fellowship Program October to December 2021

- Developed partnerships with United Way organizations in Western Australia, France, and South Africa. Wrote influential whitepaper detailing lessons learned and recommendations for future programs.

Workplace Campaign Manager August 2012 to March 2015

- Managed $830,000 workplace campaign portfolio representing 120+ accounts with 130 volunteers. Secured new sponsorships and developed sponsorship campaign materials.

ASSOCIATION MEMBERSHIPS

- Oceanview Young Professionals, Government Affairs Council, 2020 to 2021
- Florida Public Relations Society of America, 2019 to present
- University of East Florida Forensics and Debate Team, 2020 to present

HONORS & AWARDS

- Top 3 finalist in Florida's State Collegiate Debate and Forensic Competition, 2020 and 2021
- Toastmasters Regional Impromptu Speaking Contest, 1st place 2021
- Dean's and President's List every semester since entering college in 2020

FIGURE 10.2 Sample Resume

career services center or type "create resume" into your favorite search engine to see various options and tips for creating them.

Follow Application Instructions

There are stories of job candidates who do something so unusual that it catches a hiring manager's attention, but it's more likely that failing to follow instructions will create a bad impression. For example, if the announcement says "no phone calls," then don't call. As one recruiter puts it: "If you can't follow clear, simple directions, how can I trust that you will be able to give great attention to the details of your job?"[44]

Keep Organized Records of Your Interactions

Maintain a list of everyone with whom you communicate. Include contact information, when the message was sent or received, and what it was about. Save copies of all written correspondence for future reference.

10.4 Taking Part in a Job Interview

A successful job interview can be a life-changing experience. It's worth preparing for it carefully. This section presents experts' suggestions for engaging in a **selection interview**—a question and answer session during which a candidate's qualifications are considered. Employment interviews are the most common type of selection interviews, but the same principles apply if you're being considered for a promotion or reassignment, an award, a scholarship, admission to a graduate program, or another opportunity.

> **selection interview**
> A conversation aimed at evaluating a candidate for a job or other opportunity.

Preparing for an Interview

Excelling during a job interview is a process that begins before you arrive or sign on. Here are some ways you can prepare in advance.

Do Research Displaying your knowledge of an organization and industry is a terrific way to show potential employers that you're a motivated and savvy person. In some organizations, failure to demonstrate familiarity with the organization or job is an automatic disqualifier. Along with remembering what you've learned from informational interviews, do advance research. Here are some strategies you might use:

- Read the organization's website thoroughly. Most include an "About Us" section where you can learn about the organization's history, mission, core values, strategic initiatives, size, locations, and more.[45]
- Follow and study the organization's social media feeds.
- Search for media coverage about the organization, industry trends, and market shifts.
- Search for the organization online and see what others (including news organizations) are saying about it.
- If possible, talk to people who are familiar with the organization, such as employees or clients.
- Research the organization's key competitors using the tips just listed.

Based on what you learn, prepare questions to pose during the interview. For example, you might ask:

> I understand that you're piloting a location flexibility program, allowing some employees to work either remotely or on-site. I read in the *Human Resources Bulletin* that this approach has worked better for some jobs than others. Do you plan to allow this option for everybody in the company, or just for specific jobs?

You're likely to learn a lot and impress the employer by asking insightful questions.

Prepare for Likely Questions Regardless of the organization and job, most interviewers have similar concerns, which they explore with questions such as "What is your greatest weakness?" The "Communication Strategies" feature with that title suggests some ways you might respond.

Some interviewers also ask nontraditional questions to see how well candidates think on their feet, how they handle problems, and how creative they are.[46] Some examples include:

> If you could have any superpower, what would it be?
>
> Name five uses for a stapler with no staples in it.
>
> If you were a sweater, what kind would you be?

If asked one of these questions, maintain your composure and use your answers to demonstrate qualifications you would like to showcase. For example, one response to the sweater question might be, "I would be appealing but not flashy, so I would

---

COMMUNICATION STRATEGIES

Answering "What Is Your Greatest Weakness?"

It's a common question in job interviews and one of the toughest to answer. However you respond, try to show how awareness of your flaws makes you a desirable person to hire. Here are four ways you might respond, but there are endless possibilities. Ensure that your answers reflect your own experiences.

Discuss a Weakness That Can Also Be Viewed as a Strength

"When I'm involved in a big project I tend to work too hard, and I can wear myself out."

Discuss an Unrelated Weakness and Then Focus on the Job

"I'm not very interested in accounting. I'd much rather work with people selling a product I believe in."

Discuss a Weakness the Interviewer Already Knows About

"I don't have a lot of experience in multimedia design at this early stage of my career. But, based on my experience in computer programming and my internship in graphic arts, I know that I can learn quickly."

Reference a Weakness You've Been Working to Remedy

"I know that being bilingual is important for this job. That's why I have enrolled in a Spanish course."

have value over time. I would be flexible, so I could be useful in many different situations." Notice that the attributes of the sweater would make the candidate a good person to hire.

Know the Law "You're sitting in the interview for your dream job, and it's going great," says human resources specialist Angela Smith. "You and the interviewer are really hitting it off. Then, out of the blue, she asks, 'Are you planning on having kids?'"[47] It's illegal to ask this question during a job interview. What should you do?

First, it's important to understand the law and your options as an applicant. Most laws about the questions employers can ask boil down to two simple principles. One is that employers cannot legally ask questions about a person's race, religion, gender, sexual orientation, disabilities unrelated to the job, national origin, marital status, parenting status, or age. Because it's illegal to judge a candidate based on these qualities, they should never be asked. The second principle is that questions must be related to what the U.S. government's Equal Employment Opportunity Commission (EEOC) calls *bona fide occupational qualifications*. In other words, employers may only ask about topics that are related to the job at hand. An interviewer may legally ask "Are you legally authorized to work in the U.S.?" and "Do you have a college degree?" but they aren't allowed to ask where you are from or when you were born. The first might hint at race or ethnicity, and the latter indicates your age. It's legal for an interviewer to ask what languages you speak if the position requires frequent interactions with non-English speakers, but the same question is illegal if that isn't a job requirement.

Despite the law, there's a chance that interviewers will ask illegal questions. They're probably uninformed rather than malicious. Still, it's a good idea to prepare in advance for how you might respond. Here are several options:

- *Respond without objecting.* You might choose to answer a question even though you know it's unlawful. Recognize, though, that this could open the door for other illegal questions—and perhaps even discrimination in hiring decisions.

- *Point out the irrelevance.* You might say, "I'm not sure how my marital status relates to my ability to do this job."

- *Redirect.* Shift the focus away from a question that isn't job related to one that is. For example, if someone asks "Are you planning on having kids?" you might say, "You know, I'm not quite there yet. But I am very interested in career paths at your company. Can you tell me more about that?"[48]

- *Refuse to answer.* You have the option to explain politely but firmly that you will not provide the information requested. You might say, "I'd rather not talk about my religion. That's a very private and personal matter for me" or "My understanding is that this question is illegal since it's not relevant to the job."

- *Withdraw.* Another choice is to end the interview immediately and leave, stating your reasons firmly but professionally. You might say, "I'm uncomfortable with these questions about my personal life, and I don't see a good fit between me and this organization. Thank you for your time."

There's no single correct way to handle illegal questions. The option you choose will depend on several factors, including the apparent intent of the interviewer, the nature of the questions, your desire for the job—and finally, your "gut level" of comfort with the whole situation. Knowing your options going in may help you make an effective split-second decision if the need arises.

COMMUNICATION STRATEGIES

Creating a Job Interview Presentation

Ask in advance if a brief presentation about your qualifications and experience would be welcome, and if so, follow the advice here to present yourself most favorably.[49,50]

Consider Your Audience

Ask who will be present at the interview and what information is most important to them.

Plan Ahead for Technology Needs

Inquire if a projector (and speakers, if necessary) are available in the interview room, or if you might bring your own. Many university media centers and libraries will allow you to check out portable projectors. If the interview will occur via live video, investigate how you might best share your audiovisual presentation that way.

Follow the Principles of Good Public Speaking

Before putting content together, read Chapters 12–15 for guidance on creating a professional-quality presentation, such as beginning with an attention getter, developing clear points, presenting strong evidence, and ending memorably.

Focus on Relevant Accomplishments

The bottom line is how well your talents and accomplishments fit the needs of the organization. Without embellishing, demonstrate with clear evidence how you embody desired qualifications.

Be Brief

Your presentation should occupy a small fraction of the time available for the interview. Avoid giving the impression that you are a stage hog. Instead, present yourself clearly and concisely, with professionalism and enthusiasm.

Make Your Visuals Simple

Keep wording to a minimum and use photos, graphics, and video (if applicable) that are professional and simple. If the position involves projects, you might post a photo of you engaged in a professional-quality project overlaid with the words "Project Management Experience." Use your spoken words to convey the details.

Practice

Rehearse many times in advance, using a mirror and/or recording device to see how you sound and look to others. Practicing will allow you to speak without notes in a conversational tone during the actual presentation.

If Allowed, Create a Presentation About Yourself It's increasingly common for employers to ask (or allow) candidates to present a PowerPoint, Prezi, or other type of digital presentation about themselves during employment interviews. See the feature "Creating a Job Interview Presentation" for tips on creating your own.

Participating in a Job Interview

An interview is your chance to shine. Present yourself in a positive and confident light. Here are some suggestions for doing your best.

Know When and Where to Go Don't sabotage an interview before it begins by showing up late. If the interview is in person, be sure you're clear about the time and location of the meeting. Research parking or public transportation to be sure you aren't held up by delays. There's virtually no good excuse for showing up late. Even if the interviewer is forgiving, a bad start is likely to shake your confidence and impair your performance.

Reframe Any Anxiety as Enthusiasm Feeling anxious about an employment interview is natural. Managing your feelings calls for many of the same strategies as managing your apprehension while giving a speech (Chapter 12). Realize that a certain amount of anxiety is understandable. If you can reframe those feelings as *excitement* about the prospect of landing a great job, the energy can work to your advantage.

What you wear to a job interview suggests to a prospective employer how seriously you take the job.

Considering a job you would like to have one day, which of these outfits would be most likely to make a positive impression? Which would seem inappropriate and why?

Dress for Success Most interviewers form opinions about applicants within the first few minutes, so it makes sense to look your best. No matter what the job or company, be well groomed and neatly dressed, and don't overdo it with makeup or accessories. The proper style of clothing can vary from one type of job or organization to another, so do some research to find out what the standards are. Avoid wearing anything that seems flashy, flirtatious, or frivolous. Being underdressed or appearing sloppy can be taken as a sign that you don't take the job or the interview seriously.

Bring Copies of Your Resume and Portfolio If you are interviewing in person, arrive with materials that will help the interviewer learn more about why you are ready, willing, and able to do the job. Bring extra copies of your resume. If appropriate, also bring copies of your past work, such as reports you have helped prepare, performance reviews by former employers, drawings or designs you have created for work or school, letters of recommendation, and so on.

If you are interviewing via technology, consider in advance how you can provide these materials to interviewers. You might say, "I have some work samples and references you might like to see. Is it okay if I email these to you in advance?" If the interview will be conducted via videoconference, you might also ask in advance if you can show a few work samples via screen-share. (You'll learn additional tips for interviewing via video or phone in a moment.)

Whether in person or from a distance, providing evidence of your work showcases your qualifications and demonstrates that you know how to sell yourself.

Mind Your Manners It's essential to demonstrate proper business etiquette from the moment you arrive for an interview or sign on. In person, "you may be riding on the elevator with the head of your interview team," advises one business etiquette expert.[51] Turn off your phone before you enter the building, smile at people, put your shoulders back and head up, and don't fiddle with your clothing, hair, or belongings. In short, behave at all times as the sort of engaged, professional, and attentive coworker everyone wants on their team.

Whether you are in-person or on a videoconference, when you meet people, focus your attention on them and demonstrate an attentive listening posture—shoulders parallel to the speaker's and facial expressions that show you are paying attention. If multiple people are present, be sure to greet all of them and then focus on everyone when you answer a question, not just on the person who asked it.

Follow the Interviewer's Lead   Let the interviewer set the tone of the session. Along with topics and verbal style, pay attention to nonverbal cues described in Chapter 7, such as the interviewer's posture, gestures, and vocal qualities. If they are informal, you can loosen up a bit too, but if they are formal and proper, you should act the same way.

Keep Your Answers Succinct and Specific   It's easy to ramble in an interview, either out of enthusiasm, a desire to show off your knowledge, or nervousness. But in most cases, long answers are not a good idea. Generally, it's better to keep your responses concise, but provide specific examples to support your statements.

Describe Relevant Challenges, Actions, and Results   Most sophisticated employers realize that past performance can be the best predictor of future success. For that reason, many of them engage in **behavioral interviews**—question-and-answer sessions that focus on the applicant's past performance (behavior) as it relates to the job at hand. Typical behavioral questions include the following:

> Describe an experience in which you needed to work as part of a team.
>
> Tell us about a time you had to think on your feet to handle a challenging situation.
>
> Describe a situation in which you were faced with an ethical dilemma, and discuss how you handled it.

When asked behavioral questions, answer in a way that shows the prospective employer how your past performance demonstrates your ability to handle the job you are now seeking. The feature "Responding to Common Interview Questions" offers strategies that may help you stand out as a job applicant.

Ask Good Questions of Your Own   Near the end of the interview, you'll probably be asked if you have any questions. You might feel as if you already know the important facts about the job, but as mentioned before, asking questions can yield useful information and show the interviewer that you have done your research. In addition to queries about the organization and industry, here are some questions that often work well:

> What would you most like to see the person in this position accomplish in the first 30, 60, and 90 days on the job?[53]
>
> How would you describe the management style I could expect from supervisors?
>
> I understand that one of your strategic initiatives this year is to . . . . If chosen for this position, how might I use my background in . . . to help reach this goal?

Generally speaking, experts recommend that applicants not ask about emotionally intense or controversial topics during an interview. So what might you do if you're concerned about a recent scandal at the company, or if you've heard rumors that sexist or racist attitudes are tolerated there? First, conduct advance research to find out as much as you can. You're likely to find out more from people familiar with the organization (and in some cases, news coverage) than you would during

---

behavioral interview   A question and answer session that focuses on an applicant's past performance (behavior) as it relates to the job at hand.

COMMUNICATION STRATEGIES

Responding to Common Interview Questions

Following are some of the most challenging interview questions asked by potential employers. Review experts' tips for answering them effectively and then brainstorm how you might respond based on your own expectations and career goals.

"Tell me something about yourself."

This broad opening question gives the interviewer a chance to know you better. Be genuine and let your personality show. You might say something like "I love a challenge. I participated in debate in high school and learned how to think quickly in high-pressure situations" or "I have lived in five different cities, which has given me a powerful appreciation for diversity" or "I developed an interest in . . . when I . . ." Within your answer, you might describe qualities (e.g., enthusiasm, self-motivation, good listening skills) that can help the employer. Be honest, but don't ramble. This isn't a good time to talk about irrelevant hobbies or pet peeves.

"What makes you think you're qualified to work for this company?"

This question may sound like an attack, but it's really another way of asking, "How can you help us?" In answering, show how your skills and interests fit with the company's goals. Prepare in advance by making a table with three columns. In one column, list your main qualifications; in the next, list specific examples of each qualification; and in the last, explain how these qualifications would benefit the employer.

"What accomplishments have given you the most satisfaction?"

The achievements you describe should demonstrate qualities that would help you succeed in the job, such as creativity, perseverance, self-control, and dependability.

"Why do you want to work for us?"

Employers are impressed by candidates who have done their homework about the organization. This offers you the chance to demonstrate your knowledge of the organization and to show how your talents fit with the organization's goals.

"Where do you see yourself in five years?"

This often-posed query is really asking: "How ambitious are you?" "How well do your plans fit with this company's goals?" and "How realistic are you?" Answer in a way that shows you understand the industry and the company. Share your ambitions, but also make it clear that you're willing to work hard to achieve them.

"What major challenges have you faced, and how have you dealt with them?"

What (admirable) qualities have you demonstrated as you have grappled with problems in the past? Curiosity? Hard work? Reliability? The specific problems aren't as important as the way you responded to them. You may even choose to describe a problem you didn't handle well to show how the lessons you learned can help you in the future.

"What are your greatest strengths?"

Link what you say to the job. "I'm a pretty good athlete" isn't persuasive in itself, but you might talk about being a team player, having competitive drive, or having the ability to work hard and not quit in the face of adversity.

"What are your salary requirements?"

This is a tricky but important question. Shooting too high might dissuade an employer, whereas shooting too low can cost you dearly. Do research in advance to determine the prevailing compensation rates in the industry and region. (Salaries and cost of living differ considerably from one region to another.) Also consider your qualifications. Then name a salary range and back up your numbers with reasons you think you would be a valuable member of the team. For you example, you might say "Considering my skills and experience in addition to salary averages for the position, I would like to receive between $60,000 and $65,000 for this position."[52]

an interview. Second, ask questions during the interview in a way that feels non-threatening, calm, and appropriate to the moment. Keep in mind that coming off as confrontational is likely to turn off interviewers, even if they agree with your viewpoints. You might ask:

What are your goals for the company in the next five years?

What are your goals concerning diversity, equity, and inclusion?

What are you most proud of in regard to your work toward these goals so far?

Know What *Not* to Ask   Some questions are likely to create a bad impression, even if you mean well. Job search expert Susan P. Joyce advises candidates *not* to ask the following:[54]

- *What are the job duties?* Assuming that these are provided in the job listing, asking this question shows that the applicant hasn't done their homework. Moreover, it's unconvincing for the candidate to argue that they're right for the job when they don't know what it involves. If the job duties aren't clear, it's best to ask about them in advance rather than during an interview.
- *Who are your main competitors?* Some fairly easy research could yield answers to this question.
- *Do you check references?* or *Is a drug test required?* Questions such as these suggest that the applicant has something to hide.
- *How soon can I get a raise? How soon can I take vacation? Is there a company discount?* Although pay and benefits are important, most experts say it's bad form to ask about them during a selection interview because it may seem presumptuous or greedy. Employers would like to know that your number-one consideration is the work you will be doing. You'll have a chance to negotiate pay and perks if an offer is made. (You'll learn more about negotiating terms in a moment.)

Follow Up After the Interview   Send a prompt, sincere, and personalized note of thanks. A thoughtful and well-written thank-you message can set you apart from other candidates, whereas failing to send a thank-you note within a day of your interview can eliminate you from consideration. Here is what you might include:

- Express your appreciation for the chance to become acquainted with the interviewer(s) and the organization.
- Explain why you see a good fit between you and the job, highlighting your demonstrated skills.
- Let the interviewer know that the conversation has left you excited about the chance of being associated with the organization.

Since employment decisions may be made quickly, send a gracious thank-you email the same day as your interview. In some circumstances, you might also send a handwritten message of thanks. Carefully reread your message several times before sending it, and if possible, have a skilled proofreader review it as well. One job seeker ruined her chances of employment by mentioning the "report" (instead of "rapport") that she felt with the interviewer.

Negotiating the Terms of a Job Offer Debra's delight over beginning a new job dimmed a bit when she realized that another new hire had negotiated to have the company pay her moving expenses. "I wish I had asked for that, too," thought Debra. Here are some suggestions on how to negotiate effectively when a job is offered to you:

- *Do your homework in advance.* Ask trusted mentors in the field what salary and benefits are reasonable, considering the job you hope to get. That way, you'll be ready if an offer is extended.

- *Consider more than money.* Monetary compensation is not the only perk that might be available if you ask for it. Others might include new technology or software, the option to work at home one or more days a week, moving expenses, paid travel to attend conferences, and so on. Your request is likely to be taken most seriously if you link it to job performance, as in "To serve you best as social media director, I'll need to use Sprout Social and Hootsuite. Can you provide those software programs if you don't have them already?"

- *Build on common interests.* To emphasize that you and your (potential) new employer are on the same side, you might say something like "I'm thrilled about this offer and I really want to work with you. I know you've interviewed dozens of people and chosen me, so it's great to be in a place where *we both want the same thing*," suggests Jonah Berger, the author of *The Catalyst: How to Change Anyone's Mind*.[55,56]

- *Make the ask without an ultimatum.* Unless you really are ready to walk without a pay boost or additional perks, don't present them as must-haves. Rather than saying "I would accept this job if you . . ." you might say "My decision would be easier if you . . ."

- *Stay calm if they say no.* There's no shame in having asked for something you don't get. "If they shut you down, reply in a non-defensive way, such as, 'Oh, that's interesting. Can you tell me more about the reasoning behind that?'" says workplace expert Kristi DePaul.[57]

So far, this discussion has mostly involved interviews conducted in person. But—in an age when budgets are tight, communication technology is pervasive, and work teams are geographically distributed—it's no surprise that a growing number of interviews are conducted via technology. For helpful suggestions about those interviews, see the feature "Interviewing by Phone or Video."

10.5 Adapting to a New Work Environment

Congratulations! You've landed the job. Now it's time to build the career you want. No matter what field you are in, communicating effectively is a key to success. The first step is understanding the culture of your new work environment, including spoken and unspoken expectations, patterns of interaction, and communication do's and don'ts.

COMMUNICATION STRATEGIES

Interviewing by Phone or Video

Interviewing at a distance involves the same guidelines as in-person meet-ups. In addition, the following tips can help you succeed when communication technology is involved.

Present a Professional Identity

Your screen name, if you're using one, should be professional and appropriate, not provocative or edgy. Likewise, pay attention to what you wear if it's a video conference. Even if you're at home, dress professionally for the interview. In addition, think about what will show behind you. A neutral backdrop (real or virtual) without distractions is ideal. Also minimize background noise. A barking dog, crying child, or noisy roommate might interfere with the interview.

Practice with Technology in Advance

Tech problems can end a distance interview before it begins. Follow these tips to ensure that you are ready:

- Make sure you have the right software and are comfortable using it.
- Confirm that you have a solid internet connection with enough speed to handle the conversation.
- Double-check your camera, microphone, and speakers to confirm that they function properly.
- Make sure lighting is sufficient to allow people to see you clearly. The light source should shine on your face rather than on the back of your head (which will cast you in shadow.)
- Find a space with an uncluttered background and no unwanted noise.
- If you're using a phone or tablet, use a stand to avoid distracting jiggles.
- Make sure the camera is at eye level. Looking down is unflattering and can make it seem that you have your eyes closed or are focused on something else.

Ensure That You Have the Right Time for the Interview

There's nothing worse than being an hour late. Confirm the time in advance, especially when different time zones are involved. (If you're not sure, search the web for "world clock.") You might send a message such as "I'm looking forward to speaking with you Tuesday, March 12, at 8 a.m. Pacific/11 a.m. Eastern."

Ask in Advance How Long the Interview Will Last

"Long distance interviews are sometimes meant to be a brief candidate introduction, not a thorough vetting session," says one job search coach. She advises, "If this is the case, be prepared to make the most of this brief first impression!"[58]

Look at the Camera, Not at the Screen

Looking at your monitor or another device may be tempting, but it will create the impression that you're not making eye contact with the interviewer(s). Instead, look directly at the camera. Also remember to smile.

Conduct a Dress Rehearsal

Practicing is the best way to ensure that you are prepared. Recruit a trusted friend (or, even better, someone at your school's career center) to play the role of interviewer. Be sure to practice under the actual circumstances of the interview—remotely and with the same equipment and services you'll use for the real thing. Besides ironing out potential glitches, rehearsals should leave you feeling more confident.

Don't Panic If Technology Fails

Even when you've done everything possible to prepare, technological difficulties can arise. In that case, don't be unnerved. If your online connection is interrupted and you can't get back to the meeting, immediately call or email the interviewer to explain what is happening. And take heart: Your calm demeanor and quick response may signal to the employer that you will be effective in other tough situations as well.

Culture in the Workplace

As you learned in Chapter 4, organizations have cultures of their own made up of shared beliefs and patterns of behavior. Cultural expectations suggest what is frowned upon in the organization, and conversely, what is likely to garner approval. For example, colleagues in some organizations are rewarded for competing with one another, and personal conversations are largely considered a waste of time. In other organizations, people may consider friendly interactions to be productive and enjoyable. At the tech company Google, creativity and innovation are core values. Employees are encouraged to have fun and interact with each other. "You get the freedom to work in the cafeteria, lounge area, even on large bean bag chairs," says a writer familiar with the company.[59] Few organizations may take things as far as that, but as we mentioned previously, it's a good idea to investigate a company's culture before you apply for a job there.

Most experts agree that there is value in having consistent values and practices. A strong culture can unite organizational members in upholding key values and pursuing shared goals. Strong values can also be good for business. About 7 in 10 consumers say they prefer to do business with organizations that are socially and environmentally conscious.[60] However, the danger in a strong culture is that people who challenge the status quo may be marginalized for not "fitting in," even if they have good ideas to share. Organizations in which people are only rewarded if they think and act alike tend to lack innovation and lose step with changes in the environment.[61] The best bet is a balance between core values and openness to new ideas.

Even when you have researched an organization in advance, expect a period of adjustment once you become part of it. **Acculturation** (also called socialization) is the process of adapting to a culture.[62] If you have ever adjusted to a new job, a new school, or moved to a different city, you know that acculturation isn't a simple process. "It's common to experience culture shock when starting a new position," says one career coach, who compares it to the exciting but nerve-wracking experience of visiting another country.[63] Don't be discouraged if you feel out of place for a while, overwhelmed, and nostalgic for your previous position.[64] Those feelings usually lessen as you become accustomed to the new environment. In the meantime, here are a few approaches that might help:

- *Get to know people.* Although you may feel shy or awkward at first, put on a smile and reach out to others. "Jump right in and meet the new people you will be working with," recommend career advisors.[65] Relationships will make you feel more at home and boost your ability to be a great team member.
- *Take care of yourself.* Adapting to a new situation can be exhausting. "It's important to get enough sleep, maintain a healthy diet, and exercise regularly," advises career coach Edythe Richards, who adds, "For me, it's also been very important to have a support system of people I can 'vent' to when needed."[66]
- *Be patient.* It's okay if you have to ask a lot of questions and you don't get everything right at first. Everything you learn is a step toward becoming more proficient.

> **acculturation** The process of adapting to a culture.

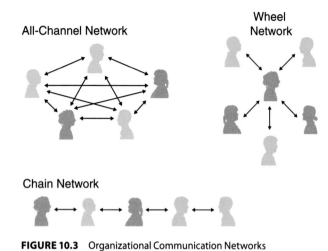

FIGURE 10.3 Organizational Communication Networks

On the flip side, when newcomers enter your workplace, go out of your way to be kind and helpful, involve them in conversations and social gatherings, and be understanding when they make inevitable missteps while adjusting.

In reality, acculturation is a two-way street. Not only do new members adjust to the culture. It also adjusts to them, as they bring novel ways of thinking and behaving. For example, as members of the Millennial generation embark on careers, they tend to favor employers who offer flexible working conditions and who are civically engaged and socially responsible.[67] Business articles abound that offer advice on attracting and keeping Millennial job candidates, mostly by being less rigid, less hierarchical, and more inclusive of diversity.[68,69] The odds are that Millennials in the career world will adapt to corporate culture, but it may also change in response to them.

Patterns of Interaction

One aspect of organizational culture involves who communicates with whom and how (Figure 10.3). Some organizations lend themselves to an **all-channel network** in which all group members share information with everyone else on the team. Emails are a handy way to accomplish this. Transparency is typically high in an all-channel network because everyone gets the same information. One drawback can be information overload. As you probably know from experience, it's nice to be in the loop, but too much information can be overwhelming.

Another option is a **chain network**, in which information moves sequentially from one member to another. Chains are an efficient way to deliver simple verbal messages or to circulate written information. However, chains aren't reliable for lengthy or complex verbal messages because the content often changes as it passes from one person to another. A simple statement such as "Cristiana is leaving Friday" might morph into "Cristiana has been fired" when the speaker really meant that she's going to a conference at the end of the week. The danger of message distortion along a chain means that you should be skeptical about rumors unless you can confirm that the information is true.

Another communication pattern is the **wheel network**, in which one person acts as a clearinghouse, receiving and relaying messages to other members. Like chains, wheel networks are sometimes a practical choice, especially if the group is small and one member is available to communicate with others all or most of the time. The central person, called the **gatekeeper**, can become the informational hub who keeps track of messages and people. If the gatekeeper is a skilled communicator, these mediated messages may help the group function effectively. But if they consciously or unconsciously distort messages to suit personal goals or play members against one another, the group is likely to suffer.

Communication and Workplace Etiquette

Patricia is so furious with members of her project group that she sends out an angry group message calling them the worst team she has ever been part of. Later, she

all-channel network A communication network pattern in which all group members share information with one another.

chain network A communication network in which information passes sequentially from one member to another.

wheel network A communication network in which a gatekeeper regulates the flow of information to and from all other members.

gatekeeper A person who acts as a hub for receiving messages and sharing (or not sharing) them with others.

wishes she had handled the situation differently. Lapses in communication etiquette can damage your reputation and relationships. The good news is that practicing good workplace etiquette is fairly easy if you follow some simple rules. Here are some communication strategies to enhance your career success and some that should be on your never-do list at work.

Cultivate Strong Relationships The benefits of networking don't end once you start a career, and getting to know people with similar interests is one of the best parts about having a job.[70] "Grabbing coffee or lunch with your coworkers, attending happy hour or other company functions, and simply making yourself available can go very far," advises one career professional,[71] who adds this word of caution: "It's important not to forget that, while you should be yourself, you're still among office mates who you'll be working side-by-side with tomorrow."[72] The rule of workplace etiquette extend beyond the physical work space.

Respect Cultural Differences Many Americans display what researchers call "instant intimacy."[73] They often address even new acquaintances, elders, and authority figures by first name. They also engage in a great deal of eye contact, touch their conversational partners, and ask personal questions. To people from different backgrounds, these behaviors may seem disrespectful. An Australian exchange student in the United States reflected on her experience this way: "There seemed to be a disproportionate amount of really probing conversations. Things I normally wouldn't chat about on a first conversation."[74] Review Chapter 4 for more guidance on being culturally sensitive.

Always Do Your Best You may have heard the phrase "don't sweat the small stuff." In fact, making a good impression requires paying attention to every detail. Show up for work looking as good as you did in the job interview. Another method for standing out is to do more than is required. For example, you might finish a job sooner than anticipated or take the initiative to suggest new ideas or launch a project that keeps getting delayed. The time that jobs like this take may be well worth the good reputation they earn you. The caveat is to pace yourself. Doing so much that you end up depleted or burnt out helps no one.

Remain Calm Losing control under pressure can jeopardize work relationships, your reputation, and your career. "You can't put lava back in the volcano," advises consultant Mark Jeffries.[75] Even if you are usually calm and polite, no one is likely to forget the time you stormed out of a meeting or raised your voice. And if your freak-out takes the form of an email or text, there's a record of it that may never go way. To stay collected when you feel yourself getting agitated, take a few deep breaths or a break, and stop to listen and ask questions before responding.[76] And remember this cardinal rule: Never vent about work-related matters on social media.

What you *should* do on the job is only part of the story. It's also important to consider what *not* to do. Here are some communication blunders to avoid at work.

Never Make Fun of Others Some people learn this lesson the hard way. For example, when two women walked into the diner where he worked, Rik wisecracked to a coworker: "Oh look! Fat old ladies in baseball caps!"[77] Then he realized that they were his coworker's mother and aunt. Rik's colleague forgave him, but he learned a valuable lesson and swore off insensitive humor forever. Wisecracks at someone else's expense

When it comes to gossip, the golden rule applies: Don't say anything behind a colleague's back that you wouldn't want said about you.

How would you feel if you knew colleagues were saying negative things about you behind your back?

can be hurtful in any situation. At work, they can be cause for a reprimand, dismissal, or even a lawsuit.

Don't Overshare It may seem important to "be yourself," but there are times when disclosing information about your personal life can damage your chances for professional success—or at the very least, annoy people. The "don't overshare" rule applies to online communication, too. Photos of your wild vacation aren't likely to impress the boss, clients, or colleagues. The best rule is to disclose cautiously, especially if the topic is a sensitive one such as religion, political views, or romantic relationships. A trusted colleague may be able to offer advice about how much to share.

Don't Gossip Communicating with integrity isn't always easy. The culture in some organizations involves gossiping, bad-mouthing, and even lying about others. Nevertheless, effective communication means following your own principles. It may be helpful to know if someone was promoted, reprimanded, or fired, and why—but malicious gossip can mark you as untrustworthy and can damage team spirit. One executive proposed this test: Before you start talking, stop and ask yourself, "Is it kind?"[78]

Don't Get Distracted by Personal Communication Human resource managers list mobile phone use and internet browsing as the two top "productivity killers" at work.[79] To minimize distractions and demonstrate that you're focused on workplace goals, it's a good policy to engage in personal communication only during

Working from home can present challenges.

What steps might you take to create a home workspace that is as quiet and free of distractions as possible?

breaks and lunch periods. Furthermore, make sure others can tell that you are on a break, as by stepping away from your desk or taking a brief walk. Even if you are legitimately "off the clock" for a few minutes, someone who walks by your desk and sees you chatting on Instagram may assume you're goofing off when you should be working.

Don't Fixate on Mistakes What if you accidentally say "I love you" while ending a call with your boss? Or your eyes fill with tears during a stressful business meeting? Minor lapses in professionalism are bound to occur, even among people who have been in the workplace for many years. You can usually recover your dignity and your reputation by following these four steps: don't panic, acknowledge the gaffe, apologize, and return to life as usual.[80] For example, you might say, "Sorry about that! I'm in the habit of saying 'I love you' when I talk to my family. Obviously, I didn't mean to end our call that way. I'm sorry for the slip-up! I'll be more careful in the future." It's okay to laugh if the other person does, but don't dwell on the mistake. You want other people's opinion of you to focus on your impressive performance, not on your goof.

Working Remotely

"Remote work is here to stay," according to business leaders in the United States.[81] That's good news, in some ways. Only about 2 in 10 workers who transitioned to remote work during the COVID-19 pandemic said they would like to return completely to working in person.[82] Most would like the flexibility to work at home sometimes or most of the time. The most popular aspects of remote work are flexible work hours and the lack of a commute. The least popular aspects are social isolation and overworking.[83] Here are some communication strategies to maximize the benefits and minimize the downsides of working remotely.

Create a Dedicated Work Space   Having a quiet, comfortable area dedicated to work activities can help you avoid distractions, signal to others in the household that you are "at work," and allow you to (literally) walk away from work when you are done each day. Even though you are at home, "entering your workspace will help you turn 'on' at the beginning of the day and get down to work," observes editor Regina Borsellino, and "on the flipside, leaving your workspace will also help you turn 'off' at the end of the day and fully disengage."[84]

Establish Ground Rules   With colleagues, the rules might involve working hours, how quickly you are expected to reply to messages, what software and equipment you should use, and data security procedures. With those who share a home with you, it may be helpful to have ground rules about noise, interruptions, and sharing home office space with others.

Keep Regular Work Hours   Establish a clear start and end time for work, take breaks, and let others know what your hours are. Resist the mindset that you are always at work. "If you feel yourself extending your work hours because you aren't doing anything in the evening . . . tell yourself it's time to put work away, recharge, and start tomorrow with a fresh mind. The work will be there in the morning," says career coach Heather Yurovsky.[85]

Dress for Work   Some people are able to work well in their pajamas or sweatpants, but for many, dressing professionally sends a clear message (to themselves and others) that they are in work mode. Work clothes make a better impression if a last-minute Zoom meeting or video chat comes up.

Respond Quickly   Let clients and colleagues know that you are attentive and on the job by replying to calls and emails quickly during work hours.

Don't Show Off   You've seen the memes—a laptop by the pool or on the balcony of a luxury resort, with the catch phrase "Working Remotely." It can feel like a dream come true to work anywhere you have Wi-Fi. But posting messages such as these might suggest to others (including potential future employers) that you don't take remote work seriously. And it can cause resentment among those who can't work off-site. "Everyone on my team was able to work remotely but me," says one frustrated professional, "and seeing those memes was like a slap in the face."

Communicate Often and Clearly   There are two reasons to communicate with colleagues more than usual when you're working remotely: one is to fend off social isolation and the other is to minimize misunderstandings. Video chat apps such as Zoom and Slack can be useful in reinforcing a spirit of teamwork and minimizing loneliness. While you're at it, use video chats, email, and other means to be especially clear about needs, expectations, and work details. Everyone can learn a lesson from the professional team who forgot that a colleague was on vacation for two days while frustrated clients awaited responses from her. "If we had been in person, it would have been obvious that Chelsea wasn't there, and we would have picked up the slack," explains a coworker, "but with everyone working from home, it was easy to forget."

Toot Your Own Horn   Your accomplishments may be less visible when you're working remotely, which can reduce your chances of raises and promotions. About two-thirds of supervisors in one study considered remote workers to be more dispensable than in-person ones.[86] Don't be shy about sharing project updates and accomplishments with colleagues and supervisors, and make it a point to engage in videoconferences with the team so you are literally seen. Being "out of sight, out of mind" is a real danger when telecommuting.[87]

As this chapter illustrates, communication skills are an "absolute must-have" for job seekers and team members.[88] And in today's high-tech world, they are more important than ever. Chapter 11 continues the theme of workplace communication by exploring what it takes to succeed as a follower, team member, and leader.

MAKING THE GRADE

OBJECTIVE 10.1   Describe the reasons why good communicators are likely to excel in the workplace.

- Good communicators work well in teams.
- Good communicators build public awareness.
- Good communicators enhance customer satisfaction.
- Good communicators make good leaders.
  - > Describe the best team you have ever been part of. What role did communication play in your success?
  - > How might you use public speaking skills to enhance your success in a career that interests you?
  - > Describe an instance in which you received great customer service. How was communication involved?

OBJECTIVE 10.2   Engage in communication behaviors conducive to becoming an appealing job candidate.

- The work habits you exhibit now will influence whether people let you know about career opportunities and recommend you for them.

- It's important to build an online identity even before you enter the job market. Present yourself in a favorable and honest light by showcasing your strengths and goals and avoiding behavior that might lead to embarrassing posts.
- To network effectively, view everyone as a networking prospect, engage in online networking, seek referrals, conduct informational interviews, and show appreciation to the people who help you.
  - > Make a list of several people you might interview to learn more about your dream job and brainstorm a list of questions you would ask them.
  - > What might people conclude about you based on your work habits as a student? Are you prompt or tardy? Well prepared or not? Well groomed or sloppy? Committed to doing excellent work or just enough to get by? How well do your answers reflect what employers are looking for in a job candidate?

OBJECTIVE 10.3   Create application materials that make it easy for prospective employers to see how your skills and aspirations match their needs.

- An appealing cover letter acts as a personal sales pitch that encourages employers to learn more about your qualifications.

- Resumes are most impressive if they are clear, powerful, honest, and specific.
- It's important to follow application instructions and keep a record of everyone with whom you communicate.
  > Find a posting for a job you would like to have one day. What talents and experiences make you a good candidate for the position?
  > Brainstorm what you might include in your resume and cover letter if you were applying for that job.

OBJECTIVE 10.4 **Practice communication strategies to make a good impression during employment interviews.**

- Do advanced research, practice answering likely questions, and plan how you will dress and what you will bring to the interview.
- Know the law regarding employment interview questions and devise strategies for how you might respond to them.
- Be specific but succinct in how you respond to interview questions.
- If allowed, create a brief media presentation that showcases your accomplishments relevant to the job.
  > Create a list of questions you might ask a prospective employer to show that you understand the industry and organization and are a good candidate for your dream job.
  > Brainstorm how you might respond to the following questions in a way that redirects the interview to more productive (and legal) topics:

  "That's a beautiful name. Where are you from?"

  "I went to State College, too. When were you a student there?"

  "Do you have children?"

  "You look familiar. Do you go to Olive Baptist Church?"

  "How would you feel about working in an all-male (or all-female) work environment?"

  > Ask a friend to pose as a job interviewer. Practice introducing yourself to them while making eye contact. Are you happy with how clear and confident you seem? In what ways would you like to improve?
  > Rehearse how you might respond to the following questions in a job interview: *What are your greatest strengths and weaknesses? Why should we hire you? What have been your greatest challenges and victories? What do you want to be doing in five years?*
  > Imagine you are being interviewed via videoconference for a job you really want. Rehearse how you might respond to the question, "Tell us more about yourself." Remember to pay attention to your posture, facial expressions, and where your gaze falls. Do you feel well prepared for a video interview? Why or why not?

OBJECTIVE 10.5 **Demonstrate skills to communicate effectively as a new employee in person and remotely.**

- Organizations have cultures of their own.
- Understanding who communicates with whom can help you be successful in the workplace.
- Good workplace etiquette involves strong relationships, respect for cultural differences, doing your best, and remaining calm
- Some of the communication gaffes to avoid at work include belittling people, oversharing, gossiping, and becoming distracted by personal communication.
- Working remotely presents communication opportunities and challenges. Clear ground rules, quick responses, and open communication can boost success.
  > Recall a time when you were in a new school, neighborhood, or job. In what ways did you adapt to the new culture? In what ways did you impact it?
  > Describe some ways you might graciously exit a conversation involving gossip about a coworker.
  > Have you ever had a working-from-home or remote-classroom moment you'd like to take back? If so, what advice would you give others?

KEY TERMS

acculturation p. 279
all-channel network p. 280
behavioral interview p. 274
chain network p. 280
digital dirt p. 260
eportfolio p. 264
gatekeeper p. 280
informational interview p. 262

networking p. 261

portfolio p. 264

selection interview p. 269

wheel network p. 280

PUBLIC SPEAKING PRACTICE

Create a brief statement about yourself that reflects your career goals and why you are (or will be) an appealing candidate for employers. Prepare to share your statement in an oral presentation.

ACTIVITIES

1. Search for your name on Google, Yahoo!, and several other search engines. Do you feel that the photos and information revealed by the search would impress prospective employers? If not, how might you change your online image to be more professional? If your search leads to information about other people with the same name as you, how might you help prospective employers find the real you? If nothing much shows up about you online, how might you cultivate a greater presence?

2. Draft a cover letter and resume for a job you would like to have. Share it with a mentor or representative at your university career center to get their feedback.

3. Conduct mock job interviews in which you take turns being the interviewer and the job candidate. Reflect on what you do well and how you might like to improve.

Teamwork and Leadership

CHAPTER OUTLINE

11.1 Communicating Well as a Follower 290
- Be Proactive
- Seek Feedback
- Support Others
- If Something Isn't Right, Speak Up
- Handle Challenges Calmly

11.2 Communicating in Groups and Teams 294
- What Makes a Group a Team?
- Motivational Factors
- Rules and Norms in Small Groups
- Individual Roles

11.3 Making the Most of Group Interaction 298
- Recognize Stages of Team Development
- Enhance Cohesiveness
- Manage Meetings Well
- Use Meeting Technology Effectively
- Use Discussion Formats Strategically

11.4 Group Problem Solving 307
- Advantages of Group Problem Solving
- A Structured Problem-Solving Approach

11.5 Communicating Effectively as a Leader 311
- Leadership Can Be Learned
- Power Comes in Many Forms
- Leadership Approaches Vary
- Good Leadership Is Situational
- Transformational Leadership

11.6 Leaving a Job Graciously 316

MAKING THE GRADE 318

KEY TERMS 320

PUBLIC SPEAKING PRACTICE 321

ACTIVITIES 321

LEARNING OBJECTIVES

11.1 Practice communication skills involved in being a collaborative, proactive follower.

11.2 Identify factors that influence communication in small groups and either help or hinder their success.

11.3 Strategize ways to build cohesiveness and communicate effectively during group discussions and meetings.

11.4 Assess the advantages and stages of group problem solving, and practice applying Dewey's group problem-solving model.

11.5 Demonstrate effective leadership skills based on the situation, goals, and team members' needs.

11.6 Describe the steps involved in resigning from a job graciously.

Julie Zhuo was the first engineering intern to work at Facebook, back when it was still a speck in the market. At the time, Facebook users were almost all high school and college students. "In some ways I was the perfect candidate," recalls the engineering graduate. "I mean, who knew Facebook's audience better than a recent grad like myself?"[1] In her book *The Making of a Manager*, Zhuo says she quickly realized that people skills were as important as technical dexterity when it came to success.[2] She worked hard to be a good follower and team member, and many of the same skills made her an appealing candidate for leadership roles. Zhuo eventually served as vice president of Facebook before launching her own tech advisory company.

This chapter considers lessons about communication that Zhuo and others have learned and how you might use them to succeed in school and at work. It begins with communication strategies for being an influential follower and a valued team member. Next, you'll learn how to use those skills and others to be an effective leader. The chapter concludes with advice on how to segue from one career or position to another when that's the wisest choice for you.

11.1 Communicating Well as a Follower

Even a great coach can't win championships without an outstanding team. The same is true in every group effort—whether it's a class project, a neighborhood committee, or a staff who works together every day. Effective followership is essential to success. To consider what followers (and you, specifically) have to offer, complete the self-assessment "How Good a Follower Are You?"

UNDERSTANDING YOUR COMMUNICATION

How Good a Follower Are You?

Check all of the following that apply to you in your role as a follower.

- ☐ I think for myself.
- ☐ I go above and beyond job requirements.
- ☐ I'm supportive of others.
- ☐ I'm goal oriented.
- ☐ I focus on the end goal and help others stay focused as well.
- ☐ I take the initiative to make improvements.
- ☐ I realize that my ideas and experiences are essential to the success of the group.
- ☐ I take the initiative to manage my time.
- ☐ I frequently reflect on the job I am doing and how I can improve.
- ☐ I keep learning.
- ☐ I am a champion for new ideas.

If the majority of these statements describe you, pat yourself on the back. These are the qualities of an outstanding follower, according to Robert E. Kelley, author of *The Power of Followership* (1992, Doubleday Business).

The self-assessment makes it clear that good followers aren't sheep who blindly follow the herd. To the contrary, the best followers are self-motivated, highly committed, talented, courageous, and well respected.[3] Effective followers are so important that successful organizations work hard to attract and keep them. The hotel chain Hilton, for example, ranks among the best places to work in the United States.[4] According to employees, Hilton leaders make it a priority to be effective communicators who listen to team members, invite their ideas, encourage teamwork, and honor diverse perspectives.[5] The company's success underscores the advantage of valuing followers' contributions. Here are four communication strategies to help you shine as a follower.

Be Proactive

Great followers show initiative. That often involves "being the person to seize an idea and get it off the ground—often, before it's been asked for," says one career advisor.[6] If you're thinking that initiative sounds a lot like leadership, you're right. Outstanding followers exhibit many of the same traits as good leaders. They work well with others and do a good job even when no one's watching.[7] One way to gauge your status as a follower is to consider the confidence that others have in you. For example, "If you are the first on your boss' radar for getting things done or dealing with a crisis, it means they have an enormous amount of trust in your abilities," observes business speaker Michael Kerr.[8]

Seek Feedback

Don't wait until an annual performance appraisal to find out how you're doing. Seek out constructive feedback year round. Organizational specialist Dan Rockwell suggests the following:[9]

- *Ask specific questions.* Rather than the nebulous query "How am I doing?" you might ask "Did I cover everything you wanted in the project update?" or "What do you think of my idea to host an employee appreciation event?"
- *Don't seek feedback simply to invite praise.* It feels good to get 100 percent positive feedback ("You're awesome!"), but you'll grow more if you also get ideas for improvement.
- *Avoid over-correcting.* If a colleague suggests that you talk a little less in meetings, don't go silent at the next one. Too little is usually as bad as too much, and colleagues who feel their advice backfired may not make the effort again.

Support Others

Supportive followers look out for others. You might pitch in when others are overwhelmed, offer guidance to new colleagues, and make sure no one is left out when staff members gather for lunch or after-hours events. In addition to helping others, your extra effort is likely to make your job easier and more fun. "Getting along with colleagues at work is important for inspiring teamwork and creating a healthy work environment," observes a career advisor.[10] It also means that colleagues are more likely to support you when you need it.

If Something Isn't Right, Speak Up

It may seem that your job as a follower is to do whatever your supervisor says, but that's not always the best policy. "Good followers possess the courage to support the

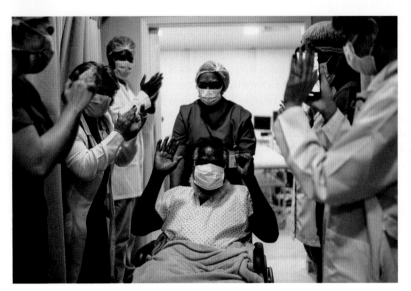

The COVID-19 pandemic underscored the value of teamwork in getting the word out and providing care for people who became sick.

What teams are you part of? How do you support other members of the team, and how do they support you?

leader as well as the courage to challenge the leader," conclude researchers who studied more than 70 years of research about followership.[11] When you're aware of ineffective, unsafe, or unethical practices, you might prevent disaster by calling attention to them. Plus, it's the right thing to do. If the situation is serious and you fear that you'll be penalized for reporting it, it's a good idea to consult with a legal expert in your organization (if there is one), a higher level of management, or even the police. To learn about federal laws that protect employees from retaliation, visit whistleblowers.gov. Most of the time, however, the issues you bring up won't be that volatile.

Courageous followership includes bringing attention to policies or situations that aren't *wrong*, they're just not ideal. The best followers understand that even good leaders sometimes make bad decisions for a variety of reasons (fatigue, lack of information, haste). If you find yourself in such a situation, try to help. "If you want to do in your boss, do everything he or she orders exactly as you are told. Sooner or later you will get a bad order and executing it will make the boss look very bad," advises Ira Chaleff, an expert on followership.[12] A better alternative is tactfully to point out potential (or real) drawbacks in an unaccusatory way. You might say, "I've talked to three clients today who are having trouble with our new web page. I think it might help if we . . ." It may seem that you're sticking your neck out, but if you approach the situation in a collaborative and positive manner, your commitment to excellence is likely to make a positive impression on others.[13]

Handle Challenges Calmly

Followers often deal with a wide range of people and situations, including some that are quite challenging. It pays to remember that just because people are difficult, it doesn't mean you are at fault, and as much as you might wish you could, you can't control the behavior of others. "Reacting negatively to someone increases your stress levels, gets you worked up, and can spoil your mood for the rest of the day," reflects one career advisor.[14] Here are a few alternatives that might help:

- If the person's behavior is abusive or frightening, seek help right away from security personnel or a trusted colleague.
- If the challenging person is rude but not threatening, try to remain calm, have compassion, and be a good listener. When things get tense, "it takes a lot of practice to be able to listen with an open mind, but doing this can actually resolve conflict faster than expected," recommends one advisor.[15]

- Show empathy. "Saying 'I'm sorry,' or 'I'm going to fix this,' can go a long way toward defusing many situations," says psychologist Barbara Markway.[16]

Unfortunately, the difficult someone might be your supervisor. See the feature "Working with a Difficult Boss" for communication strategies to manage that tricky situation.

All in all, if you'd like to succeed in the career world, the message is clear: It's essential to be a skillful and influential follower. The scenarios described here can help build your communication skills and demonstrate to others that you have the poise and confidence to be a great team member and leader.

COMMUNICATION STRATEGIES

Working with a Difficult Boss

Sooner or later you're likely to encounter a difficult boss. They may be unreasonably demanding, blast you with verbal abuse, or engage in passive aggressive behavior—perhaps being nice to you in person but sabotaging you behind your back. Or your boss may simply be incompetent. What should you do if you encounter such a manager? While every job and situation is different, here are a few strategies that may help you manage the relationship.[17]

Rise to the Challenge

Meeting your boss's expectations might make your life easier. If your boss is a micromanager, invite their input. If your boss is a stickler for detail, provide more information than you would otherwise. The extra effort may show that you care and take the job seriously.

Make Up for the Boss's Shortcomings

If your boss is forgetful, diplomatically remind them of important details. If they are disorganized, provide the necessary information before they ask for it.

Seek Advice from Others

Gratuitous complaining about your boss is a bad idea. But if other people you know have encountered similar problems, you might discover useful information by seeking their advice.

Talk with Your Boss

If your best efforts don't solve the problem, consider requesting a meeting to discuss the situation. Rather than blaming the boss, use "I" language, such as "I feel confused when two managers give me different instructions." Solicit your boss's point of view, and listen nondefensively to what they have to say. Paraphrase and use perception checking as necessary to clarify your understanding. As much as possible, seek a win–win outcome.

Adjust Your Expectations

You may not be able to change your boss's behavior, but you can control your attitude about the situation. Sometimes you must accept that there are things over which you have little control. In that case, the challenge is to decide whether you can accept working under less than ideal conditions.

Maintain a Professional Demeanor

Even if your boss has awful interpersonal skills, you will gain nothing by sinking to the same level. It's best to take the high road, practicing the professional communication skills described elsewhere in this chapter.

If Necessary, Make a Gracious Exit

If you can't fix an intolerable situation, the smartest option may be to look for more rewarding employment. If so, leave on the most positive note you can.[18] See the unit at the end of this chapter for tips on exiting a job without leaving a bad impression.

11.2 Communicating in Groups and Teams

Have you ever wondered what the perfect team would be like? The employees at Google did. In 2012, they launched Project Aristotle, a deep-dive study of hundreds of groups to pinpoint the ideal conditions for teamwork. The project—inspired by Aristotle's famous quote "the whole is greater than the sum of its parts"[19]—yielded some helpful insights.

Being a team player is central to career success. Employers rank teamwork skills among the 10 most desired traits of people they hire.[20] Employees give teamwork high priority as well. They typically say that effective teamwork makes them feel more powerful and empowered, more appreciated, more successful, closer to their colleagues, and more confident that their colleagues will support and encourage them in the future.[21]

Even though you probably believe, as Aristotle did, that teams often accomplish more than people working alone, you may groan (at least inwardly) when you find out you'll be part of a group project. You've probably been part of groups in which some members did more work than others, one or two people dominated group discussions, or members never trusted each other enough to become comfortable sharing their ideas.

On the other hand, if you've ever been part of a great team, you know that it can be one of life's most enjoyable and productive experiences. As the theologian H. E. Luccock puts it, "No one can whistle a symphony. It takes a whole orchestra to play it."[22]

Most of the time, the difference between successful and frustrating group work involves communication. This section explores the nature of small groups and the communication skills involved.

small group A limited number of interdependent people who interact with one another over time to pursue shared goals.

What Makes a Group a Team?

Groups may play a bigger role in your life than you realize. Some are informal, such as friends and family. Others are part of work and school. Project groups, sports teams, and study groups are common types, and you can probably think of more examples that illustrate how central groups are in your life.

A **small group** consists of a limited number of people who interact with one another over time to reach shared goals. More precisely, small groups embody the following characteristics:

- *Interaction.* Without interaction, a collection of people isn't a group. Students who passively listen to a lecture don't constitute a group until they begin actively to communicate with one another. This is why some students feel isolated even though they spend a great deal of time on a crowded campus.

"And so you just threw everything together? ... Matthews, a posse is something you have to *organize*."

Source: Copyright © 2019-2022 FarWorks, Inc. All rights reserved. Used with permission.

- *Interdependence.* In groups, members don't just interact—they are *interdependent.*[23] When one member behaves poorly, their actions influence how the entire group functions. On the bright side, positive actions have ripple effects, too.
- *Time.* A collection of people who interact for a few minutes doesn't qualify as a group. True groups work together long enough to develop a sense of identity and history that shapes their ongoing effectiveness.
- *Size.* How big is a small group? Most experts in the field set the lower limit at three members.[24] There is less agreement about the maximum number of people.[25] As a rule of thumb, an effective group is small enough for members to know and react to every other member, and no larger than necessary to perform the task at hand effectively.[26] Small groups usually have between 3 and 20 members.

The words *groups* and *teams* are sprinkled throughout this unit, but the terms are not synonymous. Teams share the same qualities as groups, but they take group work to a higher level. You probably know a team when you see it: Members are proud of their identity. They trust and value one another and cooperate. They seek, and often achieve, excellence. Teamwork doesn't come from *what* the group is doing, but *how* they do it. **Teams** tend to be unified by highly committed members and leaders who work well together to pursue clear, inspiring, and lofty goals.[27] High-profile examples are teams that win the Super Bowl, climb mountains, or collaborate to achieve medical or technological breakthroughs. But teamwork happens in everyday circumstances as well. Perhaps you've been part of a high-functioning team that united to help others, excel at a school assignment, or launch a campaign.

teams Groups of people who are successful at reaching goals.

If the goal is fairly simple or routine, an individual or group may accomplish it adequately. But nothing beats teamwork when the job requires a great deal of thought, collaboration, and creativity.

Motivational Factors

Two underlying motives—group and individual goals—drive small-group communication. *Group goals* are the outcomes members collectively seek by joining together. A group goal might be to have fun, win a contract or contest, make a good grade, or provide a service.

Individual goals are the personal motives of each member. Your individual goals might be to impress the boss, build your resume, or develop a new skill. Sometimes individual goals can help the larger group. For example, a student seeking a top grade on a team project will probably help the team excel. But problems arise when individual motives conflict with the group's goal. Consider a group member who monopolizes the discussion to get attention or one who engages in **social loafing**—lazy behavior some people use to avoid doing their share of the work. The feature "Getting Slackers to Do Their Share" presents some strategies for distributing the workload equitably.

social loafing The tendency of group members to do as little work as possible.

Rules and Norms in Small Groups

All groups have guidelines in the form of rules and norms that govern members' behavior. You can appreciate this by comparing the way you act in class or at work with the way you behave with your friends.

COMMUNICATION STRATEGIES

Getting Slackers to Do Their Share

Experts offer the following suggestions to make sure everyone on your team does their fair share of the work.[28,29,30]

Focus on the Endgame

True motivation arises from a sense of working together toward an important goal.

Match the Goal to the Group Size

Make sure the size and nature of the task match the size and talents of the team. Social loafing is more likely when more people than necessary are involved.

Establish Clear Goals and Responsibilities

Draft a clear action plan to make sure that group members truly know what is expected of them.

Provide Training

Make sure everyone has the training and tools to deliver.

Hold People Accountable

Ask team members to regularly share their accomplishments.

Focus on Quality

Agree on clear guidelines for high-quality work, and offer feedback at every step.

Ask Why

If team members fall behind, ask them why.

Don't Overlook Poor Performance

All team members need to pull their weight, or others may begin to slack off as well.

Guard Against Burnout

Pay attention to members' emotional states and energy levels to make sure that unrealistic demands aren't sapping their strength.

Celebrate Successes

Make sure that even low-visibility tasks are rewarded with praise and recognition.

---

rules Explicit, official guidelines that govern group functions and member behavior.

norms Typically unspoken shared values, beliefs, behaviors, and processes that influence group members.

social norms Shared expectations that influence how group members relate to one another.

Rules are official guidelines that govern what the group is supposed to do and how the members should behave. They are usually stated outright. In a classroom, rules include how absences will be treated, if late work will be accepted, and so on.

Norms are equally powerful, but they involve expectations that are typically conveyed by example rather than in words. There are three main types of small-group norms.

- **Social norms** govern how members interact with one another—what kinds of humor are considered appropriate, how much socializing is acceptable on the job, and so on.
- **Procedural norms** guide operations and decision making. Group members' actions may signal that it's okay to answer emails during a meeting, or that group discussions will continue until everyone has a chance to speak.
- **Performance norms** (also called task norms) govern what's expected of the group on a day-to-day basis. For example, one group might expect perfection right away whereas another may presume that excellence emerges over time through trial and error.

It's important to realize two things about norms. First, a group's norms don't always match what members say is ideal. Consider punctuality. The expectation may be that meetings should begin at the scheduled time, yet the norm may be to delay for about 10 minutes until everyone arrives. Second, group norms don't emerge immediately or automatically. When people first come together, it's common for them to feel unsure how to behave together. Even when groups have been together for a while, members' expectations may not match up perfectly. For example, the group norm may be for members to engage in rousing debates, but some members may wish meetings were calmer and quieter.

Several years into Project Aristotle, the Google team found that most of the groups they studied were comprised of well-intentioned people who were determined to do good work. Nevertheless, some were far more successful than others. Why? The eureka moment came when the research team realized that groups made better decisions and carried them out more effectively when the social norm was for members to treat each other with empathy and respect. As one researcher put it, the collective IQ of those teams was higher than others'.[31] The successful teams didn't all communicate in the same way. Some were orderly and calm, whereas others engaged in a free flow of ideas that seemed chaotic at times. But they had one thing in common—a shared a sense that each member was important and unique, that it was okay to disagree, and that everyone on the team deserved to be heard. Unlike less productive teams, the best teams achieved what researchers called *psychological safety*, defined by the conviction that it's okay, and even preferable, to be vulnerable and imperfect in one another's presence. As one analyst summed it up:

> To be fully present at work, to feel "psychologically safe," we must know that we can be free enough, sometimes, to share the things that scare us without fear of recriminations. We must be able to talk about what is messy or sad, to have hard conversations with colleagues who are driving us crazy. . . . We want to know that those people really hear us. We want to know that work is more than just labor.[32]

procedural norms Shared expectations that influence how a group operates or reaches decisions.

performance norms Shared expectations that influence how group members handle the job at hand.

role Pattern of behavior enacted by particular members.

Individual Roles

The next time you see people working in small groups, observe how they behave. Some may seem stuck in silence with no one willing to speak up. In others, one or two members may dominate the discussion. In still others, all members may appear to be actively engaged—sharing ideas and enjoying the give and take. Small group interaction is shaped by roles, such as those described here, that members play.

Whereas rules and norms establish expectations for how members behave overall, **roles** define patterns of behavior enacted by *particular* members. **Formal roles** are explicitly assigned by an organization or group. They usually come with a label, such as assistant coach, treasurer, or customer service representative. By contrast, **informal roles** (sometimes called "functional roles") are rarely acknowledged by the group in words.[33] Informal group roles fall into two categories: task and maintenance.

- *Task roles* help the group achieve particular outcomes, such as revising workplace policies or hosting an event. Task roles include initiator, researcher, opinion seeker, elaborator, and so on (see Figure 11.1). An initiator might ask "How about we take a different approach?" and then present an idea or propose a solution. Elaborators tend to focus on details. They might say, "Let's see . . . at 35 cents per brochure, the total cost would be $525."

formal role A role assigned to a person by group members or an organization, usually to establish order.

informal role A role usually not explicitly recognized by a group that describes functions of group members rather than their positions. These are sometimes called "functional roles."

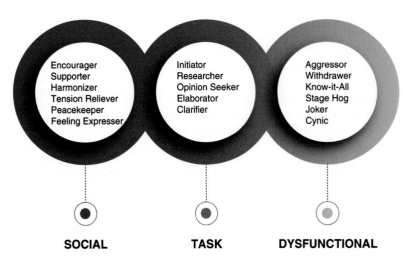

FIGURE 11.1 Roles That Team Members Play

- *Social roles* (also called maintenance roles) help the relationships among group members run smoothly. For example, an encourager might actively solicit comments by shy members, and a harmonizer might suggest compromises to help bring those with opposing viewpoints to a consensus.

Not all informal roles are constructive. Some participants may bully others (aggressors) or refuse to participate in offering opinions or ideas (withdrawers). (See Figure 11.1 for examples of informal roles in each category.)

As you might expect, groups are most effective when people fill positive social roles and no one fills dysfunctional ones.[34] Here are some strategies for using roles to benefit a group:

- *Make sure all helpful roles are filled.* For example, if key facts are missing, take on the role of researcher. If details are getting lost, it may help to be a clarifier who says, "Let me see if I can summarize what has been said so far. . . ." You can also encourage others to fill key roles.

- *Avoid role fixation.* By always occupying the same role, you prevent others from having that experience. If you're always the one to suggest new ideas, the rest of the team may stop contributing. If you always support new ideas, others may be stuck playing devil's advocate to balance things out. Another danger is that you'll fill a role out of habit, even when the situation doesn't require it. For example, you may be a world-class coordinator or critic, but these talents will only annoy others if you use them when they aren't needed.

- *Avoid dysfunctional roles.* It can be tempting to goof off or to sabotage annoying team members, but doing so may hurt the team and damage your reputation. No group needs an aggressor, an incessant joker, or so on.

11.3 Making the Most of Group Interaction

Meetings are like muffins, proclaims one beleaguered professional, "You eat too many, and you're not going to feel very good."[35] It's a common sentiment whether groups interact in person or via technology. "I don't want another Zoom call where people are bored and multitasking," bemoans a remote worker.[36]

Just about everyone knows the frustration of group interactions that are boring, unproductive, or worse. But there's a better way. The principles of effective teamwork can help make group interactions more rewarding. Here are some strategies for accomplishing that.

Recognize Stages of Team Development

Rather than being frustrated when a group doesn't immediately gel, it's helpful to recognize that teams tend to develop in stages. These are characterized as orientation (forming), conflict (storming), emerging (norming), and reinforcement (performing).[37,38] As you read about the stages, visualize how they have applied to problem-solving groups in your experience.

Orientation (Forming) Stage In the **orientation (forming) stage**, members approach the problem and one another tentatively. There is little outward disagreement at this stage. Members test out possible ideas cautiously and politely. This doesn't mean that they agree with one another. Rather, they are probably sizing up the situation before asserting themselves. Although there is little outward disagreement at this stage, it can be a calm before the storm.

Conflict (Storming) Stage After members understand the problem and become acquainted, a successful group enters the **conflict (storming) stage**. Members take strong positions and defend them. Coalitions are likely to form, and the discussion may become polarized. The conflict needn't be personal, however, and it should preserve the members' respect for one another. Conflict seems to be a necessary stage in group development. The give and take of discussion can test the quality of ideas, and weaker ones may be justly eliminated.

Emergence (Norming) Stage After a period of conflict, effective groups move to an **emergence (norming) stage** in which one idea might emerge as the best one, or the group might combine the best parts of several plans into a new solution. As they approach consensus, members back off from their dogmatic positions. Statements become more tentative again: "I guess that's a pretty good idea" or "I can see why you think that way."

Reinforcement (Performing) Stage Finally, an effective group reaches the **reinforcement (performing) stage**, at which point members not only accept the group's decision; they also endorse it. Even if members disagree with the outcome, they're not likely to voice their concerns. There is an unspoken drive toward consensus and harmony.

This isn't a one-time process. Ongoing groups can expect to move through these stages with each new issue, such that their interactions take on a cyclic pattern. They may begin discussion of a new issue tentatively, then experience conflict, emergent solutions, and reinforcement. In fact,

> **orientation (forming) stage** When group members become familiar with one another's positions and tentatively volunteer their own.

> **conflict (storming) stage** When group members openly defend their positions and question those of others.

> **emergence (norming) stage** When a group moves from conflict toward a single solution.

> **reinforcement (performing) stage** When group members endorse the decision they have made.

Source: © Original Artist. Reproduction rights obtainable from www.CartoonStock.com.

a group that deals with several issues at once might find itself in a different stage for each problem.

Knowing that these phases are natural and predictable can be reassuring. It can help curb your impatience when the group is feeling its way through the orientation stage. It may also make you feel less threatened when inevitable and necessary conflicts take place. Understanding the nature of emergence and reinforcement can help you know when it is time to stop arguing and seek consensus.

Enhance Cohesiveness

The degree to which members feel connected with and committed to a group is known as **cohesiveness**. You might think of cohesiveness as the glue that bonds individuals together, giving them a collective sense of identity. In highly cohesive groups, members spend more time interacting and express more positive feelings for one another than in those that lack cohesion. Members of cohesive groups typically report being highly satisfied and loyal. Cohesion keeps people coming back, even when the going is tough. With characteristics such as these, it's no surprise that highly cohesive groups have the potential to be productive. In fact, group cohesion is one of the strongest predictors of innovation, along with effective communication and encouragement.[39]

Despite its advantages, cohesiveness doesn't guarantee success. You've probably been part of study groups in which the members cared more about hanging out as friends than actually getting down to work. If so, your group was cohesive but not productive.

cohesiveness The degree to which members feel part of a group and want to remain in it.

COMMUNICATION STRATEGIES

Dealing with Difficult Team Members

Every now and then you will run across a team member who consistently tests your patience. Perhaps they are whiny, bossy, aloof, aggressive, overly ingratiating, or a know-it-all. Here are some tips from the experts on coping effectively.[40,41]

Keep Calm

Some people thrive on goading others and creating drama. Don't play their game.

Look for Underlying Reasons

Consider what factors might have led the person to feel ignored, hurt, or disrespected.

Surface the Issue

You might say, "I noticed that you've interrupted me several times. Do you feel that you didn't get a chance to explain your position?"

Lay Ground Rules

Establishing specific expectations will make it easier to identify and address issues. For example, you might agree that there will be no yelling, interrupting, or maintaining side conversations.

Write Down Ideas

People may be difficult because they don't feel they're being heard or respected. Writing down everyone's ideas on a board or flipchart captures what people are saying and prevents a potential source of frustration.

Clarify Repercussions

When you are coping with a difficult person whose behavior doesn't improve, it's important to have a clear understanding of the potential consequences. This may involve sharing the problem with a boss or instructor, or if you have the authority, making the repercussions clear yourself.

Groups can maximize the positive aspects of cohesiveness in the following ways:

- *Focus on shared goals.* People draw closer when they share a similar aim or when their goals can be mutually satisfied.
- *Celebrate progress.* When a group is making progress, members tend to feel highly cohesive. But when progress stops, cohesiveness often decreases.
- *Minimize competition.* Sometimes strife arises within groups. Perhaps there's a struggle over who will be the leader or decision maker. Whether the threat is real or imagined, the group must neutralize it or face the consequences of reduced cohesiveness.
- *Establish interdependence.* Groups tend to become cohesive when members realize that they must rely on each other to reach their collective goals.
- *Build relationships.* Groups often become close because the members like one another. It's a good idea to devote time and energy to building camaraderie and friendship within the group.

These suggestions work well on most teams. See the feature "Dealing with Difficult Team Members" for tips on handling especially hard cases.

Manage Meetings Well

If you've ever suffered through a meeting dominated by a few vocal or high-status members, you know that the quantity of speech doesn't equate with its quality. You can encourage the useful contributions of all members in a variety of ways.

Keep the Group Small   In groups with three or four members, participation is usually roughly equal, but after the size increases to between five and eight, there is a dramatic gap between the contributions of members.[42]

Encourage Everyone to Participate   A team isn't likely to get optimal results without everyone's input. This can mean encouraging quiet members and moderating the input of overly talkative ones. Here are a few techniques you might use:

- *Thank normally quiet people for sharing.* Although it isn't necessary to gush about a quiet person's brilliant remark, saying thanks and acknowledging the value of their idea can increase the odds that they will speak up again in the future.
- *Assign tasks strategically.* You might ask quiet members to be in charge of specific tasks. The need to report on these tasks guarantees that they will speak up.
- *Ask outright for everyone's input.* If one member is talking too much, politely express a desire to hear from others.
- *Enlist the help of talkative members.* Speak privately with verbose team members and ask them to help you encourage quieter ones to be involved.
- *Change things up.* Group members who are hesitant to speak in large groups might be more comfortable writing down their ideas, talking in pairs, or brainstorming in small groups. (You'll read about discussion formats shortly.)

Keep the Group on Topic   If the group is on a tangent and nothing else works, you might say something such as, "I'm sure last Saturday's party was awesome! But if we're going to meet the deadline, I think we'd better save those stories for after the meeting."

CHAPTER 11 Teamwork and Leadership

Teams in which everyone thinks alike often make mistakes.

How might you encourage people to challenge the status quo in groups to which you belong?

Avoid Information Underload and Overload Make sure team members know the information and nuances that bear on a problem. At the same time, recognize that too much information makes it hard to sort out what's essential from what isn't. In such cases, experts suggest parceling out areas of responsibility.[43] Instead of expecting all members to explore everything about a topic, assign groups to explore particular aspects of it and then share what they learn with the group at large.

Don't Pressure Members to Conform If you've ever supported an idea because everyone else seemed to like it or you were tired of debating the issue, you have engaged in **groupthink**—the tendency of some groups to support ideas without challenging them or providing alternatives.[44] Here are some of the reasons people might engage in groupthink:

groupthink The tendency in some groups to support ideas without challenging them or providing alternatives.

- They wish to avoid a conflict.
- They want others to see them as "team players."
- They overestimate the group's good judgment or its privileged status.
- They fear that others won't like what they have to say.
- They just want to get the discussion over with.[45]

A range of factors may make groupthink appealing, but the results can range from disappointing to downright disastrous. On a small scale, your team might downplay customer complaints that could help you improve, insisting that what you are already doing is "the best it can be" and "you'll never please everyone." Sometimes the results are more tragic. One of the most famous examples is the *Challenger* explosion in 1986, in which seven astronauts died when their spacecraft exploded shortly after launch. An investigation revealed that, in the days leading up to the event, NASA engineers who were concerned that freezing temperatures might cause a deadly malfunction were discouraged from voicing their objections.[46] More recently, in numerous high-profile cases, people didn't challenge or report child molesters because they were afraid of the repercussions.[47,48,49]

You can minimize the risk of groupthink by adopting the following practices:[50]

- *Recognize early warning signs.* If agreement comes quickly and easily, the group may be avoiding the tough but necessary search for alternatives. Considering all the options now may save time and hardship later.
- *Minimize status differences.* Group members sometimes fall into the trap of agreeing with anything the leader says. To minimize this tendency, leaders should make it clear that they encourage open debate rather than blind obedience.

They might also encourage members to conduct initial brainstorming sessions on their own, among peers.

- *Make respectful disagreement the norm.* After members recognize that questioning one another's positions doesn't signal personal animosity or disloyalty, a constructive exchange of ideas can lead to top-quality solutions.
- *Designate someone to play devil's advocate.* Before reaching a decision it can be helpful to designate a person or subgroup to bring up potential pitfalls, ask questions, and remind others about the dangers of groupthink.

Although meetings can be challenging, there is nothing so powerful as a good one. To paraphrase Julie Zhuo, whose story began this chapter, a good meeting is engaging and welcoming. It feels like a great use of your time, yields information and inspiration to help you be more effective, and leaves you with a clear sense of what to do next.[51]

Use Meeting Technology Effectively

"Love them or hate them, virtual meetings are here to stay," says a writer for *The Economist*.[52]

Even teams who always met in person before COVID-19 are likely to keep using virtual meeting technology, at least sometimes. That's partly because the nature of work has changed, perhaps forever. Experts expect that as many as 1 in 4 professional positions in the United States will be remote in the future, and many more will involve a hybrid mix of remote and on-site work.[53] As discussed in Chapter 10, the majority of workers are in favor of location flexibility. One of the main challenges, however, is maintaining collaborative communication. This section considers how to maximize the benefits of videoconference meetings and offset the drawbacks.

Develop Camaraderie Communication technology can make it easy to exchange information but hard to make a real connection. Take time to get acquainted with online team members before getting down to business. People tend to be more committed and more accountable when they know their teammates well.[54]

Learn the Technology More than a dozen videoconferencing platforms are in widespread use and more emerge all the time. Familiarize yourself with the technology in advance of a meeting. YouTube is replete with free video tutorials that can help bring you up to speed.

Practice Good Cybersecurity Imagine how Dennis Johnson felt when his online dissertation defense was disrupted by hackers who drew lewd pictures and racial slurs on the screen for everyone to see.[55] In other situations, uninvited guests may gain access to sensitive information. To keep your meetings secure, require a password to enter, create an electronic waiting room so no one is admitted until the host okays it, and turn off the "join before host" option that might let someone sneak in early.[56]

Set the Stage "Keep in mind that people aren't just seeing you, they're also seeing whatever the camera is pointed at behind you," says tech writer Sean Adams.[57] Keep the area neat and tidy, and whenever possible, limit the presence of other people and pets.

Show Your Face Whenever possible, turn on your video camera and let others see you. "In one meeting, I found myself talking to six photographs," laments Ryan, who explains, "The other participants had cameras, but they chose not to use them. I could

It can be frustrating when members of online meetings don't mute their microphones, or when they turn off their cameras so no one can see them.

What are some of your pet peeves regarding online meetings? Conversely, in your experience, what communication techniques work well?

only wonder if they were actually listening or not." Showing your face is supportive of others, and it shows that you are attentive and engaged. (Make sure lighting is adequate for people to see you clearly.)

Dress Head to Toe for the Camera Don't be like the telecommuter who logged onto a videoconference wearing a business jacket and pajama pants and then realized—once everyone was watching her—that she had left important papers on the other side of the room. Dress head to toe for video conferences as you would for in-person meetings.

Decide Who Talks When Businessperson Mita Mallick recalls the bedlam of a recent video conference. "I'm interrupted, like, three times and then I try to speak again and then two other people are speaking at the same time interrupting each other," she says.[58] Part of the problem is that turn-taking cues are less visible online than in person. One strategy is to have participants use a "raised hand" icon or physically hold up a hand when they have something to say. A facilitator can monitor these signals and call on people one by one. The team might also make use of virtual breakout rooms for small group discussions.

Use the Mute Function Wisely Even relatively quiet sounds add up when everyone's microphone is on. Use the mute function to minimize background noise when you aren't speaking. However, be sure to unmute yourself when you speak or laugh. One meeting participant remembers the silence after she told a joke during an online meeting: "No one laughed. Or maybe they did, but everyone was muted. At any rate, I won't do that again."

Pay Attention Although it may be tempting to get other work done during a long-distance meeting, looking away suggests that you're a poor listener and not interested. Position your device so that you can look directly at the camera and display the same listening posture (head up, shoulders back, facing forward) that you would in person.

Don't Over-Rely on Meetings So-called "Zoom fatigue" is real. It's exhausting to sit still and look at a computer screen for long periods of time.[59] Nevertheless, by some accounts, people now engage in more meetings virtually than they used to in person.[60] The increase may be because remote work offers fewer opportunities to talk casually in the hall or pop into a colleague's office with an idea. It's tempting to suggest "let's meet on Zoom" whenever an issue or opportunity emerges, but it's worth considering other options as well. Automattic, the company that created WordPress and other online platforms, has a long history with remote work involving staff members all over the globe. They communicate more or less continually through a company-wide blog system similar to Yammer. Anyone can review the posts, comment on them, and use them as an archive of information. "Everyone, from intern to CEO, can weigh in on anything,"[61] observes an industry analyst. The system supports transparency, frequent interaction, open debates, and shared learning—in short, dynamic communication 24 hours a day, even when team members are in different time zones. Your organization may consider similar ways to keep the lines of communication open without continually scheduling meetings.

Use Discussion Formats Strategically

Groups meet in a variety of settings and for a wide range of reasons. The formats they use are also varied. The following list is not exhaustive, but it provides a sense of how a group's structure can shape the type and quality of communication that members share.

Brainstorming During a **brainstorming** session, the goal is to think of as many ideas as possible about a specified topic without stopping to judge what is suggested or to rule anything out. The feature "Making the Most of a Brainstorming Session" presents strategies for succeeding at this.

> **brainstorming** A group activity in which members share as many ideas about a topic as possible without stopping to judge or rule anything out.

COMMUNICATION STRATEGIES

Making the Most of a Brainstorming Session

Brainstorming can be useful at any phase of problem solving, from understanding an issue to coming up with solutions. Here are some ways to make the most of this creative technique.

Include Diverse Participants

As one analyst puts it, "diversity is the mother of innovation."[62] Most problems would have been solved long ago if the solution were obvious. Applying multiple perspectives can reveal options not yet considered.

Begin by Brainstorming as Individuals

"When people work together, their ideas tend to converge," points out decision-making expert Art Markman.[63] By having people jot down their own ideas first, a group can harness the diversity of individual thought *and* the synergy of group work.

Choose a Neutral Facilitator

It's easy to get carried away evaluating the merits of a particular idea. A facilitator can keep the group focused and help them follow the principles of effective brainstorming.

Appoint a Recorder or Scribe

This person's job is to write all ideas on a board or flip chart where group members can see them.[64]

Forbid Criticism

Nothing stops the flow of ideas more quickly than negative evaluation. Criticism often leads to a defensive reaction, and it may inhibit people from thinking freely and sharing new ideas.

Share Whatever Comes to Mind

Even the most outlandish ideas sometimes prove workable, and an impractical suggestion might trigger a workable idea. The more ideas you generate, the more likely you are to come up with a good one.

Combine and Build Upon Ideas

Encourage members to "piggyback" by modifying and combining ideas already suggested. Also invite them to draw diagrams to show how concepts relate to one another.[65]

breakout groups A strategy used when the number of members is too large for effective discussion. Subgroups simultaneously address an issue and then report back to the group at large.

Breakout Groups When the number of members is too large for effective discussion, **breakout groups** (in person or via electronic chats rooms) can be used to maximize effective participation. In this approach, subgroups—usually consisting of five to seven members—simultaneously address an issue and then report back to the group at large. The best ideas of each breakout group are then discussed and synthesized.

Round Robins It can be useful to give everyone on the team a brief turn to talk. To initiate a **round robin** session, agree upon the time allotted to each member (perhaps 30 or 60 seconds) and the question everyone will address, such as "What do you most hope we can achieve as a group?" or "Of the ideas we have discussed, which one are you most excited about and why?" If you're meeting in person, have members sit (preferably in a circle) where they can easily see and hear everyone else. If your group is meeting virtually, agree on the order you will follow or have a facilitator call on people one by one.

round robin Session in which members each address the group for specified amount of time.

Since round robins aren't anonymous, they may not be effective at addressing highly sensitive issues. However, they allow members to get to know one another and appreciate the similarities and diversity of their ideas. Round robins give everyone a chance to share their thoughts, and as with brainstorming, participants in a round robin session are forbidden from critiquing or evaluating ideas as they are expressed. You might choose to have someone write all the ideas on a board or online slide to stimulate a follow-up discussion. Or a round robin can be a good way to close out a meeting.

Problem Census When the issue is sensitive or some members are more vocal than others, **problem census** can help equalize participation in a nonthreatening way.

problem census A technique in which members write their ideas on cards, which are then posted and grouped by a leader to generate key ideas the group can then discuss.

If you're meeting in person, have members list their thoughts on cards, one idea per card. A facilitator should collect the cards, read them to the overall group one by one, and post them on a board visible to everyone. Because the cards don't include names, issues are separated from personalities. As similar items are read, the leader posts and arranges them in clusters. After all cards are read and posted, members reflect on the ideas that have been presented.

If you're meeting online, members can maintain anonymity by changing their screen names to an agreed-upon moniker (such as Somebody) and then posting ideas via the chat function. In Zoom, their ideas should remain anonymous, even after they change their names back.[66] (Check your meeting software in advance to make sure it functions the same.) Members can then review ideas submitted by reading others' comments.

If you like, you can have individual members or breakout teams rank the ideas that have been posted in order from least preferable (1 point) to most preferable (a point value equal to the number of ideas on the list). You might then focus primarily on ideas that have garnered the most support.

Dialogue Sometimes the best way to tackle a problem is to stop trying to find a solution and, instead, listen to one another. **Dialogue** is a process in which people let go of the notion that their ideas are superior to others' and instead try to understand an issue from many perspectives.[67] For example, if the problem is that some children in your community are not being immunized, you might invite a collection of diverse people together to talk about the issue. Perhaps you had assumed that parents were being irresponsible, but you learn by listening that some of them don't have transportation, can't afford the cost, or don't understand medical information very well. You will probably proceed very differently once you realize the complexities of the issue.

dialogue A process in which people let go of the notion that their ideas are more correct or superior to others' and instead seek to understand an issue from many different perspectives.

In a genuine dialogue, members acknowledge that everything they "know" and believe is an assumption based on their own unavoidably limited experiences. People engage in curious and open-minded discussion about that assumption. The goal is to understand one another better, not to reach a decision or debate an issue.

11.4 Group Problem Solving

Dismayed to see small businesses going out of business during the COVID-19 pandemic, groups formed in many communities to help save them. In Detroit, bankers united to assist local entrepreneurs in purchasing inexpensive vacant lots.[68] One group of friends bought a small, inner-city plot for $100 and launched East Eats, an outdoor restaurant at which diners eat in transparent igloo-style tents that are heated in winter and shaded in summer.[69] "People might think that we're crazy, but once they see it, we think that they'll really get it," said Lloyd M. Talley, one of the owners.[70] He was right. The restaurant has won awards for its cuisine and become a neighborhood gathering space that also allows for social distancing when necessary.

Efforts such as these call attention to the problem-solving potential of groups. Unfortunately, groups don't always live up to this potential. When is group work a good option? What makes some groups succeed and others fail? Researchers have spent decades asking these questions. To discover their findings, read on.

Advantages of Group Problem Solving

Here are several reasons why groups are often more effective than individuals working alone:[71]

- *Groups have more resources.* Imagine trying to raise money for an important cause. Together, a group is likely to know far more potential contributors than any one person does. And if you have ever had the opportunity to take a test or quiz as a group, the odds are that you scored higher than any one member would have. Other resources may include space, equipment, time, and more.

- *Groups are more likely to catch errors.* At one time or another, everyone makes stupid mistakes, like the man who built a boat in his basement and then wasn't able to get it out the door. Working in a group can help prevent errors and oversights.

- *Diversity is an asset.* Although people tend to think in terms of "lone geniuses" who make discoveries and solve the world's problems, most breakthroughs are actually the result of collective creativity—people working together to create options that no one would have thought of alone.[72] When team members have different backgrounds and perspectives, the benefits are multiplied. To make the most of multiculturalism and avoid some of the common pitfalls, see the feature "Maximizing the Effectiveness of Multicultural Teams."

- *People tend to buy into solutions they helped to generate.* Besides coming up with superior solutions, groups may also generate a higher commitment to carrying them out. Nearly any professor will tell you that students cooperate much more willingly when they help develop a policy or project for themselves than when it is imposed on them.

COMMUNICATION STRATEGIES

Maximizing the Effectiveness of Multicultural Teams

Multicultural teams are typically more creative than homogeneous groups or individuals.[73] As one analyst puts it, "diversity makes us smarter."[74] So it's worth the effort to maximize the benefits and minimize the pitfalls of multicultural teams.[75] Here are some ways to do that.

Allow More Time Than Usual

When members have different backgrounds and perspectives, it can take patience and extra time to understand and appreciate where each person is coming from.

Agree on Clear Guidelines

To minimize misunderstandings, negotiate mutually acceptable ground rules for discussions, participation, and decision making.

Use a Variety of Communication Formats

People may be more or less comfortable speaking to the entire group, putting their thoughts in writing, speaking one on one, and so on. Variety will help everyone have a voice.

Educate Team Members

People are less likely to make unwarranted assumptions (that a person is lazy, disinterested, overbearing, or so on) if they understand the cultural patterns at play.

Be Open-Minded

It's easy to reject unfamiliar ideas without giving them proper consideration, but that short circuits the advantage of diverse perspectives. Make it a point to be curious and open-minded when a new idea is raised.

Despite the potential advantages of group work, it isn't always the quickest way to accomplish a task or make a decision. Following are some questions to consider when deciding whether a challenge is best addressed by a group or an individual. If you answer "yes" to the following conditions, collaborative problem solving will probably yield the best results.

- Is there more than one solution and no easy answer?
- Does the issue present implications for a large number of people?
- Is there the potential for disagreement?
- Will a good solution require more resources (e.g., time, money, supplies) than one person can provide?
- Would it be costly to make an error in judgment?
- Is buy-in by multiple people important to success?

By contrast, a problem with only one solution won't take full advantage of a group's talents. For example, phoning merchants to get price quotes and looking up a series of books in the library don't require much creative thinking. Jobs such as those can be handled by one or two people working alone. Of course, it may take a group meeting to decide how to divide the work to get the job done most efficiently.

A Structured Problem-Solving Approach

Working through a problem systematically involves a number of steps can that help a group stop and think, share ideas, and consider solutions they might not have thought of otherwise.

As early as 1910, John Dewey introduced his famous **reflective thinking method** as a systematic, multi-step approach to solving problems.[76] Since then, other experts have suggested modifications, although it is still generally known as Dewey's method. Although no single approach is best for all situations, a structured procedure usually produces better results than "no pattern" discussions.[77]

Consider the steps described here to be general guidelines, not a precise formula. Depending on the nature of a problem, you may want to focus on some steps more than others.

reflective thinking method A systematic approach to solving problems. Also known as Dewey's reflective thinking method.

Identify the Problem   Sometimes a group's problem is easy to pinpoint. Many times, however, the challenge isn't so clear. For example, if a group is meeting to discuss a low-performing employee, it may be helpful to ask why that person is underperforming. It may be because this employee has personal problems, feels unappreciated by members, or hasn't been challenged. The best way to understand a problem is to look below the surface and identify the range of factors that may be involved. It may also be helpful at this stage to take stock of group goals and the individual goals of everyone on the team.

Analyze the Problem   After you've identified the general nature of the problem, examine it in more detail. Here are some ways to do that:

- *Word the problem as a broad, open question.* For example, if your group is trying to understand why employee turnover is high, you might ask yourselves "What can we do to hire and retain good employees?" Open-ended questions like this encourage people to contribute ideas and to work cooperatively. They may also help the group identify the criteria for a successful solution.

- *Identify criteria for success.* Once you know what you're trying to achieve, you have a better chance of creating goal-oriented solutions and measuring your success. To continue the previous example, after analyzing the problem, you might set the goal of keeping employee turnover equal to or lower than the norm for your industry.

- *Gather relevant information.* It's foolish to choose a solution before you know all the options and factors at play. In this stage, you might seek answers to questions such as: How does our turnover compare to similar companies in this community? How can we measure employee satisfaction? What can we learn from current employees and those who have recently left? and What can we learn from companies with less turnover?

- *Identify supporting and restraining forces.* List the forces in favor of a desired outcome and those that will probably make it difficult to achieve. (This is called a *forcefield analysis*.[78]) For example, forces that can help you retain good employees might include offering higher pay and better training. Challenges might include a transient workforce and the inherent difficulty of the job.

Develop Creative Solutions   Considering more than one solution is important because the first solution, although the most obvious or most familiar, may not be the best one. Review the group discussion formats described earlier in the chapter for ideas on how to engage the team.

Select an Option   A good way to evaluate solutions is to ask the following questions: Which will best produce the desired changes? Which solution is most achievable? and Which solution contains the fewest serious disadvantages? As you narrow the list, review "Ways to Reach a Group Decision" to consider what option is best for you.

Implement the Plan   Everyone who makes New Year's resolutions knows the difference between making a decision and carrying it out. There are several important steps in developing and implementing a plan of action:

- *Identify specific tasks to be accomplished.* What must be done? Even a relatively simple job usually involves several steps. Now is the time to anticipate all the tasks facing the group. This may help you avoid an oversight or last-minute rush later.

> **COMMUNICATION STRATEGIES**
>
> # Ways to Reach a Group Decision
>
> Groups can use several approaches to arrive at decisions. Here are the advantages and disadvantages of each.
>
> **Reach Consensus**
>
> Consensus occurs when all members of a group support a decision. Full participation can increase the quality of the decision as well as members' willingness to support it. However, consensus building can take a great deal of time, which makes it unsuitable for emergencies.
>
> **Let the Majority Decide**
>
> Many people believe the democratic method of majority rule is always superior. That's not always true, however. A majority vote may be sufficient when the support of all members isn't necessary, but in more important matters it's risky. Even if a 51 percent majority favors a plan, 49 percent might oppose it—hardly sweeping support for any decision that needs everyone's support to work. Decisions made under majority rule are often inferior to those hashed out by a group until the members reach consensus.[79]
>
> **Rely on the Experts**
>
> If one or more group members are experts on the topic, you might give them decision-making authority. This can work well if (and only if) their judgment is truly superior.
>
> **Let a Few Members Decide**
>
> This approach works well with noncritical questions that would waste the whole group's time. But when an issue needs more support, it's best at least to have a subgroup report its findings for the approval of all members.
>
> **Honor Authority Rule**
>
> Although group decisions are usually of higher quality and gain more support from members than those made by an individual, there are times when an executive decision is in order. Sometimes there isn't time for a group to decide, or the matter is so routine that it doesn't require discussion.

- *Determine necessary resources.* Identify the equipment, material, and other resources the group will need to get the job done.
- *Define individual responsibilities.* Who will do what? Do all the members know their jobs? The safest plan here is to put everyone's duties in writing with due dates. This increases the chance of getting jobs done on time.
- *Plan ahead for emergencies.* Murphy's Law states that "whatever can go wrong will go wrong." Anyone experienced in group work knows the truth of this saying. People forget their obligations, get sick, or quit. The internet goes down when it's most needed, and so on. Whenever possible, develop contingency plans to cover foreseeable problems. Probably the single best suggestion is to plan on having all work done well ahead of the deadline so you know that, even with last-minute problems, you can still finish on time.

Follow Up on the Solution   Even the best plans usually require some modifications after they're put into practice. You can improve the group's effectiveness by meeting periodically to evaluate progress and revise the approach as necessary.

11.5 Communicating Effectively as a Leader

Julie Zhuo remembers the day she became an official leader at Facebook. Her manager said, "We need another manager, and you get along with everyone. What do you think?" At age 25, Zhuo wasn't expecting the invitation, but she felt it was a good opportunity and said yes. Later, she thought to herself, "*I got along with everyone.* Surely there was more to management than that."[80] She soon found out that there was.

Even if you don't consider yourself a leader yet, you may be surprised by how quickly the opportunity arises, and it's never too early to start building your skills. "If you want to become a leader, don't wait for the fancy title or the corner office," advises human resources expert Amy Gallo. "You can begin to act, think, and communicate like a leader long before that."[81] With that in mind, let's consider some essential points.

Leadership Can Be Learned

More than 2,000 years ago, Aristotle proclaimed, "From the hour of their birth some are marked out for subjugation, and others for command."[82] This is a radical expression of **trait theories of leadership**, which suggests that some people are born to be leaders while others are not. In fact, evidence shows that leaders of *any* personality can be effective, and the leadership skills people acquire are typically more important than anything they are born with.[83]

Julie Zhuo

Some leaders are outspoken, whereas others are quiet; some are bold, while others are tentative, and so on. The diversity is evident by considering leaders ranging from environmental activist Greta Thunberg, entrepreneur/actor/philanthropist Oprah Winfrey, civil rights hero Mahatma Gandhi, actor/composer Lin-Manuel Miranda, and corporate titan Bill Gates, to elected official Kamala Harris.

Of course, most leaders aren't famous. And many don't have official titles. People tend to gain influence gradually, even before they're aware of it. Think of a leaderless group to which you've belonged. Who did members look to for guidance and direction? **Emergent leaders** gain influence without being appointed by higher-ups. If the emergent leader of your group was respected and effective as time went on, they probably exhibited many of the following characteristics, mostly involving good communication skills:[84]

- good listening
- open to innovation
- able to work well with teams
- skillful at facilitating change
- appreciative of diversity
- honest and ethical

> **trait theories of leadership** A school of thought based on the belief that some people are born to be leaders and others are not.

> **emergent leader** A member who assumes leadership without being appointed by higher-ups.

COMMUNICATION STRATEGIES

Demonstrating Your Leadership Potential

The following behaviors are effective ways to show others that you would be a good leader.

Stay Engaged

Getting involved won't guarantee that you'll be recognized as a leader, but failing to speak up will almost certainly knock you out of the running.

Demonstrate Competence

Make sure your comments identify you as someone who can help the team succeed. Talking for its own sake will only antagonize other members.

Be Assertive, Not Aggressive

It's fine to have a say, but don't try to overpower others. Treat every member's contributions respectfully, even if they differ from yours.

Provide Solutions

How can the team obtain necessary resources? Resolve a disagreement? Meet a deadline? Members who provide answers to problems such as these are likely to rise to positions of authority.

---

power The ability to influence others' thoughts and/or actions.

legitimate power The ability to influence group members based on one's official position. Synonymous with being a **nominal leader**.

expert power The ability to influence others by virtue of one's perceived expertise on the subject in question.

connection power The ability to influence others based on having relationships that might help the group reach its goal.

The feature "Demonstrating Your Leadership Potential" suggests some ways you might communicate to earn a reputation as someone who can influence others for the better.

Power Comes in Many Forms

You can probably think of coworkers whose great attitude and talent influence everyone around them. Their presence is a good reminder that **power** (the ability to influence others) comes in many forms. Here are six types of power common in the workplace.[85,86]

Legitimate Power   Sometimes called position power, **legitimate power** arises from the title a person holds, such as supervisor, professor, or coach. People with legitimate power are said to be **nominal leaders**. Nominal comes from the Latin word for *name*, meaning that these leaders have been officially named to leadership positions. You can increase your chances of being selected for a leadership role by speaking up without dominating others, demonstrating competence, showing that you respect the group's norms and customs, and gaining the visible support of influential team members.

Expert Power   People have **expert power** when others perceive that they have valuable talents or knowledge. If you're lost in the woods, it makes sense to follow the advice of a group member who has wilderness experience. If your computer crashes at a critical time, you might turn to the team member with information technology expertise. To gain expert power, make sure that others are aware of your qualifications, be certain to convey accurate information, and don't act as if you're superior to others.

Connection Power   As its name implies, **connection power** comes from an individual's ability to develop relationships that help a group reach its goals. For example, a team seeking guest speakers for a seminar might rely on a well-connected member to line up candidates. To gain connection power, seek out opportunities to meet new people, nurture relationships through open and regular communication, and don't allow petty grievances to destroy valued relationships.

Reward Power A person with the ability to grant or promise desirable consequences has **reward power**. Rewards might involve raises, promotions, and the appreciation you show others. Even if you don't have a high-power job title, you can have reward power. For example, you might offer sincere, positive feedback to a classmate about a presentation they made in class.

> **reward power** The ability to influence others by granting or promising desirable consequences.

Coercive Power The threat or imposition of unpleasant consequences gives rise to **coercive power**. Bosses can coerce members via the threat of a demotion, an undesirable task, or even loss of a job. But peers also have coercive power. Working with an unhappy, unmotivated teammate can be punishing. For this reason, it's important to keep members feeling satisfied without compromising the team's goals. As a general rule, use rewards as a first resort and punishment as a last resort, make rewards and punishments clear in advance, and be generous with praise.

> **coercive power** The ability to influence others by threatening or imposing unpleasant consequences.

Referent Power The basis of **referent power** is the respect, liking, and trust others have for a person. If you have high referent power, you may be able to persuade others to follow your lead because they believe in you or because they are willing to do you a favor. Members acquire referent power by being genuinely likable and behaving in ways that others in the group admire. To gain referent power, listen to others' ideas and honor their contributions, do what you can to be liked and respected without compromising your principles, and present your ideas clearly and effectively.

> **referent power** The ability to influence others by virtue of being liked or respected.

After our look at various ways members influence one another, three important characteristics of power in groups become clearer.[87]

- *Power is given.* Power isn't something an individual possesses. Instead, it is conferred by the group. You may know a great deal about the subject being considered, but if the other members don't think you're knowledgeable, you don't have expert power. By the same token, you might try to reward other people by praising their contributions, but if they don't value your compliments all the praise in the world won't influence them.

- *Power is distributed among group members.* Power rarely belongs to just one person. Even when a group has an official leader, other members usually have some power to affect what happens. This influence can be positive, as when it arises from information, expertise, or social reinforcement. It can also be negative, as when a member criticizes others or withholds contributions the group needs to succeed. You can appreciate how power is distributed among members by considering the effect just one member can have by not showing up for meetings or failing to carry out their part of the job.

- *Power isn't an either–or concept.* It's incorrect to assume that power is something that members either have or lack. Rather, it's a matter of degree. Instead of talking about someone as "powerful" or "powerless," it's more accurate to talk about how much influence they have in various ways.

Leadership Approaches Vary

Early scholars identified three basic leadership approaches. The first, **authoritarian leadership**, is based on a leader's position and ability to offer rewards or punishment. It's typically a "power over" approach in which relatively powerless team

> **authoritarian leadership** A style in which a designated leader uses coercive and reward power to dictate the group's actions.

democratic leadership A style in which a leader invites the group's participation in decision making.

laissez-faire leadership A style in which a leader takes a hands-off approach, imposing few rules on a team.

servant leadership A style based on the idea that a leader's job is mostly to recruit outstanding team members and provide the support they need to do a good job.

members are expected to obey an all-powerful leader. The second approach, **democratic leadership**, describes leaders who encourage others to share in decision making. It is a "power with" dynamic in which people work together. The third approach, **laissez-faire leadership**, reflects a leader's willingness to allow team members to function independently and to make decisions on their own. Some theorists now add a fourth approach—**servant leadership**, based on the perspective that a leader's job is mostly to recruit outstanding team members and provide the support they need to do a good job.[88] Unlike laissez-faire leaders, who tend to have a hands-off approach, servant leaders are often highly involved with team members and processes. It is another example of "power with" leadership.

Research shows that each of these approaches is effective in some situations:

- Morale tends to be higher in teams with servant leaders than in those with authoritarian leaders.[89] However, an authoritarian approach sometimes produces faster results, which can be useful in a crisis.
- Satisfaction is typically high in teams led by democratic leaders, although inclusive decision making can be time consuming.[90]
- Highly experienced members may appreciate the hands-off approach of a laissez-faire leader, but for many teams, the ambiguity involved can create added stress.[91] A Gallup survey of millions of American workers revealed that nearly half of them don't have a clear sense of what their bosses expect of them.[92]
- Servant leadership often enhances team members' satisfaction and leads them to feel more self-confident and optimistic than they would otherwise.[93]

Good Leadership Is Situational

Civil rights leader Nelson Mandela was probably right when he said, "The mark of great leaders is the ability to understand the context in which they are operating and act accordingly."[94] Most contemporary scholarship supports the principle of **situational leadership**, which holds that a leader's style should change with the circumstances.[95]

situational leadership A theory that argues that the most effective leadership style varies according to the circumstances involved.

Those who exercise situational leadership consider the nature of the task, including how prepared team members are to accomplish it, and the team involved, including their relationships with each other and with the leader.[96] The **managerial grid** developed by Robert Blake and Jane Mouton (Figure 11.2) portrays leadership on the basis of these two considerations: low to high emphasis on people, and low to high emphasis on tasks.[97,98] Here are the management styles portrayed in the model.

managerial grid A two-dimensional model that identifies leadership styles as a combination of concern (in varying degrees) for people and/or for the task at hand.

Impoverished Managers This type of manager has little interest in either tasks or relationships. If you've worked with a supervisor who overlooked poor performance and seemed oblivious to team members' needs, you know how frustrating it can be. There are a number of reasons leaders may take an impoverished approach. They may lack confidence and assertive communication skills. Or they may be experiencing burnout. Once-concerned leaders who feel depleted or discouraged may largely give up. And, some leaders are so concerned with looking good personally that they try do nearly everything themselves, giving others the impression that they don't care much about the team or how it functions.[99]

It's easy to see that impoverished leadership isn't ideal. Some people equate it with a laissez-faire approach. Indeed, some laissez-faire leaders truly don't care about task fulfillment or relationships. However, other laissez-faire leaders care a great deal about both, but they take a hands-off approach for other reasons, such as that they trust the team, or they want them to develop confidence and independence. If that's their motivation, it's important that teams know it. Impoverished leaders take heed: Even when leaders take a hands-off approach, teams perform best when they perceive that leaders are invested in them personally and in their success.[100]

Country Club Managers These leaders exhibit high regard for relationships but little emphasis on accomplishing tasks. They tend to smile a lot, listen, and offer praise and encouragement. However, they may let misbehavior slide because they dread confrontations, and they typically have a hard time making difficult decisions. In many cases, people led by country club leaders like them personally but feel frustrated because their teams stagnate and underperform.[101] Country club management can foster an unproductive environment, as you know if you have been part of a project team in which meetings feel like social hours during which little actual work gets done.

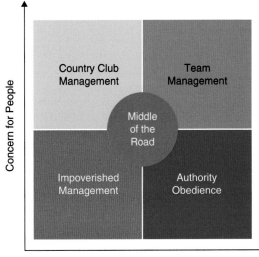

FIGURE 11.2 The Managerial Grid

Authority–Obedience Managers At the other extreme are managers who focus almost entirely on tasks and very little on relationships. Authority–obedience leaders exhibit the traits of authoritarian leadership described earlier in that they tend to focus on commands and quotas rather than investing in people. As leadership coach Bob Weinstein puts it, such leaders often come off as "hard-nosed, tough, demanding perfectionist[s]" or "unyielding control freaks" who micromanage purely for the sense of being in control.[102] These managers may engage in what some call "swoop and poop" maneuvers. To paraphrase one professional, rather than listening to team members and getting to know them, these managers tend to swoop in out of nowhere (a lot like a seagull) "make a lot of noise, dump on your work, and then fly back out, leaving a mess for others to fix."[103]

Clearly, few of us would want to work for long under those conditions. Another disadvantage is that change happens so quickly in today's environment that centralized decision-making is often too slow to be effective.[104]

Middle-of-the-Road Managers In the center of Blake and Mouton's grid is an approach characterized by a moderate interest in both tasks and relationships. If you've worked with a leader you consider not horrible or great, but more or less "okay," you've probably experienced this approach firsthand.

Team Managers These leaders are typically the most successful of all. They exhibit high regard for both tasks and relationships. This approach has a great deal in common with the model of transformational leadership, which is described next.

Transformational Leadership

Transformational leadership is defined by a leader's devotion to helping teams fulfill their potential.[105] Here are the central assumptions of transformational leadership.

> **transformational leadership** Defined by a leader's devotion to help a team fulfill an important mission.

- *People want to make a difference.* Transformational leaders believe that if the right people are on board, they will be motivated to accomplish important goals. Therefore, these leaders cultivate strong teams, actively listen to members, consider their feelings, and honor their contributions.

- *Empowerment is essential.* Transformational leaders aren't micromanagers who feel they have all the answers. Instead, they know that the best results come from well-prepared team members with the talent, training, and authority to make most decisions for themselves. The motto of transformational leaders could be, "It's not about me. It's about the *team* and what we accomplish together."
- *Mission is the driving force.* Transformational leaders expect 100 percent effort from everyone on the team because, otherwise, teams cannot live up to their full potential or accomplish their mission.
- *Transparency is key.* Although transformational leaders empower team members to make decisions for themselves as much as possible, when a tough decision is needed, these leaders aren't afraid to make it.[106] In those circumstances, they listen to diverse viewpoints, weigh all the factors, and when they announce a decision, they explain *why* they made it.[107]

It's probably clear that transformational leadership isn't easy. It requires putting your ego aside and focusing on the team and the mission. Because they are willing to make tough calls, transformational leaders aren't popular with everyone all the time. But even in tough times, team members typically hold these leaders in high regard as being effective, trustworthy, and fair.[108] Take the "What's Your Leadership Style?" self-assessment quiz to see which approach best describes you.

11.6 Leaving a Job Graciously

It's a fact of life that not all positions last forever. On average, workers in the United States hold about 12 different jobs by the time they are 54.[109] Even once you leave a position, your former colleagues and supervisors are in a position to recommend you to others and maintain your good reputation. To help ensure that you leave on good terms, follow these steps:

- *Put it in writing.* Write a brief, gracious resignation letter. Include the date you will leave (allow at least two weeks for the transition), a diplomatic explanation for why you are leaving (". . . new opportunities for growth"), and a statement of appreciation for what you have learned on the job.
- *Deliver the news personally.* Let the boss know you're leaving before you tell anyone else. If possible, schedule a face-to-face meeting, just the two of you. Give your manager the resignation letter and be professional and calm, even if you are leaving under less-than-ideal conditions.
- *Share the news graciously.* Unless instructed otherwise, let your coworkers know you are leaving. When you deliver the news, don't engage in criticism or complaints.
- *Make the change as easy as possible.* Help during the transition. You may be asked to finish key projects, create to-do lists and guides, or train new staff members. Do these things graciously and to the best of your ability as time allows.
- *Stay positive.* Even after you leave, don't complain about your employer. Bad-mouthing your old boss or the company you used to work for won't improve anything, and it's likely to make new colleagues wonder if you might criticize them in the future.

UNDERSTANDING YOUR COMMUNICATION

What's Your Leadership Style?

Choose the item in each group that best characterizes your beliefs as a leader.

1. I believe a leader's most important job is to:
   a. make sure people stay focused on the task at hand.
   b. take a hands-off approach so team members can figure things out on their own.
   c. make sure the workplace is a friendly environment.
   d. help team members build strong relationships so they can accomplish a lot together.

2. When it comes to being an employee, I believe that people:
   a. accomplish most when leaders set clear expectations.
   b. should do their work and let leaders do theirs.
   c. are most productive when they are enjoying themselves.
   d. have a natural inclination to work hard and do good work.

3. As a leader, when a problem arises, I am most likely to:
   a. announce a new policy or procedure to avoid the same problem in the future.
   b. ignore it; it will probably work itself out.
   c. try to smooth things over so no one feels upset about it.
   d. ask team members' input on how to solve it.

4. If team members were asked to describe me in a few words, I hope they would say I am:
   a. competent and results oriented.
   b. removed enough to make decisions without letting my emotions get in the way.
   c. pleasant and friendly.
   d. respectful and innovative.

5. When I see team members talking in the hallway, I am likely to:
   a. feel frustrated that they are goofing off.
   b. close my door so I can work without interruption.
   c. share my latest joke with them.
   d. feel encouraged that they get along so well.

INTERPRETING YOUR RESPONSES

For insight about your leadership style, consider which of the following best describes your answers.

Authority–Obedience

If most of your answers were "a," you feel that people should stay focused on the job at hand, and you are frustrated by inefficiency and signs that people are "wasting time." The danger is that in your zeal to get the job done, you will overlook relationships. This can be counterproductive in the long run since teams often accomplish more than individuals working alone.

Impoverished

If the majority of your answers were "b," you tend to take a hands-off approach as a leader, investing neither in relationships nor in tasks. You may take pride that you aren't a micromanager, but you're probably going too far in the opposite direction. Team members often need guidance. And even those who work well without much supervision probably crave your attention and appreciation.

Country Club

If you selected "c" more than other options, your focus on strong relationships and a pleasant work environment is likely to make you likeable as a person. However, team members may be frustrated by less-than-optimal results. A focus on both relationships *and* tasks may ultimately be more rewarding for everyone involved.

Team or Transformational

If you chose "d" most often, you balance an emphasis on results with a respect for the people involved. Although your expectations are high, your support and empowerment are likely to bring out the best in people. Most people consider this to be the ideal leadership style.

From landing a job to succeeding at it, to potentially moving on, communication is essential to career success.

Reflecting on what she has learned about followership, teamwork, and leadership, Julie Zhuo says that much of it comes down to communication. "There were so many instances where I didn't feel experienced enough, farsighted enough, empathetic enough, determined enough, or patient enough," she says.[110] Once she accepted that perfection isn't necessary, or even possible, it fueled her enthusiasm for learning from others and sharing her mistakes as well as her ideas. She concludes, "I've found that the more frequently and passionately I talk about what's important to me—including my missteps and what I've learned through them—the more positively my team responds."[111]

MAKING THE GRADE

OBJECTIVE 11.1 Practice communication skills involved in being a collaborative, proactive follower.

- Good followers do not passively follow orders. Instead, they are self-motivated, committed, talented, and courageous.
- Communications strategies that help followers succeed include presenting new ideas, seeking feedback, supporting others, and speaking up when things could be improved.

   > Think of a person you know who is not a good follower. What communication behaviors do they exhibit?

   > Describe a time when you showed initiative as a follower. How did you communicate? How were your efforts received by others?

   > Why do you think followership often gets less attention than leadership? Present an argument giving three compelling reasons to appreciate the important roles that followers play.

OBJECTIVE 11.2 Identify factors that influence communication in small groups and either help or hinder their success.

- Group work involves interaction and interdependence over time among a limited number of participants with the purpose of achieving one or more goals.

- Some groups achieve the status of teams, which embody a high level of shared goals and identity, commitment to a common cause, and high ideals.
- Social loafing is a common frustration in group work, but there are ways to help ensure that all group members feel accountable.
- Group norms suggest how members should interact with one another, how the group will do business, and how they will perform particular tasks.
- Members' goals fall into two main categories: task related and social.
- Group members play task roles (such as initiator, researcher, opinion seeker) and social roles (such as harmonizer and tension reliever). Some roles are helpful to the group, whereas others (such as aggressor and withdrawer) can damage performance and member relationships.

   > Describe the best and worst groups you have ever been part of, and then describe how communication differed in those groups.

   > What functional (and possibly dysfunctional) roles do you usually play in small groups? How do those roles help (or hinder) the groups' effectiveness? Would you be a more productive member if you modified these roles?

   > Think of a group you belong to, then make three lists: (1) your individual goals as a group member, (2) the individual goals of another group member, and (3) the group goals. How do the three lists compare? Are any of the individual goals you listed at odds with the team goals? If so, how?

OBJECTIVE 11.3 Strategize ways to build cohesiveness and communicate effectively during group discussions and meetings.

- Strong teams move through the following stages as they solve a problem: orientation (forming), conflict (storming), emergence (norming), and reinforcement (performing).
- Effective groups strive to build a sense of cohesiveness among members.
- Members of effective teams make sure that they participate equally by encouraging the contributions of quiet members and by keeping more talkative people on topic.
- Groups function best when they get the information they need without feeling overloaded.
- Effective team members guard against groupthink by minimizing pressure on members to conform for the sake of harmony or approval.
- Because face-to-face meetings can be time consuming and difficult to arrange, virtual teamwork is a good alternative for some group tasks. Making the most of them requires special rules for getting to know one another, turn-taking, and listening.
- Groups use a variety of discussion formats when solving problems, and each format has advantages and disadvantages depending on the size of the group, the behavior of members, and the nature of the problem.
  > Summarize the development of a team that is familiar to you, from orientation through reinforcement, giving an example of communication in each stage.
  > Imagine you have been asked to lead a new committee responsible for redesigning your school's grading policy. What will you do to help ensure that the committee functions to its highest potential?
  > Is a cohesive team always productive? Why or why not? Describe at least four methods of building team cohesiveness that can help a team get its job done.
  > Think of a team meeting you have been part of. Assess the productivity of that meeting on a 1 (low) to 10 (high) scale. Using the tips for managing meetings effectively in this chapter, describe the reasons you think the team either was, or was not, very productive.
  > Recall a group in which one or more members talked too much and others said very little. Describe an approach you might take to ensure more equal participation in the future.
  > Imagine that you and five other students have been asked to engage in a team project to benefit a nonprofit organization of the team's choice. What discussion formats would you use to decide on a project and host organization? If we expanded the project to include the entire class (let's say 30 people), would your choice of formats change? Why or why not? How might your communication strategies differ if members met virtually rather than in person?
  > Pretend that you are advising a newly formed team whose members will interact with each other online. What advice from this chapter would you share with them for developing trust at a distance?

OBJECTIVE 11.4 Assess the advantages and stages of group problem solving, and practice applying Dewey's group problem-solving model.

- Groups often have greater resources, including diverse ideas, than do either individuals or collections of people working in isolation.
- Teamwork can result in greater accuracy than individual efforts, and people may feel more committed to solutions they have helped to produce.
- Problem-solving groups should begin by identifying and analyzing a problem. The next step is to develop possible solutions, taking care not to stifle creativity by evaluating any of them prematurely. Finally, a group should implement the plan and follow up on the solution.
  > Describe a group experience in which you felt it would have been more productive to work alone. What factors caused you to feel that way?
  > Use your memory or imagination to envision a team of people from several different cultures. How might they differ from one another in terms of proposing new ideas, engaging in debate, and deferring (or not) to the highest-status member of the team?
  > Think of a group you were part of that effectively dealt with a problem or opportunity. What factors helped the group succeed?

OBJECTIVE 11.5 Demonstrate effective leadership skills based on the situation, goals, and team members' needs.

- Most research suggests that people can learn the skills that contribute to effective leadership.
- Leaders often emerge through a process of elimination, which suggests that, whether or not they know it, they begin "auditioning" for leadership roles as soon as they join a group.
- Power is given not taken, and it may be based on a variety of factors ranging from how likeable a person is to the position they hold.
- No one leadership approach works well in all circumstances. Instead, leaders who understand the relative strengths of various styles are most likely to succeed.
- The managerial grid developed by Robert Blake and Jane Mouton portrays leadership on the basis of these two considerations: low to high emphasis on people, and low to high emphasis on tasks
- For the most part, leaders who focus on the overall mission, relationships, and task fulfillment accomplish more than those who are motivated by the desire to achieve personal glory or maintain harmony at all costs.
  > Think of leaders (people you know or public figures) who embody each of the following leadership styles: autocratic, democratic, laissez-faire, and servant leadership. In your opinion, which of these leaders has been most effective and why?
  > What types of power do you have? How might you use that power responsibly to boost your chance of career success?
  > Imagine that the CEO of the retail clothing company where you work has challenged your team to increase sales by 20 percent. Describe how your team leader might respond to this challenge differently using each of the five leadership approaches included in Blake and Mouton's managerial grid.
  > Think about the last time you took on a new leadership role. Did you focus more on tasks or on relationships, or equally on both? How did this affect the way you communicated as a leader?
  > Transformational leaders help people make a significant and valuable contribution in business, science, civil rights, or another arena. Describe a goal that is important to you, and explain how you might embody the qualities of transformational leadership to help people achieve it.

OBJECTIVE 11.6 Describe the steps involved in resigning from a job graciously.

- Most people will leave multiple jobs in their lifetime. Resigning graciously involves sharing the news in written form and in person with the boss first and then with the team, making sure to remain positive and helpful until you leave.
- Even after you leave an organization, avoid bad-mouthing your former employer.
  > Draft a letter resigning from a hypothetical job, making sure that it meets the objectives outlined in the chapter.
  > Write down a few notes about what you might say to the boss and then to your coworkers.

KEY TERMS

authoritarian leadership p. 313
brainstorming p. 305
breakout groups p. 306
coercive power p. 313
cohesiveness p. 300
conflict (storming) stage p. 299
connection power p. 312
democratic leadership p. 314
dialogue p. 306
emergence (norming) stage p. 299
emergent leader p. 311
expert power p. 312
formal role p. 297
groupthink p. 302
informal role p. 297
laissez-faire leadership p. 314
legitimate power p. 312
managerial grid p. 314
nominal leader p. 312
norms p. 296
orientation (forming) stage p. 299
performance norms p. 296
problem census p. 306
procedural norms p. 296
power p. 312
referent power p. 313
reflective thinking method p. 308

reinforcement (performing) stage p. 299
reward power p. 313
role p. 297
round robin p. 306
rules p. 296
servant leadership p. 314
situational leadership p. 314
small group p. 294
social loafing p. 295
social norms p. 296
team p. 295
trait theories of leadership p. 311
transformational leadership p. 315

PUBLIC SPEAKING PRACTICE

Think of a leader (either good or bad) who has influenced you in a powerful way. How would you describe that person's communication style and leadership philosophy? How have they influenced your own leadership approach? Prepare an oral presentation in which you share your answers.

ACTIVITIES

1. Think about two groups to which you belong.
   a. What are your task-related goals in each?
   b. What are your social goals?
   c. Are your personal goals compatible or incompatible with those of other members?
   d. Are they compatible or incompatible with the group goals?
   e. What effect does the compatibility or incompatibility of goals have on the effectiveness of the group?
2. Describe the desirable norms and explicit rules you would like to see established in the following new groups, and describe the steps you could take to see that they are established.
   a. A group of classmates formed to develop and present a class research project
   b. A group of neighbors meeting for the first time to persuade the city to install a stop sign at a dangerous intersection
   c. A group of 8-year-olds you will coach in a team sport
   d. A group of fellow employees who will share new office space
3. Explain which of the following tasks would best be managed by a group:
   a. Collecting and editing a list of films illustrating communication principles
   b. Deciding what the group will eat for lunch at a one-day meeting
   c. Choosing the topic for a class project
   d. Finding which of six companies had the lowest auto insurance rates
   e. Designing a survey to measure community attitudes toward a subsidy for local artists

Preparing and Presenting Your Speech

CHAPTER OUTLINE

12.1 Getting Started 324
Choosing Your Topic
Defining Your Purpose
Writing a Purpose Statement
Stating Your Thesis

12.2 Analyzing the Speaking Situation 327
The Listeners
The Occasion

12.3 Gathering Information 332
Online Research
Library Research
Interviewing
Survey Research

12.4 Managing Communication Apprehension 335
Facilitative and Debilitative Communication Apprehension
Sources of Debilitative Communication Apprehension
Overcoming Debilitative Communication Apprehension

12.5 Presenting Your Speech 339
Choosing an Effective Type of Delivery
Practicing Your Speech

12.6 Guidelines for Delivery 341
Visual Aspects of Delivery
Auditory Aspects of Delivery

12.7 Sample Speech 344

MAKING THE GRADE 347

KEY TERMS 348

PUBLIC SPEAKING PRACTICE 348

ACTIVITIES 348

LEARNING OBJECTIVES

12.1
Describe the importance of topic, purpose, and thesis in effective speech preparation.

12.2
Analyze both the audience and occasion in any speaking situation.

12.3
Gather information on your chosen topic from a variety of sources.

12.4
Assess and manage debilitative speaking apprehension.

12.5
Make effective choices in the delivery of your speech.

Photo courtesy of Hailey Hardcastle

University of Oregon student Hailey Hardcastle describes herself as a happy, healthy, and successful student.

And yet, Hailey has suffered from anxiety and depression since her earliest school days. These challenges reached a peak in her junior year of high school, when college application deadlines combined with both the usual stressors of high school and problems at home. Sitting in class one day, Hailey started to suffer the symptoms of a panic attack: racing heart, shortness of breath, an overwhelming sense of terror. Hailey managed to regain her footing but continued to struggle on her own.

As Hailey explains in the sample speech at the end of this chapter,

> . . . I had gotten pretty good at managing my own mental health. I was a successful student, and I was president of the Oregon Association of Student Councils. But it was around this time that I began to realize mental health was a much bigger problem than just for me personally. Unfortunately, my hometown was touched by multiple suicides during my first year in high school. I saw those tragedies shake our entire community, and as the president of a statewide group, I began hearing more and more stories from students where this had also happened in their town.[1]

Hailey and a team she assembled helped change the laws of her state. Because of their activism, Oregon students are now allowed by law to take mental health breaks away from school, just as with physical illness. Hailey's sample speech at the end of this chapter tells the story of why that's important and how she made it happen.

This chapter describes the process of creating and delivering an effective speech. Your classroom speech might not be as personal as Hailey Hardcastle's, but it might pave the way for important moments in your life. Someday you might find yourself advocating for social change, making a job-related presentation, or speaking on a special occasion, such as a wedding or funeral. You might find yourself speaking in favor of a civic-improvement project in your hometown or trying to persuade members of your club to work toward solving local problems such as poverty, religious conflicts, or environmental threats.

Despite the potential benefits of effective speeches, the prospect of standing before an audience terrifies many people. In fact, giving a speech seems to be one of the most anxiety-producing things we can do: When asked to list their common fears, research subjects mention public speaking more often than they do insects, heights, accidents, and even death.[2]

Even if you don't love the idea of giving speeches, we promise to give you the tools to speak in a way that is clear, interesting, and effective. And it's very likely that, as your skills grow, your confidence will too. This chapter covers major steps in that process, through careful speech planning.

12.1 Getting Started

Your first tasks are generally choosing a topic, determining your purpose, and writing a purpose statement.

Choosing Your Topic

The first question many student speakers face is, "What should I talk about?" When you need to choose a topic, you should try to pick one that is right for you, your audience, and the situation. Hailey Hardcastle chose her topic because of its importance to her. If you take the same approach, your story will be unique—one that only you can tell.

Of course, you can also pick a topic based on simple interest and curiosity. Your motivation there would be to discover as much about it as you can and share that information with your classmates. Whatever your motivation for choice of topic, a key word of advice is to decide on your topic as early as possible. Waiting until the last moment will leave you without enough time to research, outline, and practice your speech effectively. That's a recipe for anxiety and poor outcomes.

Defining Your Purpose

No one gives a speech—or expresses *any* kind of message—without having a reason to do so. Your first step in focusing your speech is to formulate a clear and precise statement of that purpose.

Writing a Purpose Statement

Your **purpose statement** should be expressed in the form of a complete sentence that describes your **specific purpose**—exactly what you want your speech to accomplish. It should stem from your **general purpose**, which might be to inform, persuade, or entertain. Beyond that, though, there are three criteria for an effective purpose statement:

1. **A purpose statement should be realistic.** It's fine to be ambitious, but you need to design a purpose that has a reasonable chance of success. You can appreciate the importance of having a realistic goal by looking at some unrealistic ones, such as "My purpose is to convince my audience to make federal budget deficits illegal." Unless your audience happens to be a joint session of Congress, it won't have the power to change U.S. fiscal policy. But any audience can write its congressional representatives or sign a petition.

    If you were giving a speech on college students and depression, it would be a tall order to provide every audience member with the tools to handle their current or future bouts of depression. A better purpose statement for this speech might sound something like this:

    My speech will give my audience some tools to recognize and deal with depression.

2. **A purpose statement should be results oriented.** Having a *results orientation* means that your purpose is focused on the outcome you want to accomplish with your audience members. For example, if you were giving an informative talk on how depression affects undergraduates, this would be an inadequate purpose statement:

    My purpose is to tell my audience about depression in college students.

    As that statement is worded, your purpose is "to tell" an audience something, which suggests that the speech could be successful even if no one

> **purpose statement** A complete sentence that describes precisely what a speaker wants to accomplish.

> **specific purpose** The precise effect that the speaker wants to have on an audience. It is expressed in the form of a purpose statement.

> **general purpose** One of three basic ways a speaker seeks to affect an audience: to entertain, inform, or persuade.

listened. A results-oriented purpose statement should refer to the response you want from your audience: It should tell what the audience members will know or be able to do after listening to your speech:

After listening to my speech, audience members will be more willing to acknowledge feelings of depression and take steps to manage it.

3. **A purpose statement should be specific.** To be effective, a purpose statement should include enough details so that, after your speech, you would be able to measure or test your audience to see if you had achieved your purpose. Getting students to admit to feeling depressed is a worthy goal, but your speech might be improved if you make that goal measurable:

After listening to my speech, my audience members will be able to list three ways to deal with depression when they experience it.

Consider the following sets of purpose statements:

LESS EFFECTIVE	MORE EFFECTIVE
To have professional wrestling banned as a sport. (Not realistic)	After listening to my speech, my audience will agree that kids who imitate professional wrestlers can be seriously hurt.
To get my audience thinking about the benefits of prison reform. (Not results oriented)	After listening to my speech, my audience members will vote yes to my referendum to set up a tutoring program for local prisoners.
To tell my audience about gun control. (Not specific)	After listening to my speech, the audience will sign my petition calling for universal background checks.

A specific purpose statement usually is a tool to keep you focused on your goal as you plan your speech. You probably won't include it word for word in your actual speech.

Stating Your Thesis

After defining your purpose, you're ready to start planning what is arguably the most important sentence in your entire speech. The **thesis statement** tells your listeners the central idea of your speech. It's the one idea that you want your audience members to remember after they have forgotten everything else you had to say. The thesis statement for a speech about your local recycling program might sound like this:

"Our local MRF (Materials Recovery Facility) dramatically reduces the waste we dump into the environment."

Unlike your purpose statement, your thesis statement is almost always delivered directly to your audience. The thesis statement is usually formulated later in the speech-making process, after you have done some research on your topic.

thesis statement A complete sentence describing the central idea of a speech.

The progression from topic to purpose to thesis is another part of your focusing process, as you can see in the following example:

Topic: Organ donation

Specific Purpose: After listening to my speech, audience members will recognize the importance of organ donation and will sign an organ donor's card for themselves.

Thesis: By choosing to become an organ donor you can prevent one, or even several needless deaths.

12.2 Analyzing the Speaking Situation

There are two components to analyze in any speaking situation: the audience and the occasion. To be successful, every choice you make in putting together your speech—your purpose, topic, and all the material you use to develop your speech—must be appropriate to both of these components.

The Listeners

Audience analysis involves identifying and adapting your remarks to the most pertinent characteristics of your listeners.

Audience Purpose Just as you have a purpose for speaking, audience members have a reason for gathering. Sometimes virtually all the members of your audience will have the same, obvious goal. Expectant parents at a natural childbirth class are all seeking a healthy delivery, and people attending an investment seminar are all looking for ways to increase their net worth.

There are other times, however, when audience purpose can't be so easily defined. In some instances, different listeners will have different goals, some of which might not be apparent to the speaker. Consider a church congregation, for example. Whereas most members might listen to a sermon with the hope of applying religious principles to their lives, a few might be interested in being entertained or in merely appearing pious. In the same way, the listeners in your speech class probably have a variety of motives for attending. Becoming aware of as many of these motives as possible will help you predict what will interest them. You can ask individual audience members about their point of view or simply listen carefully when they express themselves in class. As you do so, you can start to make some judgments about audience demographics.

Demographics Your audience has a number of characteristics that you may be able to observe in advance. These factors, known as **demographics**, include cultural differences, age, gender, group membership, number of people, and so on. Demographic characteristics might affect your speech planning in a number of ways.[3] For example:

- **Cultural diversity**. Do audience members differ in terms of race, religion, or national origin? The guideline here might be, *Do not exclude or offend any portion of your audience on the basis of cultural differences*. If there is a dominant cultural group represented, you might decide to speak to it, but remember that the point is to analyze, not stereotype, your audience. If you talk down to any segment of your listeners, you have probably stereotyped them.

> **audience analysis** A consideration of characteristics, including the type, goals, demographics, beliefs, attitudes, and values of listeners.

> **demographics** Audience characteristics that can be analyzed statistically, such as age, gender, education, and group membership.

- **Gender.** Although masculine and feminine stereotypes are declining, it is still important to think about how gender can affect the way you choose and approach a topic. Every communication professor has a horror story about a student getting up in front of a class composed primarily, but not entirely, of women and speaking on a subject such as "Adapting to Motherhood."
- **Age.** Our interests vary and change with our age. These differences may run relatively deep; our approach to literature, films, finance, health, and long-term success may change dramatically over just a few years, perhaps from graphic novels to serious literature, from punk to classical music, or from hip-hop to epic poetry.
- **Group membership.** Groups generally form around shared interests. By examining the groups to which your audience members belong, you can potentially surmise their political leanings, religious beliefs, or occupation. Group membership is often an important consideration in college classes. Consider the difference between a daytime class and a class that meets in the evening. At many colleges, the evening students are generally older and tend to belong to civic groups, church clubs, and the local chamber of commerce. Daytime students are more likely to belong to sororities and fraternities, sports clubs, and social action groups.[4]
- **Number of people.** Topic appropriateness varies with the size of an audience. With a small audience, you can be less formal and more intimate; you can, for example, talk more about your feelings and personal experiences. If you gave a speech before 5 people as impersonally as if they were a standing-room-only crowd in a lecture hall, they would probably find you stuffy. On the other hand, if you talked to 300 people about your unhappy childhood, you'd probably make them uncomfortable.

You have to decide which demographics of your audience are important for a particular speech. For example, when Britton Ody, a student at Berry College in Georgia, gave a speech on the lack of treatment for hepatitis C in state prisons, he knew he had to broaden the appeal of his topic beyond the prison demographic referred to in his speech. He adapted to his broader audience this way:

Within prisons, hepatitis C is not being treated, allowing for the disease to attack new hosts uncontended. It's a virus feeding ground. And it continues to spread beyond prisons. According to the Bureau of Justice's statistics, 95% of inmates in the United States are being released back into the community, and hepatitis C with them.[5]

To keep your audience engaged, it's essential to tailor your remarks to their demographic profile, interests, beliefs, and knowledge.

COMMUNICATION STRATEGIES

Adapting with Integrity

A student asks: "Is it truly ethical for me to adapt to the audience, even if I believe something different? I hate it when politicians do that. . . ."

As a speaker, you have to decide how far to go when adapting to your audience. You should never take a position you do not sincerely hold. Audience adaptation is more a matter of recognizing where your listeners' beliefs differ from yours and then finding common ground. Former U.S. Secretary of Labor Robert Reich outlined several strategies toward this end, including the following:[6]

Seek Out Alternative Views

Don't avoid conversations with people who are likely to disagree with you. To the contrary, seek them out.

Zero In on Key Issues

Instead of raising abstract points, begin by focusing on issues that affect your listeners directly. For example, Do audience members think that ineffective government is related to big money in politics, rather than Democrat or Republican deeds?

Make It Personal

Invite your audience to think about their own experiences and stories. Share yours. Point out common ground between their positions and yours.

Acknowledge Listeners' Beliefs

Invite the audience to consider your thoughts and how your position might address their concerns.

Use Humor

Humor can be a great connector. For example, "the Supreme Court says corporations are people. Well, you'll believe they're people when Texas executes a corporation."

Avoid Causing Defensiveness

Remember, the point isn't to convince them you're right and they're wrong, which will only make them defensive.

These five demographic characteristics are important examples, but the list goes on. Other demographic characteristics that might be important in a college classroom include the following:

- Educational level
- Economic status
- Hometown
- Year in school
- Major subject
- Ethnic background

A final factor to consider in audience analysis concerns members' attitudes, beliefs, and values.

Attitudes, Beliefs, and Values Audience members' feelings about you, your subject, and your intentions are central issues in audience analysis. One way to approach these issues is through a consideration of attitudes, beliefs, and values.[7] Attitudes, beliefs, and values all deal with the way people think about different topics and how they will respond to that topic. These three traits reside in human consciousness like layers of an onion (see Figure 12.1). **Attitudes**, which are closest to the surface, are most easily

> **attitude** The predisposition to respond to an idea, person, or thing favorably or unfavorably.

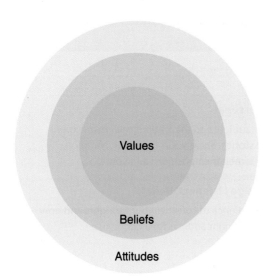

FIGURE 12.1 Attitudes, Beliefs, and Values

belief An underlying conviction about the truth of an idea, often based on cultural training.

value A deeply rooted belief about a concept's inherent worth.

observed. Sometimes we can just look at a person and say, "He's got an attitude." That's because attitudes are directly reflected in things a person says and does. **Beliefs** lie a little deeper and deal with the truth of something. It takes a little longer for people to reveal the beliefs that lie beneath their attitudes. Often, beliefs are divulged only when someone is challenged on an attitude. **Values** are the most deeply rooted of all. Individuals might not be able to express or explain their values coherently, but those values still impact the way people see the world. Values tend to be part of an individual's identity. We often speak of an individual's belief system or value system, suggesting that these factors are part of a larger group's meaningful ideas.[8]

You can begin to appreciate the usefulness of these concepts by considering an example. Suppose you wanted to give a speech on the concept of political correctness. Consider how audience analysis would help you design the most promising approach: *Attitudes.* How do your listeners feel about political correctness movements, such as Me Too and Black Lives Matter? Do they feel that such movements perform a useful and reasonable function, or do they sense that such movements are unnecessary and have gone too far? If they recognize the importance of such movements, you can proceed confidently, knowing they'll probably want to hear what you have to say. On the other hand, if they are vaguely disgusted by even thinking about the topic, you will need to dig deeper.

Beliefs. Does your audience accept the relationship between political correctness and deeper held beliefs, such as treating people equally and acting civilly in society? Or do you need to inform them about how these concepts are related?

Values. Which underlying values matter most to your listeners, and how does political correctness relate to them? For example, how would the tenets of their religion, or the lessons that their parents taught them about life, relate to the social justice movements that are part of political correctness?

Experts in audience analysis, such as professional speechwriters, often try to concentrate on values. As one team of researchers pointed out, "values have the advantage of being comparatively small in number, and owing to their abstract nature, are more likely to be shared by large numbers of people."[9] Stable American values include the ideas of good citizenship, a strong work ethic, tolerance of differing political views, individualism, and justice for all. Brianna Mahoney, a student at the University of Florida, appealed to her audience's values when she wanted to make the point that anti-homelessness laws were inhumane:

> Extreme poverty in the United States is a shockingly overlooked issue, making the homeless one of our most vulnerable populations. Recent legislation has worsened this by denying the homeless the ability to help themselves. People experiencing homelessness exist in every community; it's time to stop accusing them of becoming a burden when they are already struggling to carry their own.[10]

Mahoney pointed out that discriminating against homeless people was unfair and impractical; her analysis had suggested that the value of fairness would be important to this audience. She also surmised that they would be offended by unfairness combined with impracticality, as in this piece of evidence she used:

> Ninety-year-old Arnold Abbott made international headlines when he was arrested in Fort Lauderdale, Florida, and faced 60 days in prison for feeding hot meals to people who were homeless. When police arrived they grabbed trays of food and shoved them into the trash, while lines of hungry people were forced to just look on.[11]

You can often make an inference about audience members' attitudes by recognizing the beliefs and values they are likely to hold. In this example, Brianna knew that her audience, made up mostly of idealistic college students and professors, would dislike the idea of unfair and impractical discrimination.

Evaluation of hidden psychological states can be extremely helpful in audience analysis. For example, a religious group might hold the value of "obeying God's word." For some fundamentalists, this might lead to the belief, based on their religious training, that women are not meant to perform the same functions in society as men. This belief, in turn, might lead to the attitude that women ought not to pursue careers as firefighters, police officers, or construction workers.

You can also make a judgment about one attitude your audience members hold based on your knowledge of other attitudes they maintain. If your audience is made up of undergraduates who have a positive attitude toward liberation movements, it is a good bet they also have a positive attitude toward civil rights and ecology. If they have a negative attitude toward collegiate sports, they may also have a negative attitude toward fraternities and sororities. This should suggest not only some appropriate topics for each audience but also ways that those topics could be developed.

The Occasion

The second phase in analyzing a speaking situation focuses on the occasion. The occasion of a speech is determined by the circumstances surrounding it. Three of these circumstances are time, place, and audience expectations.

Time Your speech occupies an interval of time that is surrounded by other events. For example, other speeches might be presented before or after yours, or comments might be made that set a certain tone or mood. External events such as elections, the start of a new semester, or even the weather can color the occasion in one way or another. The date on which you give your speech might have some historical significance. If that historical significance relates in some way to your topic, you can use it to help build audience interest.

The time available for your speech is also an essential consideration. You should choose a topic that is broad enough to say something worthwhile but brief enough to fit your limits. "Wealth," for example, might be an inherently interesting topic to some college students, but it would be difficult to cover such a broad topic in a 10-minute speech and still say anything significant. However, a topic like "The problem of income inequality in America today" could conceivably be covered in 10 minutes in enough depth to be of some value. All speeches have limits, whether or not they are explicitly stated. If you are invited to say a few words, and you present a few volumes, you might not be invited back.

Place Your speech also occupies a physical space. The beauty or squalor of your surroundings and the noise or stuffiness of the room should all be taken into

It can be fun to go for a quick laugh when making a speech. But keep in mind the impression that will linger.

Have you ever made remarks inappropriate for the occasion? What can you do to avoid this kind of mistake in the future?

consideration. These physical surroundings can be referred to in your speech if appropriate. If you were talking about world poverty, for example, you could compare your surroundings to those that might be found in a poor country.

Audience Expectations Finally, your speech is surrounded by audience expectations. A speech presented in a college class, or a TED Talk such as the one that appears at the end of this chapter, is usually expected to reflect a high level of thought and intelligence. This doesn't necessarily mean that it has to be boring or humorless; wit and humor are, after all, indicative of intelligence. But it does mean that you have to put a little more effort into your presentation than if you were discussing the same subject with friends over coffee.

When you are considering the occasion of your speech, it pays to remember that every occasion is unique. Although there are obvious differences among the occasions of a college class, a church sermon, and a bachelor party "roast," there are also many subtle differences that will apply only to the circumstances of each unique event.

12.3 Gathering Information

This discussion about planning a speech purpose and analyzing the speech situation makes it apparent that it takes time, interest, and knowledge to develop a topic well. Setting aside a block of time to reflect on your own ideas is essential. However, you will also need to gather information from outside sources.

By this time you are probably familiar with both web searches and library research as forms of gathering information. Sometimes, however, speakers overlook interviewing, personal observation, and survey research as equally effective methods of gathering information. Let's review all these methods here and perhaps provide a new perspective on one or more of them.

Online Research

The ease of using search engines like Google has made them the popular favorite for speech research. But students are sometimes so grateful to have found a website dealing with their topic that they forget to evaluate it. Like any other written sources you would use, websites should be accurate and rational. Beyond that, there are four specific criteria that you can use to evaluate the quality of a website: *credibility*, *objectivity*, *currency*, and *functionality* (see the feature "Evaluating Websites").

In the case of some specialized search engines, like Google Scholar, the criteria of credibility, objectivity, and currency will be practically guaranteed. However, these guidelines are especially important when accessing information from Wikipedia, the popular online encyclopedia. Because anyone can edit a Wikipedia article at any time, many professors forbid the use of it as a primary resource. Others allow it to be used for general information and inspiration. Most will allow its use when articles have references to external sources (whether online or not) and the student reads the references and checks whether they really do support what the article says.

Library experts help you make sense of and determine the validity of the information you find, whether online or in print. And a library can be a great environment for concentration, a rare quiet place with minimal distractions.

Library Research

Libraries, like people, tend to be unique. Although many of your library's resources will be available online through your school's website, it can be extremely rewarding to get to know your library in person, to see what kind of special collections and services it offers, and just to find out where everything is. A few resources are common to most libraries, including the library catalog, reference works, periodicals, nonprint materials, and databases.

Databases, which can be particularly useful, are computerized collections of highly credible information from a wide variety of sources. One popular collection of databases is LexisNexis, which contains millions of articles from news services, magazines, scholarly journals, conference papers, books, law journals, and other sources. Other popular databases include ProQuest, Factiva, and Academic Search

> **database** A computerized collection of information that can be searched in a variety of ways to locate information that the user is seeking.

COMMUNICATION STRATEGIES

Evaluating Websites

Consider the following four criteria when choosing a website for online research:

Credibility

Anyone can establish a website, so it is important to evaluate where your information is coming from. Who wrote the page? Are their names and contact information listed? Anonymous sources should not be used. If the sources *are* listed, are their credentials listed? What institution publishes the document? Remember that although an attractive site design doesn't guarantee high-quality information, obvious mistakes such as misspellings are clear signs of low quality.

Objectivity

What opinions (if any) does the site express? Are these opinions backed up with facts, or are they purely based on inferences? Is the site trying to sell something, including a candidate or a political idea?

Currency

When was the page produced? When was it updated? Are the links working? If any of the links are dead, the information might not be current. And information that was true 10 years ago might not be true today.

Functionality

Is the site easy to use, so that you can locate information you are looking for? Are options to return to the home page and tops of pages provided? Is the site searchable?

For more tips on evaluating online information, see the feature "Evaluating (Mis)information" in Chapter 2.

Premier, and there are dozens of specialized databases, such as Communication and Mass Media Complete. Database searches are slightly different from web searches; they generally don't respond well to long strings of terms or searches worded as questions. With databases it is best to use one or two key terms with a connector such as AND, OR, or NOT.[12] Once you learn this technique and a few other rules (perhaps with a librarian's help), you will be able to locate dozens of articles on your topic in just a few minutes.

Interviewing

An information-gathering interview allows you to view your topic from an expert's perspective and take advantage of that expert's experience, research, and thought. You can also use an interview to stimulate your own thinking. Often the interview will save you hours of Internet or library research and allow you to present ideas that you could not have uncovered any other way. And because an interview is an interaction with an expert, many ideas that otherwise might be unclear can become more understandable through questions and answers. Interviews can be conducted face to face, by telephone, or by email.

Allison McKibban, a student at Rice University, used highly effective excerpts from an interview in her speech on criminalizing rape survivors. She began that speech like this:

> As I sat down in the visitor's center of the Harris County jail, my hands shaking and heart pounding, Alycia bluntly asked, "Who are you?" It's never an easy question—but I started in with college and speech and when Alycia began to open up, our similarities were overwhelming. We grew up middle class, doting mothers, swim team, straight A's, bound for college—but it all changed at 17. Her senior year of high school, Alycia detailed in a personal interview from December 2nd of this year, was the first time her stepfather's friend sold her for sex.
>
> Two years later, after being trafficked through Texas against her will for money she never received, Alycia was freed from her abuser—and sentenced to 27 months in Texas prison for prostitution.[13]

Survey Research

One advantage of **survey research**—the distribution of questionnaires for people to respond to—is that it can give you up-to-date answers concerning "the way things are" for a specific audience. Survey responses can be collected via group chats on apps like GroupMe or WhatsApp, if you have one set up for your class. If not, you can ask your professor if you can hand out questionnaires a week or so in

survey research Information gathering in which the responses of a population sample are collected to disclose information about the larger group.

Online and library research are essential to speech preparation. In addition, talking to credible sources can provide information that makes a speech more interesting and compelling.

Who can you interview to enrich the content of your speech?

advance. If you did that, and you were presenting a speech on the possible dangers of body piercing, you could present information like this in your speech:

> According to a survey I conducted last week, 90 percent of the students in this class believe that body piercing is basically safe. Only 10 percent are familiar with the scarring and injury that can result from this practice. Two of you, in fact, have experienced serious infections from body piercing: one from a pierced tongue and one from a simple pierced ear.

That statement would be of immediate interest to your audience members because *they* were the ones who were surveyed. Another advantage of conducting your own survey is that it is one of the best ways to find out about your audience: It is, in fact, *the* best way to collect the demographic data mentioned earlier. The one disadvantage of conducting your own survey is that, if it is used as evidence, it might not have as much credibility as published evidence found in the library. But the advantages seem to outweigh the disadvantages of survey research in public speaking.

No matter how you gather your information, remember that it is the *quality* rather than the quantity of the research that is most important. The key is to determine carefully what type of research will answer the questions you need to have answered. Sometimes only one type of research will be necessary; at other times every type mentioned here will have to be used. Generally, you will collect far more information than you'll use in your speech, but the winnowing process will ensure that the research you do use is of high quality.

Along with improving the quality of what you say, effective research will also minimize the anxiety of actually giving a speech. Let's take a close look at that form of anxiety.

12.4 Managing Communication Apprehension

The terror that strikes the hearts of so many beginning speakers is commonly known as *stage fright* or *speech anxiety,* or what communication scholars call *communication apprehension*.[14] Whatever term you choose, the important point to realize is that fear about speaking can be managed in a way that works for you rather than against you.

Facilitative and Debilitative Communication Apprehension

Although communication apprehension is a very real problem for many speakers, it can be overcome. The first step in feeling less apprehensive about speaking is to realize that a certain amount of nervousness is not only natural but also facilitative. That is, **facilitative communication apprehension** is a factor that can help improve your performance. Just as totally relaxed actors or musicians aren't likely to perform at the top of their potential, speakers think more rapidly and express themselves more energetically when their level of tension is moderate. In fact, experts suggest that you relabel communication apprehension as something positive: "The crackle of excitement" or "a surge of extra energy."[15]

facilitative communication apprehension A moderate level of anxiety about speaking before an audience that helps improve the speaker's performance.

Academy Award–winning actress Jennifer Lawrence suffers from communication apprehension. Her tip for reducing anxiety: "I just try to acknowledge that this scrutiny is stressful, and that anyone would find it stressful."

How can you apply the information in this chapter to manage—and benefit from—your natural apprehension?

If you can't talk yourself out of it, **debilitative communication apprehension** tends to inhibit effective self-expression. Intense fear causes trouble in two ways. First, the strong emotion keeps you from thinking clearly.[16] This has been shown to be a problem even in the preparation process: Students who are highly anxious about giving a speech will find the preliminary steps, including research and organization, to be more difficult.[17] Second, intense fear leads to an urge to do something, anything, to make the problem go away. This urge to escape often causes a speaker to speed up delivery, which results in a rapid, almost machine-gun style. As you can imagine, this boost in speaking rate leads to even more mistakes, which only add to the speaker's anxiety. Thus, a relatively small amount of nervousness can begin to feed on itself until it grows into a serious problem.

debilitative communication apprehension An intense level of anxiety about speaking before an audience, resulting in poor performance.

Sources of Debilitative Communication Apprehension

Before we describe how to manage debilitative communication apprehension, let's consider why people are afflicted with the problem in the first place.[18]

Previous Negative Experience People often feel apprehensive about speech giving because of unpleasant past experiences. Most of us are uncomfortable doing *anything* in public, especially if it is a type of performance in which our talents and abilities are being evaluated. An unpleasant experience in one type of performance can cause you to expect that a future similar situation will also be unpleasant.[19] These expectations can be realized through the self-fulfilling prophecies discussed in Chapter 3. A traumatic failure at an earlier speech and low self-esteem from critical parents during childhood are common examples of experiences that can cause later communication apprehension.

You might object to the idea that past experiences cause communication apprehension. After all, not everyone who has bungled a speech or had critical parents is debilitated in the future. To understand why past experiences affect some people more strongly than others, we need to consider another cause of communication apprehension.

irrational thinking Beliefs that have no basis in reality or logic; one source of debilitative communication apprehension.

Irrational Thinking Cognitive psychologists argue that it is not events that cause people to feel nervous but rather the beliefs they have about those events. Certain irrational beliefs leave people feeling unnecessarily apprehensive. Psychologist Albert Ellis lists several such beliefs, or examples of **irrational thinking**, which we will call "fallacies" because of their illogical nature.[20]

- **Catastrophic failure.** People who succumb to the **fallacy of catastrophic failure** operate on the assumption that if something bad can happen, it probably will. Their thoughts before a speech resemble these:

    "As soon as I stand up to speak, I'll forget everything I wanted to say."

    "Everyone will think my ideas are stupid."

    "Somebody will probably laugh at me."

    Although it is naive to imagine that all your speeches will be totally successful, it is equally naive to assume they will all fail miserably. One way to escape the fallacy of catastrophic failure is to take a more realistic look at the situation. Would your audience members really hoot you off the stage? Will they really think your ideas are stupid? Even if you did forget your remarks for a moment, would the results be a genuine disaster? It helps to remember that nervousness is more apparent to the speaker than to the audience.[21] Beginning public speakers, when congratulated for their poise during a speech, are apt to say, "Are you kidding? I was dying up there."

- **Perfection.** Speakers who succumb to the **fallacy of perfection** expect themselves to behave flawlessly. Whereas such a standard of perfection might serve as a target and a source of inspiration (like the desire to make a hole in one while golfing), it is totally unrealistic to expect that you will write and deliver a perfect speech, especially as a beginner. It helps to remember that audiences don't expect you to be perfect.

- **Approval.** The mistaken belief called the **fallacy of approval** is based on the idea that it is vital—not just desirable—to gain the approval of everyone in the audience. It is rare that even the best speakers please everyone, especially on topics that are at all controversial. To paraphrase Abraham Lincoln, you can't please all the people all the time, and it is irrational to expect you will.

- **Overgeneralization.** The **fallacy of overgeneralization** might also be labeled the fallacy of exaggeration because it occurs when a person blows one experience out of proportion. Consider these examples:

    "I'm so stupid! I mispronounced that word."

    "I completely blew it—I forgot one of my supporting points."

    "My hands were shaking. The audience must have thought I was crazy."

    A second type of exaggeration occurs when a speaker treats occasional lapses as if they were the rule rather than the exception. This sort of mistake usually involves extreme labels, such as "always" or "never."

    "I always forget what I want to say."

    "I can never come up with a good topic."

    "I can't do anything right."

> **fallacy of catastrophic failure** The irrational belief that the worst possible outcome will probably occur.

> **fallacy of perfection** The irrational belief that a worthwhile communicator should be able to handle every situation with complete confidence and skill.

> **fallacy of approval** The irrational belief that it is vital to win the approval of virtually every person a communicator deals with.

> **fallacy of overgeneralization** Irrational beliefs in which conclusions (usually negative) are based on limited evidence or communicators exaggerate their shortcomings.

Overcoming Debilitative Communication Apprehension

There are five strategies that can help you manage debilitative communication apprehension:

- **Use nervousness to your advantage.** Paralyzing fear is obviously a problem, but a little nervousness can actually help you deliver a successful speech. Being completely calm can take away the passion that is one element of a good speech. Control your anxiety, but don't try to completely eliminate it.

- **Understand the difference between rational and irrational fears.** Some fears about speaking are rational. For example, you ought to be worried if you haven't properly prepared for your speech. But fears based on the fallacies you just read about aren't constructive. It's not realistic to expect that you'll deliver a perfect speech, and it's not rational to indulge in catastrophic fantasies about what might go wrong.

- **Maintain a receiver orientation.** Paying too much attention to your own feelings—even when you're feeling good about yourself—will take energy away from communicating with your listeners. Concentrate on your audience members rather than on yourself. Focus your energy on keeping them interested and on making sure they understand you.

- **Keep a positive attitude.** Build and maintain a positive attitude toward your audience, your speech, and yourself as a speaker. Some communication consultants suggest that public speakers should concentrate on three statements immediately before speaking:

  > I'm glad I have the chance to talk about this topic.
  >
  > I know what I'm talking about.
  >
  > I care about my audience.

  Repeating these statements (until you believe them) can help you maintain a positive attitude.

  Another technique for building a positive attitude is known as **visualization**,[22] an approach that has been used successfully with athletes. It requires you to use your imagination to visualize the successful completion of your speech. Visualization can help make the self-fulfilling prophecy discussed in Chapter 3 work in your favor.

- **Be prepared!** Preparation is the most important key to controlling communication apprehension. You can feel confident if you know from practice that your remarks are well organized and supported and that your delivery is smooth. Researchers have determined that the highest level of communication apprehension occurs just before speaking; the second highest level at the time the assignment is announced and explained; and the lowest level during the time you spend preparing your speech.[23] You should take advantage of this relatively low-stress time to work through the problems that would tend to make you nervous during the actual speech. For example, if, on one hand, your anxiety is based on a fear of forgetting what you are going to say, make sure that your note cards are complete and effective, and that you have practiced your speech thoroughly (we'll go into speech practice in more

> **visualization** A technique for rehearsal using a mental visualization of the successful completion of a speech.

> **UNDERSTANDING YOUR COMMUNICATION**
>
> ## Speech Anxiety Symptoms
>
> To what degree do you experience the following anxiety symptoms while speaking?
>
> 1. Sweating
>   a. Nonexistent
>   b. Moderate
>   c. Severe
>
> 2. Rapid breathing
>   a. Nonexistent
>   b. Moderate
>   c. Severe
>
> 3. Difficulty catching your breath
>   a. Nonexistent
>   b. Moderate
>   c. Severe
>
> 4. Rapid heartbeat
>   a. Nonexistent
>   b. Moderate
>   c. Severe
>
> 5. Restless energy
>   a. Nonexistent
>   b. Moderate
>   c. Severe
>
> 6. Forgetting what you wanted to say
>   a. Nonexistent
>   b. Moderate
>   c. Severe
>
> **EVALUATING YOUR RESPONSES**
>
> Give yourself one point for every "a," two points for every "b," and three points for every "c." If your score is:
>
> **6 to 9** You have nerves of steel. You're probably a natural public speaker, but you can always improve.
>
> **10 to 13** You are the typical public speaker. Practice the strategies discussed in this chapter to improve your skills.
>
> **14 to 18** You tend to have significant apprehension about public speaking. You need to consider each strategy in this chapter carefully. Although you will benefit from the tips provided, you should keep in mind that some of the greatest speakers of all time have considered themselves highly anxious.

detail in a moment). If, on the other hand, your great fear is "sounding stupid," then getting started early with lots of research and advance thinking is the key to relieving your communication apprehension.

12.5 Presenting Your Speech

Once you have done all the planning and analysis that precedes speechmaking, you can prepare for your actual presentation. Your tasks at this point include choosing an effective type of delivery, formulating a plan for practicing your speech, and thinking carefully about the visual and auditory choices you will make.

Choosing an Effective Type of Delivery

There are four basic types of delivery: extemporaneous, impromptu, manuscript, and memorized. Each type creates a different impression and is appropriate under different conditions. Any speech may incorporate more than one of these types of delivery. For purposes of discussion, however, it is best to consider them separately.

> **extemporaneous speech** A speech that is planned in advance but presented in a direct, conversational manner.

1. An **extemporaneous speech** is planned in advance but presented in a direct, spontaneous manner. Extemporaneous speeches are conversational in tone, which means that they give the audience members the impression that you are talking to them, directly and honestly. Extemporaneous speaking is the most common type of delivery in both the classroom and the "outside" world. In fact, extemporaneous speaking is one of the factors that researchers believe led to Donald Trump's election win in 2016. Trump's adherents found his off-the-cuff remarks compelling compared with the carefully prepared speeches of more typical politicians.[24]

> **impromptu speech** A speech given "off the top of one's head," without preparation.

2. An **impromptu speech** is given off the top of one's head, without preparation. This type of speech is spontaneous by definition, but it is a delivery style that is necessary for informal talks, group discussions, and comments on others' speeches. It is also a highly effective training aid that teaches you to think on your feet and to organize your thoughts quickly.

> **manuscript speech** A speech that is read word for word from a prepared text.

3. **Manuscript speeches** are read word for word from a prepared text. They are necessary when you are speaking for the record, such as at a legal proceeding, or when presenting scientific findings. The greatest disadvantage of a manuscript speech is, of course, the lack of spontaneity.

4. **Memorized speeches**—those learned by heart—are the most difficult and often the least effective. They often seem excessively formal. However, like manuscript speeches, they may be necessary on special occasions. They are used in oratory contests and as training devices for memory, which many experts feel is needed in today's media-saturated world.[25] One guideline holds for each type of speech: Practice.

> **memorized speech** A speech learned and delivered by rote without a written text.

Practicing Your Speech

A smooth and natural delivery is the result of extensive practice. Get to know your material until you feel comfortable with your presentation. One way to do that is to go through some or all of the steps listed in the feature "Practicing Your Presentation." In each of these steps, critique your speech according to the guidelines that follow.

COMMUNICATION STRATEGIES

Practicing Your Presentation

Be sure to give yourself plenty of time to practice. Here are a few suggestions.

- ☐ First, present the speech to yourself. Talk through the entire speech, including your examples and forms of support. Don't skip parts by using placeholders. Make sure you have a clear plan for presenting your statistics and explanations.

- ☐ Record the speech on your phone, and listen to it. Because we are sometimes surprised at what we sound like and how we appear, video recording has been shown to be an especially effective tool for rehearsals.[28]

- ☐ Present the speech in front of a small group of friends or relatives.[29]

- ☐ Present the speech to at least one listener in the room where you will present the final speech (or, if that room is not available, a similar room).

12.6 Guidelines for Delivery

Let's examine some nonverbal aspects of presenting a speech. As you read in Chapter 7, nonverbal behavior can change, or even contradict, the meaning of the words a speaker utters. If audience members want to interpret how you feel about something, they are likely to trust your nonverbal communication more than the words you speak. If you tell them, "It's great to be here today," but you stand before them slouched over with your hands in your pockets and an expression on your face like you're about to be shot, they are likely to discount what you say. This might cause your audience members to react negatively to your speech, and their negative reaction might make you even more nervous. This cycle of speaker and audience reinforcing each other's feelings can work for you, though, if you approach a subject with genuine enthusiasm. Enthusiasm is shown through both the visual and auditory aspects of your delivery.

Visual Aspects of Delivery

Visual aspects of delivery include appearance, movement, posture, facial expression, and eye contact.

Appearance This is not a presentation variable as much as a preparation variable. Some communication consultants suggest new clothes, new glasses, and new hairstyles for their clients. In case you consider any of these grooming aids, be forewarned that you should be attractive to your audience but not flashy. Research suggests that audiences like speakers who are similar to them, but they prefer the similarity to be shown conservatively.[26] Speakers, it seems, are perceived to be more credible when they look businesslike. Part of looking businesslike, of course, is looking like you took care in the preparation of your wardrobe and appearance.

Movement The way you walk to the front of your audience will express your confidence and enthusiasm. And after you begin speaking, nervous energy can cause your body to shake and twitch, and that can be distressing both to you and to your audience. One way to control involuntary movement is to move voluntarily when you feel the need to move. Don't feel that you have to stand in one spot or that all your gestures need to be carefully planned. Simply get involved in your message, and let your involvement create the motivation for your movement. That way, when you move, you will emphasize what you are saying in the same way you would emphasize it if you were talking to a group of friends.

Movement can also help you maintain contact with all members of your audience. Those closest to you will feel the greatest contact. This creates what is known as the "action zone" in the typical classroom, within the area of the front and center of the room. Movement enables you to extend this action zone, to include in it people who would otherwise remain uninvolved. Without overdoing it, you should feel free to move toward, away from, or from side to side in front of your audience.

Remember: Move with the understanding that it will add to the meaning of the words you use. It is difficult to bang your fist on a podium or take a step without conveying emphasis. Make the emphasis natural by allowing your message to create your motivation to move.

Posture Generally speaking, good posture means standing with your spine relatively straight, your shoulders relatively squared off, and your feet angled out to keep your body from falling over sideways. In other words, rather than standing at military attention, you should be comfortably erect.

Good posture can help you control nervousness by allowing your breathing apparatus to work properly; when your brain receives enough oxygen, it's easier for you to think clearly. Good posture also increases your audience contact because the audience members will feel that you are interested enough in them to stand formally, yet relaxed enough to be at ease with them.

Facial Expression The expression on your face can be more meaningful to an audience than the words you say. Try it yourself with a mirror. Say, "You're a terrific audience," for example, with a smirk, with a warm smile, with a deadpan expression, and then with a scowl. It just doesn't mean the same thing. But don't try to fake it. Like your movement, your facial expressions will reflect your genuine involvement with your message.

Eye Contact Eye contact is perhaps the most important nonverbal facet of delivery.[27] Eye contact increases your connection with your audience, and at the same time it helps you control your nervousness. Direct eye contact is a form of reality testing. The most frightening aspect of speaking is the unknown. How will the audience react? Direct eye contact allows you to test your perception of your audience as you speak. Usually, especially in a college class, you will find your audience is more "with" you than you think. By deliberately establishing eye contact with audience members, you might engage them and find they are more interested than they previously appeared.

To maintain eye contact, you could try to meet the eyes of each member of your audience squarely at least once during any given presentation. After you have made definite eye contact, move on to another audience member. You can learn to do this quickly, so you can visually latch on to every member of a good-sized class in a relatively short time.

The characteristics of appearance, movement, posture, facial expression, and eye contact are visual, nonverbal facets of delivery. Now consider the auditory nonverbal messages that you might send during a presentation.

Auditory Aspects of Delivery

As you read in Chapter 6, your paralanguage—the way you use your voice—says a good deal about you, especially about your sincerity and enthusiasm. Like eye contact, using your voice well can help you control your nervousness. It's another cycle: Controlling your vocal characteristics will decrease your nervousness, which will enable you to control your voice even more. But this cycle can also work in the opposite direction. If your voice is out of control, your nerves will probably be in the same state. Controlling your voice is mostly a matter of recognizing and using appropriate volume, rate, pitch, and articulation.

Video conferencing in platforms like Zoom and Webex changed the dynamics of audience analysis and audience contact by forcing speakers to really look at individual audience members as individuals.

Volume The loudness of your voice is determined by the amount of air you push past the vocal folds in your throat. The key to controlling volume, then, is controlling the amount of air you use. The key to determining the right volume is audience contact. Your delivery should be loud enough so that your audience members can hear everything you say but not so loud that they feel you are talking to someone in the next room. Too much volume is seldom the problem for beginning speakers. Usually, they either are not loud enough or tend to fade off at the end of a thought. Sometimes, when they lose faith in an idea in midsentence, they compromise by mumbling the end of the sentence so that it isn't quite coherent.

Rate There is a range of personal differences in speaking speed, or **rate**. Daniel Webster, for example, is said to have spoken at around 90 words per minute, whereas one actor who is known for his fast-talking commercials speaks at about 250. Normal speaking speed, however, is between 120 and 150 words per minute. If you talk much more slowly than that, you may tend to lull your audience to sleep. Faster speaking rates are stereotypically associated with speaker competence,[30] but if you speak too rapidly, you will tend to be unintelligible. Once again, your involvement in your message is the key to achieving an effective rate.

> **rate** The speed at which a speaker utters words.

Pitch The highness or lowness of your voice—**pitch**—is controlled by the frequency at which your vocal folds vibrate as you push air through them. Because taut vocal folds vibrate at a greater frequency, pitch is influenced by muscular tension. This explains why nervous speakers have a tendency occasionally to "squeak," whereas relaxed speakers seem to be in more control. Pitch will tend to follow rate and volume. As you speed up or become louder, your pitch will tend to rise. If your range in pitch is too narrow, your voice will have a singsong quality; if it is too wide, you may sound overly dramatic. You should control your pitch so that your listeners believe you are talking with them rather than performing in front of them. Once again, your involvement in your message should take care of this problem naturally for you.

> **pitch** The highness or lowness of one's voice.

When considering volume, rate, and pitch, keep emphasis in mind. Remember that a change in volume, pitch, or rate will result in emphasis. If you pause or speed up, your rate will suggest emphasis. Words you whisper or scream will be emphasized by their volume.

Articulation The final auditory nonverbal behavior, articulation, is perhaps the most important. For our purposes here, **articulation** means pronouncing all the parts of all the necessary words and nothing else.

> **articulation** The process of pronouncing all the necessary parts of a word, with all its distinct syllables.

It is not our purpose to condemn regional or ethnic dialects within this discussion. It is true that a considerable amount of research suggests that regional dialects can cause negative impressions,[31] but our purpose here is to suggest careful, not standardized, articulation. Incorrect articulation is usually nothing more than careless articulation. It is caused by (1) leaving off parts of words (deletion), (2) replacing parts of words (substitution), (3) adding parts to words (addition), or (4) overlapping two or more words (slurring).

Deletion The most common mistake in articulation is **deletion**, or leaving off part of a word. As you are thinking the complete word, it is often difficult to recognize that you are saying only part of it. The most common deletions occur at the ends of words, especially *-ing* words. *Going*, *doing*, and *stopping* become *goin'*, *doin'*, and *stoppin'*. Parts of words can be left off in the middle, too, as in *terr'iss* for *terrorist*, *Innernet* for *Internet*, and *asst* for *asked*.

> **deletion** An articulation error that involves leaving off parts of spoken words.

substitution The articulation error that involves replacing part of a word with an incorrect sound.

Substitution *Substitution* takes place when you replace part of a word with an incorrect sound. The ending *-th* is often replaced at the end of a word with a single *t*, as when *with* becomes *wit*. The *th-* sound is also a problem at the beginning of words, as *this*, *that*, and *those* tend to become *dis*, *dat*, and *dose*. (This tendency is especially prevalent in many parts of the northeastern United States.)

addition The articulation error that involves adding extra parts to words.

Addition The articulation problem of **addition** is caused by adding extra parts to words, such as *incentative* instead of *incentive*, *athalete* instead of *athlete*, and *orientated* instead of *oriented*. Sometimes this type of addition is caused by incorrect word choice, as when *irregardless* is used for *regardless*.

Another type of addition is the use of "tag questions," such as *you know?* or *you see?* or *right?* at the end of sentences. To have every other sentence punctuated with one of these barely audible superfluous phrases can be annoying.

Probably the worst type of addition, or at least the most common, is the use of *uh* and *anda* between words. *Anda* is often stuck between two words when *and* isn't even needed. If you find yourself doing that, you might just want to pause or swallow instead.[32]

slurring The articulation error that involves overlapping the end of one word with the beginning of the next.

Slurring The articulation error of **slurring** is caused by trying to say two or more words at once—or at least overlapping the end of one word with the beginning of the next. Word pairs ending with *of* are the worst offenders in this category. *Sort of* becomes *sorta*, *kind of* becomes *kinda*, and *because of* becomes *becausa*. Word combinations ending with *to* are often slurred, as when *want to* becomes *wanna*. Sometimes even more than two words are blended together, as when *that is the way* becomes *thatsaway*. Careful articulation means using your lips, teeth, tongue, and jaw to bite off your words, cleanly and separately, one at a time.

12.7 Sample Speech

Hailey Hardcastle, the University of Oregon student whose profile began this chapter, presented the following speech at a TED event in 2020. Her general purpose was to inform the audience about her progress in establishing a law in Oregon that allows students to take a mental health day off from school, the same way one would a sick day.

"I have three younger sisters who are in middle school right now . . . and part of the reason I do this is so high school and beyond will be even easier for them than it was for me," she said.[33]

Hailey's speech tells the story of why she chose her topic and how she determined her purpose, so we'll let her speak for herself on that. Her thesis statement might be summarized like this:

Mental health affects us all, so we should have solutions that are accessible to all of us.

In analyzing her audience, Hailey had to take into consideration students, parents, and a general audience that included everyone from journalists to legislators to fellow activists. The live audience was relatively small, but the virtual audience that would be able to access this speech on TED's website was huge and would extend over time.

SAMPLE SPEECH — Hailey Hardcastle

The Case for Student Mental Health Days[34]

1. When I was a kid, my mom and I made this deal. I was allowed to take three mental health rest days every semester as long as I continued to do well in school. This was because I started my mental health journey when I was only six years old. I was always what my grade-school teachers would call "a worrier," but later on we found out that I have trauma-induced anxiety and clinical depression.

2. This made growing up pretty hard. I was worried about a lot of things that other kids weren't, and school got really overwhelming sometimes. This resulted in a lot of breakdowns, panic attacks—sometimes I was super productive, and other days I couldn't get anything done.

3. This was all happening during a time when mental health wasn't being talked about as much as it is now, especially youth mental health. Some semesters I used all of those rest days to the fullest. Others, I didn't need any at all. But the fact that they were always an option is what kept me a happy, healthy and successful student.

4. Now I'm using those skills that I learned as a kid to help other students with mental health challenges. I'm here today to offer you some insight into the world of teenage mental health: what's going on, how did we get here and what can we do?

5. But first you need to understand that while not everyone has a diagnosed mental illness like I do, absolutely everyone—all of you—have mental health.

6. All of us have a brain that needs to be cared for in similar ways that we care for our physical well-being. Our head and our body are connected by much more than just our neck after all. Mental illness even manifests itself in some physical ways, such as nausea, headaches, fatigue and shortness of breath. So since mental health affects all of us, shouldn't we be coming up with solutions that are accessible to all of us?

7. That brings me to my second part of my story. When I was in high school I had gotten pretty good at managing my own mental health. I was a successful student, and I was president of the Oregon Association of Student Councils. But it was around this time that I began to realize mental health was a much bigger problem than just for me personally.

8. Unfortunately, my hometown was touched by multiple suicides during my first year in high school. I saw those tragedies shake our entire community, and as the president of a statewide group, I began hearing more and more stories from students where this had also happened in their town.

9. So in 2018 at our annual summer camp, we held a forum with about 100 high school students to discuss teenage mental health. What could we do? We approached this conversation with an enormous amount of empathy and honesty, and the results were astounding. What struck me the most was that every single one of my peers had a story about a mental health crisis in their school, no matter if they were from a tiny town in eastern Oregon or the very heart of Portland.

She begins with a personal anecdote, establishing this as a speech that only she can tell.

She establishes the importance of her topic to her general audience, which includes students, parents, and everyone who needs to understand mental health problems...

...And then continues her personal story.

10. This was happening everywhere. We even did some research, and we found out that suicide is the second leading cause of death for youth ages 10 to 24 in Oregon. The second leading cause. We knew we had to do something.

11. So over the next few months, we made a committee called Students for a Healthy Oregon, and we set out to end the stigma against mental health. We also wanted to prioritize mental health in schools. With the help of some lobbyists and a few mental health professionals, we put forth House Bill 2191. This bill allows students to take mental health days off from school the same way you would a physical health day. Because oftentimes that day off is the difference between feeling a whole lot better and a whole lot worse—kind of like those days my mom gave me when I was younger.

Here she projects the following visual aid:
STUDENTS FOR A HEALTHY OREGON

12. So over the next few months, we lobbied and researched and campaigned for our bill, and in June of 2019 it was finally signed into law.

Here Hailey pauses for effect, essentially giving her audience time to react. They do, with cheers and applause.

13. This was a groundbreaking moment for Oregon students. Here's an example of how this is playing out now. Let's say a student is having a really hard month. They're overwhelmed, overworked, they're falling behind in school, and they know they need help. Maybe they've never talked about mental health with their parents before, but now they have a law on their side to help initiate that conversation.

Here she uses a hypothetical example to show how the bill will impact students...

14. The parent still needs to be the one to call the school and excuse the absence, so it's not like it's a free pass for the kids, but most importantly, now the school has that absence recorded as a mental health day, so they can keep track of just how many students take how many mental health days. If a student takes too many, they'll be referred to the school counselor for a check-in.

...But reminds parents that they will still be in charge.

15. This is important because we can catch students who are struggling before it's too late. One of the main things we heard at that forum in 2018 is that oftentimes stepping forward and getting help is the hardest step. We're hoping that this law can help with that. This not only will start teaching kids how to take care of themselves and practice self-care and stress management, but it could also literally save lives.

16. Now students from multiple other states are also trying to pass these laws. I'm currently working with students in both California and Colorado to do the same, because we believe that students everywhere deserve a chance to feel better.

Here she extends her message to those who might be able to encourage similar changes nationwide.

17. Aside from all the practical reasons and technicalities, House Bill 2191 is really special because of the core concept behind it: that physical and mental health are equal and should be treated as such. In fact, they're connected. Take health care for example. Think about CPR. If you were put in a situation where you had to administer CPR, would you know at least a little bit of what to do?

18. Think to yourself—most likely yes because CPR trainings are offered in most schools, workplaces and even online. We even have songs that go with it. But how about mental health care? I know I was trained in CPR in my seventh-grade health class. What if I was trained in seventh grade how to manage my mental health or how to respond to a mental health crisis?

When mentioning CPR training, Hailey doesn't have time for extended details. In a different speech, she might explain that the songs that go with CPR training include "Stayin' Alive" by the Bee Gees and "Dancing Queen" by ABBA, both of which have the recommended 100 beats per minute that are needed for chest compressions!

19. I'd love to see a world where each of us has a toolkit of skills to help a friend, coworker, family member or even stranger going through a mental health crisis.

And these resources should be especially available in schools because that's where students are struggling the most.

20 The other concept that I sincerely hope you take with you today is that it is always OK to not be OK, and it is always OK to take a break. It doesn't have to be a whole day; sometimes that's not realistic. But it can be a few moments here and there to check in with yourself.

21 Think of life like a race . . . like a long-distance race. If you sprint in the very beginning you're going to get burnt out. You may even hurt yourself from pushing too hard. But if you pace yourself, if you take it slow, sometimes intentionally, and you push yourself other times, you are sure to be way more successful.

22 So please, look after each other, look after the kids and teens in your life especially the ones that look like they have it all together. Mental health challenges are not going away, but as a society, we can learn how to manage them by looking after one another. And look after yourself, too. As my mom would say, "Once in a while, take a break." Thank you.

Hailey begins to establish her conclusion by reminding her audience about her thesis and main points—first for her general audience . . .

. . . And then, specifically for students.

MAKING THE GRADE

OBJECTIVE 12.1 Describe the importance of topic, purpose, and thesis in effective speech preparation.

- Choose a topic that is right for you, your audience, and the situation.
- Formulating a clear purpose statement serves to keep you focused while preparing your speech.
- A straightforward thesis statement helps the audience understand your intent.
  > Why is a carefully worded purpose statement essential for speech success?
  > Why do you believe your topic has the potential to be effective for you as a speaker?
  > How would you word your thesis statement for your next speech?

OBJECTIVE 12.2 Analyze both the audience and occasion in any speaking situation.

- When analyzing your audience, you should consider the audience's purpose, demographics, attitudes, beliefs, and values.
- When analyzing the occasion, you should consider the time (and date) when your speech will take place, the location, and audience expectations given the occasion.
  > What are the most important aspects of analyzing your audience?
  > What are the most important aspects of analyzing the occasion for your next speech?
  > How will you adapt your next speech to both your audience and occasion?

OBJECTIVE 12.3 Gather information on your chosen topic from a variety of sources.

- When researching a speech, students usually think first of online searches.
- Also consider searching the collections or databases at a library, interviewing, making personal observations, and conducting survey research.
  > Why are multiple forms of research important in speech preparation?
  > How do you analyze the reliability of information you find online?
  > What is the most important piece of research you need for your next speech?

OBJECTIVE 12.4 Assess and manage debilitative speaking apprehension.

- Sources of debilitative communication apprehension often include irrational thinking, such as a belief in one or more fallacies.
- To help overcome communication apprehension, remember that nervousness is natural, and use it to your advantage.
- Other methods of overcoming communication apprehension involve being rational, receiver oriented, positive, and prepared.
  > What is the relationship between self-talk and communication apprehension?
  > Which types of self-defeating thoughts create the greatest challenge for you?
  > What forms of rational self-talk can you use to overcome self-defeating thoughts?

OBJECTIVE 12.5 Make effective choices in the delivery of your speech.

- Choose an effective type of delivery, or combinations of extemporaneous, impromptu, manuscript, and memorized speeches.
- Practice your speech thoroughly.
- Consider both visual and auditory aspects of delivery.
  > What are the primary differences among the main types of speeches?
  > In your last speech, why did you make the delivery choices that you did? Were they as effective as they could have been?
  > What visual and auditory aspects of delivery will be most important in the next speech you present?

KEY TERMS

addition p. 344
articulation p. 343
attitude p. 329
audience analysis p. 327
belief p. 330
database p. 333
debilitative communication apprehension p. 336
deletion p. 343
demographics p. 327
extemporaneous speech p. 340
facilitative communication apprehension p. 335
fallacy of approval p. 337
fallacy of catastrophic failure p. 337
fallacy of overgeneralization p. 337
fallacy of perfection p. 337
general purpose p. 325
impromptu speech p. 340
irrational thinking p. 336
manuscript speech p. 340
memorized speech p. 340
pitch p. 343
purpose statement p. 325
rate p. 343
slurring p. 344
specific purpose p. 325
substitution p. 344
survey research p. 334
thesis statement p. 326
value p. 330
visualization p. 338

PUBLIC SPEAKING PRACTICE

Prepare a brief, one-paragraph explanation of your choice of topic for your next classroom speech. Discuss it with your classmates.

ACTIVITIES

1. **Formulating Purpose Statements** Write a specific purpose statement for each of the following speeches:
   a. An after-dinner speech at an awards banquet in which you will honor a team that has a winning, but not championship, record. (You pick the team. For example: "After listening to my speech, my audience members will appreciate the individual sacrifices made by the members of the chess team.")
   b. A classroom speech in which you explain how to do something. (Again, you choose the topic: "After listening to my speech, my audience members will know at least three ways to maximize their comfort and convenience on an economy class flight.")
   c. A campaign speech in which you support the candidate of your choice. (For example: "After

listening to my speech, my audience members will consider voting for Alexandra Rodman in order to clean up student government.")

d. Answer the following questions about each of the purpose statements you make up: Is it result oriented? Is it precise? Is it attainable?

2. **Formulating Thesis Statements** Turn each of the following purpose statements into a statement that expresses a possible thesis. For example, if you had a purpose statement such as this:

   After listening to my speech, my audience will recognize the primary advantages and disadvantages of home teeth bleaching.

   you might turn it into a thesis statement such as this:

   Home bleaching your teeth can significantly improve your appearance, but watch out for injury to the gums and teeth.

   a. At the end of my speech, the audience members will be willing to sign my petition supporting the local needle exchange program for drug addicts.

   b. After listening to my speech, the audience members will be able to list five disadvantages of tattoos.

   c. During my speech on the trials and tribulations of writing a research paper, the audience members will show their interest by paying attention and their amusement by occasionally laughing.

3. **Communication Apprehension: A Personal Analysis** To analyze your own reaction to communication apprehension, think back to your last public speech, and rate yourself on how rational, receiver oriented, positive, and prepared you were. How did these attributes affect your anxiety level?

Speech Organization and Support

CHAPTER OUTLINE

13.1 Building Your Speech 352
- Your Preliminary Notes
- Your Working Outline
- Your Formal Outline
- Your Full-Sentence Outline
- Your Speaking Notes

13.2 Principles of Outlining 359
- Standard Symbols
- Standard Format
- The Rule of Division
- The Rule of Parallel Wording

13.3 Organizing Your Outline into a Logical Pattern 360
- Time Patterns
- Space Patterns
- Topic Patterns
- Problem-Solution Patterns
- Cause-Effect Patterns
- Monroe's Motivated Sequence

13.4 Beginnings, Endings, and Transitions 364
- The Introduction
- The Conclusion
- Transitions

13.5 Supporting Material 368
- Functions of Supporting Material
- Types of Supporting Material
- Styles of Support: Narration Versus Citation
- Plagiarism Versus Originality

13.6 Sample Speech 375
- Speech Outline
- Annotated Bibliography

MAKING THE GRADE 379

KEY TERMS 380

PUBLIC SPEAKING PRACTICE 381

ACTIVITIES 381

LEARNING OBJECTIVES

13.1 Describe different types of speech outlines and their functions.

13.2 Construct an effective speech outline using the organizing principles described in this chapter.

13.3 Choose an appropriate organizational pattern for your speech.

13.4 Develop a compelling introduction and conclusion, and use effective transitions at key points in your speech.

13.5 Choose supporting material that will help make your points clear, interesting, memorable, and convincing.

CHAPTER 13 Speech Organization and Support

Photo courtesy Dan O'Connor

Dan O'Connor is currently a student at Seton Hall University. When he was in high school, he worked in a local grocery store as a cashier.

At the cash register, he greeted each customer the same way.

"Hi, did you find everything okay?"

And then one day, a customer gave him no response.

He thought, "Well, someone's in a bad mood."

But then something went wrong with her order, and as they struggled to communicate Dan realized she was deaf.

Dan felt bad about assuming the customer was just being grouchy, and he felt that he had failed to accommodate one of her basic human needs—communication.

He began to learn American Sign Language. It opened him up to what he now calls the "amazing culture and history of the deaf." He says, "I fell in love with ASL, a full and truly beautiful language."

Months later, the same deaf customer came through the store. As she gathered her bags, Dan managed to sign, "Have a nice day!" She looked surprised, then grateful as she signed, "Thank you." The connection he made with this stranger motivated him to study more ASL.

By the end of his freshman year at Seton Hall University, Dan had become fairly competent in ASL. He was also starting to learn about the prison system, and when he was asked to come up with a speech topic for a tournament speech, it occurred to him that deaf people might face significant challenges in prison. As he researched that topic, he began organizing the ideas he found, and finding information to support the points he wanted to make. The resulting speech is the sample speech at the end of this chapter.

As Dan O'Connor prepared his speech, he learned that organization and structure are essential in both the development and effectiveness of public speaking. In the following pages, you will learn methods of organizing and supporting your thoughts effectively.

13.1 Building Your Speech

As you read in Chapter 3, people tend to arrange their perceptions in some meaningful way in order to make sense of the world. Being clear to your audience, however, isn't the only benefit of good organization; structuring a message clearly will help you refine your own ideas and construct more effective messages.

Your Preliminary Notes

You should start crafting your speech from the moment you decide on a topic and purpose. Your initial notes will be extremely rough: probably including some basic ideas, materials you have already collected, and potential places you might look for further research. Figure 13.1 represents the kind of preliminary notes that Dan O'Connor eventually whipped into shape. The notes there are a snapshot of Dan's

Plan for speech:
1. Problems faced by deaf inmates
2. Causes
3. Solutions

Over 150,000 deaf and hard-of-hearing inmates and prisoners in the USA

Inhumane abuse of deaf inmates in almost all US prisons. (Quote Jerry Coen, National Association of the Deaf 3/19/2019)

American Disability Act (ADA) guarantees equitable access to communication for deaf inmates.

There's "Systemic abuse" of deaf inmates—A "complete failure of state and federal prisons to adhere to the ADA." (Advocacy group HEARD 6/16/2020)

"...we end the daily, constant suffering faced by America's deaf inmates, and finally realize the decades-old obligations of the ADA.)

It's time we stop turning a blind eye to America's deaf inmates. (My thesis)

CAUSES

Outdated technology

Most deaf inmates communicate via American Sign Language (ASL). Grammar and syntax are different than spoken/written English. (Show examples?)

The minority of prisons that do provide "accommodations" for deaf communication use Teletype. Users must type in English—which many deaf inmates are not proficient in. (Chronicle Herald 3/20/2020)

This creates a barrier for both hearing and deaf users—meaningful conversation can become impossible. (US Department of Education 3/29/2020)

Average deaf or hard of hearing American can read and write at a third-grade level. (Is this relevant? If not, drop)

Teletype in modern form invented in its modern form in 1964! It's drastically outdated and outmoded—like giving prisoners only a telegraph and saying it's their problem if they don't know Morse code. (CREEC Law Center 6/29/2019)

Deaf inmates left using technology from a time when even a handheld calculator was science fiction—never mind personal computers. (Keep in or drop?)

Lack of interpreters

Demand for interpreters outweighs supply. Only ~15,000 ASL interpreters for the entire USA . . . Deaf and hard of hearing population of 11 million. (Registry of Interpreters for the Deaf) Shortage of interpreters is driven by low salary. (Bureau of Labor Statistics 9/1/2020) (Include this or not?)

EFFECTS

Poor physical health

James Woody story—4 years in a Georgia state prison. Doctors and nurses didn't understand ASL, so no diagnosis. After release he was diagnosed with cancer. (Find more statistics)

Mental health

Communication with outside is important factor in protecting mental wellbeing of inmates. (Prison Policy Initiative 2/9/2019) Deaf inmates are left cut off from their loved ones, their lawyers, and the entire world beyond their cell.

Recidivism

When deaf inmates can't connect, rehab unlikely. (Find evidence)

Safety risks

Deaf and disabled individuals are more than twice as likely to experience sexual abuse. "Deaf inmates are often left unable to say 'me too.'" (Maryland Coalition against Sexual Assault reports on 10/3/2019)

FIGURE 13.1 Preliminary Notes

> **Unjust punishments**
>
> Deaf inmates miss meals, showers, and important announcements. This leads to punishment by prison guards & reduces chances of sentence reductions for good behavior. (ACLU 6/28/2020)
>
> Deaf inmates can't attend rehabilitative courses and programs, participation in which is often a factor in a convict getting parole. Deaf prisoners often cannot access doctors, mental healthcare, the administration, or even other inmates. Jeremy Sanders was sent to prison due to his painkiller addiction. And without any ability to communicate with the prison medical staff.
>
> Florida's Department of Corrections was required by a 2017 settlement to provide videophones and interpreters for the deaf. However, two years later, the state's prisons still fail to provide these essential services. Lawsuits have demonstrated that videophones are required in prisons by the ADA, but we cannot wait for judges in every state to tell prisons what we already know.
>
> **SOLUTIONS**
>
> **Better technology**
>
> The pandemic has demonstrated the value of videoconferencing. Videophones are more valuable for the deaf, who depend on visual cues. (CBS Reports, 9/30/2019)
>
> **More interpreters**
>
> To honor the ADA/civil rights acts/Bill of Rights, Department of Justice must create a program to reimburse students studying ASL interpretation.
>
> **Visit prisoners**
>
> Learn sign language and visit inmates (following state law). Just asking "how are you?" can make a difference.
>
> **Request (Demand?) government action**
>
> Show memo audience can send to local jails and state department of corrections explaining how they can help provide adequate and equal communication for the deaf.
>
> (Modified from National Association of the Deaf model)

FIGURE 13.1 (Continued)

ideas at a single point in time. Earlier versions were messier, and later versions more complete and clear.

Don't worry about making your preliminary notes look perfect. They're a space for holding all your ideas before you try to organize them into anything approaching the remarks you'll eventually deliver.

Some students use their favorite word processing to develop their notes. You could also work with pen and paper. Another option is to jot your ideas onto a set of file cards or sticky notes that you can shift around once you start organizing your random thoughts.

Your Working Outline

Just as an architect creates a blueprint before construction begins, your next step should be to start organizing your ideas before you begin to think about the actual words you want to say. This involves creating a **working outline**: a rough organizational scheme you'll modify as you build your speech.

Chapter 12 showed you how to begin this planning by formulating a purpose, analyzing your audience, and conducting research. You apply that information to the structure of the speech through outlining. Like any other plan, a speech outline is the framework on which your message is built. It contains your main ideas and shows how they relate to one another and to your thesis.

working outline A constantly changing organizational aid used in planning a speech.

Virtually every speech you outline ought to follow a **basic speech structure** that looks like this:

I. INTRODUCTION
  A. Attention Getter
  B. Thesis
  C. Preview of main points
II. BODY (2–5 main points)
  A. Main point 1
  B. Main point 2
  C. Main point 3
  D. Etc.
III. CONCLUSION
  A. Review of thesis and main points
  B. Conclusion

> **basic speech structure** The division of a speech into introduction, body, and conclusion.

This structure demonstrates the old aphorism for speakers: "Tell what you're going to say, say it, and then tell what you said." Although this sort of repetition may seem redundant, the research on listening cited in Chapter 5 demonstrates that receivers forget much of what they hear. The clear, repetitive nature of the basic speech structure reduces the potential for memory loss because audiences tend to listen more carefully during the beginning and ending of a speech.[1] Your outline will reflect this basic speech structure.

Your working outline should follow the basic speech structure, even in rough form. It's for your eyes only, and you'll probably create several drafts as you refine your ideas. As your ideas solidify, your outline will change accordingly, becoming more polished as you go along.

Dan O'Connor's working outline was an expansion of his preliminary notes and ultimately led to the final version that he used for the sample speech at the end of this chapter.

Your Formal Outline

A **formal outline** uses a consistent format and set of symbols to identify the structure of ideas. It's a document you will create once you have finalized what you plan to say during your speech.

> **formal outline** A consistent format and set of symbols used to identify the structure of ideas.

A formal outline serves several purposes. In simplified form, it can be displayed as a visual aid or distributed as a handout. It can also serve as a record of a speech that was delivered; many organizations send outlines to members who miss meetings at which presentations were given. Finally, in speech classes, instructors often use speech outlines to analyze student speeches. When one is used for that purpose, it is usually a full-sentence outline and includes the purpose, thesis, and topic or title. Some speakers use a formal outline while delivering their remarks, although your connection with the audience will probably be stronger if you use the kind of speaking notes described in this chapter.

Most instructors also require a bibliography of sources at the end of the outline. The bibliography should include full research citations, the correct form for which can be found in any style guide, such as *The Craft of Research* by Wayne Booth et al.[2] There are at least six standard bibliographic styles. Whichever style you use, you should be consistent in form and remember the two primary functions of a bibliographic citation: to demonstrate the credibility of your source and to enable the readers—in this case, your professor or your fellow students—to find the source if they want to check its accuracy or explore your topic in more detail.

Another person should be able to understand the basic ideas included in your speech by reading the formal outline. In fact, that's one test of the effectiveness of your outline. The formal outline for the sample speech at the end of the chapter appears on pp. 375–376. Figure 13.2 is a formal outline for a speech on physician suicide.

Your Full-Sentence Outline

While a formal outline shows the basic skeleton of your speech, a full-sentence outline can be used as the final step in making sure that all of your ideas are expressed clearly and that you have covered your topic completely. A handy checklist appears in the feature "Building a Full-Sentence Speech Outline."

Your Speaking Notes

Many teachers suggest that speaking notes should be in the form of a brief keyword outline, with just enough information listed to jog your memory but not enough to get lost in. They also might suggest that you fit your notes on one side of one 3- by

---

Introduction

I. Attention-Getter

II. Preview of Main Points

Body

I. Physician suicide has two main causes.
   - **A.** Hazing Culture
   - **B.** Assembly-Line Medicine

II. Physician suicide affects us all.
   - **A.** Disguised Mental Anguish
   - **B.** Physician Burnout

III. Physician suicide can be solved.
   - **A.** Public Action
     1. We can educate the public.
     2. We can use the documentary, *Do No Harm*.
   - **B.** Individual Action
     1. We must care for our doctors.
     2. We must thank them when they help us.

Conclusion

I. Review

II. Final Remarks

FIGURE 13.2 Formal Outline

5-inch note card. Other teachers recommend that you also have your introduction and conclusion on note cards, and still others recommend that your longer quotations be written out on note cards, as in Figure 13.3. When you use note cards for your speaking notes, they can be in any format that works for you, because they are for your eyes only.

One variation on speaking notes is to use them as a visual aid for your speech, projected so your audience can see them (see pp. 395–402 for more on the use of visual aids). When you do that, they need to be in the form of the formal outline in Figure 13.2, so that all members of your audience can make sense of them.

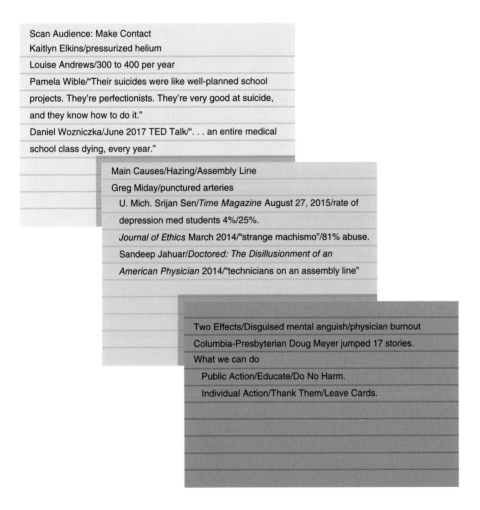

FIGURE 13.3 Speaking Notes

COMMUNICATION STRATEGIES

Building a Full-Sentence Speech Outline

As you formulate your own speech outline, you might want to consult this guide used by students at the University of Colorado at Colorado Springs. Your outline doesn't necessarily need all these elements, and you might want to add some others of your own, but this can serve as a handy reminder of possible elements

(Continued)

that you might want to include. It specifies the critical elements of a speech, including the title, general purpose, specific purpose, thesis statement, introduction, main points, subpoints, transitions, visual aids, citations, conclusion, and bibliography.

General Purpose: To inform (or persuade, etc.)

Specific Purpose: After listening to my speech, my audience will . . .

Practice Time Average: (Average all your timed practices—cut if long, add if short)

Title

I. Introduction
   A. The introduction of a speech should contain the opening statements of a speech.
   B. The introduction also includes the thesis of the speech—a kind of preview.
   C. Typically, the introduction will comprise 20% of delivery time of the speech.
   D. It is also best to use "attention-getters" in this portion of the speech.

***At this point there should be a transition from the introduction to the body.

II. Body
   A. The body of the speech can contain three to four main points about the central idea being presented.
      1. For each main idea, subpoints may be included to help discuss the topic further.
      2. You may have citations to lend credibility or quotes to draw in your audience.

Transition

   B. The bulk of the speech is usually centered in the body of the speech.
      1. Thus, typically comprising 70% of the delivery time for the speech.
      2. You should note what visual aids you plan to use and when you will incorporate them into the speech.

Transition

   C. This would be the third main point; remember to add visuals to keep the audience engaged.
      1. Cue visuals
      2. Check time

***At this point, there should be a transition from the body to the conclusion.

III. Conclusion
   A. The closing statements of a speech will sum up the three main points of the speech.
   B. Typically, this aspect of the speech will comprise 10% of delivery time.

IV. Bibliography (This can be an additional outline number, an additional page, or an additional paragraph with the title *Bibliography* centered at the top.)

> **A.** In this portion of the outline, a list of all sources used in the preparation of the speech will be included.
> **B.** APA format

Source: Adapted from "Outlining the Speech," Speech Program, University of Colorado at Colorado Springs. Available at https://www.uccs.edu/Documents/eberhardt/Outlining%20a%20Speech.doc

13.2 Principles of Outlining

Over the years, a series of rules or principles for the construction of outlines has evolved. These rules are based on the use of the standard symbols and format discussed next.

Standard Symbols

A speech outline generally uses the following symbols:

I. Main point (roman numeral)
   A. Subpoint (capital letter)
      1. Sub-subpoint (standard number)
         a. Sub-subsubpoint (lowercase letter)

In the examples in this chapter, the major divisions of the speech—introduction, body, and conclusion—are not given symbols. They are listed by name, and the roman numerals for their main points begin anew in each division. An alternative form is to list these major divisions with roman numerals, main points with capital letters, and so on.

Standard Format

In the sample outlines in this chapter, notice that each symbol is indented a number of spaces from the symbol above it. Besides keeping the outline neat, the indentation of different-order ideas is actually the key to the technique of outlining; it enables you to coordinate and order ideas in the form in which they are most comprehensible to the human mind. If the standard format is used in your working outline, it will help you create a well-organized speech. If it is used in speaking notes, it will help you remember everything you want to say.

Proper outline form is based on a few rules and guidelines, the first of which is the rule of division.

The Rule of Division

In formal outlines, main points and subpoints always represent a division of a whole. Because it is impossible to divide something into fewer than two parts, you always have at least two main points for every topic. Then, if your main points are divided, you will always have at least two subpoints, and so on. Thus, the rule for formal outlines is as follows: Never a "I" without a "II," never an "A" without a "B," and so on.

Three to five is considered to be the ideal number of main points. It is also considered best to divide those main points into three to five subpoints, when necessary and possible.

The Rule of Parallel Wording

Your main points should be worded in a similar, or "parallel," manner. For example, if you are developing a speech against capital punishment, your main points might look like this:

I. Capital punishment is not effective: it is not a deterrent to crime.

II. Capital punishment is not constitutional: it does not comply with the Eighth Amendment.

III. Capital punishment is not civilized: it does not allow for a reverence for life.

Whenever possible, subpoints should also be worded in a parallel manner. For your points to be worded in a parallel manner, they should each contain one, and only one, idea. (After all, they can't really be parallel if one is longer or contains more ideas than the others.) This will enable you to completely develop one idea before moving on to another one in your speech. If you were discussing cures for indigestion, your topic might be divided incorrectly if your main points looked like this:

I. "Preventive cures" help you before eating.

II. "Participation cures" help you during and after eating.

You might actually have three ideas there and thus three main points:

I. Prevention cures (before eating)

II. Participation cures (during eating)

III. Postparticipation cures (after eating)

13.3 Organizing Your Outline into a Logical Pattern

An outline should reflect a logical order for your points. You might arrange them from newest to oldest, largest to smallest, best to worst, or in a number of other ways that follow. The organizing pattern you choose ought to be the one that best develops your thesis.

Time Patterns

time pattern An organizing plan for a speech based on chronology.

Arrangement according to **time patterns**, or chronology, is one of the most common patterns of organization. The period of time could be anything from centuries to seconds. In a speech on airline food, a time pattern might look like this:

I. Early airline food: a gourmet treat

II. The middle period: institutional food at 30,000 feet

III. Today's airline food: the passenger starves

Arranging points according to the steps that make up a process is another form of time patterning. The topic "Recording a Hit Song" might use this type of patterning:

I. Record the demo.

II. Post a YouTube video.

III. Get a recording company to listen and view.

Time patterns are also the basis of **climax patterns**, which are used to create suspense. For example, if you wanted to create suspense in a speech about military intervention, you could chronologically trace the steps that eventually led us into Afghanistan or Iraq in such a way that you build up your audience's curiosity. If you detail these steps through the eyes of a soldier who entered military service right before one of those wars, you will be building suspense as your audience wonders what will become of that soldier.

The climax pattern can also be reversed. When it is reversed, it is called *anticlimactic* organization. If you started your military intervention speech by telling the audience that you were going to explain why a specific soldier was killed in a specific war, and then you went on to explain the things that caused that soldier to become involved in that war, you would be using anticlimactic organization. This pattern is helpful when you have an essentially uninterested audience and you need to build interest early in your speech to get the audience to listen to the rest of it.

> **climax pattern** An organizing plan for a speech that builds ideas to the point of maximum interest or tension.

Space Patterns

Speech organization by physical area is known as a **space pattern**. The area could be stated in terms of continents or centimeters or anything in between. If you were discussing the Great Lakes, for example, you could arrange them from west to east:

I. Superior
II. Michigan
III. Huron
IV. Erie
V. Ontario

> **space pattern** An organizing plan in a speech that arranges points according to their physical location.

Topic Patterns

A topical arrangement or **topic pattern** is based on types or categories. These categories could be either well known or original; both have their advantages. For example, a division of college students according to well-known categories might look like this:

I. Freshmen
II. Sophomores
III. Juniors
IV. Seniors

> **topic pattern** An organizing plan for a speech that arranges points according to logical types or categories.

Well-known categories are advantageous because audiences quickly understand them. But familiarity also has its disadvantages. One disadvantage is the "Oh, this again" syndrome. If the members of an audience feel they have nothing new to learn about the components of your topic, they might not listen to you. To avoid this, you could invent original categories that freshen up your topic by suggesting an original analysis. For example, original categories for "college students" might look like this:

I. Grinds: Students who go to every class and read every assignment before it is due.
II. Renaissance students: Students who find a satisfying balance of scholarly and social pursuits.
III. Burnouts: Students who have a difficult time finding the classroom, let alone doing the work.

Sometimes topics are arranged in the order that will be easiest for your audience to remember. To return to our Great Lakes example, the names of the lakes could be arranged so that their first letters spell the word "HOMES." Words used in this way are known as *mnemonics*. Carol Koehler, a professor of communication and medicine, uses the mnemonic "CARE" to describe the characteristics of a caring doctor:

C stands for *concentrate*. Physicians should pay attention with their eyes and ears . . .

A stands for *acknowledge*. Show them that you are listening . . .

R stands for *response*. Clarify issues by asking questions, providing periodic recaps . . .

E stands for *exercise emotional control*. When your "hot buttons" are pushed . . .[3]

Problem-Solution Patterns

Describing what's wrong and proposing a way to make things better is known as a **problem-solution pattern**. It is usually (but not always) divisible into two distinct parts, as in this example:

I. The Problem: Addiction (which could then be broken down into addiction to cigarettes, alcohol, prescribed drugs, and street drugs)

II. The Solution: A national addiction institute (which would study the root causes of addiction in the same way that the National Cancer Institute studies the root causes of cancer)

We will discuss this pattern in more detail in Chapter 15.

Cause-Effect Patterns

Cause-effect patterns are similar to problem-solution patterns in that they are basically two-part patterns: First you discuss something that happened, and then you discuss its effects.

A variation of this pattern reverses the order and presents the effects first and then the causes. Persuasive speeches often have effect-cause or cause-effect as the first two main points. Elizabeth Hallum, a student at Arizona State University, organized the first two points of a speech on "workplace revenge"[4] like this:

I. The effects of the problem
   A. Lost productivity
   B. Costs of sabotage
II. The causes of the problem
   A. Employees feeling alienated
   B. Employers' light treatment of incidents of revenge

The third main point in this type of persuasive speech is often "solutions," and the fourth main point is often "the desired audience behavior." Hallum's final points were as follows:

III. Solutions: Support the National Employee Rights Institute.
IV. Desired Audience Response: Log on to www.disgruntled.com.

Cause-effect and problem-solution patterns are often combined in various arrangements. One extension of the problem-solution organizational pattern is Monroe's Motivated Sequence.

---

problem-solution pattern An organizing pattern for a speech that describes an unsatisfactory state of affairs and then proposes a plan to remedy the problem.

cause-effect pattern An organizing plan for a speech that demonstrates how one or more events result in another event or events.

Monroe's Motivated Sequence

The Motivated Sequence was proposed by a scholar named Alan Monroe in the 1930s.[5] In this persuasive pattern, the problem is broken down into an attention step and a need step, and the solution is broken down into a satisfaction step, a visualization step, and an action step. In a speech on "random acts of kindness,"[6] the Motivated Sequence might break down like this:

I. The attention step draws attention to your subject. ("Just the other day Ron saved George's life with a small, random, seemingly unimportant act of kindness.")[6]

II. The need step establishes the problem. ("Millions of Americans suffer from depression, a life-threatening disease.")

III. The satisfaction step proposes a solution. ("One random act of kindness can lift a person from depression.")

UNDERSTANDING YOUR COMMUNICATION

Main Points and Subpoints

To get an idea of your ability to distinguish main points from subpoints, set the "timer" function on your mobile phone and see how long it takes you to fit the following concepts for a speech entitled "The College Application Process" into outline form:

CONCEPTS	RECOMMENDED OUTLINE FORM
Participation in extracurricular activities	I.
Visit and evaluate college websites	A.
Prepare application materials	B.
Career ambitions	II.
Choose desired college	A.
Letters of recommendation	B.
Write personal statement	C.
Visit and evaluate college campuses	III.
Choose interesting topic	A.
Test scores	B.
Include important personal details	1.
Volunteer work	2.
Transcripts	3.

You can score yourself as follows:
   A minute or less: Congratulations, organization comes naturally to you.
   61–90 seconds: You have typical skills in this area.
   More than 90 seconds: Give yourself extra time while building your speech outline.

IV. The visualization step describes the results of the solution. ("Imagine yourself having that kind of effect on another person.")

V. The action step is a direct appeal for the audience to do something. ("Try a random act of kindness today!")

Chapter 15 has more to say about the organization of persuasive speeches.

13.4 Beginnings, Endings, and Transitions

The introduction and conclusion of a speech are vitally important, although they usually will occupy less than 20 percent of your speaking time. Listeners form their impression of a speaker early, and they remember what they hear last; it is, therefore, vital to make those few moments at the beginning and end of your speech work to your advantage. It is also essential that you connect sections within your speech using effective transitions.

The Introduction

The **introduction** of a speech has four functions: to capture the audience's attention, preview the main points, set the mood and tone of the speech, and demonstrate the importance of the topic.

Capturing Attention There are several ways to capture an audience's attention, including the following:

- **Refer to the audience.** This technique is especially effective if it is complimentary: "Zoe's speech about how animals communicate was so interesting that I decided to explore a related topic."
- **Refer to the occasion.** "Given our assignment, it seems appropriate to talk about something you might have wondered."
- **Refer to the relationship between the audience and the subject.** "It's fair to say that all of us here believe it's important to care for our environment."
- **Refer to something familiar to the audience.** "Most of us have cared for a plant at some point in our lives."
- **Cite a startling fact or opinion.** "New scientific evidence suggests that plants appreciate human company, kind words, and classical music."
- **Ask a question.** "Have you ever wondered why some people have a green thumb, whereas others couldn't make a weed grow?"
- **Tell an anecdote.** "The other day, while taking a walk near campus, I saw a man talking quite animatedly to a sunflower."
- **Use a quotation.** "The naturalist Max Thornton recently said, 'Psychobiology has shown that plants can communicate.'"
- **Tell an (appropriate) joke.** "We once worried about people who talked to plants, but that's no longer the case. Now we only worry if the plants talk back."

---

introduction (of a speech) The first structural unit of a speech, in which the speaker captures the audience's attention and previews the main points to be covered.

Previewing Main Points   After you capture the attention of the audience, an effective introduction will almost always state the speaker's thesis and give the listeners an idea of the upcoming main points. Katharine Graham, the former publisher of the *Washington Post*, addressed a group of businessmen and their wives in this way:

> I am delighted to be here. It is a privilege to address you. And I am especially glad the rules have been bent for tonight, allowing so many of you to bring along your husbands. I think it's nice for them to get out once in a while and see how the other half lives. Gentlemen, we welcome you.
>
> Actually, I have other reasons for appreciating this chance to talk with you tonight. It gives me an opportunity to address some current questions about the press and its responsibilities—whom we are responsible to, what we are responsible for, and generally how responsible our performance has been.[7]

Thus, Graham previewed her main points:

1. To explain who the press is responsible to
2. To explain what the press is responsible for
3. To explain how responsible the press has been

Sometimes your preview of main points will be even more straightforward:

"I have three points to discuss: They are _____, _____, and _____."

Sometimes you will not want to refer directly to your main points in your introduction. Your reasons for not doing so might be based on a plan calling for suspense, humorous effect, or stalling for time to win over a hostile audience. In that case, you might preview only your thesis:

"I am going to say a few words about _____."

"Did you ever wonder about _____?"

"_____ is one of the most important issues facing us today."

Setting the Mood and Tone of Your Speech   Notice, in the example just given, how Katharine Graham began her speech by joking with her audience. She was a powerful woman speaking before an all-male organization; the only women in the audience were the members' wives. That is why Ms. Graham felt it necessary to put her audience members at ease by joking with them about women's traditional role in society. By beginning in this manner, she assured the men that she would not berate them for the sexist bylaws of their organization. She also showed them that she was going to approach her topic with wit and intelligence. Thus, she set the mood and tone for her entire speech. Imagine how different that mood and tone would have been if she had begun this way:

> Before I start today, I would just like to say that I would never have accepted your invitation to speak here had I known that your organization does not accept women as members. Just where do you Cro-Magnons get off, excluding more than half the human race from your little club?

Demonstrating the Importance of Your Topic to Your Audience   Your audience members will listen to you more carefully if your speech relates to them as individuals.

Based on your audience analysis, you should state directly *why* your topic is of importance to your audience members. This importance should be related as closely as possible to their specific needs at that specific time.

For example, when Anna Claire Tucker, a student at Berry College, presented a speech about medical waste, she began her introduction with a striking example of how medical practices result in overcharging patients.

> When you need milk, a walk down the dairy aisle provides you with a variety of options: pint, quart, gallon. With so many choices, you wouldn't buy a gallon if you only needed a glass. But when it comes to thousand-dollar chemotherapy treatments, patients aren't given the same choice. Shaun Recchi didn't get to choose between small, medium, or large vials of rituximab. Manufacturers made that decision for him when they produced it in one size. Their greed cost Shaun $13,000 for a drug he only needed half of, and it only gets worse. While an open jug of milk could be saved, Shaun's medication couldn't, so the remaining $6,500 worth was simply thrown in the trash.[8]

That example caught her audience's attention, and Anna followed it up immediately with a reminder that her problem was a problem for everyone in the audience:

> From overprescribed pills to unused surgical tools, medical waste is a $765 billion problem. According to the National Academy of Medicine, it accounts for one quarter of all healthcare spending. Manufacturers and hospitals perpetuate the problem, but as Senator Amy Klobuchar decries, "It's the American taxpayers footing the bill."[9]

Thus, she reminded the audience that they all, as potential patients and definitely as taxpayers, were affected by the topic of her speech.

Establishing Credibility One final consideration for your introduction is to establish your credibility to speak on your topic. One way to do this is to be *well prepared*. Another is to *appear confident* as soon as you face your audience. A third technique is to *tell your audience about your personal experience* with the topic, in order to establish why it is important to you. We will discuss credibility, and how to establish it, in greater detail in Chapter 15.

conclusion (of a speech) The final part of a speech, in which the main points are reviewed and final remarks are made to motivate the audience or help listeners remember key ideas.

The Conclusion

Like the introduction, the **conclusion** is an especially important part of your speech. It has three essential functions: to restate the thesis, to review your main points, and to provide a memorable final remark.

You can review your thesis either by repeating it or by paraphrasing it. Or you might devise a striking summary statement for your conclusion to help your audience remember your thesis. Grant Anderson, a student at

The introduction and conclusion are extremely important to the impact of your speech.
What would grab the attention of your audience at the beginning and end of your next speech?

Minnesota State University, gave a speech against the policy of rejecting blood donations from homosexuals. He ended his conclusion with this statement: "The gay community still has a whole host of issues to contend with, but together all of us can all take a step forward by recognizing this unjust and discriminatory measure. So stand up and raise whatever arm they poke you in to draw blood and say 'Blood is Blood' no matter who you are."[10]

Grant's statement was concise but memorable.

Your main points can also be reviewed artistically. For example, first look back at that example introduction by Katharine Graham, and then read her conclusion to that speech:

> So instead of seeking flat and absolute answers to the kinds of problems I have discussed tonight, what we should be trying to foster is respect for one another's conception of where duty lies, and understanding of the real worlds in which we try to do our best. And we should be hoping for the energy and sense to keep on arguing and questioning, because there is no better sign that our society is healthy and strong.

Let's take a closer look at how and why this conclusion was effective. Graham posed three questions in her introduction. She dealt with those questions in her speech and reminded her audience, in her conclusion, that she had answered the questions.

PREVIEW (FROM INTRODUCTION OF SPEECH)	REVIEW (FROM CONCLUSION)
1. To whom is the press responsible?	1. To its own conception of where its duty lies
2. What is the press responsible for?	2. For doing its best in the "real world"
3. How responsible has the press been?	3. It has done its best

---

COMMUNICATION STRATEGIES

Effective Conclusions

You can make your final remarks most effective by avoiding the following mistakes:

- ☐ **Don't end abruptly.** Make sure your conclusion accomplishes everything it is supposed to accomplish. Develop it fully. You might want to use signposts such as, "Finally . . . ," "In conclusion . . . ," or "To sum up what I've been talking about here . . ." to let your audience know that you have reached the conclusion of the speech.

- ☐ **Don't ramble either.** Prepare a definite conclusion, and never, ever end by mumbling something like, "Well, I guess that's about all I wanted to say . . . ."

- ☐ **Don't introduce new points.** The worst kind of rambling is, "Oh, yes, and something I forgot to mention is . . ."

- ☐ **Don't apologize.** Don't make statements such as "I'm sorry I didn't have more time to research this subject." You will only highlight the possible weaknesses of your speech, which may have been far more apparent to you than to your audience. It's best to end strong. You can use any of the attention-getters suggested for the introduction to make the conclusion memorable, or you can revisit your attention-getting introduction.

Transitions

To keep your message moving forward, **transitions** perform the following functions:

They tell how the introduction relates to the body of the speech.
They tell how one main point relates to the next main point.
They tell how your subpoints relate to the points they are part of.
They tell how your supporting points relate to the points they support.

> **transition** A phrase that connects ideas in a speech by showing how one relates to the other.

To be effective, transitions should refer to the previous point and to the upcoming point, showing how they relate to each other and to the thesis. They usually sound something like this:

"Like [previous point], another important consideration in [topic] is [upcoming point]."

"But _____ isn't the only thing we have to worry about. _____ is even more potentially dangerous."

"Yes, the problem is obvious. But what are the solutions? Well, one possible solution is . . ."

Sometimes a transition includes an internal review (a restatement of preceding points), an internal preview (a look ahead to upcoming points), or both:

"So far we've discussed _____, _____, and _____. Our next points are _____, _____, and _____."

It isn't always necessary to provide a transition between every set of points. You have to choose when one is necessary for your given audience to follow the progression of your ideas. You can find several examples of transitions in the sample speech at the end of this chapter.

13.5 Supporting Material

It is important to organize ideas clearly and logically. But clarity and logic by themselves won't guarantee that you'll interest, enlighten, or persuade others; these results call for the use of supporting materials. These materials—the facts and information that back up and prove your ideas and opinions—are the flesh that fills out the skeleton of your speech.

Functions of Supporting Material

Supporting material has four functions.

To Clarify   As explained in Chapter 4, people of different backgrounds tend to attach different meanings to words. Supporting material can help you overcome this potential source of confusion by helping you to clarify key terms and ideas. For example, when the biologist Prosanta Chakrabarty talks about the confusion inherent in the idea of evolutionary theory, he explains it like this:

> For instance, we're taught to say "the theory of evolution." There are actually many theories, and just like the process itself, the ones that best fit the data are the ones

that survive to this day. The one we know best is Darwinian natural selection. That's the process by which organisms that best fit an environment survive and get to reproduce, while those that are less fit slowly die off. And that's it. Evolution is as simple as that, and it's a fact.

Evolution is a fact as much as the "theory of gravity." You can prove it just as easily. You just need to look at your bellybutton that you share with other placental mammals, or your backbone that you share with other vertebrates, or your DNA that you share with all other life on earth. Those traits didn't pop up in humans. They were passed down from different ancestors to all their descendants, not just us.[11]

To Prove A second function of support is to be used as evidence, to prove the truth of what you are saying. If you were giving a speech on how college fraternities are changing today, you might want to quote from a researcher who has been looking into that topic.

Americans demonize fraternities as bastions of toxic masculinity where young men go to indulge their worst impulses. Universities have cracked down: Since November 2017, more than a dozen have suspended all fraternity events. But

---

COMMUNICATION STRATEGIES

Organizing Business Presentations

When top business executives plan an important speech, they often call in a communication consultant to help organize their remarks. Even though they are experts, executives are so close to the topic of their message that they may have difficulty arranging their ideas so others will understand or be motivated by them.

Consultants stress how important organization and message structure are in giving presentations. Seminar leader and corporate trainer T. Stephen Eggleston sums up the basic approach: "Any presentation . . . regardless of complexity . . . should consist of the same four basic parts: an opening, body, summary and closing."[13]

Ethel Cook, a Massachusetts consultant, is very specific about how much time should be spent on each section of a speech. "In timing your presentation," she says, "an ideal breakdown would be:

Opening—10 to 20 percent

Body—65 to 75 percent

Closing—10 to 20 percent"[14]

Business coach Vadim Kotelnikov gives his clients a step-by-step procedure to organize their ideas within the body of a presentation. "List all the points you plan to cover," he advises. "Group them in sections and put your list of sections in the order that best achieves your objectives. Begin with the most important topics."[15] Toastmasters International, an organization that runs training programs for business professionals, suggests alternative organizational patterns:

To organize your ideas into an effective proposal, use an approach developed in the field of journalism—the "inverted pyramid." In the "inverted pyramid" format, the most important information is given in the first few paragraphs. As you present the pitch, the information becomes less and less crucial. This way, your presentation can be cut short, yet remain effective.[16]

Imagine a business presentation you might have to make in your future career. Why would organization be important in such a presentation?

Alexandra Robbins, a researcher who writes about the lives of young Americans, spent more than two years interviewing fraternity members nationwide for a book about what college students think it means to "be a man," and what she learned was often heartening. Contrary to negative headlines and popular opinion, many fraternities are encouraging brothers to defy stereotypical hypermasculine standards and to simply be good people.[12]

To Make Interesting   A third function of support is to make an idea interesting or to catch your audience's attention. For example, when the marketing executive Aparna Mehta got up to speak about the problem online retailers have with returns, audience members who were not in her field might have been tempted to run out for a snack. But before they could, Mehta began telling the following anecdote:

> Hi. My name is Aparna. I am a shopaholic—and I'm addicted to online returns.
>
> Well, at least I was. At one time, I had two or three packages of clothing delivered to me every other day. I would intentionally buy the same item in a couple different sizes and many colors, because I did not know what I really wanted. So I overordered, I tried things on, and then I sent what didn't work back. Once my daughter was watching me return some of those packages, and she said, "Mom, I think you have a problem."
>
> I didn't think so. I mean, it's free shipping and free returns, right? I didn't even think twice about it, until I heard a statistic at work that shocked me.
>
> You see, I'm a global solutions director for top-tier retail, and we were in a meeting with one of my largest customers, discussing how to streamline costs. One of their biggest concerns was managing returns. Just this past holiday season alone, they had 7.5 million pieces of clothing returned to them.
>
> I could not stop thinking about it. What happens to all these returned clothes? So I came home and researched. *And I learned that every year, four billion pounds of returned clothing ends up in the landfill. That's like every resident in the US did a load of laundry last night and decided to throw it in the trash today.*[17]

To Make Memorable   A final function of supporting materials, related to the preceding one, is to make a point memorable. We have already mentioned the importance of "memorable" statements in a speech conclusion; use of supporting material in the introduction and body of the speech provides another way to help your audience retain important information. For example, when Nathan Hill, a student at the University of Akron, spoke about how certain local ordinances discriminated against victims of abuse, he started with this example:

> Overjoyed. That's the word single mother Lakesha Briggs used to describe how she felt when she finally found a safe home for herself and her two daughters. That is, until Lakesha's boyfriend, a guest in the home, began to abuse her daily. . . . When the authorities responded, the family was startled when the police had a stark warning . . . for Ms. Briggs: "Strike One." The officers told her that if the police were called again to the property, they would make sure her family's landlord prepared an eviction. The police were referring to a local "Nuisance Property Ordinance," which labels a property as a nuisance when it is the site of a designated number of calls to police . . . making no distinction between perpetrators and victims of crimes.[18]

Types of Supporting Material

As you may have noted, each function of support could be fulfilled by several different types of material. Let's take a look at these different types of supporting material.

Definitions It's a good idea to give your audience members definitions of your key terms, especially if those terms are unfamiliar to them or are being used in an unusual way. A good definition is simple and concise, and might even encapsulate the main points of your speech. For example, when Reagan Williams, a student at Arkansas State University, gave a speech on Rape Culture, she defined her key term as follows:

> This term refers to a society or culture that normalizes sexual violence through patriarchal myths about the rape, the nature of a reaction to the crime, and rigid gender constructs that normalize violence.[19]

Examples An example is a specific case that is used to demonstrate a general idea. Examples can be either factual or hypothetical, personal or borrowed. In Reagan Williams's speech, she used the following **factual example** to demonstrate how rapes are often perpetrated by someone the victim knows:

> On January 8, 2012, Daisy Coleman's mother found her lying on the front lawn, hair frozen to the ground. Daisy was fourteen. Her mother drove her to the hospital, where she underwent a rape kit. When the result came back, their worst fears were confirmed.
>
> The evening before, the senior quarterback asked Daisy and her friend to join him at his home. When they arrived, they were encouraged to drink vodka from the "bitch cup." They were inexperienced with alcohol, and as the night went on, everything got blurry. They both blacked out. Less than an hour later, five senior boys left a fourteen-year-old girl lying on the ground in below freezing temperature. Years later, Daisy would learn that her 13-year-old friend had been raped in the room right next door.[20]

factual example A true, specific case that is used to demonstrate a general idea.

Hypothetical examples, which ask audience members to imagine something, can often be more powerful than factual examples because they create active participation in the thought. Stephanie Wideman of the University of West Florida used a hypothetical example to start off her speech on oil prices:

> The year is 2025. One day you are asked not to come into work, not because of a holiday, but instead because there is not enough energy available to power your office. You see, it is not that the power is out, but that they are out of power.[21]

hypothetical example An example that asks an audience to imagine an object or event.

Statistics Data organized to show that a fact or principle is true for a large percentage of cases are known as **statistics**. These are actually collections of examples, which is why they are often more effective as proof than are isolated examples. Here's the way a newspaper columnist used statistics to demonstrate a point about gun violence:

> I had coffee the other day with Marian Wright Edelman, president of the Children's Defense Fund, and she mentioned that since the murders of Robert Kennedy and the Rev. Martin Luther King Jr. in 1968, well over a million Americans have been killed by firearms in the United States. That's more than the combined U.S. combat deaths in all the wars in all of American history. "We're losing eight children and teenagers a day to gun violence," she said.[22]

statistics Numbers arranged or organized to show how a fact or principle is true for a large percentage of cases.

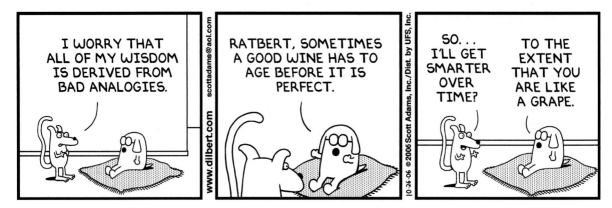

Source: DILBERT © 2006 Scott Adams. Used by permission of UNIVERSAL UCLICK. All rights reserved.

Because statistics can be powerful support, you have to follow certain rules when using them. You should make sure that the statistics make sense and that they come from a credible source. You should also cite the source of the statistic when you use it. A final rule is based on effectiveness rather than ethics. You should reduce the statistic to a concrete image if possible. For example, $1 billion in $100 bills. Using concrete images such as this will make your statistics more than "just numbers" when you use them. For example, one observer expressed the idea of Bill Gates's wealth this way:

> Examine Bill Gates' wealth compared to yours: Consider the average American of reasonable but modest wealth. Perhaps he has a net worth of $100,000. Mr. Gates' worth is 400,000 times larger. Which means that if something costs $100,000 to him, to Bill it's as though it costs 25 cents. So for example, you might think a new Lamborghini Diablo would cost $250,000, but in Bill Gates dollars that's 63 cents.[23]

analogies An extended comparison that can be used as supporting material in a speech.

Analogies/Comparison-Contrast We use **analogies**, or comparisons, all the time, often in the form of figures of speech, such as similes and metaphors. A simile is a direct comparison that usually uses *like* or *as*, whereas a metaphor is an implied comparison that does not use *like* or *as*. So if you said that the rush of refugees from a war-torn country was "like a tidal wave," you would be using a simile. If you used the expression "a tidal wave of refugees," you would be using a metaphor.

Analogies are extended metaphors. They can be used to compare or contrast an unknown concept with a known one. For example, here's how one writer made her point against separate Academy Awards for men and women:

> Many hours into the Academy Awards ceremony this Sunday, the Oscar for best actor will go to Morgan Freeman, Jeff Bridges, George Clooney, Colin Firth, or Jeremy Renner. Suppose, however, that the Academy of Motion Picture Arts and Sciences presented separate honors for best white actor and best non-white actor, and that Mr. Freeman was prohibited from competing against the likes of Mr. Clooney and Mr. Bridges. Surely, the Academy would be derided as intolerant and

out of touch; public outcry would swiftly ensure that Oscar nominations never again fell along racial lines.

Why, then, is it considered acceptable to segregate nominations by sex, offering different Oscars for best actor and best actress?[24]

Anecdotes An **anecdote** is a brief story with a point, often (but not always) based on personal experience. (The word *anecdote* comes from the Greek, meaning "unpublished item.") Alyssa Gieseck, a student at the University of Akron, used the following anecdote in her speech about the problems some deaf people encounter with police:

> Jonathan Meister, a deaf man, was retrieving his personal belongings from a friend's home when police arrived, responding to a report of suspicious behavior. Trying to sign to the police officers, Meister was seen as a threat, which is when the officers decided to handcuff him. Handcuffing a deaf person is equivalent to putting duct tape over a hearing person's mouth. Meister initially pulled away to sign that he was deaf, but one police officer pushed him up against a fence, kneed him twice in the abdomen, put him in a choke hold, and tasered him. A second officer repeatedly punched him in the face, tasering him again. But this wasn't the end of the assault. He was then shoved to the ground, kicked, elbowed, tasered for a third time, and put in a second chokehold, which left him unconscious.[25]

> **anecdote** A brief, personal story used to illustrate or support a point in a speech.

Quotations/Testimonies Using a familiar, artistically stated quotation will enable you to take advantage of someone else's memorable wording. For example, if you were giving a speech on personal integrity, you might quote Mark Twain, who said, "Always do right. This will gratify some people, and astonish the rest." A quotation like that fits Alexander Pope's definition of "true wit": "What was often thought, but ne'er so well expressed."

You can also use quotations as **testimony** to prove a point by using the support of someone who is more authoritative or experienced on the subject than you are. When Julia Boyle, a student at Northern Illinois University, wanted to prove that spyware stalking was a serious problem, she used testimony this way:

> Michella Cash, advocate for the Women's Service network, noted, "Spyware technology is the new form of domestic violence abuse that enables perpetrators to exert round the clock control over their victims."[26]

> **testimony** Supporting material that proves or illustrates a point by citing an authoritative source.

Sometimes testimony can be paraphrased. For example, when one business executive was talking on the subject of diversity, he used a conversation he had with Jesse Jackson Sr., an African American political activist, as testimony:

> At one point in our conversation, Jesse talked about the stages of advancement toward a society where diversity is fully valued. He said the first stage was emancipation—the end of slavery. The second stage was the right to vote and the third stage was the political power to actively participate in government—to be part of city hall, the governor's office and Capitol Hill. Jesse was clearly focused, though, on the fourth stage—which he described as the ability to participate fully in the prosperity that this nation enjoys. In other words, economic power.[27]

Styles of Support: Narration Versus Citation

Most of the forms of support discussed in the preceding section could be presented in either of two ways: through narration or through citation. **Narration** involves telling a story with your information. You put it in the form of a small drama, with a beginning, middle, and end. For example, Evan McCarley of the University of Mississippi narrated the following example in his speech on the importance of drug courts:

> Oakland contractor Josef Corbin has a lot to be proud of. Last year his firm, Corbin Building Inc., posted revenue of over 3 million dollars after funding dozens of urban restoration projects. His company was ranked as one of the 800 fastest-growing companies in the country, all due to what his friends call his motivation for success. Unfortunately, Corbin used this motivation to rob and steal on the streets of San Francisco to support a heroin and cocaine habit. But when he was charged with possession, Josef was given the option to participate in a state drug court, a program targeted at those recently charged with drug use, possession, or distribution. The drug court offers offenders free drug treatment, therapy, employment, education, and weekly meetings with a judge, parole officer and other accused drug offenders.[28]

Citation, unlike narration, is a simple statement of the facts. Citation is shorter and more precise than narration, in the sense that the source is carefully stated. Citation will always include such phrases as, "According to the July 25, 2016, edition of *Time* magazine," or, "As Mr. Smith made clear in an interview last April 24." Evan McCarley cited statistics later in his speech on drug courts:

> Fortunately, Corbin's story, as reported in the May 30th *San Francisco Chronicle*, is not unique, since there are currently over 300 drug courts operating in 21 states, turning first-time and repeat offenders into successful citizens with a 70% success rate.[29]

Some forms of support, such as anecdotes, are inherently more likely to be expressed as narration. Statistics, on the other hand, are nearly always cited rather than narrated. However, when you are using examples, quotation/testimony, definitions, and analogies, you often have a choice.

Plagiarism Versus Originality

Some experts believe that social media is redefining how students understand the concept of authorship and originality. After all, the internet is the home of file sharing that allows us to download music, movies, and TV programs without payment. Also, Google and Wikipedia are our main portals to random free information. It all seems to belong to us, residing on our computer as it does. Information wants to be free.

According to one expert on the topic, "Now we have a whole generation of students who've grown up with information that just seems to be hanging out there in cyberspace and doesn't seem to have an author. It's possible to believe this

---

narration The presentation of speech supporting material as a story with a beginning, middle, and end.

citation A brief statement of supporting material in a speech.

information is just out there for anyone to take."³⁰ Other experts beg to differ. They say students are fully aware of what plagiarism is, online or off, and they know it's cheating. It's just that it's so easy to copy and paste online material, and students like to save time wherever they can.

Public speaking instructors are on the front lines of those fighting plagiarism because it's so important for successful student speakers to speak from the heart, in their own words and with their own voice. In addition, citing research enhances credibility. Plagiarism in public speaking isn't just cheating, it's ineffective.

The general rule for the digital age is as follows: Thou shalt not cut and paste into a final draft—not for a paper and not for a speech. Cutting and pasting is fine for research, but everything that's cut and pasted should be placed in a separate "research" file, complete with a full citation for the website in which you found it. Then switch to your "draft" file to put everything in your own words, and go back to the research file to find the attribution information when you need to cite facts and ideas that you got from those sources.

13.6 Sample Speech

The following speech was given by Dan O'Connor, whose profile opened this chapter.

Speech Outline

A formal outline for Dan's speech might appear like the following one. This is a refinement of the working outline shown on pages 353–354. Numbers in parentheses correspond to the numbered paragraphs of the speech.

INTRODUCTION

  I. Attention-Getter: Jerry Coen example (1)
 II. Thesis: Prisons unnecessarily abuse deaf inmates. (2)
III. Preview of Main Points: Today, we look within our prisons' walls, to examine the causes of this abuse, their effects, and finally explore solutions. (3)

Example, transition to 1st main point (4)

BODY

  I. The abuse of deaf inmates has two main causes. (5–9)
   A. Outdated technology (5–6)
      1. Teletype is the only technology used (5)
      2. Teletype is outmoded (6)
      3. Teletype is difficult for deaf prisoners to use (6)
   B. Lack of interpreters. (7–9)
      1. There aren't enough ASL interpreters. (7)
      2. Prisons often use unavailability as an excuse. (8)
      3. Even where there are interpreters, prisons don't bother with them. (9)

Example, transition to 2nd main point (10)

  II. The abuse of deaf inmates has two main effects (10–15)

    A. Deaf prisoners are cut off from support systems in the outside world. (11–12)

      1. Isolation eliminates due process (11–12)

      2. Isolation devastates relationships (11–12)

      3. Isolation negatively affects recidivism (11–12)

    B. Deaf prisoners are cut off from important activities within the prison. (13–15)

      1. This deprives deaf inmates of human contact, (13–15)

      2. And leads to longer sentences, (13–15)

      3. And leaves little hope for change. (13–15)

Transition to 3rd main point (16)

  III. The abuse of deaf inmates requires two main solutions (17–21)

    1. We need to provide videophones (17–19)

    2. We need to fill the interpretation gap. (20–21)

CONCLUSION (22)

  I. Review (22)

  II. Final Remarks (22)

A brief excerpt of Dan's annotated bibliography shows the thoroughness of his research (the complete bibliography is too lengthy to reproduce here):

Annotated Bibliography

"The message would be typed in ASL, which would be very difficult for whoever is receiving the call to understand, because ASL is not a written form. It's a visual language," he said. "The receiver would be reading disjointed English and they wouldn't understand . . . that creates a barrier for both the hearing people receiving the call and don't know how to respond, and for the deaf population who is faced with the receiver who is not able to facilitate the conversation."

WUSF: Florida DOC sued for deaf descrimination, again. December 5, 2019

The Guardian: Deaf man commits suicide after hearing aids taken, no interpreters for suicide watch January 22, 2020

https://www.theguardian.com/society/2020/jan/22/prison-death-tyrone-givans-deaf-hmp-pentonville

"a catalogue of errors."

"an 'absence of meaningful contact' with him, as well as a failure to consider his disability, or to document, let alone act on, Givans' concerns that he was under threat."

Because of the COVID-19 pandemic, Dan presented his speech via Zoom. He takes advantage of the medium at several points in his speech.

SAMPLE SPEECH  **Dan O'Connor**

Deaf Justice

1. At his arraignment, Jerry Coen thought he signed a parole agreement. In truth, he signed on to ten years in prison. The miscommunications didn't stop there. In prison, he couldn't call a lawyer, talk to his family, or even get to know his cellmates. It was a decade of isolation, fear, and pain. The reason? Coen is deaf.

    Dan uses a striking example to gain audience attention

2. According to the National Association of the Deaf, nearly all American prisons are perpetrating systems of constant oppression of deaf communities. The advocacy group HEARD recently revealed the systemic abuse of deaf inmates, and a complete failure of state and federal prisons to adhere to the Americans with Disabilities Act—the ADA. The most recent data from the Bureau of Justice Statistics suggests that there are over 150,000 deaf and hard-of-hearing Americans in prison, nearly all of whom face constant suffering.

    He uses information and statistics from recent articles to establish his thesis.

3. Today, we look within our prisons' walls, to examine the causes of this abuse, their effects, and finally explore solutions. To make this speech accessible, I will use simultaneous communication, English alongside signs from American Sign Language, ASL. Because in our national discourse on criminal justice, we have brushed aside deaf issues for far too long. It's time to put their struggle front and center.

    He previews his main points and explains that he will act as his own interpreter for deaf audience members.

4. James Woody spent four years in a Georgia prison, where the doctors and nurses had no knowledge of ASL. His health worsened. Upon leaving prison, he was diagnosed with cancer. Woody's story demands that we examine the causes of prisons' inability to accommodate deafness. First, outdated technology, and second, a systemic lack of interpreters.

    Dan uses another example as part of his transition to his first main point.

5. First, outdated technology. *The Chronicle Herald* on March 20, 2020 explains that Teletype, a machine that sends written text, is many prisons' only attempt at accommodations for the deaf. But the Department of Education on June 29, 2020 reveals that the average deaf or hard-of-hearing American reads at a third-grade level. Teletype forces ASL, with its own grammar and syntax, into an unnatural, often unfamiliar written medium. Conversation is impossible.

    He uses research to establish his point.

6. CREEC Law Center on June 21, 2019 explains, the teletype, invented in 1964, is drastically outdated and outmoded. It's like giving prisoners a telegraph, and saying it's their problem if they don't know Morse code. Despite over half a century of invention in telecommunications, deaf inmates are left with technology from a time when a handheld calculator was science fiction.

    Here he uses an analogy comparing the teletype with the outmoded technology of the telegraph and Morse code

7. Second, a systemic lack of interpreters. According to the Registry of Interpreters for the Deaf's most recent data, there are about 15,000 ASL interpreters in the United States, for an estimated deaf and hard-of-hearing population of eleven million.

    Here he uses statistics as support.

8. *The Oregonian* of September 26, 2019 reveals that prisons often use unavailability as an excuse, but even where there are interpreters, prisons don't bother with them. Prison guards, who use fear and intimidation to control inmates, don't call for interpreters, and rarely care to learn ASL.

    Here he paraphrases an article . . .

9. Jeremy Sanders' painkiller addiction landed him behind bars, where no one bothered to sign. He struggled, unable to ask for help, repeatedly attempting suicide.

    . . . and introduces another example.

Combining minimal incentive to interpret with an unfeeling prison system abandons deaf inmates, depriving them of basic, essential interactions.

10. Tyrone Givans, a partner, father, and deaf inmate, committed suicide just weeks into his sentence. Prison officials had confiscated his hearing aids and failed to provide interpreters, leading to a complete absence of meaningful human contact. The effects on deaf prisoners are excruciating. Let's examine two: A barrier to the outside, and a failure of understanding inside.

Another example is used as part of his transition to his second main point.

11. First, a barrier to the outside. The Prison Policy Initiative on February 9, 2019 explains, the ability to communicate beyond prison is crucial to the mental well-being of inmates. Contact with a support structure massively affects recidivism. When deaf inmates can't connect, it not only damages their health, it undermines prison's supposed role as a rehabilitative force in society.

Again, Dan supports his idea with evidence based on research ...

12. The Maryland Coalition against Sexual Assault reports that deaf prisoners are more than twice as likely to experience sexual abuse. And while their hearing counterparts have access to abuse hotlines, deaf inmates are unable to report abuse. Our prisons' inability to provide communication leaves the deaf cut off from lawyers, family, and the entire world, eliminating due process and devastating their relationships.

... including this authoritative source.

13. Second, a failure of understanding inside. The ACLU reported in 2020 that without interpreters, deaf inmates miss meals, showers, and announcements. This leads to punishment by prison guards and reduces the chances of sentence reductions for good behavior.

He moves directly to his next sub-point ...

14. Prison Legal News reported recently that many deaf inmates can't attend rehabilitative programs, which are often a requirement for parole. After one California inmate attended without an interpreter, the parole board still denied him—citing his lack of participation. He couldn't apply again for five years.

... and backs it up with another example.

15. We are forcing deaf inmates to live in solitary confinement every day. This torturous system deprives the deaf of human contact, leads to longer sentences, and leaves little hope for change.

He uses an analogy to compare deafness with solitary confinement ...

16. A 2017 settlement required Florida's Department of Corrections to provide videophones and interpreters for the deaf. Three years later, the state's prisons still fail. Lawsuits have demonstrated that change is required, but we cannot wait for judges to tell prisons what we already know. We need solutions. First, to provide videophones, and second, to fill the interpretation gap.

... and another example to introduce his final main point.

17. First, videophones. The pandemic has made videoconferencing ubiquitous. Videophones, like this, can be uniquely beneficial for the deaf, allowing direct and fluid communication.

Here he gestures toward his own camera, which is recording his speech via Zoom conference ...

18. The National Association of the Deaf has created a memo you can send to your local jails and state department of corrections, informing them of their failures, and what's needed to fix them. However, the NAD calls for teletypes, despite their clear inadequacy. So, I have created a modified version of this memo, which I will send in a folder after this speech.

... and pastes a link for further information in the chat function.

19. We have the most up to date, comprehensive advocacy information available. It's vital that we share. By telling our government to make changes the law

He explains why audience members should access the link.

already requires, we can help rapidly and drastically improve the lives of deaf inmates.

20 Second, we must fill the interpretation gap. The ADA, Civil Rights Acts, and the Bill of Rights demand the federal government protect us from such blatant discrimination and violations of due process. For prisons to live up to this obligation, we must leave them no excuse. We need more interpreters. We must create a federal system of tuition reimbursement for students pursuing a career in ASL interpretation, as we have done to incentivize other careers. In the resource folder is information for your representatives. It also contains resources for learning ASL.

He introduces his second solution...

21 I learned this beautiful language to connect with others. So try it out. Consider becoming an interpreter. You might just fall in love with ASL and improve equity in the process. Through individual and societal action, we can allow our deaf inmates to finally return from isolation.

...and develops the point by directly engaging the audience

22 Today we have examined the causes, the effects, and solutions to America's failure of deaf communities. Jerry Coen's prison sentence is up. But there are still thousands of other deaf people suffering silently behind bars. It is crucial that we use our privileged voices to speak out.

...and wraps up his speech with a conclusion that summarizes his main points and gives the audience a final thought to take with them.

MAKING THE GRADE

OBJECTIVE 13.1 Describe different types of speech outlines and their functions.

- A working outline is used to map out the structure of your speech, in rough form.
- A formal outline uses a standard format and a consistent set of symbols.
- Speaking notes are used to jog your memory while giving a speech. Like a working outline, they are for your eyes only.
- Explain the primary differences among working outlines, formal outlines, and speaking notes.
- Which type of outline would be most important for your next speech?
- What style of speaking notes would you use for your next speech?

OBJECTIVE 13.2 Construct an effective speech outline using the organizing principles described in this chapter.

- Principles for the effective construction of outlines are based on the use of standard symbols and a standard format.
- The rule of division requires at least two divisions of every point or subpoint.
- The rule of parallel wording requires that points at each level of division be worded in a similar manner whenever possible.
- What are the standard symbols used in a formal outline?
- Which principle of outlining is least intuitive for you? Which one is most intuitive?
- How would you divide your next speech into main points and subpoints?

OBJECTIVE 13.3 Choose an appropriate organizational pattern for your speech.

- The organization of ideas should follow a pattern, such as time, space, topic, problem-solution, cause-effect, or Monroe's Motivated Sequence.
- Each pattern of organization has its own advantages in helping your audience understand and remember what you have to say.
- Why is it important to organize your ideas according to a pattern?
- When should you use each of the six patterns of organization discussed in this chapter?
- Which pattern of organization would be best for your next speech?

OBJECTIVE 13.4 Develop a compelling introduction and conclusion, and use effective transitions at key points in your speech.

- The main idea of the speech is established in the introduction, developed in the body, and reviewed in the conclusion.
- The introduction will also gain the audience's attention, preview the main points, set the mood and tone of the speech, and demonstrate the importance of the topic to the audience.
- The conclusion will review your main points and supply the audience with a memory aid in the form of compelling final remarks.
- Effective transitions keep your message moving forward and demonstrate how your ideas are related.
  > Why are the beginning and end of a speech so important? Why are transitions?
  > What would be the most important function of the introduction of your next speech?
  > What idea would you like to leave your audience with in your next speech?

OBJECTIVE 13.5 Choose supporting material that will help make your points clear, interesting, memorable, and convincing.

- Supporting materials are the facts and information you use to back up what you say.
- Types of support include *definitions, examples, statistics, analogies, anecdotes, quotations,* and *testimony*.
- Support may be narrated or cited.
  > Give examples of five forms of support that could be used for a speech on financing a college education.
  > Which form of support do you find to be most effective for most speeches?
  > What are the main forms of support that you would use for your next speech?

KEY TERMS

analogy p. 372
anecdote p. 373
basic speech structure p. 355
cause-effect pattern p. 362
citation p. 374
climax pattern p. 361
conclusion (of a speech) p. 366
factual example p. 371
formal outline p. 355
hypothetical example p. 371
introduction (of a speech) p. 364
narration p. 373
problem-solution pattern p. 362
space pattern p. 361
statistics p. 371
testimony p. 373
time pattern p. 360

topic pattern p. 361
transition p. 368
working outline p. 354

PUBLIC SPEAKING PRACTICE

Prepare a brief, one-paragraph explanation of your choice of organization pattern for your next classroom speech. Discuss it with your classmates.

ACTIVITIES

1. **Dividing Ideas** For practice in the principle of division, divide each of the following into three to five subcategories:
   a. Clothing
   b. Academic studies
   c. Crime
   d. Health care
   e. Fun
   f. Charities

2. **Organizational Effectiveness** Take any written statement at least three paragraphs long that you consider effective. This statement might be an editorial in your local newspaper, a short magazine article, or even a section of one of your textbooks. Outline this statement according to the rules discussed here. Was the statement well organized? Did its organization contribute to its effectiveness?

3. **The Functions of Support** For practice in recognizing the functions of support, identify three instances of support in each of the speeches at the end of Chapters 13 and 14. Explain the function of each instance of support. (Keep in mind that any instance of support *could* perform more than one function.)

Informative Speaking

CHAPTER OUTLINE

14.1 Types of Informative Speaking 386
By Content
By Purpose

14.2 Informative Versus Persuasive Topics 387
Type of Topic
Speech Purpose

14.3 Techniques of Informative Speaking 388
Define a Specific Informative Purpose
Create Information Hunger
Make It Easy to Listen
Use Clear, Simple Language
Use a Clear Organization and Structure

14.4 Using Supporting Material Effectively 392
Emphasizing Important Points
Generating Audience Involvement
Using Visual Aids
Using Presentation Software
Alternative Media for Presenting Graphics
Rules for Using Visual Aids

14.5 Sample Speech 402

MAKING THE GRADE 410

KEY TERMS 411

PUBLIC SPEAKING PRACTICE 411

ACTIVITIES 411

LEARNING OBJECTIVES

14.1
Distinguish among the main types of informative speaking.

14.2
Describe the differences between informative and persuasive speaking.

14.3
Outline techniques that increase the effectiveness of informative speeches.

14.4
Explain how you might use visual aids appropriately and effectively.

Photo courtesy Adam Grant

Adam Grant is a super-professor. He teaches MBA and undergraduate courses at the University of Pennsylvania's Wharton School of Business, where his students consistently rate him among the best professors.

Grant does cutting-edge research in organizational psychology. His *New York Times* bestselling books have sold millions of copies, his TED Talks have more than 25 million views, and his podcast, *WorkLife with Adam Grant*, tops the charts. He has been recognized by *Forbes* as one of "The World's Ten Most Influential Business Thinkers," highlighted as a *Fortune* "40 Under 40," and received distinguished scientific achievement awards from the American Psychological Association and the National Science Foundation.

On top of all that, Grant is a former magician who uses his conjuring skills in lectures whenever possible. He was a junior Olympic springboard diver.

One of Grant's many talents is the ability to translate complex ideas into clear language for lay audiences. One of his explanations, "Frogs in Hot Water," is the sample speech at the end of this chapter.

Informative speaking skills like Adam Grant's are extremely important today because there is a huge amount of information competing for everyone's attention, and that amount is increasing exponentially. One expert estimates that every two days now we create as much information as we did from the dawn of civilization up until 2003.[1] To make matters worse, much of the information that is produced is purposely inaccurate and misleading. This is especially true of the disinformation that circulates online, on cable TV, in politics, and in advertising.

Social scientists describe a form of psychological stress known as *information overload*: the difficulty of sorting through all the information available to us.[2] Some experts use the term "**information anxiety**" for the state of being overwhelmed by information; this is one of the symptoms of the information overload mentioned in Chapter 11.

information anxiety
The psychological stress of dealing with too much information.

As an informative speaker, your responsibility goes beyond providing new information. When it's done effectively, informative speaking relieves information overload by helping to make a crazy world make sense. This is done by converting information to knowledge. *Information* is the raw material, the sometimes contradictory statements and competing claims you encounter. *Knowledge* results from being able to make sense of that raw material. Effective public speakers filter, organize, and illustrate information to reach audiences with tailored messages, in an environment where they can see if the audience is understanding. If they aren't, the speaker can adjust as needed.

Informative speaking goes on all around you: in your professors' lectures or in a mechanic's explanation of how to keep your car from breaking down. You engage in this type of speaking frequently whether you realize it or not. Sometimes it's formal, as when you give a report in class. At other times, it's more casual, as when you tell a friend how to prepare your favorite dish. The main objective of this chapter is to give you the skills you need to enhance all your informative speaking.

UNDERSTANDING YOUR COMMUNICATION

Are You Overloaded?

Problems in informative speaking are often the result of information overload—on both the speaker's and the audience's parts. For each statement to the right, select "often," "sometimes," or "seldom."

1. I forget information I need to know.
   OFTEN   SOMETIMES   SELDOM

2. I have difficulty concentrating on important tasks.
   OFTEN   SOMETIMES   SELDOM

3. When I go online, I feel anxious about the work I don't have time to do.
   OFTEN   SOMETIMES   SELDOM

4. I have email messages sitting in my inbox that are more than 2 weeks old.
   OFTEN   SOMETIMES   SELDOM

5. I constantly check my online services because I am afraid that if I don't, I will never catch up.
   OFTEN   SOMETIMES   SELDOM

6. I find myself easily distracted by things that allow me to avoid work I need to do.
   OFTEN   SOMETIMES   SELDOM

7. I feel fatigued by the amount of information I encounter.
   OFTEN   SOMETIMES   SELDOM

8. I delay making decisions because of too many choices.
   OFTEN   SOMETIMES   SELDOM

9. I make wrong decisions because of too many choices.
   OFTEN   SOMETIMES   SELDOM

10. I spend too much time seeking information that is *nice to know* rather than information I *need to know*.
    OFTEN   SOMETIMES   SELDOM

Scoring: Give yourself 3 points for each "often," 2 points for each "sometimes," and 1 point for each "seldom." If your score is:

10–15: Information overload is not a big problem for you. However, it's probably still a significant problem for at least some members of your audience, so try to follow the guidelines for informative speaking outlined in this chapter.

16–24: You have a normal level of information overload. The guidelines in this chapter will help you be a more effective speaker.

25–30: You have a high level of information overload. Along with observing the guidelines in this chapter, you might also want to search online for guidelines to help you overcome this problem.

14.1 Types of Informative Speaking

There are several types of informative speaking. The primary types have to do with the content and purpose of the speech.

By Content

Informative speeches are generally categorized according to their content and include the following types.

Speeches About Objects This type of informative speech is about anything that is tangible (that is, capable of being seen or touched). Speeches about objects might include an appreciation of the Grand Canyon (or any other natural wonder) or a demonstration of the newest smartphone (or any other product).

Speeches About Processes A process is any series of actions that leads to a specific result. If you spoke on the process of aging, the process of learning to juggle, or the process of breaking into a social networking business, you would be giving this type of speech.

Speeches About Events You would be giving this type of informative speech if your topic dealt with anything notable that happened, was happening, or might happen: an upcoming protest against hydraulic fracturing ("fracking"), for example, or the prospects of your favorite baseball team winning the national championship.

Speeches About Concepts Concepts include intangible ideas, such as beliefs, theories, ideas, and principles. The sample speech at the end of this chapter deals with such a concept. If you gave an informative speech about postmodernism, vegetarianism, or any other "ism," you would be giving this type of speech. Other topics would include everything from New Age religions to theories about extraterrestrial life to rules for making millions of dollars.

By Purpose

We also distinguish among types of informative speeches depending on the speaker's purpose. We ask, "Does the speaker seek to describe, explain, or instruct?"

Descriptions A speech of **description** is the most straightforward type of informative speech. You might introduce a new product like a wearable computer to a group of customers, or you might describe what a career in nursing would be like. Whatever its topic, a descriptive speech uses details to create a "word picture" of the essential factors that make that thing what it is.

Explanations **Explanations** clarify ideas and concepts that are already known but not understood by an audience. For example, your audience members might already know that a U.S. national debt exists, but they might be baffled by the reasons why it has become so large. Explanations often deal with the question of *why* or *how*. Why do we have to wait until the age of 21 to drink legally? How did China evolve from an impoverished economy to a world power in a single generation? Why did tuition need to be increased this semester?

Instructions **Instructions** teach something to the audience in a logical, step-by-step manner. They are the basis of training programs and orientations. They often deal

---

description A type of speech that uses details to create a "word picture" of something's essential factors.

explanations Speeches or presentations that clarify ideas and concepts already known but not understood by an audience.

instructions Remarks that teach something to an audience in a logical, step-by-step manner.

with the question of *how to*. This type of speech sometimes features a demonstration or a visual aid. Thus, if you were giving instructions on "how to promote your career via social networking sites," you might demonstrate by showing the social media profile of successful people. For instructions on "how to perform CPR," you could use a volunteer or a dummy.

These types of informative speeches aren't mutually exclusive. As you'll see in the sample speech at the end of this chapter, there is considerable overlap, as when you give a speech about a complex concept that has the purpose of explaining it. Still, even this imperfect categorization demonstrates how wide a range of informative topics is available. One final distinction we need to make, however, is the difference between an informative and a persuasive speech topic.

14.2 Informative Versus Persuasive Topics

There are many similarities between an informative and a persuasive speech. In an informative speech, for example, you are constantly trying to "persuade" your audience to listen, understand, and remember. In a persuasive speech, you "inform" your audience about your arguments, your evidence, and so on. Nonetheless, two basic characteristics differentiate an informative topic from a persuasive topic.

Type of Topic

In an informative speech, you generally do not present information that your audience is likely to disagree with. Again, this is a matter of degree. For example, you might want to give a purely informative talk on the differences between hospital births and home-based midwife births by simply describing what the practitioners of each method believe and do. By contrast, a talk either boosting or criticizing one method over the other would clearly be persuasive. The sample speech at the end of this chapter is almost entirely informative, although the argument could be made that Adam Grant is trying to convince us to change our behavior by rethinking as well as thinking.

The noncontroversial nature of informative speaking doesn't mean that your speech topic should be uninteresting to your audience; rather, it means that your approach to it should not engender conflict. You could speak about the animal rights movement, for example, by explaining the points of view of both sides in an interesting but objective manner.

Speech Purpose

The informative speaker does seek a response (such as attention and interest) from the listener and does try to make the topic important to the audience. But the speaker's primary intent isn't to change attitudes or to make the audience members *feel* differently about the topic. For example, an informative speaker might explain how facial recognition software works but will not try to change attitudes about whether this technology is a boon or a threat.

The speaker's intent is best expressed in a specific informative purpose statement, which brings us to the first of our techniques of informative speaking.

14.3 Techniques of Informative Speaking

The techniques of informative speaking are based on a number of principles covered in earlier chapters. The most important principles to apply to informative speaking include those that help an audience understand and care about your speech. Let's look at how these principles apply to specific techniques.

Define a Specific Informative Purpose

As Chapter 12 explained, any good speech must be based on a purpose statement that is audience oriented, precise, and attainable. When you are preparing an informative speech, it is especially important to define in advance, for yourself, a clear informative purpose. An **informative purpose statement** will generally be worded to stress audience knowledge, ability, or both:

> *After listening to my speech, my audience will be able to recall the three most important questions to ask when shopping for a smartphone.*
>
> *After listening to my speech, my audience will be able to identify the four reasons that online memes go viral.*
>
> *After listening to my speech, my audience will be able to discuss the pros and cons of using drones in warfare.*

Notice that in each of these purpose statements a specific verb such as *to recall*, *to identify*, or *to discuss* points out what the audience will be able to do after hearing the speech. Other key verbs for informative purpose statements include these:

Accomplish	Contrast	Integrate	Perform
Analyze	Describe	List	Review
Apply	Explain	Name	Summarize
Choose	Identify	Operate	

A clear purpose statement will lead to a clear thesis statement. As you remember from Chapter 12, a thesis statement presents the central idea of your speech. Sometimes your thesis statement for an informative speech will just preview the central idea:

> *Today's smartphones have so many features that it is difficult for the uninformed consumer to make a choice.*
>
> *Understanding how memes go viral could make you very wealthy someday.*
>
> *Soldiers and civilians have different views on the morality of drones.*

At other times, the thesis statement for an informative speech will delineate the main points of that speech:

> *When shopping for a smartphone, the informed consumer seeks to balance price, dependability, and user friendliness.*

informative purpose statement A complete statement of the objective of a speech, worded to stress audience knowledge and/or ability.

The four basic principles of aerodynamics—lift, thrust, drag, and gravity—can explain why memes go viral.

Drones can save warrior lives but cost the lives of civilians.

Setting a clear informative purpose will help keep you focused as you prepare and present your speech.

Create Information Hunger

An effective informative speech creates **information hunger**: a reason for your audience members to want to listen to and learn from your speech. To do so, you can use the analysis of communication functions discussed in Chapter 1 as a guide. You read there that communication of all types helps us meet our physical needs, identity needs, social needs, and practical needs. In informative speaking, you could tap into your audience members' physical needs by relating your topic to their survival or to the improvement of their living conditions. If you gave a speech on food (eating it, cooking it, or shopping for it), you would be dealing with that basic physical need. In the same way, you could appeal to identity needs by showing your audience members how to be respected—or simply by showing them that you respect them. You could relate to social needs by showing them how your topic could help them be well liked. Finally, you can relate your topic to practical audience needs by telling your audience members how to succeed in their courses, their job search, or their quest for the perfect outfit.

> **information hunger**
> Audience desire, created by a speaker, to learn information.

Make It Easy to Listen

Keep in mind the complex nature of listening, discussed in Chapter 5, and make it easy for your audience members to hear, pay attention, understand, and remember. This means first that you should speak clearly and with enough volume to be heard by all your listeners. It also means that as you put your speech together, you should take into consideration techniques that recognize the way human beings process information.

Limit the Amount of Information You Present Remember that you probably won't have enough time to transmit all your research to your audience in one sitting. It's better to make careful choices about the three to five main ideas you want to get across and then develop those ideas fully. Remember, too much information leads to overload, anxiety, and lack of attention on the part of your audience.

Use Familiar Information to Increase Understanding of the Unfamiliar Move your audience members from familiar information (on the basis of your audience analysis) to your newer information. For example, if you are giving a speech about how the stock market works, you could compare the daily activity of a broker with that of a salesperson in a retail store, or you could compare the idea of capital growth (a new concept to some listeners) with interest earned in a savings account (a more familiar concept).

Use Simple Information to Build Up Understanding of Complex Information Just as you move your audience members from the familiar to the unfamiliar, you can move them from the simple to the complex. An average college audience, for example, can understand the complexities of genetic modification if you begin with the concept of inherited characteristics.

In the sample speech at the end of this chapter, Adam Grant uses simple, familiar examples to explain the psychology of "escalation of commitment to a losing course of action":

> Escalation of commitment explains why you might have stuck around too long in a miserable job, why you've probably waited for a table way too long at a restaurant and why you might have hung on to a bad relationship long after your friends encouraged you to leave.

Use Clear, Simple Language

Another technique for effective informative speaking is to use clear language, which means using precise, simple wording and avoiding jargon. As you plan your speech, consult online dictionaries such as Dictionary.com to make sure you are selecting precise vocabulary. Remember that picking the right word seldom means using a word that is unfamiliar to your audience; in fact, just the opposite is true. Important ideas do not have to sound complicated. Along with simple, precise vocabulary, you should also strive for direct, short sentence structure. For example, when Warren Buffett, one of the world's most successful investors, wanted to explain the impact of taxes on investing, he didn't use unusual vocabulary or complicated sentences. He explained it like this:

> Suppose that an investor you admire and trust comes to you with an investment idea. "This is a good one," he says enthusiastically. "I'm in it, and I think you should be, too." Would your reply possibly be this? "Well, it all depends on what my tax rate will be on the gain you're saying we're going to make. If the taxes are too high, I would rather leave the money in my savings account, earning a quarter of 1 percent." So let's forget about the rich and ultrarich going on strike and

---

COMMUNICATION STRATEGIES

Techniques of Informative Speaking

- ☐ **Define a specific informative purpose.**
- ☐ **Create information hunger by relating to audience needs.**
- ☐ **Make it easy for audience members to listen.**
  - Limit the amount of information presented.
  - Use familiar information to introduce unfamiliar information.
  - Start with simple information before moving to more complex ideas.
- ☐ **Use clear, simple language.**
- ☐ **Use clear organization and structure.**
- ☐ **Support and illustrate your points.**
  - Provide interesting, relevant facts and examples, citing your sources.
  - Use visual aids that help make your points clear, interesting, and memorable.
- ☐ **Emphasize important points.**
  - Repeat key information in more than one way.
  - Use signposts: words or phrases that highlight what you are about to say.
- ☐ **Generate audience involvement.**
  - Personalize the speech.
  - Use audience participation.
  - Use volunteers.
  - Have a question-and-answer period at the end.

stuffing their ample funds under their mattresses if—gasp—capital gains rates and ordinary income rates are increased. The ultrarich, including me, will forever pursue investment opportunities.[3]

Each idea within that explanation is stated directly, using simple, clear language.

Use a Clear Organization and Structure

Because of the way humans process information (that is, in a limited number of chunks at any one time),[4] organization is extremely important in an informative speech. Rules for structure may be mere suggestions for other types of speeches, but for informative speeches they are ironclad.

Chapter 12 discusses some of these rules:

- Limit your speech to three to five main points.
- Divide, coordinate, and order those main points.
- Use a strong introduction that previews your ideas.
- Use a conclusion that reviews your ideas and makes them memorable.
- Use transitions, internal summaries, and internal previews.

The repetition that is inherent in strong organization will help your audience members understand and remember those points. This will be especially true if you use a well-organized introduction, body, and conclusion.

The Introduction The following principles of organization from Chapter 12 become especially important in the introduction of an informative speech:

1. Establish the importance of your topic to your audience.
2. Preview the thesis, the one central idea you want your audience to remember.
3. Preview your main points.

For example, Kevin Allocca, the trends manager at YouTube, began his TED Talk "Why Videos Go Viral" with the following introduction:

> I professionally watch YouTube videos. It's true. [Attention getter.] So we're going to talk a little bit today about how videos go viral and then why that even matters. [Previews the thesis that videos go viral for specific reasons, that can be emulated.] Web video has made it so that any of us or any of the creative things that we do can become completely famous in a part of our world's culture. Any one of you could be famous on the Internet by next Saturday. [Importance of topic to audience.] But there are over 48 hours of video uploaded to YouTube every minute. And of that, only a tiny percentage ever goes viral and gets tons of views and becomes a cultural moment. So how does it happen? Three things: tastemakers, communities of participation, and unexpectedness.[5] [Previews main points]

The Body In the body of an informative speech, the following organizational principles take on special importance:

1. Limit your division of main points to three to five subpoints.
2. Use transitions, internal summaries, and internal previews.
3. Order your points in the way that they will be easiest to understand and remember.

Kevin Allocca followed these principles for organizing his speech on why some videos go viral and some do not. He developed his speech with the following three main points:

1. Tastemakers: Tastemakers like Jimmy Kimmel introduce us to new and interesting things and bring them to a larger audience.
2. Communities of participation: A community of people who share this big inside joke start talking about it and doing things with it.
3. Unexpectedness: In a world where more than two days of video get uploaded every minute, only those that are truly unique can go viral.

The Conclusion Organizational principles are also important in the conclusion of an informative speech:

1. Review your main points.
2. Remind your audience members of the importance of your topic to them.
3. Provide your audience with a memory aid.

This is how Kevin Allocca concluded his speech on viral videos:

> Tastemakers, creative participating communities, complete unexpectedness, [Review of main points] these are characteristics of a new kind of media and a new kind of culture where anyone has access and the audience defines the popularity. [Reminds audience of importance of the topic to them.] One of the biggest stars in the world right now . . . got his start on YouTube. [Memory aid.] No one has to green-light your idea. And we all now feel some ownership in our own pop culture. And these are not characteristics of old media, and they're barely true of the media of today, but they will define the entertainment of the future.

14.4 Using Supporting Material Effectively

Another technique for effective informative speaking has to do with the supporting material discussed in Chapter 12. All of the purposes of support (to clarify, to prove, to make interesting, to make memorable) are essential to informative speaking. Therefore, you should be careful to support your thesis in every way possible. Notice the way Adam Grant uses solid supporting material like visual aids in the sample speech at the end of this chapter.

Emphasizing Important Points

One specific principle of informative speaking is to stress the important points in your speech through repetition and the use of signposts.

Repetition Repetition is one of the age-old rules of learning. Human beings are more likely to comprehend information that is stated more than once. This is especially true in a speaking situation, because, unlike a written paper, your audience members cannot go back to reread something they have missed. If their minds have wandered the first time you say something, they just might pick it up the second time.

Of course, simply repeating something in the same words might bore the audience members who actually are paying attention, so effective speakers learn to say the same thing in more than one way, sometimes by adding an additional example. Shayla Cabalan, a student at the University of Indianapolis, used this technique in her speech on child marriage:

> Parents who marry off their minor children are often motivated by the need to control sexuality and unwanted behavior, prevent unsuitable relationships, protect family honor and perceived religious or cultural ideals, or achieve financial gain. The *Washington Post* of February 10, 2017, details the story of 15-year-old Sara Siddiqui of Nevada, whose father married her off to a man 13 years her senior, simply because she was seeing someone of a different cultural background.[6]

Redundancy can be effective when you use it to emphasize important points.[7] It is ineffective only when (1) you are redundant with obvious, trivial, or boring points or (2) you run an important point into the ground. There is no sure rule for making certain you have not overemphasized a point. You just have to use your best judgment to make sure that you have stated the point enough that your audience members get it without repeating it so often that they want to give it back.

Signposts Another way to emphasize important material is by using **signposts**: words or phrases that emphasize the importance of what you are about to say. You can state, simply enough, "What I'm about to say is important," or you can use some variation of that statement: "But listen to this . . . ," or "The most important thing to remember is . . . ," or "The three keys to this situation are . . . ," and so on.

signpost A phrase that emphasizes the importance of upcoming material in a speech.

Generating Audience Involvement

The final technique for effective informative speaking is to get your audience involved in your speech. **Audience involvement** is the level of commitment and attention that listeners devote to a speech. Educational psychologists have long known that the best way to teach people something is to have them do it; social psychologists have added to this rule by proving, in many studies, that involvement in a message increases audience comprehension of, and agreement with, that message.

audience involvement The level of commitment and attention that listeners devote to a speech.

There are many ways to encourage audience involvement in your speech. One way is to follow the rules for good delivery by maintaining enthusiasm, energy, eye contact, and so on. Other ways include personalizing your speech, using audience participation, using volunteers, and having a question-and-answer period.

Personalize Your Speech One way to encourage audience involvement is to give audience members a human being to connect to. In other words, don't be afraid to be yourself and to inject a little of your own personality into the speech. If you happen to be good at storytelling, make a narration part of your speech. If humor is a personal strength, be funny. If you feel passion about your topic, show it. Certainly, if you have any experience that relates to your topic, use it.

In this chapter's sample speech, Adam Grant uses several experiences from his life to personalize his speech. He tells stories about mountain climbing in Panama (a near-death experience) and his attempts at various high school sports. Both stories illustrate his ideas about his topic.

Another way to personalize your speech is to link it to the experience of audience members . . . maybe even naming one or more.

audience participation
Listener activity during a speech; a technique to increase audience involvement.

Use Audience Participation Having your listeners actually do something during your speech—**audience participation**—is another way to increase their involvement in your message. For example, if you were giving a demonstration on isometric exercises (which don't require too much room for movement), you could have the entire audience stand up and do one or two sample exercises. If you were explaining how to fill out a federal income tax form, you could give each class member a sample form to fill out as you explain it. Outlines and checklists can be used in a similar manner for just about any speech.

Use Volunteers Some points or actions are more easily demonstrated with one or two volunteers. Selecting volunteers from the audience will increase the psychological involvement of all audience members because they will tend to identify with the volunteers.

Kathryn Schulz, author of the best-seller *Being Wrong*, subtly enlisted volunteers when she wanted to impress an important point on her audience. She began by addressing a rhetorical question to her entire audience but then directed it to a few individuals in the front row:

> So let me ask you guys something—or actually, let me ask you guys something, because you're right here: How does it feel—emotionally—how does it feel to be wrong?

Schulz then listened to the responses and repeated them for the rest of the audience:

> Dreadful. Thumbs down. Embarrassing. . . . Thank you, these are great answers, but they're answers to a different question.
>
> You guys are answering the question: How does it feel to *realize* you're wrong? When we're wrong about something—not when we realize it, but before that—*it feels like being right*.[8]

Have a Question-and-Answer Period One way to increase audience involvement that is nearly always appropriate if time allows is to answer questions at the end of your speech. You should encourage your audience to ask questions. Solicit questions and be patient waiting for the first one. Often no one wants to ask the first question. When the questions do start coming, the following suggestions might increase your effectiveness in answering them:

1. Listen to the substance of the question. Don't zero in on irrelevant details; listen for the big picture, the basic, overall question that is being asked. If you are not really sure what the substance of a question is, ask the questioner to paraphrase it. Don't be afraid to let the questioners do their share of the work.

2. Paraphrase confusing or quietly asked questions. Use the active listening skills described in Chapter 5. You can paraphrase the question in just a few words: "If I understand your question, you are asking _____. Is that right?"

Rubes® By Leigh Rubin

"In order to adequately demonstrate just how many ways there are to skin a cat, I'll need a volunteer from the audience."

Source: By permission of Leigh Rubin and Creators Syndicate, Inc.

3. Avoid defensive reactions to questions. Even if the questioner seems to be calling you a liar or stupid or biased, try to listen to the substance of the question and not to the possible personal attack.

4. Answer the question briefly. Then check the questioner's comprehension of your answer by observing his or her nonverbal response or by asking, "Does that answer your question?"

Using Visual Aids

Visual aids are graphic devices used in a speech to illustrate or support ideas. Although they can be used in any type of speech, they are especially important in informative speeches. For example, they can be extremely useful when you want to show how things look (photos of your trek to Nepal or the effects of malnutrition) or how things work (a demonstration of a new ski binding or a diagram of how seawater is made drinkable). Visual aids can also show how things relate to one another (a graph showing the relationships among gender, education, and income).

There is a wide variety of types of visual aids. The most common types include the following.

Objects and Models Sometimes the most effective visual aid is the actual thing you are talking about. This is true when the thing you are talking about is easily portable and simple enough to use during a demonstration before an audience (e.g., a lacrosse racket). **Models** are scaled representations of the object you are discussing and are used when that object is too large (the new campus arts complex) or too small (a DNA molecule) or simply doesn't exist anymore (a *Tyrannosaurus rex*).

Photos, Videos, and Audio Files Speaking venues, including college classrooms, are often set up with digital projectors connected to online computers, DVD players, and outlets for personal laptops. In these cases, a wealth of photos, video clips, and audio files are available to enhance your speech. If you can project photos, you can use them to help explain geological formations, for example, or underwater habitats. Videos can be used to play a visual **sound bite** (a brief recorded excerpt) from an authoritative source, or to show a process such as plant growth. Using audio files can help you compare musical styles or demonstrate the differences in the sounds of gas and diesel engines.

Photos can stay up as long as you are referring to them, but videos and audio files should be used sparingly because they allow audience members to receive information passively. You want your audience to actively participate in the presentation. The general rule when using videos and sound files is *Don't let them get in the way of the direct, person-to-person contact that is the primary advantage of public speaking.*

Videos and sound files should be carefully introduced, controlled, and summarized at the end. They should also be very brief, and they should include what media

> **visual aids** Graphic devices used in a speech to illustrate or support ideas.
>
> **model (in speeches and presentations)** A replica of an object being discussed. It is usually used when it would be difficult or impossible to use the actual object.
>
> **sound bite** A brief recorded excerpt from a longer statement.

Sometimes everyday objects can be used to illustrate ideas.
If you were comparing the health benefits of an apple and a cheeseburger, would these objects be effective visual aids?

clip wraparound A brief introduction before a visual aid is presented, accompanied by a brief conclusion afterward.

clip intro A brief explanation or comment before a visual aid is used.

clip outro A brief summary or conclusion after a visual aid has been used.

diagram A line drawing that shows the most important components of an object.

word chart A visual aid that lists words or terms in tabular form in order to clarify information.

number chart A visual aid that lists numbers in tabular form to clarify information.

professionals call **clip wraparounds** or **clip intros** and **clip outros**: careful introductions and closing summaries of each clip. Using these devices enables the speakers to remain in control of the message. Adam Grant uses extensive clips and photos in his sample speech at the end of this chapter.

Other types of visual aids include charts and graphs for representing facts and data. They include the following.

Diagrams   A **diagram** is any kind of line drawing that shows the most important properties of an object. Diagrams do not try to show everything but just those parts of a thing that the audience most needs to be aware of and understand. Blueprints and architectural plans are common types of diagrams, as are maps and organizational charts. A diagram is most appropriate when you need to simplify a complex object or phenomenon and make it more understandable to the audience. Figure 14.1 is a diagram that shows the progression of different types of "connecting the dots" in inference-making. (In a speech you would help your audience follow along by first showing the "Data" figure and then steadily adding subsequent images.)

Word and Number Charts   **Word charts** and **number charts** are visual depictions of key facts or statistics. Figure 14.2 combines both words and numbers to

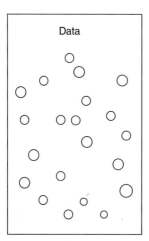
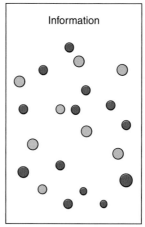
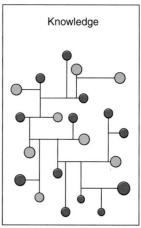
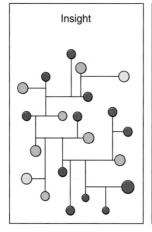
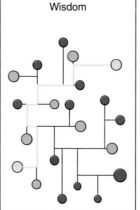

FIGURE 14.1 Diagram

14.4 Using Supporting Material Effectively

demonstrate how many ancestors every person has. (Again, you would start with the "2 parents" line and gradually add the following ones to help your audience appreciate the point.) Your audience will understand and remember facts and numbers like these better if you show them than if you just talk about them. Many speakers arrange the main points of their speech, often in outline form, as a word chart. Other speakers list their main statistics. An important guideline for word and number charts is, *Don't read them to your audience; use them to enhance what you are saying and help your audience remember key points.*

Pie Charts Circular graphs with dividing wedges are known as **pie charts**. They are used to show divisions of any whole: where your tax dollars go, the percentage of the population involved in various occupations, and so on. Pie charts are often made up of percentages that add up to 100. Usually, the wedges of the pie are organized from largest to smallest. The pie chart in Figure 14.3 represents one person's perception of "how cartoon princesses spend their time," and Figure 14.4 shows how the U.S. government adapted a pie chart for a new nutrition diagram. Coincidentally, Figure 14.4 is also a **pictogram**, which is a visual aid that conveys its meaning through images of an actual object.

Ancestral Mathematics
In order to be born, you needed:

2 parents
4 grandparents
8 great-grandparents
16 second great-grandparents
32 third great-grandperents
64 fourth great-grandperents
128 fifth great-grandperents
256 sixth great-grandperents
512 seventh great-grandperents
1,024 eighth great-grandperents
2,048 ninth great-grandperents

For you to be born today from 12 previous generations. you needed a total of 4,094 ancestors over the last 400 years.

FIGURE 14.2 Word and Number Chart

> **pie chart** A visual aid that divides a circle into wedges, representing percentages of the whole.

> **pictogram** A visual aid that conveys its meaning through an image of an actual object.

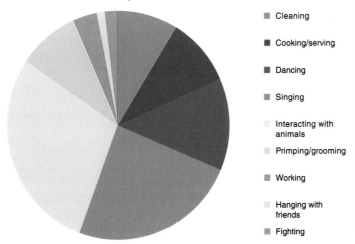

FIGURE 14.3 Pie Chart

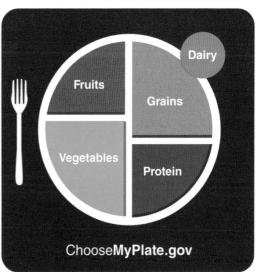

FIGURE 14.4 Adaptation of Pie Chart

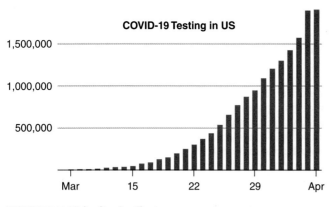

FIGURE 14.5 Misleading Bar Chart
Source: Presidential press conference, April 2020

bar chart A visual aid that compares two or more values by showing them as elongated horizontal rectangles.

column chart A visual aid that compares two or more values by showing them as elongated vertical rectangles.

line chart A visual aid consisting of a grid that maps out the direction of a trend by plotting a series of points.

flow chart A diagram that depicts the steps in a process with shapes and arrows.

Bar and Column Charts Figures 14.5 and 14.6 are both **bar charts**, a type of chart that compares two or more values by stretching them out in the form of horizontal rectangles. **Column charts** perform the same function as bar charts but use vertical rectangles.

Line Charts A **line chart** maps out the direction of a moving point; it is ideally suited for showing changes over time. The time element is usually placed on the horizontal axis so that the line visually represents the trend over time. Figure 14.7 is a line chart that maps out how tactful people become as they mature.

Flow Charts A **flow chart** is a diagram that depicts the steps in a process with shapes and arrows. Often, it branches according to yes/no decisions (if this, then this). Figure 14.8 represents one way to explain "mansplaining."

The Ethics of Charts and Graphs Ethical communication requires speakers to make sure that each visual aid they use accurately represents the points they are claiming. For example, at a press briefing in 2020, members of the White House Coronavirus Task Force presented a graph titled "COVID-19 Testing in the U.S." (see Figure 14.5) to illustrate the point that testing was "going up at a rapid rate."

This graph seems to support the assertion. However, it actually shows the total cumulative number of tests performed over months, not the number of new tests each day. Figure 14.6 shows the same graph with the number of new tests in a separate color, making it clear that the number of COVID-19 tests performed between March and April did increase, but not rapidly.

You need to consider graphic design in this way, and you also need to consider where and how the data were collected. For example, when *Fox Business News* host Lou Dobbs aired a chart about President Trump's handling of COVID-19 (Figure 14.9), he

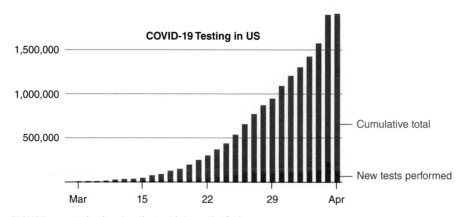

FIGURE 14.6 Misleading Bar Chart with Data Clarified
Source: Our World in Data, April 2020, https://ourworldindata.org/coronavirus-testing

14.4 Using Supporting Material Effectively 399

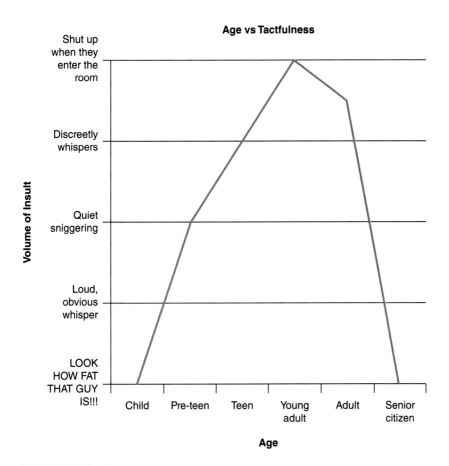

FIGURE 14.7 Line Chart
Source: Adapted from GraphJam

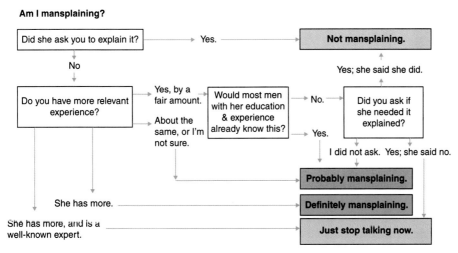

FIGURE 14.8 Flow Chart: Am I Mansplaining?

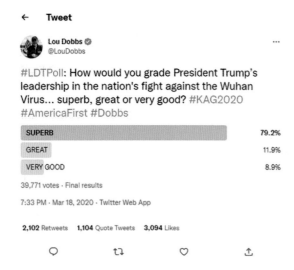

FIGURE 14.9 Bar Chart with Only Positive Options

gave his TV and social media audiences only three options: "Superb," "Great," and "Very Good." The resulting chart was very impressive, and very misleading. Experts in data visualization explain that this example is just "one of many when important information was not properly understood or well communicated."[9]

Using Presentation Software

Several specialized programs exist just to produce visual aids. Among the most popular of these programs are Microsoft PowerPoint, Apple's Keynote, and Prezi.

In its simplest form, presentation software lets you build an effective slide show out of your basic outline. You can choose color-coordinated backgrounds and consistent formatting that match the tone and purpose of your presentation. Most presentation software programs contain a clip art library that allows you to choose images to accompany your words. They also allow you to import images from outside sources and to build your own charts.

If you would like to learn more about using PowerPoint, Keynote, and Prezi, you can easily find several web-based tutorial programs by typing the name of your preferred program into your favorite search engine.

Alternative Media for Presenting Graphics

When a digital projector and/or screen are unavailable, you can use other methods to present visual aids.

Chalkboards, Whiteboards, and Polymer Marking Surfaces The major advantage of these write-as-you-go media is their spontaneity. With them you can create your visual aid as you speak, including items generated from audience responses. Along with the odor of whiteboard markers and the squeaking of chalk, a major disadvantage of these media is the difficulty of preparing visual aids on them in advance, especially if several speeches are scheduled in the same room at the same hour.

Flip Pads and Poster Board Flip pads are like oversized writing tablets attached to a portable easel. Flip pads enable you to combine the spontaneity of the chalkboard (you can write on them as you go) with portability, which enables you to prepare them in advance. If you plan to use your visuals more than once, you can prepare them in advance on rigid poster board and display them on the same type of easel.

Despite their advantages, flip pads and poster boards are bulky, and preparing professional-looking exhibits on them requires a fair amount of artistic ability.

Handouts The major advantage of handouts is that audience members can take away the information they contain after your speech. For this reason, handouts are excellent memory and reference aids. The major disadvantage is that they are distracting when handed out during a speech: First, there is the distraction of passing them out, and second, there is the distraction of having them in front of the audience members while you have gone on to something else. It's best, therefore, to pass them out at the end of the speech, so that audience members can use them as take-aways.

Rules for Using Visual Aids

It's easy to see that each type of visual aid and each medium for its presentation have their own advantages and disadvantages. No matter which type you use, however, there are a few rules to follow.

Simplicity Keep your visual aids simple. Your goal is to clarify, not confuse. Use only key words or phrases, not sentences. The "rule of seven" states that each exhibit you use should contain no more than seven lines of text, each with no more than seven words. Keep all printing horizontal. Omit all nonessential details.

Size Visual aids should be large enough for your entire audience to see them at one time but portable enough for you to get them out of the way when they no longer pertain to the point you are making.

Attractiveness Visual aids should be visually interesting and as neat as possible. If you don't have the necessary artistic or computer skills, try to get help from a friend or at the computer or audiovisual center on your campus.

Appropriateness Visuals must be appropriate to all the components of the speaking situation—you, your audience, and your topic—and they must emphasize the

---

COMMUNICATION STRATEGIES

The Pros and Cons of Presentation Software

PowerPoint is by far the most popular form of work presentation today.[11] As with any software, however, it is not without its drawbacks.

The Pros

Proponents say that PowerPoint slides can focus the attention of audience members on important information at the appropriate time. In addition, the slides can help listeners appreciate the relationship between different pieces of information. The software may also help speakers organize their thoughts in advance. But its primary benefit may be in providing a second, visual channel of information. One psychology professor puts it this way: "If you zone out for 30 seconds—and who doesn't?—it is nice to be able to glance up on the screen and see what you missed."[12]

The Cons

For all its popularity, PowerPoint has received some bad press.[13,14] One particularly strong criticism came from Edward R. Tufte, an influential author of several books on the effective design of visual aids.[15] According to Tufte, the use of low-content PowerPoint slides trivializes important information. It encourages oversimplification by asking the presenter to summarize key concepts in as few words as possible—the ever-present bullet points.

Tufte also insists that PowerPoint makes it easier for a speaker to hide lies and logical fallacies.

Perhaps most seriously, opponents of PowerPoint say it is an enemy of interaction. One expert argued, "Instead of human contact, we are given human display."[16]

The Middle Ground

PowerPoint proponents say that it is just a tool, one that can be used effectively or ineffectively. They are the first to admit that a poorly done PowerPoint presentation can be boring and ineffective. In the infamous "triple delivery," the same text is seen on the screen, spoken aloud, and printed on the handout in front of you. Effective speakers know that PowerPoint should not be allowed to overpower a presentation—it should be just one element, not the whole thing.

point you are trying to make. Don't make the mistake of using a visual aid that looks good but has only a weak link to the point you want to make—such as showing a map of a city transit system while talking about the condition of the individual cars.

Reliability   You must be in control of your visual aid at all times. Test all electronic media (projectors, computers, and so on) in advance, preferably in the room where you will speak. Just to be safe, have nonelectronic backups ready in case of disaster. Be conservative when you choose demonstrations: Wild animals, chemical reactions, and gimmicks meant to shock a crowd can often backfire.

When it comes time for you to use the visual aid, remember one more point: Talk to your audience, not to your visual aid. Some speakers become so wrapped up in their props that they turn their backs on their audience and sacrifice all their eye contact.

14.5 Sample Speech

This speech was presented by Adam Grant, whose profile opened this chapter. He presented it at a 2021 TED Institute event that was given in partnership with Boston Consulting Group, a management consulting firm. At more than 15 minutes, it is considerably longer than the typical classroom speech, and, because it was presented during the pandemic era, it has the advantage of video post-production.

Grant was talking to a diverse audience, including business executives, professors, students and others who were interested in new ideas, as well as more than a million viewers who would eventually see the online version of the talk.

Grant's speech provides the explanation of a concept that is at once highly technical and extremely common: Thinking, or more specifically, rethinking. He uses a collection of well-chosen examples, graphs, video clips and even a magic trick as supporting material.

His informative purpose statement might be:

After listening to my speech, my audience will be able to identify and cite examples of failures in rethinking.

His thesis statement could be worded:

Failing to rethink is like reacting to a slow-boiling pot: We don't realize emerging problems until it's too late.

He organizes and structures his speech carefully. His outline would look like this (numbers in parenthesis correspond to paragraphs of the speech):

INTRODUCTION (1–3)

  I. Attention-Getter: Frog myth (1)
 II. Thesis Statement: Our world is full of slow-boiling pots. (2)
III. Preview of Main Points (3)

BODY

I. Why We Don't Rethink Effectively (4–17)

    A. One cause is the "'I'm Not Biased' Bias." (4)

    B. Another cause is "Escalation of Commitment to a Losing Course of Action." (5–7)

        1. We make a bad choice, but instead of rethinking it, we double down. (7)

        2. We tell ourselves, "If I just try harder, I can turn this around." (8–9)

        3. We say, "Never give up," but that doesn't mean "Keep doing the thing that's failing." (10)

    C. Another cause is "Identity Foreclosure." (11–15)

        1. Your concept of identity can cause a kind of shortsightedness. (11)

        2. Opening your mind to new identities can open new doors. (14–15)

    D. Another cause is "Cognitive Entrenchment." (16–17)

        1. You can get stuck in the way you've always done things.

        2. Emotion regulation: As you gain perspective, you can rethink and revise what you feel.

II. Solutions to our Failures to Rethink (18–26)

    A. Regulate Emotions (18–19)

    B. Do a Checkup (20)

        1. Career Checkup (20)

        2. Relationship Checkup (20)

        3. Identity Checkup (20)

    C. Be Open to Criticism (21–24)

    D. Be Confidently Humble (2–26)

        1. Be secure enough in your strengths to acknowledge your weaknesses. (26)

        2. Believe that the best way to prove yourself is to improve yourself. (26)

        3. Know that weak leaders silence their critics and make themselves weaker. (26)

        4. Know that strong leaders engage their critics and make themselves stronger. (26)

        5. Say "I don't know" instead of pretending to have all the answers. (26)

        6. Say "I was wrong," instead of insisting you were right. (26)

        7. Listen to ideas that make you think hard. (26)

        8. Surround yourself with people who challenge your thought process. (26)

CONCLUSION (27–28)

Final Remarks: Frogs in hot water is a myth. (27)
Review Main Idea: What's not a myth is that we have to be quick to rethink. (28)

Throughout his speech, Grant makes it easy for his audience to listen by using clear, simple language, and by personalizing his speech with humor and stories about his own life. It serves as a model for your own speeches because it demonstrates a wide range of informative techniques. You'll be able to pick and choose those that are best for your own presentations.

SAMPLE SPEECH — Adam Grant

What Frogs in Hot Water Can Teach Us About Thinking Again[10]

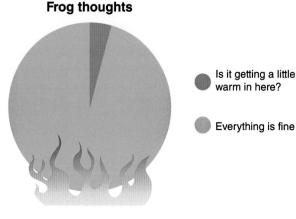

FIGURE 14.10 Frog Thoughts

1. You might have heard that if you drop a frog in a pot of boiling water, it will jump out right away, but if you put it in lukewarm water, and then slowly heat it up, the frog won't survive. The frog's big problem is that it lacks the ability to rethink the situation. It doesn't realize that the warm bath is becoming a death trap—until it's too late.

Attention-Getter
This diagram is shown quickly, giving the audience just enough time to register the idea.

2. Humans might be smarter than frogs, but our world is full of slow-boiling pots. Think about how slow people were to react to warnings about a pandemic, climate change or a democracy in peril. We fail to recognize the danger because we're reluctant to rethink the situation.

Statement of thesis, with examples to illustrate the idea.

3. We struggle with rethinking in all kinds of situations. We expect our squeaky brakes to keep working, until they finally fail on the freeway. We believe the stock market will keep going up, even after we hear about a real-estate bubble. And we keep watching "Game of Thrones" even after the show jumps the shark. Rethinking isn't a hurdle in every part of our lives. We're happy to refresh our wardrobes and renovate our kitchens. But when it comes to our goals, identities and habits, we tend to stick to our guns. And in a rapidly changing world, that's a huge problem.

Jumping the shark: When a TV series runs out of ideas for episodes.

4. I'm an organizational psychologist. It's my job to rethink how we work, lead and live. But that hasn't stopped me from getting stuck in slow-boiling pots, so I started studying why. I learned that intelligence doesn't help us escape; sometimes, it traps us longer. Being good at thinking can make you worse at rethinking. There's evidence that the smarter you are, the more likely you are to fall victim to the "I'm not biased" bias. You can always find reasons to convince yourself you're on the right path, which is exactly what my friends and I did on a trip to Panama.

He establishes his credibility in terms of both professional and personal experience.

5. I worked my way through college, and by my junior year, I'd finally saved enough money to travel. It was my first time leaving North America. I was excited for my first time climbing a mountain, actually an active volcano, literally a slow-boiling pot. I set a goal to reach the summit and look into the crater.

Setting up the story...

6. So, we're in Panama, we get off to a late start, but it's only supposed to take about two hours to get to the top. After four hours, we still haven't reached the top. It's a little strange that it's taking so long, but we don't stop to rethink whether we should turn around. We've already come so far. We have to make it to the top. Do not stand between me and my goal. We don't realize we've read the wrong map. We're on Panama's highest mountain, it actually takes six to eight hours to hike to the top. By the time we finally reach the summit, the sun is setting. We're stranded, with no food, no water, no cell phones, and no energy for the hike down.

He tells his story with rising tension, in traditional story-telling form.

7. There's a name for this kind of mistake. It's called "escalation of commitment to a losing course of action." It happens when you make an initial investment of time or money, and then you find out it might have been a bad choice, but instead of rethinking it, you double down and invest more. You want to prove to yourself and everyone else that you made a good decision. Escalation of commitment explains so many familiar examples of businesses plummeting. Blockbuster, BlackBerry, Kodak. Leaders just kept simmering in their slow-boiling pots, failing to rethink their strategies. Escalation of commitment explains why you might have stuck around too long in a miserable job, why you've probably waited for a table way too long at a restaurant and why you might have hung on to a bad relationship long after your friends encouraged you to leave.

Definition of key term, customized for both his general and business audiences.
His examples are also customized for his audiences.

8. It's hard to admit that we were wrong and that we might have even wasted years of our lives. So we tell ourselves, "If I just try harder, I can turn this around."

9. We live in a culture that worships at the altar of hustle and prays to the high priest of grit. But sometimes, that leads us to keep going when we should stop to think again. Experiments show that gritty people are more likely to overplay their hands in casino games and more likely to keep trying to solve impossible puzzles. My colleagues and I have found that NBA basketball coaches who are determined to develop the potential in rookies keep them around much longer than their performance justifies. And researchers have even suggested that the most tenacious mountaineers are more likely to die on expeditions, because they're determined to do whatever it takes to reach the summit.

FIGURE 14.11 Climbing a Mountain

10. In Panama, my friends and I got lucky. About an hour into our descent, a lone pickup truck came down the volcano and rescued us from our slow-boiling pot. There's a fine line between heroic persistence and stubborn stupidity. Sometimes the best kind of grit is gritting your teeth and packing your bags. "Never give up" doesn't mean "keep doing the thing that's failing." It means "don't get locked into one narrow path, and stay open to broadening your goals. The ultimate goal is to make it down the mountain, not just to reach the top. Your goals can give you tunnel vision, blinding you to rethinking the situation.

11 And it's not just goals that can cause this kind of shortsightedness, it's your identity too. As a kid, my identity was wrapped up in sports. I spent countless hours shooting hoops on my driveway, and then I got cut from the middle school basketball team, all three years. I spent a decade playing soccer, but I didn't make the high school team. At that point, I shifted my focus to a new sport, diving. I was bad, I walked like Frankenstein, I couldn't jump, I could hardly touch my toes without bending my knees, and I was afraid of heights.

12 But I was determined. I stayed at the pool until it was dark, and my coach kicked me out of practice. I knew that the seeds of greatness are planted in the daily grind, and eventually, my hard work paid off. By my senior year, I made the All-American list, and I qualified for the Junior Olympic Nationals. I was obsessed with diving. It was more than something I did, it became who I was. I had a diving sticker on my car, and my email address was "DiverAG@aol.com." Diving gave me a way to fit in and to stand out. I had a team where I belonged and a rare skill to share. I had people rooting for me and control over my own progress.

Short clips of home videos of Adam as a boy practicing diving and becoming successful at diving.

13 But when I got to college, the sport that I loved became something I started to dread. At that level, I could not beat more talented divers by outworking them. I was supposed to be doing higher dives, but I was still afraid of heights, and 6 AM practice was brutal. My mind was awake, but my muscles were still asleep. I did back smacks and belly flops and my slow-boiling pot this time was a freezing pool. There was one question, though, that stopped me from rethinking. "If I'm not a diver, who am I?" In psychology, there's a term for this kind of failure to rethink—it's called "identity foreclosure." It's when you settle prematurely on a sense of who you are and close your mind to alternative selves.

14 You've probably experienced identity foreclosure. Maybe you were too attached to an early idea of what school you'd go to, what kind of person you'd marry, or what career you'd choose. Foreclosing on one identity is like following a GPS that gives you the right directions to the wrong destination. After my freshman year of college, I rethought my identity. I realized that diving was a passion, not a purpose. My values were to grow and excel, and to contribute to helping my teammates grow and excel. Grow, excel, contribute. I didn't have to be a diver to grow, excel and contribute.

15 Research suggests that instead of foreclosing on one identity, we're better off trying on a range of possible selves. Retiring from diving freed me up to spend the summer doing psychology research and working as a diving coach. It also gave me time to concentrate on my dorkiest hobby, performing as a magician. I'm still working on my sleight of hand. Opening my mind to new identities opened new doors. Research showed me that I enjoyed creating knowledge, not just consuming it. Coaching and performing helped me see myself as a teacher and an entertainer. If that hadn't happened, I might not have become a psychologist and a professor, and I probably wouldn't be giving this TED talk.

Here he demonstrates a card trick ... and here, he brings his story back to the present, and the topic of his speech.

16 See, I'm an introvert, and when I first started teaching, I was afraid of public speaking. I had a mentor, Jane Dutton, who gave me some invaluable advice. She said, "You have to unleash your inner magician." So I turned my class into a live show. Before the first day, I memorized my students' names and backgrounds, and then, I mastered my routine. Those habits served me well. I started to relax more and I started to get good ratings. But just like with goals and identities, the routines that help us today can become the ruts we get trapped in tomorrow.

(To watch him speak today, you'd be surprised that he had a fear of public speaking.)

17. One day, I taught a class on the importance of rethinking, and afterward, a student came up and said, "You know, you're not following your own principles." They say feedback is a gift, but right then, I wondered, "How do I return this?" I was teaching the same material, the same way, year after year. I didn't want to give up on a performance that was working. I had my act down. Even good habits can stand in the way of rethinking. There's a name for that too. It's called "cognitive entrenchment," where you get stuck in the way you've always done things. Just thinking about rethinking made me defensive. And then, I went through the stages of grief.

Stages of feedback grief

Denial	That's not true!
Anger	You're an idiot
Shame	I'm an idiot
Guilt	I was being unreasonable
Acceptance	Sometimes I have the self-control of a toddler
Growth	I'll do better next time

FIGURE 14.12 Stages of Grief

18. I happened to be doing some research on emotion regulation at the time, and it came in handy. Although you don't always get to choose the emotions you feel, you do get to pick which ones you internalize and which ones you express. I started to see emotions as works in progress, kind of like art. If you were a painter, you probably wouldn't frame your first sketch. Your initial feelings are just a rough draft. As you gain perspective, you can rethink and revise what you feel.

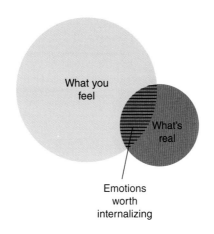

FIGURE 14.13 Venn Diagram

19. So that's what I did. Instead of defensiveness, I tried curiosity. I wondered, "What would happen if I became the student?" I threw out my plan for one day of class, and I invited the students to design their own session. The first year, they wrote letters to their freshman selves, about what they wish they'd rethought or known sooner. The next year, they gave passion talks. They each had one minute to share something they loved or cared about deeply. And now, all my students give passion talks to introduce themselves to the class. I believe that good teachers introduce new thoughts, but great teachers introduce new ways of thinking. But it wasn't until I ceded control that I truly understood how much my students had to teach one another, and me. Ever since then, I put an annual reminder in my calendar to rethink what and how I teach.

A sampling of student assignment cover pages.

FIGURE 14.14 Cover Pages

FIGURE 14.15 Word Cloud

Here, a word cloud of passion talk titles.

20 It's a checkup. Just as you go to the doctor for an annual checkup when nothing seems to be wrong, you can do the same thing in the important parts of your life. A career checkup to consider how your goals are shifting. A relationship checkup to re-examine your habits. An identity checkup to consider how your values are evolving. Rethinking does not have to change your mind—it just means taking time to reflect and staying open to reconsidering. A hallmark of wisdom is knowing when to grit and when to quit, when to throw in the towel on an old identity and dive into a new one, when to walk away from some old habits and start scaling a new mountain. Your past can weigh you down, and rethinking can liberate you.

21 Rethinking is not just a skill to master personally, it's a value we need to embrace culturally. We live in a world that mistakes confidence for competence, that pressures us to favor the comfort of conviction over the discomfort of doubt, that accuses people who change their minds of flip-flopping, when in fact, they might be learning. So let's talk about how to make rethinking the norm. We need to invite it and to model it.

22 A few years ago, some of our students at Wharton challenged the faculty to do that. They asked us to record our own version of Jimmy Kimmel's Mean Tweets. We took the worst feedback we'd ever received on student course evaluations, and we read it out loud.

23 *Angela Duckworth: "It was easily one of the worst three classes I've ever taken . . . one of which the professor was let go after the semester."*

Mohamed El-Erian: "The number of stories you tell give 'Aesop's Fables' a run for its money. Less can be more." Ouch.

Adam Grant: "You're so nervous you're causing us to physically shake in our seats."

Mae McDonnell: "So great to finally have a professor from Australia. You started strong but then got softer. You need tenure, so toughen up with these brats." I'm from Alabama.

Michael Sinkinson: "Prof Sinkinson acts all down with pop culture but secretly thinks Ariana Grande is a font in Microsoft Word."

Here, he shows clips of faculty members reading their Mean Tweets.

24 After I show these clips in class, students give more thoughtful feedback. They rethink what's relevant. They also become more comfortable telling me what to think, because I'm not just claiming I'm receptive to criticism. I'm demonstrating that I can take it. We need that kind of openness in schools, in families, in businesses, in governments, in nonprofits.

25 A couple of years ago, I was working on a project for the Gates Foundation, and I suggested that leaders could record their own version of Mean Tweets. Melinda Gates volunteered to go first, and one of the points of feedback that she read said "Melinda is like Mary effing Poppins. Practically perfect in every way." And then, she started listing her imperfections. People at the Gates Foundation who saw that video ended up becoming more willing to recognize and overcome their own limitations. They were also more likely to speak up about problems and solutions. What Melinda was modeling was confident humility.

Another example from his professional experiences serves as a transition to the next idea.

26 Confident humility is being secure enough in your strengths to acknowledge your weaknesses. Believing that the best way to prove yourself is to improve yourself, knowing that weak leaders silence their critics and make themselves weaker, while strong leaders engage their critics and make themselves stronger. Confident humility gives you the courage to say "I don't know," instead of pretending to have all the answers. To say "I was wrong," instead of insisting you were right. It encourages you to listen to ideas that make you think hard, not just the ones that make you feel good, and to surround yourself with people who challenge your thought process, not just the ones who agree with your conclusions. And sometimes, it even leads you to challenge your own conclusions, like with the story about the frog that can't survive the slow-boiling pot.

27 I found out recently that's a myth. If you heat up the water, the frog will jump out as soon as it gets uncomfortably warm. Of course it jumps out, it's not an idiot.

Again, he shows a quick diagram.

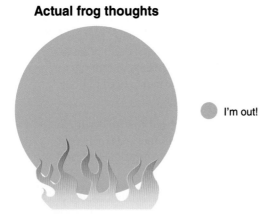

FIGURE 14.16 Actual Frog Thoughts

28 The problem is not the frog, it's us. Once we accept the story as true, we don't bother to think again. What if we were more like the frog, ready to jump out if the water gets too warm? We need to be quick to rethink.

Repeats his main idea with this memorable image.

Thank you.

MAKING THE GRADE

OBJECTIVE 14.1 Distinguish among the main types of informative speaking.

- Informative speeches can be classified by the type of content, including speeches about objects, processes, events, and concepts.
- Informative speeches can also be classified according to their purpose, including descriptions, explanations, and instructions.
  > Explain the primary differences among types of informative speeches.
  > Which type of informative speech is most important in your own life?
  > Which type of speech will you use for your next informative speech?

OBJECTIVE 14.2 Describe the differences between informative and persuasive speaking.

- Two basic characteristics differentiate an informative topic from a persuasive topic.
- In an informative speech, you generally do not present information that your audience is likely to disagree with.
- The speaker's primary intent is not to change attitudes or to make the audience members *feel* differently about the topic.
  > Explain the primary difference between informative and persuasive speaking.
  > How are informative and persuasive speeches similar?
  > Why is it important to distinguish between informative and persuasive speaking?

OBJECTIVE 14.3 Outline techniques that increase the effectiveness of informative speeches.

- Decide on a specific purpose statement.
- Create information hunger by tapping into audience needs.
- Make it easy to listen by limiting the amount of information and by using familiar information,

straightforward organization, clear language, and effective supporting material.

- Involve your audience through audience participation, the use of volunteers, and a question-and-answer period.
  > List at least three techniques for increasing informative effectiveness.
  > Which is the most important technique in informative speaking?
  > Which technique do you need to work on for your next speech?

OBJECTIVE **14.4** **Explain how you might use visual aids appropriately and effectively.**

- Visual aids include objects and models, photos, videos, audio files, charts, and graphs.
- Media for the presentation of visual aids include digital projectors with presentation software, chalkboards, whiteboards, flip pads, and handouts.
- Each type of visual aid has its own advantages and disadvantages.
- Keep visual aids simple, large enough for the audience to see, and visually interesting.
  > List at least three types of visual aids, giving examples of when to use them.
  > Describe the advantages and disadvantages of the visual aids on your list.
  > Which types of visual aids will you use for your next informative speech?

KEY TERMS

audience involvement p. 393
audience participation p. 394
bar chart p. 398
clip intro p. 396
clip outro p. 396
clip wraparound p. 396
column chart p. 398
description p. 386
diagram p. 396
explanations p. 386
flow chart p. 398
information anxiety p. 384
information hunger p. 389
informative purpose statement p. 388
instructions p. 386
line chart p. 398
model (in speeches and presentations) p. 395
number chart p. 396
pictogram p. 397
pie chart p. 397
signpost p. 393
sound bite p. 395
visual aids p. 395
word chart p. 396

PUBLIC SPEAKING PRACTICE

Prepare a brief, one-paragraph description of the visual aids that you would use for your next informative speech. Discuss your choices with your classmates.

ACTIVITIES

1. **Informative Purpose Statements** For practice in defining informative speech purposes, reword the following statements so that they specifically point out what the audience will be able to do after hearing the speech.
   a. My talk today is about building a wood deck.
   b. My purpose is to tell you about vintage car restoration.
   c. I am going to talk about toilet training.
   d. I'd like to talk to you today about sexist language.
   e. There are six basic types of machines.
   f. The two sides of the brain have different functions.
   g. Do you realize that many of you are sleep deprived?
2. **Effective Repetition** Create a list of three statements, or use the three that follow. Restate each of these ideas in three different ways.
   a. The magazine *Modern Maturity* has a circulation of more than 20 million readers.
   b. Before buying a used car, you should have it checked out by an independent mechanic.
   c. One hundred thousand pounds of dandelions are imported into the United States annually for medical purposes.
3. **Using Clear Language** For practice in using clear language, select an article from any issue of a professional journal in your major field. Using the suggestions in this chapter, rewrite a paragraph from the article so that it will be clear and interesting to a layperson.
4. **Inventing Visual Aids** Take any sample speech. Analyze it for where visual aids might be effective. Describe the visual aids that you think will work best. Compare the visuals you devise with those of your classmates.

Persuasive Speaking

CHAPTER OUTLINE

15.1 Characteristics of Persuasion 414
Persuasion Is Not Coercive
Persuasion Is Usually Incremental
Persuasion Is Interactive
Persuasion Can Be Ethical

15.2 Categorizing Persuasive Attempts 418
By Type of Proposition
By Desired Outcome
By Directness of Approach
By Type of Appeal: Aristotle's Ethos, Pathos, and Logos

15.3 Creating a Persuasive Message 422
Set a Clear Persuasive Purpose
Structure the Message Carefully
Use Solid Evidence
Avoid Fallacies

15.4 Adapting to the Audience 428
Establish Common Ground
Organize According to the Expected Response
Neutralize Potential Hostility

15.5 Building Credibility as a Speaker 430
Competence
Character
Charisma

15.6 Sample Speech 433

MAKING THE GRADE 437
KEY TERMS 438
PUBLIC SPEAKING PRACTICE 438
ACTIVITIES 439

LEARNING OBJECTIVES

15.1
Identify the primary characteristics of persuasion.

15.2
Compare and contrast different types of persuasion.

15.3
Apply the guidelines for persuasive speaking to a speech you will prepare.

15.4
Explain how to best adapt a specific speech to a specific audience.

15.5
Improve your credibility in your next persuasive speech.

CHAPTER 15 Persuasive Speaking

Saeed Malami. Photo courtesy of Saeed Malami.

Growing up in Nigeria, Saeed Malami learned early that he had to use persuasive skills to make his way through the world. Nigeria is the most populous country in Africa, made up of more than 250 ethnic groups speaking around 500 different languages. In that world of diversity, he needed to find common ground with anyone he dealt with in school, in his neighborhood, and in his extended multi-ethnic family.

When he was 13, Saeed persuaded the admissions committee at King's College, Lagos to accept him as a scholarship student. Then he convinced his parents to let him make the 450-mile journey to the prestigious prep school. Once there, he learned to negotiate his way through the levels of command at the largely student-run school. As Saeed explains it, King's College was like "*Lord of the Flies* meets Eton":

> All activities outside of the classroom were student-run, and in a boarding school, extracurricular activities include eating and sleeping. The third-years were in charge of the second-years, who were in charge of the first-years, and there was no fighting back. You learned quickly that if you were going to get what you needed, everything you asked for had to be in the form of well thought out persuasive appeals.[1]

By the time he was in his second year, Saeed was the school captain, the top student leader who had to negotiate a variety of issues between the students and the all-powerful principal. He was also the captain of the debate club, and by the time he was thinking about applying to a college he was a Nigerian national champion in debate.

When an admissions counselor from Lafayette College visited the U.S. embassy in Lagos, Saeed presented the best arguments he could muster for a full scholarship to the small but well-respected U.S. school. He won that scholarship, and then convinced several members of his extended family to chip in for his travel expenses to Easton, Pennsylvania.

Saeed arrived at Lafayette in the fall of 2016, just as the deep political divisions in the U.S. were becoming blatant. He joined the speech team and started presenting persuasive speeches at intercollegiate tournaments. One of those speeches is the sample speech at the end of this chapter.

Like Saeed, you probably aim to persuade others all the time—in personal relationships, at work, and perhaps even in impersonal settings. You know that persuasion can help you succeed in your career. And you probably feel strongly about both personal and social issues. Learning the art and science of persuasion can help you become more effective in bringing about results in all the areas that matter to you.

persuasion The act of motivating a listener, through communication, to change a particular belief, attitude, value, or behavior.

15.1 Characteristics of Persuasion

Persuasion is the process of motivating others through communication to change a particular belief, attitude, or behavior. Implicit in this definition are several characteristics of persuasion.

Persuasion Is Not Coercive

Persuasion is not the same thing as coercion. If you put someone in a headlock and said, "Do this, or I'll choke you," you would be acting coercively. Besides being illegal, this approach would be ineffective. As soon as the authorities took you away, the person would stop following your demands.

The failure of coercion to achieve lasting results is also apparent in less dramatic circumstances. Children whose parents are coercive often rebel as soon as they can; students who perform from fear of an instructor's threats rarely appreciate the subject matter; and employees who work for abusive and demanding employers are often unproductive and eager to switch jobs as soon as possible. Persuasion, by contrast, makes a listener *want* to think or act differently.

Persuasion Is Usually Incremental

Attitudes do not normally change instantly or dramatically. Persuasion is a process. When it is successful, it generally succeeds over time, usually in increments. So, despite your passion, the best way to change hearts and minds is to have modest but achievable goals.

Social judgment theory explains how and why a gradualist approach can be most effective.[2] This theory tells us that when members of an audience hear a persuasive appeal, they compare it to opinions they already hold. The preexisting opinion is called an **anchor**, but around this anchor there exist what are called **latitudes of acceptance**, **latitudes of rejection**, and **latitudes of noncommitment**. A diagram of any opinion, therefore, might look something like Figure 15.1.

People who care very strongly about a point of view will have a narrow latitude of noncommitment, and those who care less strongly will have a wider latitude of noncommitment. Research suggests that audience members respond least favorably to appeals that fall within their latitude of rejection. This means that persuasion in the real world takes place in a series of small movements. One persuasive speech may be but a single step in an overall persuasive campaign. The best example of this is the various communications that take place during the months of a political campaign. Candidates watch the opinion polls carefully, adjusting their appeals to fall within the latitudes of acceptance and noncommitment of uncommitted voters.

Public speakers who heed the principle of social judgment theory tend to seek realistic, if modest, goals in their speeches. For example, if you want to change audience views on the pro-life/pro-choice question, social judgment theory suggests that the first step would be to consider a range of arguments such as this:

Abortion is a sin.
Abortion should be absolutely illegal.

social judgment theory The theory that opinions will change only in small increments, and only when the target opinions lie within the receiver's latitudes of acceptance and noncommitment.

anchor The position supported by audience members before a persuasion attempt.

latitude of acceptance In social judgment theory, statements that a receiver would not reject.

latitude of rejection In social judgment theory, statements that a receiver would not accept.

latitude of noncommitment In social judgment theory, statements that a receiver would not care strongly about one way or another.

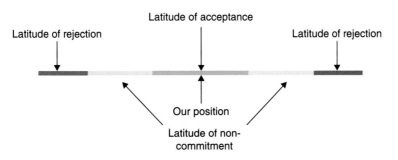

FIGURE 15.1 Latitudes of Acceptance, Rejection, and Noncommitment

Abortion should be allowed only in cases of rape and incest.

A woman should be required to have the father-to-be's permission to have an abortion.

A girl under the age of 18 should be required to have a parent's permission before she has an abortion.

Abortion should be allowed during the first three months of pregnancy.

A girl under the age of 18 should not be required to have a parent's permission before she has an abortion.

A woman should not be required to have the father-to-be's permission to have an abortion.

Abortion is a woman's personal decision.

Abortion should be discouraged but legal.

Abortion should be available anytime to anyone.

Abortion should be considered simply a form of birth control.

You could then arrange these positions on a continuum and estimate how listeners would react to each one. The statement that best represented the listeners' point of view would be their anchor. Other items that might also seem reasonable to them would make up their latitude of acceptance, and opinions they would reject would make up their latitude of rejection. The remaining statements would be the listeners' latitude of noncommitment.

Social judgment theory suggests that the best chance of changing audience attitudes would come by presenting an argument that falls somewhere within the listeners' latitude of noncommitment—even if this wasn't the position that you ultimately want them to accept. If you push too hard by arguing a position in your audience's latitude of rejection, your appeals would probably backfire, making your audience *more* opposed to you than before.

Persuasion Is Interactive

It's easy to think of public speaking as a one-way event in which you talk and the audience listens. But the transactional model of communication described in Chapter 1 is a reminder that persuasion isn't something you do *to* audience members but rather something you do *with* them. Pay attention to your listeners' nonverbal behavior during the speech. If you notice they seem confused, you might decide to elaborate on a point. If they seem annoyed, you might recognize the need to address their objections. If they're enthusiastic, you might decide to make your goals more ambitious. If time permits after you've spoken, responding to listeners' questions can be a way to address their concerns.

Persuasion Can Be Ethical

Even when you understand the difference between persuasion and coercion, you might still feel uncomfortable with the idea of persuasive speaking. You might see it as the work of high-pressure hucksters: salespeople with their feet stuck in the door, unscrupulous politicians taking advantage of beleaguered taxpayers, and so on. Indeed, many of the principles we are about to discuss have been used by unethical speakers for unethical purposes, but that's not what all—or even most—persuasion is about.

Persuasion can influence others' lives in worthwhile ways. Saying "I don't want to influence other people," really means, "I don't want to get involved with other people." Look at the good you can accomplish through persuasion: You can convince people to live healthier lives; you can get members of your community to conserve energy or to join together to refurbish a park; or you can persuade an employer to hire you for a job in which your own talents, interests, and abilities will be put to their best use.

Persuasion is considered ethical if it conforms to accepted standards. But what are the standards today? If your plan is selfish and not in the best interest of your audience members, but you are honest about your motives—is that ethical? If your plan is in the best interest of your audience members, yet you lie to them to get them to accept the plan—is that ethical? Philosophers and rhetoricians have argued for centuries over questions like these.

There are many ways to define **ethical persuasion**.[3] For our purpose, we will consider it as *communication in the best interest of the audience that does not depend on false or misleading information to change an audience's attitude or behavior.* The best way to appreciate the value of this simple definition is to consider the many strategies listed in Table 15.1 that do not fit it. For example, faking enthusiasm about a speech

> **ethical persuasion**
> Persuasion in an audience's best interest that does not depend on false or misleading information to induce change in that audience.

TABLE 15.1
Unethical Communication Behaviors

1. Committing Plagiarism
    a. Claiming someone else's ideas as your own
    b. Quoting without citing the source
2. Relaying False Information
    a. Deliberate lying
    b. Ignorant misstatement
    c. Deliberate distortion and suppression of material
    d. Fallacious reasoning to misrepresent truth
3. Withholding Information; Suppression
    a. About self (speaker); not disclosing private motives or special interests
    b. About speech purpose
    c. About sources (not revealing sources; plagiarism)
    d. About evidence; omission of certain evidence (card stacking)
    e. About opposing arguments; presenting only one side
4. Appearing to Be What One Is Not; Insincerity
    a. In words, saying what one does not mean or believe
    b. In delivery (for example, feigning enthusiasm)
5. Using Emotional Appeals to Hinder Truth
    a. Using emotional appeals as a substitute or cover-up for lack of sound reasoning and valid evidence
    b. Failing to use balanced appeals

Source: Adapted from Andersen, M. K. (1979). *An analysis of the treatment of ethos in selected speech communication textbooks* (Unpublished dissertation). University of Michigan, Ann Arbor, pp. 244–247.

> **COMMUNICATION STRATEGIES**
>
> ## You Versus the Experts
>
> Read Table 15.1 carefully. The behaviors listed there are presented in what some (but certainly not all) communication experts would describe as "most serious to least serious" ethical faults. Do you agree or disagree with the order of this list? Explain your answer and whether you would change the order of any of these behaviors. Are there any other behaviors that you would add to this list?

topic, plagiarizing material from another source, and passing it off as your own and making up statistics to support your case are clearly unethical.

Besides being wrong on moral grounds, unethical attempts at persuasion have a major practical disadvantage: If your deception is uncovered, your credibility will suffer. If, for example, prospective buyers uncover your attempt to withhold a structural flaw in the condominium you are trying to sell, they will probably suspect that the property has other hidden problems. Likewise, if your speech instructor suspects that you are lifting material from other sources without giving credit, your entire presentation will be suspect. One unethical act can cast doubt on future truthful statements. Thus, for pragmatic as well as moral reasons, honesty really is the best policy.

15.2 Categorizing Persuasive Attempts

There are several ways to categorize the persuasive attempts you will make as a speaker. What kinds of subjects will you focus on? What results will you be seeking? How will you go about getting those results? In this section, we look at each of these questions.

By Type of Proposition

Persuasive topics fall into one of three categories, depending on the type of thesis statement (referred to as a "proposition" in persuasion) you are advancing. The three categories are propositions of fact, propositions of value, and propositions of policy.

Propositions of Fact   Some persuasive messages focus on **propositions of fact**: issues in which there are two or more sides about conflicting information, in which listeners are required to choose the truth for themselves. For example:

> Global climate change has already had serious economic consequences.
>
> Early childhood vaccination is essential for public health.
>
> Ride-hailing services have contributed to traffic gridlock in many cities.

These examples show that many questions of fact require careful examination and interpretation of evidence, usually collected from a variety of sources. That's why

proposition of fact A claim bearing on issue in which there are two or more sides of conflicting factual evidence.

it's possible to debate questions of fact, and that also explains why these propositions form the basis of persuasive speeches and not informative ones.

Propositions of Value Beyond issues of truth or falsity, **propositions of value** explore the worth of some idea, person, or object. For example:

> We are/aren't obliged to obey laws that seem foolish or morally wrong.
>
> The United States is/is not justified in attacking countries that harbor terrorist organizations.
>
> The use of laboratory animals for scientific experiments is/is not cruel and immoral.

In order to deal with most propositions of value, you will have to explore certain propositions of fact. For example, you won't be able to debate whether the experimental use of animals in research is immoral—a proposition of value—until you have dealt with propositions of fact such as how many animals are used in experiments and whether experts believe they suffer.

Propositions of Policy A step beyond questions of fact or value are **propositions of policy**, which recommend a specific course of action (a "policy"). For example:

> Members of Congress should/should not be prohibited from investments about which they have insider knowledge.
>
> The Electoral College should/should not be abolished.
>
> Genetic engineering of plants and livestock is/is not an appropriate way to increase the food supply.

Looking at persuasion according to the type of proposition is a convenient way to generate topics for a persuasive speech because each type of proposition suggests different topics. Selected topics could also be handled differently depending on how they are approached. For example, a campaign speech could be approached as a proposition of fact ("Candidate X has done more for this community than the opponent"), a proposition of value ("Candidate X is a better person than the opponent"), or a proposition of policy ("We should get out and vote for Candidate X"). Remember, however, that a fully developed persuasive speech is likely to contain all three types of propositions. If you were preparing a speech advocating that college athletes should be compensated for their talents (a proposition of policy), you might want to first prove that the practice is already widespread (a proposition of fact) and that payments at some schools are unfair to athletes from other, less wealthy ones (a proposition of value).

By Desired Outcome

We can also categorize persuasion according to two major outcomes: convincing and actuating.

Convincing When you set out to **convince** an audience, you want to change the way its members think. When we say that convincing an audience changes the way its members think, we do not mean that you have to swing them from one belief or attitude to a completely different one. Sometimes audience members will already think the way you want them to, but they will not be firmly

> **proposition of value** A claim bearing on an issue involving the worth of some idea, person, or object.

> **proposition of policy** A claim bearing on an issue that involves adopting or rejecting a specific course of action.

> **convincing** A speech goal that aims at changing audience members' beliefs, values, or attitudes.

Engraved on the Statue of Liberty's pedestal are the words "Give me your tired, your poor, your huddled masses yearning to breathe free." Deciding whether those sentiments should apply to today's immigrants involves questions of values and policy.

What values guide your beliefs about immigration policy? How could you use the principles in this chapter to persuade others?

enough committed to that way of thinking. When that is the case, you reinforce, or strengthen, their opinions. For example, if your audience already believed that the federal budget should be balanced but did not consider the idea important, your job would be to reinforce members' current beliefs. Reinforcing is still a type of change, however, because you are causing an audience to adhere more strongly to a belief or attitude. In other cases, a speech to convince will begin to shift attitudes without bringing about a total change of thinking. For example, an effective speech to convince might get a group of skeptics to consider the possibility that bilingual education is/isn't a good idea.

Actuating When you set out to **actuate** an audience, you want to move its members to a specific behavior. Whereas a speech to convince might move an audience to action, it won't be any specific action that you have recommended. In a speech to actuate, you do recommend that specific action.

You can ask for two types of action—adoption or discontinuance. Adoption asks an audience to engage in a new behavior, whereas discontinuance asks an audience to stop behaving in an established way. If you gave a speech for a political candidate and then asked for contributions to that candidate's campaign, you would be asking your audience to adopt a new behavior. If you gave a speech against smoking and then asked your audience members to sign a pledge to quit, you would be asking them to discontinue an established behavior.

By Directness of Approach

We can also categorize persuasion according to the speaker's directness of approach.

Direct Persuasion Using **direct persuasion**, speakers make their purpose clear, usually by stating it outright early in the speech. This is the best strategy to use with a friendly audience, especially when you are asking for a response the audience is likely to give you. Direct persuasion is the kind we hear in most academic situations.

Indirect Persuasion In using **indirect persuasion**, the speaker's purpose is disguised or deemphasized in some way. The question, "Is a season ticket to the symphony worth the money?" (when you intend to prove it is), is based on indirect persuasion, as is any strategy that does not express the speaker's purpose at the outset.

Indirect persuasion is sometimes easy to spot. A television commercial that shows us attractive young men and women romping in the surf on a beautiful day and then flashes the product name on the screen is indisputably indirect persuasion. Political oratory also is sometimes indirect persuasion, and it can be more difficult to identify as such. A political hopeful ostensibly might be speaking on some great social issue when the real persuasive message is, "Please remember my name, and vote for me in the next election."

In public speaking, indirect persuasion is usually disguised as informative speaking, but this approach isn't necessarily unethical. In fact, it's probably the best approach when your audience is hostile to either you or your topic. It can also be necessary to use the indirect approach to get a hearing from listeners who would tune you out if you took a more direct approach. Under such circumstances, you might want to ease into your speech slowly.[4] You might take some time to make your audience feel good about you or the social action you are advocating. If you're speaking in favor of your candidacy for city council, but you are in favor of a tax increase that would displease your audience, you might talk for a while about the

---

actuate To move members of an audience toward a specific behavior.

direct persuasion Persuasion that does not try to hide or disguise the speaker's persuasive purpose.

indirect persuasion Persuasion that disguises or deemphasizes the speaker's persuasive goal.

benefits that a well-financed city council could provide to the community. You might even want to change your desired audience response. Rather than trying to get audience members to rush out to vote for you, you might want them simply to read a policy statement that you have written or become more informed on a particular issue. The one thing you cannot do in this instance is to begin by saying, "My appearance here today has nothing to do with my candidacy for city council." That would be a false statement. It is more than indirect; it is untrue and therefore unethical.

To test the ethics of an indirect approach, ask whether you would express your persuasive purpose directly if you were asked to do so. In other words, if someone in the audience asked, "Don't you want us to vote for you for city council?," you would admit to it rather than denying your true purpose, if you were ethical.

By Type of Appeal: Aristotle's Ethos, Pathos, and Logos

Over 2,000 years ago Aristotle created the first comprehensive theory of persuasion. His *Rhetoric* is generally regarded as "the most important single work on persuasion ever written."[5] He proposed three approaches to influencing others: ethos (speaker credibility), pathos (emotional appeals), and logos (logical reasoning). Aristotle advised a balance of these three approaches, called the **Rhetorical Triad** (Figure 15.2). This chapter will examine each type of appeal.

Ethos From the Greek word for *character*, **ethos**-based proofs rely on the audience's faith in a speaker. This faith is based on judgments of the speaker's competence, character, and charisma. Later in this chapter you'll learn how to boost your persuasiveness by demonstrating each of these qualities.

ethos Appeals based on the credibility of the speaker.

Pathos In modern terms, **pathos** involves appealing to emotion. Most people claim to be rational, but a large body of research shows that emotions often shape reasoning, rather than the other way around. Social scientists have suggested that emotions are like elephants, while reason is like an elephant rider.

pathos Appeals based on emotion.

Perched atop the Elephant, the Rider holds the reins and seems to be the leader. But the Rider's control is precarious because the Rider is so small relative to the Elephant. Anytime the Elephant and the Rider disagree about which direction to go, the Rider is going to lose.[6]

As psychologist Jonathan Haidt puts it, "If you want to change someone's mind about a moral or political issue, *talk to the elephant first.*"[7]

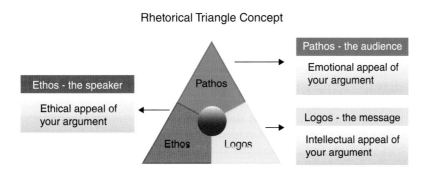

FIGURE 15.2 Aristotle's Rhetorical Triad

Effective speakers appeal to emotions in a variety of ways, including vivid descriptions, emotional stories, and powerful testimony. You can see examples of emotional appeals in Saeed Malami's speech at the end of this chapter.

Logos Later in this chapter, you'll read about boosting the persuasiveness of your message by using solid evidence and reasoning. But even the best evidence-based reasoning won't work if you antagonize your audience, even if you don't mean to. Social psychologist Jonathan Haidt uses the elephant-rider metaphor to emphasize that an argument has to be delivered with respect in order to be heard:

> When discussions are hostile, the odds of change are slight. The elephant [the audience's emotional reaction] leans away from the opponent [speaker], and the rider [the audience's logic] works frantically to rebut the opponent's charges. But if there is affection, admiration, or a desire to please the other person, then the elephant leans toward that person and the rider tries to find truth in the other person's arguments.[8]

> **logos** Appeals based on logical reasoning.

15.3 Creating a Persuasive Message

Persuasive speaking has been defined as "reason-giving discourse." Its principal technique, therefore, involves proposing claims and then backing those claims up with reasons that are true. Preparing an effective persuasive speech isn't easy, but it can be made easier by observing a few simple rules.

Set a Clear Persuasive Purpose

Remember that your objective in a persuasive speech is to move the audience to a specific, attainable attitude or behavior. In a speech to convince, the purpose statement will probably stress an attitude:

> After listening to my speech, my audience members will agree a global minimum tax on corporations should be established.

In a speech to actuate, the purpose statement will stress behavior in the form of a desired audience response. That desired audience response should be as straightforward and clear-cut as possible.

As Chapter 12 explained, your purpose statement should always be specific, attainable, and worded from the audience's point of view. "The purpose of my speech is to save the whales" is not a purpose statement that has been carefully thought out. Your audience members wouldn't be able to jump into the ocean and save the whales, even if your speech motivated them into a frenzy. They might, however, be able to support a specific piece of legislation.

A clear, specific purpose statement will help you stay on track throughout all the stages of preparation of your persuasive speech. Because the main purpose of your speech is to have an effect on your audience, you have a continual test that you can use for every idea, every piece of evidence, and every organizational structure that you think of using. The question you ask is: "Will this help me to get the audience members to think/feel/behave in the manner I have described in my purpose statement?" If the answer is "yes," you forge ahead.

Structure the Message Carefully

A sample structure of the body of a persuasive speech is outlined in Figure 15.3. With this structure, if your objective is to convince, you concentrate on the first two components: establishing the problem and describing the solution. If your objective is to actuate, you add the third component, describing the desired audience reaction.

Of course, other structures can be used for persuasive speeches. This one can be used as a basic model, however, because it's easily applied to most persuasive topics.

Describe the Problem To convince an audience that something needs to be changed, you have to show members that a problem exists. After all, if your listeners don't recognize the problem, they won't find your arguments for a solution very important.

Julia Anthon, a student at the University of Louisiana at Lafayette, presented a speech on maternal mortality in modern America, and she established her problem this way:

> According to the CDC, about 700 women die each year from pregnancy-related complications, and 60% of these deaths are completely preventable. This means that every year, 400 mothers lose their lives, not because we cannot save them, but because we do not stop them from dying. These numbers are sobering, but the situation is only getting worse. Harvard Medical School found that since 1990, the American maternal mortality rate has increased by 50%, and with the abortion bans that have swept this country in the past year, more women are going to be forced to bring their pregnancies to full term, exacerbating the problems that already exist.[9]

An effective description of the problem will answer two questions, either directly or indirectly.

Show How the Problem Affects Your Audience It's not enough to prove that a problem exists. You have to show your listeners that the problem affects them in some way. This is relatively easy in some cases: the high cost of tuition, the lack of convenient parking near campus, the quality of food in the student center. In other cases, you will

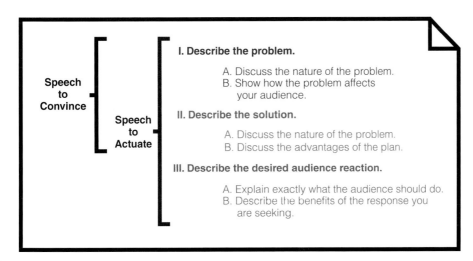

FIGURE 15.3 Sample Structure for a Persuasive Speech

In the movie *Don't Look Up*, two scientists (played by Leonardo DiCaprio and Jennifer Lawrence) try to convince a skeptical TV host (played by Tyler Perry) about a life-ending asteroid hurtling toward earth, but the host refuses to take them seriously.

What would be your first step in convincing someone that a problem exists?

need to spell out the impact to your listeners more clearly. Rebecca Yocum, a student at West Chester University in Pennsylvania, did this when she explained the extent of the problem with the 911 emergency system:

> The National Emergency Number Association website estimates that 240 million calls are made to 911 in the United States every year, with the potential for millions of cases of inaccurate dispatches. In order to mitigate this risk for each of us, we need a better system for locating those in need so that our emergency response units can do their jobs with efficiency.[10]

The problem section of a persuasive speech is often broken up into segments discussing the cause and effect of the problem. (The sample speech at the end of this chapter is an example of this type of organization.)

Describe the Solution   Your next step in persuading your audience members is to convince them that there is an answer to the problem you have just introduced. To describe your solution, you should show that the solution will work and then you should explain the advantages of this solution.

Show That the Solution Will Work   A skeptical audience might agree with the desirability of your solution but still not believe that it has a chance of succeeding. In the homeless speech discussed previously, you would need to prove that establishing a shelter can help unlucky families get back on their feet—especially if your audience analysis shows that some listeners might view such a shelter as a way of coddling people who are too lazy to work.

Explain What Advantages Will Result from Your Solution   You need to describe in specific terms how your solution will lead to the desired changes. In this step, you will paint a vivid picture of the benefits of your proposal. In a speech proposing a shelter for unhoused families, the benefits you describe would probably include the following:

1. Families will have a safe place to stay, free of the danger of living on the street.
2. Parents will have the resources that will help them find jobs: an address, telephone, clothes washers, and showers.
3. The police won't have to apply antivagrancy laws (such as prohibitions against sleeping in cars) to people who aren't the intended target of those laws.
4. The community (including your listeners) won't need to feel guilty about ignoring the plight of unfortunate citizens.

Describe the Desired Audience Response   When you want to go beyond simply convincing your audience members and impel them to follow your solution, you need to describe exactly what you want them to do. This action step, like the previous ones, should answer two questions.

Explain How the Audience Can Put Your Solution into Action   Make the behavior you are asking your audience members to adopt as clear and simple as possible for them. If you want them to vote in a referendum, tell them when and where to go to vote and how to go about registering, if necessary (some activists even provide transportation). If you're asking them to support a legislative change, don't expect them to write their congressional representative. *You* write the letter or draft a petition and ask them to sign it. If you're asking for a donation, pass the hat at the conclusion of your speech, or give audience members a stamped, addressed envelope and simple forms that they can return easily.

Demonstrate the Direct Rewards   Your solution might be important to society, but your audience members will be more likely to adopt it if you can show that they will get a personal payoff. Show them that supporting legislation to reduce greenhouse emissions will produce a wide range of benefits, from less extreme weather to lower food costs, to a better life for their children. Explain that saying "no" to a second drink before driving will not only save lives but also help your listeners avoid expensive court costs, keep their insurance rates low, and prevent personal humiliation. Show how helping to establish and staff a homeless shelter can lead to personal feelings of satisfaction and provide an impressive demonstration of community service on a job-seeking résumé.

Use Solid Evidence

All the forms of support discussed in Chapter 12 can be used to back up your persuasive arguments.[11] Your objective here is not just to find supporting material that clarifies your ideas, but rather to find the perfect example, statistic, definition, analogy, anecdote, or testimony to establish the truth of your claim in the mind of this specific audience.

Whatever type of evidence you use, cite your sources carefully. It is important that your audience know that your sources are credible, unbiased, and current. If you are quoting the source of an interview, give a full statement of the source's credentials:

> According to Sean Wilentz, Dayton–Stockton Professor of History, Director of American Studies at Princeton University, and the author of several books on this topic . . .

If the currency of the interview is important, you might add, "I spoke to Professor Wilentz just last week. . . ." If you're quoting an article, give a quick statement of the author's credentials and the full date and title of the magazine:

> According to Professor Sean Wilentz of Princeton University, in an article in the April 21, 2021, *Rolling Stone Magazine* . . .

You don't need to give the title of the article (although you may, if it helps in any way) or the page number. If you're quoting from a book, include a quick statement of the author's credentials:

> According to Professor Sean Wilentz of Princeton University, in his book *The Rise of American Democracy* . . .

Avoid Fallacies

fallacy An error in logic.

A **fallacy** (from the Latin word meaning *false*) is an error in logic. Although the original meaning of the term implied purposeful deception, most logical fallacies are not recognized as such by those who use them. Scholars have devoted lives and volumes to the description of various types of logical fallacies.[12] Here are some of the most common types to keep in mind when building your persuasive argument:[13]

***ad hominem* fallacy** A fallacious argument that attacks the integrity of a person to weaken the person's position.

Attack on the Person Instead of the Argument (*Ad Hominem*) In an ***ad hominem* fallacy** the speaker attacks the integrity of a person in order to weaken the argument. At its crudest level, an *ad hominem* argument is easy to detect. "How can you believe that fat slob?" is hardly persuasive. It takes critical thinking to catch more subtle *ad hominem* arguments, however. Consider this argument: "All this talk about 'family values' is hypocritical. Take Senator _____, who made a speech about the 'sanctity of marriage' last year. Now it turns out he was having an affair with a staffer, and his wife is suing him for divorce." Although the senator certainly does seem to be a hypocrite, his behavior doesn't necessarily weaken the merits of family values.

***reductio ad absurdum* fallacy** Fallacious reasoning that unfairly attacks an argument by extending it to such extreme lengths that it looks ridiculous.

Reduction to the Absurd (*Reductio ad Absurdum*) A ***reductio ad absurdum* fallacy** unfairly attacks an argument by extending it to such extreme lengths that it looks ridiculous. "If we allow developers to build homes in one section of this area, soon we will have no open spaces left. Fresh air and wildlife will be a thing of the past." "If we allow the administration to raise tuition this year, soon they will be raising it every year, and before we know it only the wealthiest students will be able to go to school here." This extension of reasoning doesn't make any sense: Developing one area doesn't necessarily mean that other areas have to be developed, and one tuition increase doesn't mean that others will occur. Any of these policies might be unwise or unfair, but the *ad absurdum* reasoning doesn't prove it.

either–or fallacy Fallacious reasoning that sets up false alternatives, suggesting that if the inferior one must be rejected, then the other must be accepted.

Either–Or An **either–or fallacy** sets up false alternatives, suggesting that if the inferior one must be rejected, then the other must be accepted. An angry citizen used either–or thinking to support a proposed city ordinance: "Either we outlaw alcohol in city parks, or there will be no way to get rid of drunks." This reasoning overlooks the possibility that there may be other ways to control public drunkenness besides banning all alcoholic beverages. The old saying "America, love it or leave it" provides another example of either–or reasoning. For instance, when an Asian-born college professor pointed out examples of lingering discrimination in the United States, some suggested that if she didn't like her adopted country, she should return to her native home—ignoring the fact that it is possible to admire a country and still envision ways to make it a better place.

***post hoc* fallacy** Fallacious reasoning that mistakenly assumes that one event causes another because they occur sequentially.

False Cause (*Post Hoc Ergo Propter Hoc*) A ***post hoc* fallacy** mistakenly assumes that one event causes another because they occur sequentially. An old (and not especially funny) joke illustrates the *post hoc* fallacy. Mac approaches Jack and asks, "Hey, why are you snapping your fingers?" Jack replies, "To keep the elephants away." Mac is incredulous: "What are you talking about? There aren't any elephants within a thousand miles of here." Jack smiles and keeps on snapping: "I know. Works pretty well, doesn't it?"

In real life, *post hoc* fallacies aren't always so easy to detect. For example, one critic of education pointed out that the increase in sexual promiscuity among adolescents began at about the same time that the courts prohibited prayer in public schools. A causal link in this case may exist: Decreased emphasis on spirituality could contribute to promiscuity. But it would take evidence to establish a *definite* connection between the two phenomena.

COMMUNICATION STRATEGIES

Recognizing Cultural Differences in Persuasion

Different individuals have a tendency to view persuasion differently, and often these differences are based on cultural background. Even the ability to recognize logical argument is, to a certain extent, culturally determined. Not all cultures use logic in the same way that the European American culture does. The influence of the dominant culture is seen even in the way we talk about argumentation. When we talk about "defending" ideas and "attacking our opponent's position," we are using male-oriented militaristic/aggressive terms. Logic is also based on a trust in objective reality, on information that is verifiable through our senses. As one researcher points out, such a perspective can be culturally influenced:

> Western culture assumes a reality that is materialist and limited to comprehension via the five senses. African culture assumes a reality that is both material and spiritual viewed as one and the same.[14]

The way logic is viewed also differs between Eastern and Western Hemisphere cultures. As Larry A. Samovar and Richard E. Porter point out:

> Westerners discover truth by active searching and the application of Aristotelian modes of reasoning. On the contrary, many Easterners wait patiently, and if truth is to be known it will make itself apparent.[15]

It is because of cultural differences such as these that speech experts have always recommended a blend of logical and emotional evidence.

Appeal to Authority (*Argumentum ad Verecundiam*) An *argumentum ad verecundiam* **fallacy** involves relying on the testimony of someone who is not an authority in the case being argued. Relying on experts is not a fallacy, of course. A movie star might be just the right person to offer advice on how to seem more glamorous, and a professional athlete could be the best person to comment on what it takes to succeed in organized sports. But an *ad verecundiam* fallacy occurs when the movie star promotes a political candidate or the athlete tells us why we should buy a certain kind of automobile. When considering endorsements and claims, it's smart to ask yourself whether the source is qualified to make them.

argumentum ad verecundiam fallacy Fallacious reasoning that tries to support a belief by relying on the testimony of someone who is not an authority on the issue being argued.

Bandwagon Appeal (*Argumentum ad Populum*) An *argumentum ad populum* **fallacy** is based on the often dubious notion that, just because many people favor an idea, you should, too. Sometimes, of course, the mass appeal of an idea can be a sign of its merit. If most of your friends have enjoyed a film or a new book, there is probably a good chance that you will, too. But in other cases widespread acceptance of an idea is no guarantee of its validity. In the face of almost universal belief to the contrary, Galileo reasoned accurately that the Earth is not the center of the universe, and he suffered for his convictions. The lesson here is simple to comprehend but often difficult to follow: When faced with an idea, don't just follow the crowd. Consider the facts carefully and make up your own mind.

argumentum ad populum fallacy Fallacious reasoning based on the dubious notion that because many people favor an idea, you should, too.

Other Common Fallacies There is a wide range of other common fallacies, as shown in Table 15.2. Often, dogmatic speakers don't even realize they are using faulty logic; other times, it is purposeful manipulation. How many of these do you recognize from advertising, politics, or everyday arguments? How many other fallacies can you name?

TABLE 15.2
Other Common Fallacies

FALLACY	DEFINITION	EXAMPLE
Straw Man	Setting up an argument that was not proposed and then attacking it as if it were the original argument.	"You say we should support animal rights, but many animal rights activists have supported the destruction of research facilities." (The speaker then goes on to argue that the destruction of research facilities is wrong.)
Red Herring	Shifting the focus to a tangential subject, similar to dragging a fish across a trail to distract a bloodhound.	"Bill says that buying a term paper is immoral. But what is morality, anyway?" (The speaker then goes on to discuss this philosophical question.)
Begging the Question	Repeating an argument but never providing support for a point of view.	"I can't believe people eat dog. That's just plain gross. Why? Because it's a dog, of course. How could someone eat a dog?"
Faulty Analogy	Using a comparison suggesting that two things are more alike than they really are.	"If we legalize gay marriage, next we'll legalize marriage between people and their pets."
Hasty Generalization	Reaching an unjustifiable conclusion after making assumptions or misunderstanding statistics.	"You are likely to be shot if you visit New York City." (In fact, fewer people are murdered, per capita, in New York City than in most rural American small towns.)

15.4 Adapting to the Audience

When making a persuasive speech, it's important to know as much as possible about your audience. For one thing, you should appeal to the values of your audience whenever possible, even if they are not *your* strongest values. This advice does not mean you should pretend to believe in something. According to our definition of *ethical persuasion*, pretense is against the rules. It does mean, however, that you have to stress those values that the members of your audience feel most forcefully.[16]

In addition, you should analyze your audience carefully to predict the type of response you will get. Sometimes you have to pick out one part of your audience—a **target audience**, the subgroup you must persuade to reach your goal—and aim your speech mostly at those members. Some of your audience members might be so opposed to what you are advocating that you have no hope of reaching them. Still others might already agree with you, so they do not need to be persuaded. A middle portion of your audience members might be undecided or uncommitted, and they would be the most productive target for your appeals.

Of course, you need not ignore that portion of your audience that does not fit your target. For example, if you were giving a speech about getting more exercise or eating a healthier diet, your target might be the class members who are obviously out of shape. Your main purpose would be to get them to improve their health habits, but at the same time, you could convince the relatively fit class members to encourage friends and family to be more health conscious.

target audience That part of an audience that must be influenced to achieve a persuasive goal.

All of the methods of audience analysis described in Chapter 12—surveys, observation, interviews, and research—are valuable in collecting information about your audience for a persuasive speech.

Establish Common Ground

It helps to stress as many similarities as possible between yourself and your audience members. This technique helps prove that you understand them—if not, why should they listen to you? Also, if you share a lot of common ground, it shows you agree on many things. Therefore, it should be easy to settle one disagreement: the one related to the attitude or behavior you would like them to change.

Robert Reich, a professor at the University of California Berkeley and a former Secretary of Labor, believes that common ground is essential in today's political environment. He points out that people with different political views have stopped talking with each other, and this is a huge problem because democracy depends on our capacity to deliberate together. He suggests that when people try to reach across the great divide, they should avoid mention of specific politicians or hot-button issues:

> Start instead with "kitchen table" issues like stagnant wages, shrinking benefits, the escalating costs of health care, college, pharmaceuticals, housing. . . .
>
> [And then ask,] do they think any of this has to do with big money in politics?
>
> - Is the system rigged? And if so, who's doing the rigging, and why?
> - How can average people be heard when there's so much big money in politics? Should we try to get big money out of politics?
> - And if so, how do we do it?
>
> Notice, you're not using labels. You're not talking about Democrats or Republicans, left or right, capitalism or socialism, government or free market. . . . [17]

Organize According to the Expected Response

It's much easier to get an audience to agree with you if the members have already agreed with you on a previous point. Therefore, you should arrange your points in a persuasive speech so that you develop a "yes" response. In effect, you get your audience into the habit of agreeing with you. For example, if you were giving a speech on organ donation, you might begin by asking the members of the audience if they would like to be able to get a kidney if they needed one. Then you might ask them if they would like to have a major role in curbing tragic and needless dying. The presumed response to both questions is "yes." It is only when you have built a pattern of "yes" responses that you would ask the audience to sign organ donor cards.

If audience members are already basically in agreement with you, you can organize your material to reinforce their attitudes quickly and then spend most of your time convincing them to take a specific course of action. If, on the other hand, they are hostile to your ideas, you have to spend more time getting the first "yes" out of them.

Neutralize Potential Hostility

One of the trickier problems in audience adaptation occurs when you face an audience hostile to you or your ideas. Hostile audiences are those who have a significant number of members who feel adversely toward you, your topic, or the speech situation. Members of a hostile audience could range from unfriendly to violent. Two guidelines for handling this type of audience are to (1) show that you understand their point of view, and (2) if possible, use appropriate humor.

15.5 Building Credibility as a Speaker

credibility The believability of a speaker or other source of information.

Credibility refers to the believability of a speaker. Credibility isn't an objective quality; rather, it is a perception in the minds of the audience. In a class such as the one you're taking now, students often wonder how they can build their credibility. After all, the members of the class tend to know one another well by the time the speech

---

UNDERSTANDING YOUR COMMUNICATION

Persuasive Speech

Use this self-assessment for a persuasive speech you have presented or plan to present.

1. Have you set a clear, persuasive purpose?

   I'VE DONE MY BEST.   I'VE GOT WORK TO DO.   I'VE BARELY STARTED.

2. Is your purpose in the best interest of the audience?

   I'VE DONE MY BEST.   I'VE GOT WORK TO DO.   I'VE BARELY STARTED.

3. Have you structured the message to achieve a "yes" response?

   I'VE DONE MY BEST.   I'VE GOT WORK TO DO.   I'VE BARELY STARTED.

4. Have you used solid evidence for each point?

   I'VE DONE MY BEST.   I'VE GOT WORK TO DO.   I'VE BARELY STARTED.

5. Have you used solid reasoning for each point?

   I'VE DONE MY BEST.   I'VE GOT WORK TO DO.   I'VE BARELY STARTED.

6. Have you adapted to your audience?

   I'VE DONE MY BEST.   I'VE GOT WORK TO DO.   I'VE BARELY STARTED.

7. Have you built your own credibility?

   I'VE DONE MY BEST.   I'VE GOT WORK TO DO.   I'VE BARELY STARTED.

8. Is your information true to the best of your knowledge?

   I'VE DONE MY BEST.   I'VE GOT WORK TO DO.   I'VE BARELY STARTED.

Scoring on this assessment is self-evident.

assignments roll around. This familiarity illustrates why it's important to earn a good reputation before you speak, through your class comments and the general attitude you've shown.

It is also possible for credibility to change during a speaking event. In fact, researchers speak in terms of initial credibility (what you have when you first get up to speak), derived credibility (what you acquire while speaking), and terminal credibility (what you have after you finish speaking). It is not uncommon for a student with low initial credibility to earn increased credibility while speaking and to finish with much higher terminal credibility.

Without credibility, you won't be able to convince your listeners that your ideas are worth accepting, even if your material is outstanding. On the other hand, if you can develop a high degree of credibility in the eyes of your listeners, they will likely open up to ideas they wouldn't otherwise accept. Members of an audience form judgments about the credibility of a speaker based on their perception of many characteristics, the most important of which might be called the "three Cs" of credibility: competence, character, and charisma.[18]

Competence

Competence refers to the speaker's expertise on the topic. Sometimes this competence can come from personal experience that will lead your audience to regard you as an authority on the topic you are discussing. If everyone in the audience knows you've earned big profits in the stock market, they will probably take your investment advice seriously. If you say that you lost 25 pounds from a diet-and-exercise program, most audience members will be likely to respect your opinions on weight loss.

The other way to be seen as competent is to be well prepared for speaking. A speech that is well researched, organized, and presented will greatly increase the audience's perception of the speaker's competence. Your personal credibility will therefore be enhanced by the credibility of your evidence, including the sources you cite, the examples you choose, the way you present statistics, the quality of your visual aids, and the precision of your language.

Character

Competence is the first component of being believed by an audience. The second is being trusted, which is a matter of character. *Character* involves the audience's perception of at least two ingredients: honesty and impartiality. You should try to find ways to talk about yourself (without boasting, of course) that demonstrate your integrity. You might describe how much time you spent researching the subject or demonstrate your open-mindedness by telling your audience that you changed your mind after your investigation. For example, if you were giving a speech arguing against a proposed tax cut in your community, you might begin this way:

> You might say I'm an expert on the municipal services of this town. As a lifelong resident, I owe a debt to its schools and recreation programs. I've been protected by its police and firefighters and served by its hospitals, roads, and sanitation crews.
>
> I'm also a taxpayer who's on a tight budget. When I first heard about the tax cut that's been proposed, I liked the idea. But then I did some in-depth investigation into the possible effects, not just to my tax bill but to the quality of life of our entire community. I looked into our municipal expenses and into the expenses of similar communities where tax cuts have been mandated by law.

COMMUNICATION STRATEGIES

Persuasion in the World of Sales

Persuasive skills have a range of applications in the workplace. Business consultant George Rodriguez makes it clear that developing a successful sales plan is very much like the planning involved in building a persuasive speech.

"A sales plan is basically your strategic and tactical plan for achieving your marketing objectives," Rodriguez explains. "It is a step-by-step and detailed process that will show how you will acquire new business; and how you will gain more business from your existing customer base."[19]

The process of audience analysis is as important in sales-plan development as it is in persuasive speaking. "The first step is to clearly identify your target markets," Rodriguez says. "Who are more likely to buy your product? The more defined your target market, the better. Your target market can be defined as high-income men ages 30–60 who love to buy the latest electronic gadgets; or mothers with babies 0–12 months old living in urban areas."

And don't forget that persuasion is interactive. "Prospects are more likely to purchase if you can talk to them about solving their problems," Rodriguez points out. He is far from alone in pointing out the importance of thinking in terms of problems and solutions. Business consultant Barbara Sanfilippo advises her clients to "prepare, prepare, and plan your calls. Today's customers and prospects have very little time to waste. They want solutions. A sales consultant who demonstrates a keen understanding of customers' needs and shows up prepared will earn the business."[20] Sanfilippo suggests reviewing the customer's website and interviewing key people in advance of the meeting.

Sanfilippo also points out the importance of building credibility: "How can you stand out from the pack of sales professionals and consultants all offering similar services?" she asks rhetorically. "Establish Credibility and Differentiate!" But George Rodriguez probably has the last word on the value of persuasive speaking to the sales professional. Before you make that first sales call, he says, "You may want to take courses on how to improve your confidence and presentation skills."

Charisma

Charisma is spoken about in the popular press as an almost indefinable, mystical quality. Even the dictionary defines it as "a special quality of leadership that captures the popular imagination and inspires unswerving allegiance and devotion." Luckily, communication scholars favor a more down-to-earth definition. For them, charisma is the audience's perception of two factors: the speaker's enthusiasm and likability. Whatever the definition, history and research have both shown us that audiences are more likely to be persuaded by a charismatic speaker than by a less charismatic one who delivers the same information.

Communication scholars sometimes call enthusiasm "dynamism." Your enthusiasm will mostly be perceived from the way you deliver your remarks, not from what you say. The nonverbal parts of your speech, far better than your words, will show that you believe in what you are saying. Is your voice animated and sincere? Do your gestures reflect your enthusiasm? Do your facial expression and eye contact show you care about your audience?

You can boost your likability by showing that you like and respect your audience. Insincere flattery will probably boomerang, but if you can find a way to give your listeners a genuine compliment, they'll be more receptive to your ideas.

Building your personal credibility through recognition of the roles of competence, character, and charisma is an important component of your persuasive strategy. When combined with careful consideration of audience adaptation and persuasive structure and purpose, it will enable you to formulate the most effective strategy possible.

Aristotle warned of the strength of human emotions. Extending the point, he warned that political leaders who used only emotional appeals rather than reasoning and ethical standards were dangerous. In a phenomenon researchers call **confirmation bias**, people have an emotional tendency to interpret new information as confirmation of their existing beliefs. For example, if workers already feel that regulations may cost them their jobs, they will believe this even more strongly when presented with evidence to the contrary. If people want to believe something because of the emotions involved, they will, even if it is both untrue and ultimately not in their best interests.

> **confirmation bias** The emotional tendency to interpret new information as reinforcing of one's existing beliefs.

15.6 Sample Speech

This speech was presented by Saeed Malami, whose profile began this chapter. When Saeed emigrated from Nigeria to attend school at Lafayette University, he became interested in the forensic team that competed in tournaments at other colleges. At first he shied away from persuasive speaking, because all of the topics seemed to be specialized for the stateside audience, and Saeed didn't feel that he knew enough about the new country he found himself in. When his coach suggested child slavery as a topic, it immediately resonated with Saeed. "This is a West African problem," he said, "and that's where I come from."

He came up with the following purpose statement:

> *After listening to my speech, my audience will be able to identify slave-free chocolate and choose it over the brands that are not slave-free.*

Saeed collected his material from a wide range of sources, including an interview at a local artisanal chocolate maker. He became more involved in his topic as he did more research. His coach, Scott Placke, explains, "The topic was something he cared about initially but grew to care about more deeply the more research he did on it. There's no way to be involved in the process of researching and performing and not caring."[21] Saeed adds, "The topic remains dear to me; I've since stopped eating chocolate from major brands and a lot of my friends and family have done the same."[22]

Saeed used a simple problem-effect-solution organizational structure, as seen in the following outline (numbers in parentheses correspond to paragraphs of the speech):

INTRODUCTION [1–4]

   I. Attention-Getter: Willy Wonka had slaves [1–2]

  II. Thesis: Today's big chocolate industry acquires its chocolate with child slavery. [2–3]

 III. Preview of Main points: [3–4]

     A. We must first expose the problem: Big Chocolate.

     B. We must then unravel the effects: Child Slavery.

     C. We must offer some personal solutions: Munch and Mobilize.

BODY [5–20]

   I. Big Chocolate is the problem. [5–8]

     A. Chocolate manufacturers are part of a near-invisible cocoa supply chain. [5–8]

        1. Small slave farms sell the beans to pisteurs.

        2. Pisteurs then sell the beans to a cooperative.

        3. Cooperatives then sell the beans to conglomerates

        4. Conglomerates then sell the processed beans to chocolate manufacturers.

     B. Chocolate manufacturers have plausible deniability. [9–10]

        1. The supply chain is made up of anonymous middlemen.

        2. The path from the bean to bar is untraceable.

        3. Big Chocolate maintains supposed ignorance as to how their cocoa is grown.

  II. Big Chocolate affects cocoa growers in a variety of ways. [11]

     A. Their practices create poverty.

     B. Their practices encourage child slavery.

 III. Big Chocolate can be impacted by our personal solutions.

     A. We must Munch responsibly.

     B. We must Mobilize effectively.

     C. We must Boycott Big Chocolate

CONCLUSION [21–22]

   I. While Willy Wonka may have gotten away with using slaves, we should not allow big chocolate companies to do the same.

  II. While we enjoy chocolate, we should make it less of a guilty pleasure.

As you read this speech, notice how Saeed structures his argument while he follows the various guideline for persuasive speaking that you read about in this chapter.

SAMPLE SPEECH — Saeed Malami

Child Slavery in the Chocolate Industry[23]

1. Willy Wonka's chocolate factory held the secret to his success. By some miracle, he was able to churn out the greatest chocolate in the world. Kids and adults alike loved every bite. And here's the miracle: He did all this without employing a single person. How? Oompa Loompas. [show photo] They served him and did all his bidding without complaining.

Allusion to a classic movie, with a photo as a visual aid.

2. At least that's how the story is told. Here's a retelling of the story in one word: Slavery. That's how Willy Wonka got his chocolate. And here is how the big chocolate industry gets its chocolate today in two words: child slavery.

The thesis statement is presented with a dramatic flair.

3. UNICEF first detailed in 1998 how farmers in Ivory Coast enslaved children from neighboring countries and forced them to work on cocoa farms. Now, twenty years later, this problem has grown to 2.2 million children illegally laboring in the West African cocoa industry, as revealed by the Cocoa Barometer Report. Every major chocolate manufacturer has condemned child slavery and has "executed" a plan to eliminate it. Yet this problem is as big as it has ever been. And it is growing.

This powerful item of evidence is introduced early on.

4. No Kit Kat bar, no Oreo cookie, and no Hershey's kiss is worth condemning a child to slavery. So we must first expose Big Chocolate, then unravel these companies' effects on cocoa growers, and finally offer some personal solutions. Because as *The Washington Post* of July 7, 2018 argues, for many people, like those of us in this room, "chocolate is synonymous with indulgence, a treat to be relished. But for millions . . . it's a synonym for poverty."

Preview of main points.

5. Collectively, Americans eat over 100 pounds of chocolate every second of every day, so we need to look at the causes of this problem: a near-invisible cocoa supply chain and plausible deniability.

Introduction of the first main point and two subpoints.

6. Recently *Raconteur* magazine revealed how cocoa beans are grown in thousands of small, individual slave farms across West Africa. A report by Mighty Earth explains that motorcyclists, called pisteurs, buy the beans from the farmers and transport them to a cooperative. The cooperatives are large warehouses with bulk capacity who purchase the beans in cash from the pisteurs.

As he develops his idea he bolsters his credibility with research-based evidence.

7. The cooperatives then sell the beans again, in this instance to one of three corporate conglomerates: Olam, Cargill, and Barry Callebaut. Those conglomerates in turn ship and process the raw beans into cocoa powder and butter that gets sold to the chocolate manufacturers that we know and adore.

8. The key here is the supply chain of several anonymous middlemen. Regardless of where the cocoa is made, it ends up in one "big pile." As such, the path from the bean to bar is untraceable.

9. Companies such as Mars, Hershey, and Nestlé maintain a strategy of plausible deniability as to how their cocoa is grown. Given this absurdly extensive supply chain, they claim to be unable to audit for child slavery. See no evil. Hear no evil. These major international companies even claim further to champion human rights, having all churned out reports condemning child slavery.

10. Yet their shadowy supply chains remain intact. The industry newsletter *Supply Chain Dive* explains that their sustainable cocoa promises have been empty, with a steady growth of child labor and decline in farmer income in the countries over the past 10 years. Furthermore, a *New Food Economy* article laments, "Big Chocolate has failed to move to child slavery-free cocoa production primarily because such a model is far less profitable—but deniability is free."

Supply Chain Dive is a newsletter that covers topics such as logistics, freight, and procurement for manufacturing industries.

11. There are two effects Big Chocolate companies have on individual growers: First, poverty. And second, child slavery.

Introduction of second main points, with two subpoints.

12. First, poverty is directly linked to modern slavery. The Business for Good podcast of January 1st, 2019, states that the "consensus definition of modern slavery is… exploitation and forced labor." The previously cited *Raconteur* article reports that the average cocoa-growing household in West Africa earns about 78 cents a day. That is an extremely low level of income, even by West African standards, and it keeps farmers in perpetual poverty while it keeps children out of school without the skill development they would need to earn a living for themselves in the future.

Definition as support.

13. Second, child slavery—crass, brutal, and dehumanizing—is thriving in the very lands that were once the starting point of the transatlantic slave trade. Now the masters are corporate giants. *Fortune* produced a documentary exposé entitled "Behind A Bittersweet Industry" that details an average child's workday on a chocolate plantation.

Detailed description as support.

14. Work begins on the farms at 6 am in the morning and ends well after dark. Children use chainsaws to clear the forest, and machetes to cut open the cocoa pods. Their hands are sliced with scars from several misses in swinging at the pod because every strike has the potential to slice their flesh. The children are malnourished from only eating bananas and corn paste. They sleep on wooden planks and have no access to clean water. As one child who had been working on a cocoa farm said of chocolate consumers, "They are enjoying something that I suffered to make. They are eating my flesh."

15. We must not despair. We have the ability to take this issue into our hands. For this, I have a 2-step solution that's as simple as an M&M. Munch and Mobilize.

Introduction of final main point: The solution, with a catchy mnemonic.

16. The first step is to munch. But instead of munching on that Hershey bar, let's do as the aforementioned Business For Good podcast suggests: rethink the supply chain.

17. Some socially responsible upstarts are committing to 100% child slavery-free production of cocoa by refusing to purchase from the anonymous "big pile" and instead purchasing—at a higher price—cocoa from a separate, transparent pile. A real living wage and an end to child slavery is the result. A *Forbes* magazine article from last November tells about one such company, Tony's Chocolonely. [show photo]

18. Despite slightly lower profit margins this socially conscious business model still makes money showing that other companies can do it as well. Tony's can be purchased online or from a growing number of vendors such as Whole Foods. If you would like to try a bite feel free to grab a sample of either milk or dark from me after this speech.

Chocolate samples are offered to enhance audience involvement. Saeed would later joke about "bribing the judges" with the samples.

19 So here's the deal. We also need to mobilize. On this side of this card (show card), labeled STOP, I have provided a QR Code that links you to chocolate companies who still use child slavery supply chain in their production, including Mars, Nestlé, Hershey, Cadbury, and Godiva. On the other side of the card, labeled ENJOY, is a QR code that links you to several chocolate companies that are committed to ethical cocoa production. Join these companies in taking a stand against the use of child slavery.

The cards are shown as a visual aid during the speech and given out as handouts, with the sample chocolates, at the end of the speech.

20 So I am suggesting a boycott, to create the sense of urgency needed for Big Chocolate to stop its marriage to child slavery. Search for the hashtag I created, #BOYCOTTSLAVECHOCO (show card) and join dozens of others who are already committing to end child slavery.

The hashtag leads to Saeed's Twitter feed with more information on the boycott.

21 In addition, many tournaments put out candy for competitors and judges. While this is appreciated, I have asked the organizers of this tournament to refrain from putting out slave produced chocolate. I would also ask those of you who organize tournaments to do the same. We can all do our part here in the forensics community. Munch and Mobilize. M&M!

Direct appeal for action that is relevant to this audience.

22 Today, we have exposed Big Chocolate, its effects on cocoa growers, and some solutions. So while Willy Wonka may have gotten away with enjoying the fictional inhuman labor of making chocolate using oh so dear Oompa Loompas, we should not allow big chocolate companies do the same in our real world.

Conclusion with summary of main points . . .

23 While this speech may have left a dark bittersweet taste in your mouth, it is important so when you indulge in some chocolatey pleasure, it's a little less guilty.

. . . and a final bon mot to help his audience remember his main point.

MAKING THE GRADE

OBJECTIVE 15.1 Identify the primary characteristics of persuasion.

- Persuasion is the act of moving someone, through communication, toward a belief, attitude, or behavior. Despite a sometimes bad reputation, persuasion can be both worthwhile and ethical.
- Ethical persuasion requires that the speaker be sincere and honest and avoid such behaviors as plagiarism.
- Ethical persuasion must also serve the best interest of the audience, as perceived by the speaker.
  > What are some examples of persuasive speaking in your everyday life?
  > Using the examples you identified earlier, describe the difference between ethical and unethical persuasion.
  > Describe an ethical approach to a persuasive presentation you could deliver.

OBJECTIVE 15.2 Compare and contrast different types of persuasion.

- Persuasion can be categorized according to the type of proposition (fact, value, or policy).
- Persuasion can be categorized according to the desired outcome (convincing or actuating).
- Persuasion can be categorized according to the type of approach (direct or indirect).
  > For persuasive speeches you could deliver, describe propositions of fact, value, and policy.

> In your next persuasive speech, do you intend to convince or to actuate your audience? Why have you chosen this goal?

> In your next persuasive speech, are you planning to use a direct or indirect persuasive approach? Why?

OBJECTIVE 15.3 **Apply the guidelines for persuasive speaking to a speech you will prepare.**

A persuasive strategy is put into effect through the use of several strategies. These include
- setting a specific, clear persuasive purpose,
- structuring the message carefully,
- using solid evidence (including emotional evidence),
- using careful reasoning,
- adapting to the audience, and
- building credibility as a speaker.

> For a speech from your personal experience or one that you have watched online (e.g., a TED Talk), identify its purpose, message structure, use of evidence and reasoning, audience adaptation, and enhancement of speaker credibility.

> Apply the preceding guidelines to a speech you are developing.

OBJECTIVE 15.4 **Explain how to best adapt a specific speech to a specific audience.**

In adapting to your audience, you should
- establish common ground,
- organize your speech in such a way that you can expect a "yes" response along each step of your persuasive plan, and
- take special care with a hostile audience.

> Give examples from speeches you have observed (either in person or online) that illustrate these three strategies. Explain why each strategy contributes to the success of the speech.

> Apply these strategies to a speech you are developing.

OBJECTIVE 15.5 **Improve your credibility in your next persuasive speech.**

- In building credibility, you should keep in mind the audience's perception of your competence, character, and charisma.

> For the most effective persuasive speech you can recall, describe how the speaker enhanced his or her competence, character, and charisma.

> How can you enhance your perceived competence, character, and charisma in a speech you are developing? Suggest specific improvements.

KEY TERMS

actuate p. 420
ad hominem **fallacy** p. 426
anchor p. 415
argumentum ad populum **fallacy** p. 427
argumentum ad verecundiam **fallacy** p. 427
confirmation bias p. 433
convincing p. 419
credibility p. 430
direct persuasion p. 420
either–or fallacy p. 426
ethical persuasion p. 417
ethos p. 421
fallacy p. 426
indirect persuasion p. 420
latitude of acceptance p. 415
latitude of noncommitment p. 415
latitude of rejection p. 415
logos p. 422
pathos p. 421
persuasion p. 414
post hoc **fallacy** p. 426
proposition of fact p. 418
proposition of policy p. 419
proposition of value p. 419
reductio ad absurdum **fallacy** p. 426
social judgment theory p. 415
target audience p. 428

PUBLIC SPEAKING PRACTICE

Prepare a brief, one-paragraph description of a persuasive appeal that would be effective in a speech in your class. Discuss your choice with your classmates.

ACTIVITIES

1. **Audience Latitudes of Acceptance** To better understand the concept of latitudes of acceptance, rejection, and noncommitment, formulate a list of perspectives on a topic of your choice. This list should contain 8 to 10 statements that represent a variety of attitudes, such as the list pertaining to the pro-life/pro-choice issue on pages 415–416. Arrange this list from your own point of view, from most acceptable to least acceptable. Then circle the single statement that best represents your own point of view. This will be your "anchor." Underline those items that also seem reasonable. These make up your latitude of acceptance on this issue. Then cross out the numbers in front of any items that express opinions that you cannot accept. These make up your latitude of rejection. Those statements that are left would be your latitude of noncommitment. Do you agree that someone seeking to persuade you on this issue would do best by advancing propositions that fall within this latitude of noncommitment?

2. **Personal Persuasion** When was the last time you changed your attitude about something after discussing it with someone? In your opinion, was this persuasion interactive? Not coercive? Incremental? Ethical? Explain your answer.

3. **Propositions of Fact, Value, and Policy** Which of the following are propositions of fact, propositions of value, and propositions of policy?
   a. "Three Strikes" laws that put felons away for life after their third conviction are/are not fair.
   b. Elder care should/should not be the responsibility of the government.
   c. The mercury in dental fillings is/is not healthy for the dental patient.
   d. Congressional pay raises should/should not be delayed until an election has intervened.
   e. Third-party candidates strengthen/weaken American democracy.
   f. National medical insurance should/should not be provided to all citizens of the United States.
   g. Elderly people who are wealthy do/do not receive too many social security benefits.
   h. Tobacco advertising should/should not be banned from all media.
   i. Domestic violence is/is not on the rise.
   j. Pit bulls are/are not dangerous animals.

4. **Structuring Persuasive Speeches** For practice in structuring persuasive speeches, choose one of the following topics, and provide a full-sentence outline that conforms to the outline in Figure 15.3.
   a. It should/should not be more difficult to purchase a handgun.
   b. Public relations messages that appear in news reports should/should not be labeled as advertising.
   c. Newspaper recycling is/is not important for the environment.
   d. Police should/should not be required to carry nonlethal weapons only.
   e. Parole should/should not be abolished.
   f. The capital of the United States should/should not be moved to a more central location.
   g. We should/should not ban capital punishment.
   h. Bilingual education should/should not be offered in all schools in which students speak English as a second language.

5. **Find the Fallacy** Test your ability to detect shaky reasoning by identifying which fallacy is exhibited in each of the following statements.
   > *Ad hominem*
   > *Ad absurdum*
   > Either–or
   > *Post hoc*
   > *Ad verecundiam*
   > *Ad populum*

   a. Some companies claim to be in favor of protecting the environment, but you can't trust them. Businesses exist to make a profit, and the cost of saving Earth is just another expense to be cut.
   b. Take it from me, imported cars are much better than domestic cars. I used to buy only American, but the cars made here are all junk.
   c. Rap music ought to be boycotted. After all, the number of assaults on police officers went up right after rap became popular.
   d. Carpooling to cut down on the parking problem is a stupid idea. Look around—nobody carpools!

e. I know that staying in the sun can cause cancer, but if I start worrying about every environmental risk I'll have to stay inside a bomb shelter breathing filtered air, never drive a car or ride my bike, and I won't be able to eat anything.

f. The biblical account of creation is just another fairy tale. You can't seriously consider the arguments of those Bible-thumping, know-nothing fundamentalists, can you?

6. **The Credibility of Persuaders** Identify someone who tries to persuade you via public speaking or mass communication. This person might be a politician, a teacher, a member of the clergy, a coach, a boss, or anyone else. Analyze this person's credibility in terms of the three dimensions discussed in the chapter. Which dimension is most important in terms of this person's effectiveness?

Notes

CHAPTER 1

1. Winfrey, O. (n.d.). *What Oprah knows for sure about communicating.* Oprah.com. https://www.oprah.com/spirit/what-oprah-knows-for-sure-communication. Quote appears in para. 3.
2. *Motivational speech by Oprah Winfrey* [Video]. (2021, October 15). Motivation Ark on YouTube. https://www.youtube.com/watch?v=lRrZiiMT68g. Quotes at 0:22, 0:40, and 1:18 mins, respectively.
3. Winfrey, O. (n.d.). *What Oprah knows for sure about communicating.* Oprah.com. https://www.oprah.com/spirit/what-oprah-knows-for-sure-communication. Quote appears in para. 1.
4. *Average hours per day spent on socializing and communicating by the U.S. population from 2009 to 2019 (in hours per day).* (2021). Statista. https://www.statista.com/statistics/189527/daily-time-spent-on-socializing-and-communicating-in-the-us-since-2009/.
5. Gergen, K. J. (1991). *The saturated self: Dilemmas of identity in contemporary life.* Basic Books. Quote appears on p. 158.
6. *Feeling overwhelmed—too many messages!* (2020, May). Patreon Community. https://www.patreoncommunity.com/t/feeling-overwhelmed-too-many-messages/7900. Quote appears in para. 1.
7. Barnlund, D. C. (1970). Transactional model of communication. In K. K. Sereno & C. D. Mortensen (eds.), *Language behavior: A book of readings in communication* (pp. 43–61). De Gruyter Mouton.
8. *Have you ever met someone online and then in person? What was your first experience like?* (2021, July 6). Quora. https://www.quora.com/Have-you-ever-met-someone-online-and-then-in-person-What-was-your-first-experience-like. Quote appears in para. 3 of Della Ailey post.
9. For an in-depth look at this topic, see Cunningham, S. B. (2012). Intrapersonal communication: A review and critique. In S. Deetz (Ed.), *Communication yearbook* 15 (pp. 597–620). Sage.
10. Samuels, E. (2018, August 2). Grocery store worker lets autistic teen stock shelves, causing a "'miracle in action." *The Washington Post.* https://www.washingtonpost.com/news/inspired-life/wp/2018/08/02/grocery-store-worker-lets-an-autistic-teen-stock-shelves-causing-a-miracle-in-action/. Quote appears in para. 9.
11. *Tweeting yourself out of a job—"Cisco Fatty" style.* Career Goods. https://www.careergoods.com/how-to-lose-a-job-via-twitter-the-cisco-fatty-story/. Quotes appear in paras. 5 and 8.
12. *What do you love about social media?* (2019). Quora. https://www.quora.com/What-do-you-love-about-social-media. Quotes appears in para. 1 of Kiran Chavan post.
13. Greenwell, L. (2021, April 27). *How to use social media more wisely and mindfully.* Creative Little Soul. https://creativelittlesoul.com.au/2021/04/how-to-use-social-media-more-wisely-and-mindfully/. Quote appears in item 3 under "5 Ways."
14. Manry, B. (n.d.). Escaping the social media time warp. *BM.* https://www.bomanry.com/blog/escaping-the-social-media-time-warp. Quote appears in para. 2.
15. U.S. Bureau of Labor Statistics. (2019). *Average hours per day spent in selected activities by age.* https://www.bls.gov/charts/american-time-use/activity-by-age.htm.
16. Rideout, V., & Robb, M. B. (2019). *The Common Sense census: Media use by tweens and teens.* Common Sense. https://www.commonsensemedia.org/research/the-common-sense-census-media-use-by-tweens-and-teens-2019.
17. Livingston, G. (2019, February 20). *The way U.S. teens spend their time is changing, but differences between boys and girls persist.* Pew Research Center. https://www.pewresearch.org/fact-tank/2019/02/20/the-way-u-s-teens-spend-their-time-is-changing-but-differences-between-boys-and-girls-persist/.
18. Hambleton, J. (2021, June 25). Put down your phone: Why presence is the best gift you'll ever give. *Secret Life of Mom.* https://secretlifeofmom.com/put-down-your-phone-why/. Quote appears in para. 1.
19. Blanchard, J. (n.d.). 31 ways to appreciate the present moment and feel happier right now. *Tiny Buddha.* https://tinybuddha.com/blog/31-ways-to-appreciate-the-present-moment-and-feel-happier-right-now/.
20. See Wiemann, J. M., Takai, J., Ota, H., & Wiemann, M. (1997). A relational model of communication competence. In B. Kovacic (Ed.), *Emerging theories of human communication.* SUNY Press. These goals, and the strategies used to achieve them, needn't be conscious. See Fitzsimons, G. M., & Bargh, J. A. (2003). Thinking of you: Nonconscious pursuit of interpersonal goals associated with relationship partners. *Journal of Personality and Social Psychology, 84,* 148–164.

21. Light, J., & Mcnaughton, D. (2014). Communicative competence for individuals who require augmentative and alternative communication: A new definition for a new era of communication? *Augmentative & Alternative Communication, 30*(1), 1–18.
22. Teel, C. M. (2013, March 3). *8 English words you should never use abroad*. SmarterTravel. https://www.smartertravel.com/8-english-words-you-should-never-use-abroad/.
23. Bruno, G., & Gareth, R. (2014, July). Do we notice when communication goes awry? An investigation of people's sensitivity to coherence in spontaneous conversation. *Plos ONE, 9*(7), E103182.
24. Wang, S., Hu, Q., & Dong, B. (2015). Managing personal networks: An examination of how high self-monitors achieve better job performance. *Journal of Vocational Behavior, 91*, 180–188.
25. Vikanda, P. (2018). Excessive use of Facebook: The influence of self-monitoring and Facebook usage on social support. *Kasetsart Journal of Social Sciences, 39*(1), 116–121.
26. Koenig Kellas, J., Horstman, H. K., Willer, E. K., & Carr, K. (2015). The benefits and risks of telling and listening to stories of difficulty over time: Experimentally testing the expressive writing paradigm in the context of interpersonal communication between friends. *Health Communication, 30*(9), 843.
27. Confusing sentences that actually make sense. (2017, April 7). *Grammarly*. https://www.grammarly.com/blog/confusing-sentences-actually-make-sense/. Quote appears in item 3.
28. Baker, C. (2021, December 3). *Communication is key: 4 communication strategies for leaders*. Leaders. https://leaders.com/articles/public-speaking/communication-is-key/. Quote appears in para. 1.
29. Winfrey, O. (n.d.). *What Oprah knows for sure about communicating*. Oprah.com. https://www.oprah.com/spirit/what-oprah-knows-for-sure-communication. Quote appears in para. 3.

CHAPTER 2

1. Farr, C. (2018, December 1). *I quit Instagram and Facebook and it made me a lot happier—and that's a big problem for social media companies*. CNBC News. https://www.cnbc.com/2018/12/01/social-media-detox-christina-farr-quits-instagram-facebook.html. Quote appears in para 12.
2. Farr, C. (2018, December 1). Quote appears in para 13.
3. Allcott, H., Braghieri, L., Eichmeyer, S., & Gentzkow, M. (2020). The welfare effects of social media. *American Economic Review, 110*(3), 629–676. https://doi.org/10.1257/aer.20190658.
4. Isaac, M. (2021, October 28.) Facebook renames itself Meta. *New York Times*. https://www.nytimes.com/2021/10/28/technology/facebook-meta-name-change.html.
5. Farr, C. (2018, December 1). *I quit Instagram and Facebook and it made me a lot happier—and that's a big problem for social media companies*. CNBC News. https://www.cnbc.com/2018/12/01/social-media-detox-christina-farr-quits-instagram-facebook.html. Quote appears in para 7.
6. LaFrance, A. (2020, December 15). Facebook is a doomsday machine. *The Atlantic*. https://www.theatlantic.com/technology/archive/2020/12/facebook-doomsday-machine/617384/.
7. Ghosh, D. (2021, January 7). Blame Facebook, Twitter and YouTube for the mob at the Capitol. *The Washington Post*. https://www.washingtonpost.com/outlook/2021/01/07/social-media-facebook-capitol-mob/.
8. Spangler, T. (2021, January 8). Twitter permanently bans Donald Trump. *Variety*. https://variety.com/2021/digital/news/twitter-bans-donald-trump-1234881689/.
9. Allington D., Duffy B., Wessely S., Dhavan N., & Rubin, J. (2020). Health-protective behaviour, social media usage and conspiracy belief during the COVID-19 public health emergency. *Psychological Medicine*. https://doi.org/10.1017/S003329172000224X.
10. Wells, G., Horwitz, J., & Seetharaman, D. (2021, September 14). Facebook knows Instagram is toxic for teen girls, company documents show. *The Wall Street Journal*. https://www.wsj.com/articles/facebook-knows-instagram-is-toxic-for-teen-girls-company-documents-show-11631620739.
11. Murray, M. (2021, May 9). *Poll: Nearly two-thirds of Americans say social media platforms are tearing us apart*. NBC News. https://www.nbcnews.com/politics/meet-the-press/poll-nearly-two-thirds-americans-say-social-media-platforms-are-n1266773.
12. Towey, H. (2021, July 20). The messy history of #DeleteFacebook: Why the trend is one thing the left and right are agreeing on. *Business Insider*. https://www.businessinsider.com/why-deletefacebook-is-one-thing-left-and-right-agree-on-2021-7.
13. Feezell, J. T. (2017). Agenda setting through social media: The importance of incidental news exposure and social filtering in the digital era. *Political Research Quarterly, 71*, 482–494.
14. Osman, M. (2021, January 3). Wild and interesting Facebook statistics and facts. *Kinsta*. https://kinsta.com/blog/facebook-statistics/.
15. Bellan, R. (2020, December 3). The top social media apps of 2020, according to Apptopia. *Forbes*. https://www.forbes.com/sites/rebeccabellan/2020/12/03/the-top-social-media-apps-of-2020/?sh=1ba1a2f21d4e.
16. Doyle, B. (2021, January 1). TikTok statistics–updated January 2021. *Wallaroo Media*. https://wallaroomedia.com/blog/social-media/tiktok-statistics/.
17. Warren, J. (2020, December 4). Year in review: The biggest social media moments of 2020. *Later*. https://later.com/blog/social-media-moments-2020/.
18. Morales, C. (2020, November 30). On TikTok, fans are making their own "Ratatouille" musical. *New York Times*. https://www.nytimes.com/2020/11/30/arts/tiktok-disney-ratatouille-musical.html.
19. Carras, C. (2021, January 12). "Ratatouille: The TikTok Musical" served up $2 million for COVID-19 relief. *Los Angeles Times*. https://www.latimes.com/entertainment-arts/story/2021-01-12/ratatouille-tiktok-musical-covid-19-relief.
20. Severin, W. J., & Tankard, J. W. (1997). *Communication theories: Origins, methods, and uses in the mass media* (4th ed.). Longman, pp. 197–214.

21. Whiting, A. & Williams, D. (2013). Why people use social media: A uses and gratifications approach. *Qualitative Market Research: An International Journal, 16*(4). 362–369. https://doi.org/10.1108/QMR-06-2013-0041.
22. O'Sullivan, P. B., & Carr, C. T. (2018). Masspersonal communication: A model bridging the mass-interpersonal divide. *New Media and Society, 20,* 1161–1180.
23. Trujillo, J. (2020, November 20). Viral TikTok star doggface208 buys a new house in cash: Meet the Native-Mexican that skated into the world's heart. *Hola!* https://us.hola.com/celebrities/20201119ft7yi5eht5/tiktok-star-doggface208-nathan-apodaca-buys-house.
24. Bariso, J. (2019, October 16). A Panera employee says she was fired after her video went viral: Her response is a lesson in emotional intelligence. *Inc.* https://www.inc.com/justin-bariso/a-panera-employee-says-she-was-fired-after-her-video-went-viral-her-response-is-a-lesson-in-emotional-intelligence.html.
25. Shahzeidi, A. (2020, October 28). 134 Live streaming statistics for 2021. *Uscreen.* https://www.uscreen.tv/blog/live-streaming-statistics/.
26. McLachlan, S. (2020, April 22). The ultimate guide to social media live streaming in 2020. *Hootsuite.* https://blog.hootsuite.com/social-media-live-streaming/.
27. Green, N. (2013) Wheelchair-bound gamer banned from Twitch.tv for faking disability and scamming donors. *Gamed Dynamo.* http://www.gamedynamo.com/article/showarticle/5681/en/wheelchair.
28. Otondo, R. F., Van Scotter, J. R., Allen, D. G., & Palvia, P. (2008). The complexity of richness: Media, message, and communication outcomes. *Information and Management, 45,* 21–30.
29. Surinder, K. S., & Cooper, R. B. (2003). Exploring the core concepts of media richness theory: The impact of cue multiplicity and feedback immediacy on decision quality. *Journal of Management Information Systems, 20,* 263–299.
30. Walther, J. B. (2007). Selective self-presentation in computer-mediated communication: Hyperpersonal dimensions of technology, language, and cognition. *Computers in Human Behavior. 23*(5): 2538–2557.
31. Jiang, C., Bazarova, N., & Hancock, J. (2011). The disclosure-intimacy link in computer-mediated communication: An attributional extension of the hyperpersonal model. *Human Communication Research, 37,* 58–77.
32. McEwan, B., & Zanolla, D. (2013). When online meets offline: A field investigation of modality switching. *Computers in Human Behavior, 29,* 1565–1571.
33. Rains, S. A., & Tsetsi, E. (2017). Social support and digital inequality: Does internet use magnify or mitigate traditional inequities in support availability? *Communication Monographs, 84,* 54–74.
34. Miller, D., & Madianou, M. (2012). *Migration and new media: Transnational families and polymedia.* Routledge.
35. Kalinov, K. (2017). Transmedia narratives: definition and social transformations in the consumption of media content in the globalized world. *Postmodernism Problems, 7,* 60–68. http://ppm.swu.bg/media/45765/kalinov_k_%20transmedia_narratives.pdf.
36. Ozkul, D., & Humphreys, L. (2015). Record and remember: Memory and meaning-making practices through mobile media. *Mobile Media and Communication, 3,* 351–365.
37. Merrett, R. (2018, August 23). Woman loses prized NASA internship over vulgar tweet. *People.* https://people.com/human-interest/woman-loses-nasa-internship-over-tweet/. Quote appears in para. 3.
38. Merrett, R. (2018, August 23). Quote appears in para. 5.
39. Piwek, L., & Joinson, A. (2016). What do they Snapchat about? Patterns of use in time-limited instant messaging service. *Computers in Human Behavior, 54,* 358–367.
40. Utz, S., Muscanell, N., & Khalid, C. (2015). Snapchat elicits more jealousy than Facebook: A comparison of Snapchat and Facebook use. *Cyberpsychology, Behavior, and Social Networking, 18,* 141–146. https://doi.org/10.1089/cyber.2014.0479
41. Clue & Kinsey Institute. (2017). *Technology and sexuality: Results from Clue and Kinsey's international sex survey.* https://helloclue.com/articles/sex/technology-modern-sexuality-results-from-clue-kinseys-international-sex-survey.
42. Kirkpatrick, D. (1992, March 23). Here comes the payoff from PCs. *Fortune.*
43. Anderson, M., & Jiang, J. (2018). *Teens' social media habits and experiences.* Pew Research Center. http://www.pewinternet.org/2018/11/28/teens-social-media-habits-and-experiences.
44. Iqbal, M. (2020, October 30). *Tinder revenue and usage statistics.* Business of Apps. https://www.businessofapps.com/data/tinder-statistics/. Quote appears in para. 200.
45. Iqbal, M. (2020, October 30).
46. D'Aluisio, A. (2020, May 12). Celebrity couples who met on social media. *US Weekly.* https://www.usmagazine.com/celebrity-news/pictures/celebrity-couples-who-met-on-social-media-nick-priyanka-more/.
47. Lee, E. (2019, October 30). *Dating apps are officially the most popular way to meet a spouse.* The Knot. https://www.theknot.com/content/online-dating-most-popular-way-to-meet-spouse.
48. Cacioppo, J. T., Cacioppo, S., Gonzaga, G. C., Ogburn, E. L., & VanderWeele, T. J. (2013). Marital satisfaction and break-ups differ across on-line and off-line meeting venues. *Proceedings of the National Academy of Sciences, 110*(25), 10135–10140.
49. Vitak, J. (2014). Facebook makes the heart grow fonder: Relationship maintenance strategies among geographically dispersed and communication-restricted connections. In Proceedings of the 17th ACM Conference on Computer Supported Cooperative Work and Social Computing. ACM; Walther, J. B., & Ramirez, A., Jr. (2010). New technologies and new directions in online relating. In S. W. Smith & S. R. Wilson (Eds.), *New directions in interpersonal communication research* (pp. 264–284). Sage.
50. Anderson, J. Q., & Rainie, L. (2010, July 2). *The future of social relations.* Pew Internet and American Life Project.
51. Lenhart, A. (2016, August 6). Chapter 4: Social media and friendships. In *Teens, technology and friendships.* Pew Research Center Report. http://www.pewinternet.org/2015/08/06/chapter-4-social-media-and-friendships.
52. Ezumah, B. A. (2013). College students' use of social media: Site preferences, uses and gratifications theory

revisited. *International Journal of Business and Social Science, 4*(5). http://www.ijbssnet.com/journals/Vol_4_No_5_May_2013/3.pdf.
53. Tong, S. T., & Walther, J. B. (2011.) Relational maintenance and computer-mediated communication. In K. Wright & L. Webb (Eds.) *Computer-mediated communication and personal relationships* (pp. 98–118). Hampton Press.
54. Aslam, S. (2022, January 4). *81 LinkedIn statistics you need to know in 2022*. Omnicore Agency. https://www.omnicoreagency.com/linkedin-statistics/.
55. Aslam, S. (2022, January 4).
56. Davis, J., Wolff, H., Forret, M. L., & Sullivan, S. E. (2020). Networking via LinkedIn: An examination of usage and career benefits. *Journal of Vocational Behavior, 118*, Article e103396.
57. Mogaji, E. (2019). Student engagement with LinkedIn to enhance employability. In A. Diver (Ed.), *Employability via higher education: Sustainability as scholarship* (pp. 321–330). Springer.
58. Aslam, S. (2022, January 4). *81 LinkedIn statistics you need to know in 2022*. Omnicore Agency. https://www.omnicoreagency.com/linkedin-statistics/.
59. Porter, C. E. (2006). A typology of virtual communities: A multi-disciplinary foundation for future research. *Journal of Computer-Mediated Communication, 10*, Article 3; Schwammlein, E., & Wodzicki, K. (2012). What to tell about me? Self-presentation in online communities. *Journal of Computer-Mediated Communication, 17*, 387–407.
60. Orr, E. S., Sisic, M., Ross, C., Simmering, M. G., Arseneault, J. M., & Orr, R. R. (2009). The influence of shyness on the use of Facebook in an undergraduate sample. *Cyberpsychology Behavior, 12*, 337–340.
61. Cotten, S. R., Anderson, W. A., & McCullough, B. M. (2013). Impact of internet use on loneliness and contact with others among older adults: Cross-sectional analysis. *Journal of Medical Internet Research, 15*, e39.
62. Lee, K., Noh, M., & Koo, D. (2013). Lonely people are no longer lonely on social networking sites: The mediating role of self-disclosure and social support. *Cyberpsychology, Behavior, and Social Networking, 16*, 413–418.
63. Drouin, M., Reining, L., Flanagan, M., Carpenter, M., & Toscos, T. (2018). College students in distress: Can social media be a source of social support? *College Student Journal, 52*(4). 494–504. https://www.ingentaconnect.com/content/prin/csj/2018/00000052/00000004/art00009.
64. Valenzuela, S., Halpern, D., & Katz, J. E. (2014). Social network sites, marriage well-being and divorce: Survey and state-level evidence from the United States. *Computers in Human Behavior, 36*, 94–101.
65. McClure, E. A., Acquavita, S. P., Dunn, K. E., Stoller, K. B., & Sitzer, M. L. (2014). Characterizing smoking, cessation services, and quit interest across outpatient substance abuse treatment modalities. *Journal of Substance Abuse Treatment, 46*, 194–201.
66. Luxton, D. D., June, J. D., & Kinn, J. T. (2011). Technology-based suicide prevention: Current applications and future directions. *Telemedicine and e-Health, 17*, 50–54.
67. Hawdon, J., & Ryan, R. (2012). Well-being after the Virginia Tech mass murder: The relative effectiveness of face-to-face and virtual interactions in providing support to survivors. *Traumatology, 18*, 3–12.
68. Selkie, E., Adkins, V., Masters, E., Bajpai, A., & Shummer, D. (2020). Transgender adolescents' uses of social media for social support. *Journal of Adolescent Health, 66*(3). 275–280. https://doi.org/10.1016/j.jadohealth.2019.08.011.
69. Sanford, A. A. (2010). "I can air my feelings instead of eating them": Blogging as social support for the morbidly obese. *Communication Studies, 61*, 567–584.
70. Fox, S. (2011, June). *Peer-to-peer healthcare*. Pew Internet and American Life Project; Rains, S. A., & Keating, D. M. (2011). The social dimension of blogging about health: Health blogging, social support, and well-being. *Communication Monographs, 78*, 511–553.
71. Brad, K. (2017). Social networking, survival, and healing. In Adler, R. B., & Proctor, R. F. *Looking out/looking in*. (15th ed.). Cengage, p. 46.
72. Boulianne, S., Lafancette, M., & Ilkiw, D. (2020). "School Strike 4 Climate": Social media and the international youth protest on climate change. *Media and Communication, 8*(2), 208–218. http://dx.doi.org/10.17645/mac.v8i2.2768.
73. Mundt, M., Ross, K., & Burnett, C. M. (2018). Scaling social movements through social media: The case of Black Lives Matter. *Social + Society, 4*(4). 1–14. https://doi.org/10.1177%2F2056305118807911.
74. Pressgrove, G., McKeever, B. W., & Jang, S. M. (2017). What is contagious? Exploring why content goes viral on Twitter: A case study of the ALS Ice Bucket Challenge. *International Journal of Nonprofit and Voluntary Sector Marketing, 23*(1). Article e1586. https://doi.org/10.1002/nvsm.1586.
75. DeAndrea, D. C., Tong, S. T., & Walther, J. B. (2010). Dark sides of computer-mediated communication. In W. R. Cupach & B. H. Spitzberg (Eds.), *The dark side of close relationships II* (pp. 95–118). Routledge.
76. Dunbar, R. (2010). *How many friends does one person need? Dunbar's number and other evolutionary quirks*. Harvard University Press.
77. Bryant, E. M., & Marmo, J. (2012). The rules of Facebook friendship: A two-stage examination of interaction rules in close, casual, and acquaintance friendships. *Journal of Social and Personal Relationships, 29*, 1013–1035.
78. Parks, M. R. (2007). *Personal networks and personal relationships*. Lawrence Erlbaum.
79. Dunbar, R. (2012). Social cognition on the Internet: Testing constraints on social network size. *Philosophical Transactions of the Royal Society, 367*, 2192–2201.
80. Tong, S. T., Van Der Heide, B., Langwell, L., & Walther, J. B. (2008). Too much of a good thing? The relationship between number of friends and interpersonal impressions on Facebook. *Journal of Computer-Mediated Communication, 13*, 531–549.
81. Lee, J. R., Moore, D. C., Park, E., & Park, S. G. (2012). Who wants to be "friend rich"? Social compensatory friending on Facebook and the moderating role of public self-consciousness. *Computers in Human Behavior, 28*, 1036–1043; Kim, J., & Lee, J. R. (2011). The Facebook paths to happiness: Effects of the number of Facebook friends and self-presentation on subjective well-being. *Cyberpsychology, Behavior, and Social Networking, 14*, 359–364.

82. Caplan, S. E. (2005). A social skill account of problematic Internet use. *Journal of Communication, 55*, 721–736; Schiffrin, H., Edelman, A., Falkenstein, M., & Stewart, C. (2010). Associations among computer-mediated communication, relationships, and well-being. *Cyberpsychology, Behavior, and Social Networking, 13*, 1–14; Morrison, C. M., & Gore, H. (2010). The relationship between excessive Internet use and depression: A questionnaire-based study of 1,319 young people and adults. *Psychopathology, 43*, 121–126.
83. Woods, H. C., & Scott, H. (2–16). #Sleepyteens: Social media use in adolescence is associated with poor sleep quality, anxiety, depression and low self-esteem. *Journal of Adolescence 51*, 41–49.
84. Whelan, E., Islam, N., & Brooks, S. (2020). Applying the SOBC paradigm to explain how social media overload affects academic performance. *Computers & Education, 143*. Article 103692. https://doi.org/10.1016/j.compedu.2019.103692.
85. Kleemans, M., Daalmans, S., Carbaat, I., & Anschütz, D. (2018). Picture perfect: The direct effect of manipulated Instagram photos on body image in adolescent girls. *Media Psychology, 21*(1), 93–110. https://doi.org/10.1080/15213269.2016.1257392.
86. Kim, H. M. (2020). What do others' reactions to body posting on Instagram tell us? The effects of social media comments on viewers' body image perception. *New Media & Society*. Advanced online publication. https://doi.org/10.1177/1461444820956368.
87. Vogel, E., Rose, J. Roberts, L. & Eckles, K. (2014). Social comparison, social media, and self-esteem. *Psychology of Popular Media Culture, 3*, 208–232.
88. Kelly, S. M. (2020, February 10). *Plastic surgery inspired by filters and photo editing apps isn't going away*. CNN Business. https://www.cnn.com/2020/02/08/tech/snapchat-dysmorphia-plastic-surgery/index.html.
89. Longstreet, P., & Brooks, S. (2017). Life satisfaction: A key to managing internet and social media addiction. *Technology in Society, 50*, 73–77. https://doi.org/10.1016/j.techsoc.2017.05.003.
90. Brailovskaia, J., & Margraf, J. (2017). Facebook Addiction Disorder (FAD) among German students—A longitudinal approach. *PLoS ONE 12*(12), Article e0189719. https://doi.org/10.1371/journal.pone.0189719.
91. Raypole, C. (2020). *How Facebook can become an "addiction."* Healthline. https://www.healthline.com/health/facebook-addiction#why-its-addictive.
92. Caplan, S. E. (2003). Preference for online social interaction: A theory of problematic internet use and psychosocial well-being. *Communication Research, 30*, 625–648.
93. Phu, B., & Gow, A. (2019). Facebook use and its association with subjective happiness and loneliness. *Computers in Human Behavior, 92*, 151–159.
94. Lundy, B. L., & Drouin, M. (2016). From social anxiety to interpersonal connectedness: Relationship building within face-to-face, phone and instant messaging mediums. *Computers in Human Behavior, 54*, 271–277.
95. Walther, J. B., Van Der Heide, B., Hamel, L., & Shulman, H. (2009). Self generated versus other generated statements and impressions in computer-mediated communication: A test of warranting theory using Facebook. *Communication Research, 36*, 229–253.
96. Vogels, E. A., & Anderson, M. (2020, May 8). *Dating and relationships in the digital age*. Pew Research Center. https://www.pewresearch.org/internet/2020/05/08/dating-and-relationships-in-thedigital-age/.
97. Hand, M. M., Thomas, D. B., Walter, C., Deemer, E. D., & Buyanjargal, M. (2013). Facebook and romantic relationships: Intimacy and couple satisfaction associated with online social network use. *Cyberpsychology, Behavior, and Social Networking, 16*, 8–13.
98. Arikewuyo, A. O., Lasisi, T. T., Abdulbaqi, S. S., Omoloso, A. I., & Arikewuyo, H. O. (2020). Evaluating the use of social media in escalating conflicts in romantic relationships. *Journal of Public Affairs*. Advance online publication. https://doi.org/10.1002/pa.2331.
99. Fejes-Vékássy, L., Ujhelyi, A., & Faragó, L. (2020). From #RelationshipGoals to #Heartbreak–We use Instagram differently in various romantic relationship statuses. *Current Psychology*. Advance online publication. https://doi.org/10.1007/s12144-020-01187-0.
100. Wagner, L. (2015). When your smartphone is too smart for your own good: How social media alters human relationships. *The Journal of Individual Psychology, 71*(2). 114–121. doi:10.1353/jip.2015.0009.
101. Derzakarian, A. (2017). Dark side of social media romance: Civil recourse for catfish victims. *Loyola of Los Angeles Law Review, 50*(4). https://digitalcommons.lmu.edu/cgi/viewcontent.cgi?article=3013&context=llr.
102. Wagner, L. (2015). When your smartphone is too smart for your own good: How social media alters human relationships. *The Journal of Individual Psychology, 71*(2), 114–121. doi:10.1353/jip.2015.0009.
103. Derzakarian, A. (2017). Dark side of social media romance: Civil recourse for catfish victims. *Loyola of Los Angeles Law Review, 50*(4). https://digitalcommons.lmu.edu/cgi/viewcontent.cgi?article=3013&context=llr.
104. Toma, C. L., Hancock, J. T., & Ellison, N. B. (2008). Separating fact from fiction: An examination of deceptive self-presentation in online dating profiles. *Personality and Social Psychology Bulletin, 34*, 1023–1036.
105. Lyndon, A., Bonds-Raacke, J., & Cratty, A. D. (2011). College students' Facebook stalking of ex-partners. *Cyberpsychology, Behavior, and Social Networking, 14*, 711–716.
106. Reyns, B. W., Henson, B., & Fisher, B. S. (2012). Stalking in the twilight zone: Extent of cyberstalking victimization and offending among college students. *Deviant Behavior, 33*, 1–25.
107. DreBing, H., Bailer, J., Anders, A., Wagner, H., & Gallas, C. (2014). Cyberstalking in a large sample of social network users: Prevalence, characteristics, and impact upon victims. *Cyberpsychology, Behavior, and Social Networking, 17*, 61–67.
108. Shahani, A. (September 15, 2014). *Smartphones are used to stalk, control domestic abuse victims*. All Tech Considered. https://www.npr.org/sections/alltechconsidered/2014/09/15/346149979/smartphones-are-used-to-stalk-control-domestic-abuse-victims
109. Smith, A., & Duggan, M. (2018). *Crossing the line: What counts as online harassment?* Pew Research Center. http://www.pewinternet.org/2018/01/04/crossing-the-line-what-counts-as-online-harassment.

110. Bauman, S., Toomey, R. B., & Walker, J. L. (2013). Associations among bullying, cyberbullying, and suicide in high school students. *Journal of Adolescence, 36*, 341–350; Huang, Y.-Y., & Chou, C. (2010). An analysis of multiple factors of cyberbullying among junior high school students in Taiwan. *Computers in Human Behavior, 26*, 1581–1590.
111. Anderson, M. (2018). *A majority of teens have experienced some form of cyberbullying*. Pew Research Center. http://www.pewinternet.org/2018/09/27/a-majority-of-teens-have-experienced-some-form-of-cyberbullying.
112. Cassidy, W., Faucher, C., & Jackson, M. (2013). Cyberbullying among youth: A comprehensive review of current international research and its implications and application to policy and practice. *School Psychology International, 34*, 575–612; Roberto, A. J., Eden, J., Savage, M. W., Ramos-Salazar, L., & Deiss, D. M. (2014). Prevalence and predictors of cyberbullying perpetration by high school seniors. *Communication Quarterly, 62*, 97–114.
113. Clement, J. (2020, October 12). *Cyber bullying—statistics & facts*. Statista. https://www.statista.com/topics/1809/cyber-bullying/.
114. Kempton, S. D. (2020). Erotic extortion: Understanding the cultural propagation of revenge porn. *SAGE Open, 10*(2). https://doi.org/10.1177/2158244020931850.
115. Curtis, C. (2018, October 5). *Deepfakes are being weaponized to silence women—but this woman is fighting back*. The Next. https://thenextweb.com/code-word/2018/10/05/deepfakes-are-being-weaponized-to-silence-women-but-this-woman-is-fighting-back/.
116. Ley, D. (2017, March 17). Why men post revenge porn pictures. *Psychology Today*. https://www.psychologytoday.com/us/blog/women-who-stray/201703/why-men-post-revenge-porn-pictures.
117. Bates, S. (2016). Revenge porn and mental health: A qualitative analysis of the mental health effects of revenge porn on female survivors. *Feminist Criminology, 12*(1), 22–42. https://doi.org/10.1177/1557085116654565.
118. Berger, M. (2013, November 20). *Brazilian 17-year-old commits suicide after revenge porn posted online*. BuzzFeed News. https://www.buzzfeednews.com/article/miriamberger/brazilian-17-year-old-commits-suicide-after-revenge-porn-pos.
119. Shearer, E., & Matsa, K.E. (2018). *News use across social media platforms 2018*. Pew Research Center. https://www.journalism.org/2018/09/10/news-use-across-social-media-platforms-2018/.
120. Bradshaw, S., Howard, P. N., Kollanyi, B., & Neudert, L. (2020). Sourcing and automation of political news and information over social media in the United States, 2016–2018. *Political Communication, 37*(2), 173–193. https://doi.org/10.1080/10584609.2019.1663322.
121. Song, H., Gil, H. de Zúñiga, & Boomgaarden, H. G. (2020). Social media news use and political cynicism: Differential pathways through "news finds me" perception. *Mass Communication and Society, 23*(1), 47–70. https://doi.org/10.1080/15205436.2019.1651867.
122. Song, H., Gil, H. de Zúñiga, & Boomgaarden, H. G. (2020).
123. Singh, L., Bode, L., Budak, C., Kawintiranon, K., Padden, C., & Vraga, E. (2020). Understanding high and low quality URL sharing on COVID-19 Twitter streams. *Journal of Computational Social Science, 3*, 343–366. https://doi.org/10.1007/s42001-020-00093-6.
124. Allington, D., Duffy, B., Wessely, S., Dhavan, N., & Rubin, J. (2020). Health-protective behaviour, social media usage and conspiracy belief during the COVID-19 public health emergency. *Psychological Medicine*. https://doi.org/10.1017/S003329172000224X.
125. Naeem, S. B., Bhatti, R., & Khan, A. (2020). An exploration of how fake news is taking over social media and putting public health at risk. *Health Information and Libraries Journal*. Advance online publication. https://doi.org/10.1111/hir.12320.
126. Bradshaw, S., Howard, P. N., Kollanyi, B., & Neudert, L. (2020). Sourcing and automation of political news and information over social media in the United States, 2016–2018. *Political Communication, 37*(2), 173–193. https://doi.org/10.1080/10584609.2019.1663322.
127. Masip, P., Suau, J., & Ruiz-Caballero, C. (2020). Incidental Exposure to non-like-minded news through social media: Opposing voices in Echo-Chambers' news feeds. *Media and Communication, 8*(4), 53–62. http://dx.doi.org/10.17645/mac.v8i4.3146.
128. Wamp, Z. (2022, January 24). *1 Republican argues for a narrower approach to changing a 19th century voting law* [Interview]. NPR. https://www.npr.org/transcripts/1075264794
129. Rogers, K. (2019, June 13). *The latest disinformation threat online? Old news stories*. CBC News. https://www.cbc.ca/news/science/old-news-shared-as-if-its-new-disinformation-1.5172449.
130. Rogers, K. (2019, June 13).
131. Bengani, P. (2019, December 18). Hundreds of "pink slime" local news outlets are distributing algorithmic stories and conservative talking points. *Columbia Journalism Review*. https://www.cjr.org/tow_center_reports/hundreds-of-pink-slime-local-news-outlets-are-distributing-algorithmic-stories-conservative-talking-points.php.
132. Bengani, P. (2019, December 18).
133. Sample, I. (2020, January 13). What are deepfakes and how can you spot them? *The Guardian*. https://www.theguardian.com/technology/2020/jan/13/what-are-deepfakes-and-how-can-you-spot-them.
134. Sample, I. (2020, January 13).
135. Watson, A. (2021, August 31). *Credibility of major news organizations in the U.S. 2017–2021*. Statista. https://www.statista.com/statistics/239784/credibility-of-major-news-organizations-in-the-us/.
136. Hossenei, M., & Tammimy, Z. (2016). Recognizing users' gender in social media using linguistic features. *Computers in Human Behavior, 56*, 192–197.
137. Pennebaker, J. W. (2011). *The secret lives of pronouns: What our words say about us*. Bloomsbury.
138. Tifferet, S. (2020). Gender differences in social support on social network sites: A meta-analysis. *Cyberpsychology, Behavior, and Social Networking, (23)*4. https://doi.org/10.1089/cyber.2019.0516.
139. Wenliang, S, Xiaoli, H., Hanlu, Y., Yiling, W., Wu, & Potenza, M. N. (2020). Do men become addicted to inter-

140. Wenliang, S, Xiaoli, H., Hanlu, Y., Yiling, W., Wu, & Potenza, M. N. (2020).
141. Gramlich, J. (2019, May 16). *10 facts about Americans and Facebook.* Pew Research Center https://www.pewresearch.org/fact-tank/2019/05/16/facts-about-americans-and-facebook/.
142. Pew Research Center. (2021, April 7). *Social media fact sheet.* https://www.pewresearch.org/internet/fact-sheet/social-media/.
143. Marwick, A. E. (2021). Morally motivated networked harassment as normative reinforcement. *Social Media + Society (7)* 2. https://doi.org/10.1177/20563051211021378.
144. Al-Rawi, A. Kekistanis and the meme war on social media. *The Journal of Intelligence, Conflict, and Warfare (3)*2, 1–18. https://journals.lib.sfu.ca/index.php/jicw/article/view/2360/3002.
145. Zhou, X. (2020). For better or for worse for females?: A content analysis of gender-motivated comments on social media platforms. *Advances in Social Science, Education and Humanities Research, 466,* 1192–1199.
146. Zhou, X. (2020).
147. Mahon, C. & Hevey, D. (2021) Processing body image on social media: Gender differences in adolescent boys' and girls' agency and active coping. *Frontiers in Psychology (12).* https://www.frontiersin.org/article/10.3389/fpsyg.2021.626763.
148. Oberst, U., Chamarro, A., & Renau, V. (2016). Gender stereotypes 2.0: Self-representations of adolescents on Facebook. *Cominicar: Media Education Research Journal (48)*24, 81–89.
149. Wells, G., Horwitz, J., & Seetharaman, D. (2021, September 14). Facebook knows Instagram is toxic for teen girls, company documents show. *The Wall Street Journal.* https://www.wsj.com/articles/facebook-knows-instagram-is-toxic-for-teen-girls-company-documents-show-11631620739.
150. Prensky, M. (2001). Digital natives, digital immigrants. *On the Horizon, 9,* 1–6; Rainie, L. (October 27, 2006). *Digital natives: How today's youth are different from their "digital immigrant" elders and what that means for libraries.* Pew Research Internet Project.
151. Hyman, I. (January 26, 2014). Cell phones are changing social interaction. *Psychology Today.* https://www.psychologytoday.com/us/blog/mental-mishaps/201401/cell-phones-are-changing-social-interaction.
152. Smith, A. (September 19, 2011). *Americans and text messaging.* Pew Internet and American Life Project. https://www.pewresearch.org/internet/2011/09/19/americans-and-text-messaging/
153. Kluger, J. (August 16, 2012). We never talk anymore: The problem with text messaging. *Time.* https://techland.time.com/2012/08/16/we-never-talk-anymore-the-problem-with-text-messaging/
154. Gramlich, J. (2019, May 16). *10 facts about Americans and Facebook.* Pew Research Center https://www.pewresearch.org/fact-tank/2019/05/16/facts-about-americans-and-facebook/.
155. Perrin, A. (2019). *Share of U.S. adults using social media, including Facebook, is mostly unchanged since 2018.* Pew Research Center. https://www.pewresearch.org/fact-tank/2019/04/10/share-of-u-s-adults-using-social-media-including-facebook-is-mostly-unchanged-since-2018/.
156. Alili, A. (2021, December 23). *Which social networks have the highest usage among Gen Z and Millennials?* MSN News. https://www.msn.com/en-ph/news/technology/which-social-networks-have-the-highest-usage-among-gen-z-and-millennials/ar-AAS5Goc.
157. Riorda, M. A., Kreuz, R. J., & Blair, A.N. (2018). The digital divide: Conveying subtlety in online communication. *Journal of Computers in Education, 5,* 49–66. https://doi.org/10.1007/s40692-018-0100-6
158. Riorda, M. A., Kreuz, R. J., & Blair, A. N. (2018).
159. Bennett, J. (2015, February 27). When your punctuation says it all (!). *New York Times.* https://www.nytimes.com/2015/03/01/style/when-your-punctuation-says-it-all.html.
160. Watts, S. A. (2007). Evaluative feedback: Perspectives on media effects. *Journal of Computer-Mediated Communication, 12*(2), 384–411. http://jcmc.indiana.edu/vol12/issue2/watts.html. See also Turnage, A. K. (2007). Email flaming behaviors and organizational conflict. *Journal of Computer-Mediated Communication, 13*(1), 43–59. http://jcmc.indiana.edu/vol13/issue1/turnage.html.
161. Lapidot-Lefler, N. & Barak, A. (2012). Effects of anonymity, invisibility, and lack of eye-contact on toxic online disinhibition. *Computers in Human Behavior 28*(2), 434–443.
162. Konnikova, M. (2014, March 22). The lost art of the unsent angry letter. *New York Times.* https://www.nytimes.com/2014/03/23/opinion/sunday/the-lost-art-of-the-unsent-angry-letter.html.
163. Chotpitayasunondh, V., & Douglas, K. M. (2018). The effects of "phubbing" on social interaction. *Journal of Applied Social Psychology, 48*(6), 304–316. https://doi.org/10.1111/jasp.12506.
164. Vogels, E. A., & Anderson, M. (2020, May 8). *Dating and relationships in the digital age.* Pew Research Center. https://www.pewresearch.org/internet/2020/05/08/dating-and-relationships-in-thedigital-age/.
165. Bauerlein, M. (2009, September 4). Why Gen-Y Johnny can't read nonverbal cues. *Wall Street Journal.* http://www.wsj.com/articles/SB10001424052970203863204574348493483201758.
166. Turkle, S. (2012, April 21). The flight from conversation. *New York Times.* http://www.nytimes.com/2012/04/22/opinion/sunday/the-flight-from-conversation.html?pagewanted=all&_r=0.
167. Roberts, J. J. (2017, June 5). Harvard Yanks 10 acceptance letters over offensive Facebook posts. *Fortune.* http://fortune.com/2017/06/05/harvard-acceptance-rescinded/?utm_campaign=time&utm_source=facebook.com&utm_medium=social&xid=time_socialflow_facebook, Paragraph 6.
168. Levin, D. (2020, July 2). Colleges rescinding admissions offers as racist social media posts emerge. *New York Times.* https://www.nytimes.com/2020/07/02/us/racism-social-media-college-admissions.html?searchResultPosition=10.

169. Levin, D. (2020, July 2).
170. Levine, T. R. (2014). Truth-default theory (TDT): A theory of human deception and deception detection. *Journal of Language and Social Psychology, 33*(4), 378–392. https://doi.org/10.1177/0261927X14535916.
171. Winnick, M. (2016, June 16). Putting a finger on our phone obsession. *Dscout*. https://dscout.com/people-nerds/mobile-touches#:~:text=People%20tapped%2C%20swiped%20and%20clicked,the%20less%20restrained%20among%20us.
172. Allcott, H., Braghieri, L., Eichmeyer, S., & Gentzkow, M. (2020). The welfare effects of social media. *American Economic Review, 110*(3), 629–676. https://doi.org/10.1257/aer.20190658.
173. Farr, C. (2018, December 1). *I quit Instagram and Facebook and it made me a lot happier—and that's a big problem for social media companies*. CNBC News. https://www.cnbc.com/2018/12/01/social-media-detox-christina-farr-quits-instagram-facebook.html. Quote appears in para. 34.
174. LaFrance, A. (2020, December 15). Facebook is a doomsday machine. *The Atlantic*. https://www.theatlantic.com/technology/archive/2020/12/facebook-doomsday-machine/617384/. Quote appears in para. 37.

CHAPTER 3

1. *The Mahatma. Life chronology*. (n.d.) Gandhi Ashram at Sabarmati. https://gandhiashramsabarmati.org/en/the-mahatma/life-chronology.html.
2. Birkeland, M. S., Breivik, K., & Wold, B. (2014, January). Peer acceptance protects global self-esteem from negative effects of low closeness to parents during adolescence and early adulthood. *Journal of Youth and Adolescence, 43*, 70–80.
3. Back, M., Mund, M., Finn, C., Hagemeyer, B., Zimmermann, J., & Neyer, F. J. (2015). The dynamics of self-esteem in partner relationships. *European Journal of Personality, 2*, 235–249.
4. Zhang, L., Zhang, S., Yang, Y., & Li, C. (2017). Attachment orientations and dispositional gratitude: The mediating roles of perceived social support and self-esteem. *Personality & Individual Differences, 114*, 193–197.
5. Luerssen, A., Jhita, G. J., & Ayduk, O. (2017). Putting yourself on the line: Self-esteem and expressing affection in romantic relationships. *Personality & Social Psychology Bulletin, 7*, 940–956.
6. Kille, D. R., Eibach, R. P., Wood, J. V., & Holmes, J. G. (2017). Who can't take a compliment? The role of construal level and self-esteem in accepting positive feedback from close others. *Journal of Experimental Social Psychology, 68*, 40–49.
7. Dredge, R., Gleeson, J. M., & de la Piedad Garcia, X. (2014). Risk factors associated with impact severity of cyberbullying victimization: A qualitative study of adolescent online social networking. *Cyberpsychology, Behavior & Social Networking, 17*(5), 287–291.
8. Vukasović, T., & Bratko, D. (2015). Heritability of personality: A meta-analysis of behavior genetic studies. *Psychological Bulletin, 141*(4), 769–785.
9. Sanchez-Roige, S., Gray, J. C., MacKillop, J., Chen, C.-H., & Palmer, A. A. (2018). The genetics of human personality. *Genes, Brain, and Behavior, 17*(3), e12439. https://onlinelibrary.wiley.com/doi/full/10.1111/gbb.12439.
10. Jeroen Kuntze, Henk T. van der Molen, & Marise Ph. Born. (2016). Big five personality traits and assertiveness do not affect mastery of communication skills. *Health Professions Education, 2*(1), 33–43.
11. Widisto, A. (2020, December 5). *Do you really not care about what others think of you?* Medium. https://medium.com/better-advice/do-you-really-not-care-about-what-others-think-of-you-ef1642d6ec98. Quote appears in para. 29.
12. Festinger, L. (1954). A theory of social comparison processes. *Human Relations, 7*, 117–140.
13. Gerber, J. P., Wheeler, L., & Suls, J. (2018). A social comparison theory meta-analysis 60+ years on. *Psychological Bulletin, 144*(2), 177–197.
14. Afarin. (2019, July 25). Attention: Instagram models don't actually look like that. *Weekends with Afarin. 1013 the River*. https://www.1013theriver.com/2019/07/25/attention-instagram-models-dont-actually-look-like-that/. Quote appears in title.
15. White, J. B., Langer, E. J., Yariv, L., & Welch, J. C. (2006). Frequent social comparisons and destructive emotions and behaviors: The dark side of social comparisons. *Journal of Adult Development, 1*, 36–44.
16. Robinson, A., Bonnette, A., Howard, K., Ceballos, N., Dailey, S., Lu, Y., & Grimes, T. (2019). Social comparisons, social media addiction, and social interaction: An examination of specific social media behaviors related to major depressive disorder in a millennial population. *Journal of Applied Biobehavioral Research, 24*(1), 1–14.
17. Vedantam, S. (2017, May 2). Why social media isn't always very social. *Hidden Brain*. Transcript retrieved from http://www.npr.org/2017/05/02/526514168/why-social-media-isnt-always-very-social. Quotes appear in para. 9.
18. Brown, B. (2019). *The call to courage* [documentary film]. Netflix.
19. Bailey, E. R., Matz, S. C., Youyou, W., Ivengar, S. S. (2020). Authentic self-expression on social media is associated with greater subjective well-being. *Nature Communications, 11*(1), 1–9.
20. Yang, C.-c., Holden, S. M., & Carter, M. D. K. (2017). Emerging adults' social media self-presentation and identity development at college transition: Mindfulness as a moderator. *Journal of Applied Developmental Psychology, 52*, 212–221.
21. Desta, Y. (2015, March 13). *Obsessing over the perfect social media post is ruining your life, study says*. Mashable. http://mashable.com/2015/03/13/social-media-ruining-your-life/#qbjK5m5pKaq3.
22. Holmes, J. G. (2002). Interpersonal expectations as the building blocks of social cognition: An interdependence theory perspective. *Personal Relationships, 9*, 1–26.
23. Rosenthal, R., & Jacobson, L. (1968). *Pygmalion in the classroom*. Holt, Rinehart and Winston.

24. Tatlah, I. A., Masood, S., & Amin, M. (2019). Impact of parental expectations and students' academic self-concept on their academic achievements. *Journal of Research & Reflections in Education, 13*(2), 170–182.
25. Darlow, V., Norvilitis, J., & Schuetze, P. (2017). The relationship between helicopter parenting and adjustment to college. *Journal of Child & Family Studies, 26*(8), 2291–2298.
26. Carrera, C. (2019, September 19). Bulldozer parents: Stop overparenting college students. It can do more harm than good. *USA Today.* https://www.northjersey.com/story/news/education/2019/09/19/overparenting-college-students-can-have-long-term-mental-health-effect/1899995001/. Quote appears in para. 4 of "How to Stop Bulldozing" section.
27. Agrahari, S. K., & Kinra, A. (2017). Impact of parental expectation on self-concept of adolescent. *Indian Journal of Health & Wellbeing, 8*(11), 1357–1360.
28. Müller-Pinzler, L., Czekalla, N., Mayer, A. V., Stolz, D. S., Gazzola, V., Keysers, C., Paulus, F. M., & Krach, S. (2019). Negativity-bias in forming beliefs about own abilities. *Scientific Reports, 9*(1), nonpaginated.
29. Hay, L. (n.d.). Self-criticism quotes. *AZ Quotes.* https://www.azquotes.com/quotes/topics/self-criticism.html. Quote appears fourth.
30. Robbins, M. (n.d.). *Quote catalog.* https://quotecatalog.com/quote/mel-robbins-if-you-wouldnt-zpWdkwa. Quote appears first.
31. Biali Haas, S. (2018, March 5). How to stop comparing yourself to others. *Psychology Today.* https://www.psychologytoday.com/us/blog/prescriptions-life/201803/how-stop-comparing-yourself-others. Quote appears in item 1.
32. Scuderi, R. (n.d.). The number one secret to life success: Baby steps. *Lifehack.* https://www.lifehack.org/articles/communication/secret-to-life-success-baby-steps.html. Quote appears in para. 4.
33. Tan, T. (2016, June 21). *What it's like to be friends with someone who has special needs.* Odyssey. https://www.theodysseyonline.com/friends-someone-special-needs. Quote appears in para. 2.
34. James, W. (1920). *The letters of William James* (H. James, Ed.). Boston, p. 462.
35. Elford, M. (2020, December 27). *Things I didn't appreciate about my mother until she was gone.* Welcome to the Zoo. https://welcometothezoo.ca/things-i-didnt-appreciate-about-my-mother-until-she-was-gone/. Quote appears in para. 8.
36. Trilla, I., Weigand, A., & Dziobek, I. (2021). Affective states influence emotion perception: Evidence for emotional egocentricity. *Psychological Research, 85*(3), 1005–1015.
37. Ogolsky, B. G., Monk, J. K., Rice, T. M., Theisen, J. C., & Maniotes, C. R. (2017). Relationship maintenance: A review of research on romantic relationships. *Journal of Family Theory and Review, 9*(3), 275–306.
38. Hoge, R. (2017, November 9). Don't judge me based on my speech disability. *Toronto Star.* https://www.thestar.com/life/2017/11/09/dont-judge-me-based-on-my-speech-disability.html. Quote appears in para. 26.
39. Spencer, S. J., Logel, C., & Davies, P. G. (2016, January). Stereotype threat. *Annual Review of Psychology. 67,* 415–437.
40. Santos-Díaz, S. (2019, February 28). As an underrepresented minority student, my community's support can feel like pressure. *Science.* https://www.sciencemag.org/careers/2019/02/underrepresented-minority-student-my-community-s-support-can-feel-pressure.
41. Johnson, K. (2021, April 3). If you've always wondered if you're "Black enough," you're not alone. *The Every Girl.* https://theeverygirl.com/black-enough-stereotype/. Quotes appear in para. 6.
42. Johnson (2021). Quote appears in para. 10.
43. Johnson (2021). Quote appears in para. 1.
44. Pisano, V. (2019). *What do you think people think of you when they see you in public?* Quora. https://www.quora.com/What-do-you-think-people-think-of-you-when-they-see-you-in-public. Quotes appears in paras. 4, 5, and 13, respectively.
45. *Ask Matt: Why do trans people make me feel uncomfortable?* (2013, February 25). The Transadvocate. http://transadvocate.com/ask-matt-why-do-trans-people-make-me-uncomfortable_n_8773.htm. Quote appears in third to last paragraph.
46. Marlow, S. L., Lacerenza, C. N., & Iwig, C. (2018). The influence of textual cues on first impressions of an email sender. *Business and Professional Communication Quarterly, 81*(2), 149–166.
47. Kanouse, D. E., & Hanson, L. (1972). Negativity in evaluations. In E. E. Jones, D. E. Kanouse, S. Valins, H. H. Kelley, R. E. Nisbett, & B. Weiner (Eds.), *Attribution: Perceiving the causes of behavior.* General Learning Press.
48. Rozin, P., & Royzman, E. B. (2001). Negativity bias, negativity dominance, and contagion. *Personality & Social Psychology Review, 5*(4), 296–320.
49. Williams, R. (2014, June 30). Are we hardwired to be positive or negative? *Psychology Today.* https://www.psychologytoday.com/blog/wired-success/201406/are-we-hardwired-be-positive-or-negative.
50. Black, S. L., & Johnson, A. F. (2012). Employers' use of social networking sites in the selection process. *Journal of Social Media in Society, 1*(1), 7–28.
51. For a review of these perceptual biases, see Hamachek, D. (1992). *Encounters with the self* (3rd ed.). Harcourt Brace Jovanovich. See also Bradbury, T. N., & Fincham, F. D. (1990). Attributions in marriage: Review and critique. *Psychological Bulletin, 107,* 3–33. For information on the self-serving bias, see Shepperd, J., Malone, W., & Sweeny, K. (2008). Exploring causes of the self-serving bias. *Social and Personality Psychology Compass, 2/2,* 895–908.
52. Thorndike, E. L. (1920). A constant error in psychological ratings. *Journal of Applied Psychology, 4*(1), 25–29.
53. Talamas, S. N., Mavor, K. I., & Perrett, D. I. (2016). Blinded by beauty: Attractiveness bias and accurate perceptions of academic performance. *Plos ONE, 11*(2), 1–18.
54. Thorndike, E. L. (1920). A constant error in psychological ratings. *Journal of Applied Psychology, 4*(1), 25–29.
55. Bacev-Giles, C., & Haji, R. (2017). Person perception in social media profiles. *Computers in Human Behavior, 75,* 50–57.
56. Herrick, L. (2014, June 17). *Respecting political differences: As told by a liberal living in a sea of conservatives.* Huffington Post. https://www.huffingtonpost.com/lexi-herrick/respecting-political-diff_b_5500406.html. Quote appears in para. 2.

57. Goleman, D. (2006). *Emotional intelligence [why it can matter more than IQ]*. Bantam Books.
58. Barrett, L. F. (2018, June 21). *Try these two smart techniques to help you master your emotions*. Ideas.Ted.com. https://ideas.ted.com/try-these-two-smart-techniques-to-help-you-master-your-emotions/.
59. Kozina, A., & Mleku, A. (2016). Intrinsic motivation as a key to school success: Predictive power of self-perceived autonomy, competence and relatedness on the achievement in international comparative studies. *Solsko Polje, 27*(1/2), 63–88.
60. Preece, J. (2004). Etiquette, empathy and trust in communities of practice: Stepping-stones to social capital. *Journal of Universal Computer Science, 10*(3), 294–302.
61. Stiff, J. B., Dillard, J. P., Somera, L., Kim, H., & Sleight, C. (1988). Empathy, communication, and prosocial behavior. *Communication Monographs, 55*, 198–213.
62. Chen, Q., Zhu, Y., & Chui, W. H. (2020). A meta-analysis on effects of parenting programs on bullying prevention. *Trauma Violence & Abuse*, nonpaginated online publication. https://pubmed.ncbi.nlm.nih.gov/32242506/.
63. Pounds, G., Salter, C., Platt, M. J., & Bryant, P. (2017). Developing a new empathy-specific admissions test for applicants to medical schools: A discourse-pragmatic approach. *Communication & Medicine, 14*(2), 165–180.
64. Bradberry, T. (2018, January 11). *What are some good examples of emotional intelligence?* Quora. https://www.quora.com/What-are-some-good-examples-of-emotional-intelligence.
65. Ramanauskas, K. (2016). The impact of the manager's emotional intelligence on organisational performance. *Management Theory & Studies for Rural Business & Infrastructure Development, 38*(1), 58–69.
66. Bain, E. (2021). *Who is Jenny Parrish? Woman's wine glass reflection blunder takes over TikTok!* HITC. https://www.hitc.com/en-gb/2021/03/24/jenny-parrish-wine-glass/. Quotes appear in the second to last paragraph.
67. Goffman, E. (1971). *The presentation of self in everyday life*. Doubleday; and Goffman, E. (1971). *Relations in public*. Basic Books.
68. Sezer, O., Gino, F., & Norton, M. I. (2015, April 15). *Humblebragging: A distinct—and ineffective—self-presentation strategy*. Harvard Business School Marketing Unit Working Paper No. 15–080. http://papers.ssrn.com/sol3/papers.cfm?abstract_id=2597626.
69. Appiah, K. (August 10, 2018). Go ahead, speak for yourself. *New York Times*. https://www.nytimes.com/2018/08/10/opinion/sunday/speak-for-yourself.html.
70. Benet-Martínez, V., Leu, J., Lee, F., & Morris, M. (2002). Negotiating biculturalism: Cultural frame switching in biculturals with oppositional versus compatible cultural identities. *Journal of Cross-Cultural Psychology, 33*, 492–516.
71. Toomey, A., Dorjee, T., & Ting-Toomey, S. (2013). Bicultural identity negotiation, conflicts, and intergroup communication strategies. *Journal of Intercultural Communication Research, 42*(2), 112–134. Quote appears on p. 120.
72. *12 ways to impress your first days on the job*. (2019, May 8). Executive Drafts. https://executivedrafts.com/12-ways-to-impress-your-first-days-on-the-job/. Quote appears in para. 2 of item 2.
73. Triumph, T. (n.d.). *Reverse Undercover Boss*. https://tomtriumph.com/reverse-undercover-boss/.
74. Rollender, N. (2019, January 8). I've been a manager for 10 years—here are the 5 best ways to impress your boss. *Insider*. https://www.businessinsider.com/how-to-impress-your-boss-according-to-manager-2018-10. Quote appears in para. 2 of item 5.
75. Snyder, M. (1979). Self-monitoring processes. In L. Berkowitz (Ed.), *Advances in experimental social psychology*. Academic Press; Snyder, M. (1983, March). The many me's of the self-monitor. *Psychology Today*, p. 34f.
76. Hall, J. A., & Pennington, N. (2013). Self-monitoring, honesty, and cue use on Facebook: The relationship with user extraversion and conscientiousness. *Computers in Human Behavior, 29*, 1556–1564.

CHAPTER 4

1. Vasadi, E. (n.d.). My 6 culture shocks with moving to America. [Personal blog.] https://editvasadi.com/american-culture-shock/. Quotes appear in paras. 1 and 3 of Culture Shock 3, respectively.
2. Tajfel, H., & Turner, J. C. (1986). The social identity theory of inter-group behavior. In S. Worchel & L. W. Austin (Eds.), *Psychology of intergroup relations*. Nelson-Hall.
3. Tajfel, H., & Turner, J. C. (1986). The social identity theory of inter-group behavior. In S. Worchel & L. W. Austin (Eds.), *Psychology of intergroup relations*. Nelson-Hall.
4. Flateland, S. M., Pryce-Miller, M., Skisland, A. V.-S., Tønsberg, A. F., & Söderhamn, U. (2019). Exploring the experiences of being an ethnic minority student within undergraduate nurse education: A qualitative study. *BMC [BioMed Central] Nursing, 18*, nonpaginated.
5. Song, S. (2021, April 14). *8 Asian musicians on racism and being "othered."* Paper. https://www.papermag.com/aapi-musicians-round-table-2,652,573,767.html?rebelltitem=19#rebelltitem19. Quote appears in para. 1 under "Zhu."
6. Lessard, L. M., Kogachi, K., & Juvonen, J. (2019). Quality and stability of cross-ethnic friendships: Effects of classroom diversity and out-of-school contact. *Journal of Youth and Adolescence, 48*(3), 554–566.
7. Benner, A. D., & Wang, Y. (2017). Racial/ethnic discrimination and adolescents' well-being: The role of cross-ethnic friendships and friends' experiences of discrimination. *Child Development, 88*(2), 493–504.
8. Richardson, D., & Big Debo. (n.d.) *My Black friend* [podcast]. https://player.fm/series/my-black-friend.
9. Wong, B. (2020, September 4). *Why we need more close interracial friendships (and why we're bad at them)*. HuffPost. https://www.huffpost.com/entry/close-interraacial-friendships_l_5f5122c8c5b6946f3eaed704. Quote appears in fifth to last paragraph.
10. Samson, J. (2018, February 12). *I'm an Asian American and I will never fit your stereotypes*. Odyssey. https://www.theodysseyonline.com/asians-are-not-weak. Quote appears in paras. 1, 2, and 5.

11. Kelly, N. (2013, July 30). Bad-luck numbers that scare off customers. *Harvard Business Review*. https://hbr.org/2013/07/the-bad-luck-numbers-that-scar
12. *A quick guide: Gift giving in Japan—dos and don'ts.* (n.d.). Zooming Japan. http://zoomingjapan.com/culture/gift-giving-in-japan/.
13. *Working with sign language interpreters: The DOs and DON'Ts.* (2014, September 29). Interpreting Services. http://www.signlanguagenyc.com/working-with-sign-language-interpreters-the-dos-and-donts/.
14. *How to eat in China—Chinese dining etiquette.* (n.d.). China Highlights. https://www.chinahighlights.com/travelguide/chinese-food/dining-etiquette.htm.
15. DiMeo, D. F. (n.d.). *Arabic greetings and good-byes*. Arabic for Dummies. http://www.dummies.com/languages/arabic/arabic-greetings-and-good-byes/.
16. *Body language in Arab cultures.* (n.d.). Word Press Culture Convo. https://tbell7.wordpress.com/2012/10/23/body-language-in-arab-cultures/.
17. *Traveling in a Muslim country.* (n.d.). Embassy of the United Arab Emirates. https://www.uae-embassy.org/about-uae/travel-culture/traveling-muslim-country.
18. van de Kemenade, D. (2013, September 26). Life lessons learnt from my intercultural relationship. [Personal blog.] http://www.daniellevandekemenade.com/life-lessons-learnt-from-my-intercultural-relationship/. This quote appears in para. 5 and the next in para 12.
19. Triandis, H. C. (1995). *Individualism and collectivism*. Westview.
20. Merkin, R. (2015). The relationship between individualism/collectivism. *Journal of Intercultural Communication, 39*, nonpaginated.
21. *Shinzo Abe or Abe Shinzo? A basic guide to using Asian names.* (2019, May 23). Asia Media Centre. https://www.asiamediacentre.org.nz/features/a-guide-to-using-asian-names/.
22. Babe, A. (2017, December 18). *How the South Korean language was designed to unify*. BBC. http://www.bbc.com/travel/story/20171217-why-south-koreans-rarely-use-the-word-me. Quotes appear in paras. 1 and 16, respectively.
23. *Culture shock in Hungary.* (2020, December). Expat Arrivals. https://www.expatarrivals.com/europe/hungary/culture-shock-hungary.
24. FitzGerald, C. (2015, May 11). *9 culture shocks Americans will have in Hungary*. Matador Network. https://matadornetwork.com/abroad/9-culture-shocks-americans-will-hungary/. Quote appears in item 6.
25. Daeley, S. (2015, September 22). *16 culture shocks every American will experience in Japan*. Matador Network. https://matadornetwork.com/bnt/16-culture-shocks-every-american-will-experience-japan/. Quote appears in item 3.
26. *10 things to know about U.S. culture.* (n.d.). InterExchange. https://www.interexchange.org/articles/career-training-abroad/10-things-to-know-about-u-s-culture/. Quote appears in "Competition" section.
27. Merkin, R. (2015). The relationship between individualism/collectivism. *Journal of Intercultural Communication, 39*, 4.
28. Chopik, W. J., O, B. E., & Konrath, S. H. (2017). Differences in empathic concern and perspective taking across 63 countries. *Journal of Cross-Cultural Psychology, 48*(1), 23–38.
29. Steinhage, A., Cable, D., & Wardley, D. (2017). The pros and cons of competition among employees. *Harvard Business Review Digital Articles*, 1–5.
30. Biçer, C. (2020). The hedgehog's dilemma in organizations: The pros and cons of workplace friendship. *Hitit Üniversitesi Sosyal Bilimler Enstitüsü Dergisi, 13*(1), 201–217.
31. Moghaddam, M. M. (2017). Politeness at the extremes: Iranian women's insincere responses to compliments. *Language & Dialogue, 7*(3), 413–431. Quotes appear on pp. 422 and 421, respectively.
32. Sarhan, N. M., Harb, A., Shrafat, F. D., & Alshishany, A. (2019). The impact of individualism and collectivism on communication apprehension: A study of university academic staff. *Journal of Institutional Research South East Asia, 17*(2), 71–85.
33. Merkin, R. (2015). The relationship between individualism/collectivism. *Journal of Intercultural Communication, 39*, 4.
34. Cai, D. A., & Fink, E. L. (2002). Conflict style differences between individualists and collectivists. *Communication Monographs, 69*, 67–87.
35. Croucher, S. M., Galy-Badenas, F., Jäntti, P., Carlson, E., & Cheng, Z. (2016). A test of the relationship between argumentativeness and individualism/collectivism in the United States and Finland. *Communication Research Reports, 33*(2), 128–136.
36. Hall, E. T. (1959). *Beyond culture*. Doubleday.
37. Chlopicki, W. (2017). Communication styles—an overview. *Styles of Communication, 9*(2), 9–25.
38. Vasadi, E. (n.d.). My 6 culture shocks with moving to America. [Personal blog.] https://editvasadi.com/american-culture-shock/.
39. Chung, J. H. J. (2021). "We participate, silently": Explicating Thai university students' perceptions of their classroom participation and communication. *Qualitative Research in Education, 10*(1), 62–87.
40. Hahn, M., & Molinsky, A. (2016, March 25). Having a difficult conversation with someone from a different culture. *Harvard Business Review*. https://hbr.org/2016/03/having-a-difficult-conversation-with-someone-from-a-different-culture.
41. Hahn & Molinsky (2016). Quote appears in para. 4.
42. Chen, Y.-S., Chen, C.-Y. D., & Chang, M.-H. (2011). American and Chinese complaints: Strategy use from a cross-cultural perspective. *Intercultural Pragmatics, 8*, 253–275.
43. Croucher, S., Bruno, A., McGrath, P., Adams, C., McGahan, C., Suits, A., & Huckins, A. (2012). Conflict styles and high-low context cultures: A cross-cultural extension. *Communication Research Reports, 29*(1), 64–73.
44. Hofstede, G. (2001). *Culture's consequences: Comparing values, behaviors, institutions, and organizations across nations* (2nd ed.). Sage.
45. Zhao, H.-Y., Kwon, J.-W., & Yang, O.-S. (2016, October). Updating Hofstede's cultural model and tracking changes in cultures indices. *Journal of International Trade & Commerce, 12*(5), 85–106.
46. Schinkel, S., Schouten, B. C., Kerpiclik, F., Van Den Putte, B., & Van Weert, J. C. M. (2019). Perceptions of barriers to patient participation: Are they due to language, culture, or

46. discrimination? *Health Communication, 34*(12), 1469–1481. Quote appears on p. 1477.
47. Sweetman, K. (2012, April 10). In Asia, power gets in the way. *Harvard Business Review.* https://hbr.org/2012/04/in-asia-power-gets-in-the-way. Quote appears in para. 1.
48. Zhao, H.-Y., Kwon, J.-W., & Yang, O.-S. (2016, October). Updating Hofstede's cultural model and tracking changes in cultures indices. *Journal of International Trade & Commerce, 12*(5), 85–106.
49. Spear, J., & Matusitz, J. (2015). Doctor-patient communication styles: A comparison between the United States and three Asian countries. *Journal of Human Behavior in the Social Environment, 25*(8), 871–884.
50. Zhao, H.-Y., Kwon, J.-W., & Yang, O.-S. (2016, October). Updating Hofstede's cultural model and tracking changes in cultures indices. *Journal of International Trade & Commerce, 12*(5), 85–106.
51. Dailey, R. M., Giles, H., & Jansma, L. L. (2005). Language attitudes in an Anglo-Hispanic context: The role of the linguistic landscape. *Language & Communication, 25*(1), 27–38.
52. Basso, K. (2012). "To give up on words": Silence in Western Apache culture. In L. Monogahn, J. E. Goodman, & J. M. Robinson (Eds.), *A cultural approach to interpersonal communication: Essential readings* (2nd ed., pp. 73–83). Blackwell. Quote appears on p. 84.
53. Bowleg, L. (2008). When black + lesbian + woman ≠ black lesbian woman: The methodological challenges of qualitative and quantitative intersectionality research. *Sex Roles, 59*(5/6), 312–325. Quote appears on p. 312.
54. DeFrancisco, V. P., & Palczewski, C. H. (2014). *Gender in communication.* Sage.
55. *The state of the gender pay gap in 2021.* (2021). Payscale. https://www.payscale.com/data/gender-pay-gap.
56. *Racial gap for men.* (2019, May 7). Payscale. https://www.payscale.com/data/racial-wage-gap-for-men.
57. Jones, J. (2021, March 19). 5 facts about the state of the gender pay gap. *U.S. Department of Labor Blog.* https://blog.dol.gov/2021/03/19/5-facts-about-the-state-of-the-gender-pay-gap.
58. Examining racial gap in graduate school enrollments in the United States. (2019, October 14). *The Journal of Blacks in Higher Education.* https://www.jbhe.com/2019/10/examining-the-racial-gap-in-graduate-school-enrollments-in-the-united-states/.
59. National Center for Educational Statistics. (2021, May). *Graduate degree fields.* https://nces.ed.gov/programs/coe/indicator/ctb.
60. Sheth, S., Hoff, M., Ward, M., & Tyson, T. (2021, March 24). These 8 charts show the glaring gap between men's and women's salaries in the US. *Business Insider.* https://www.businessinsider.com/gender-wage-pay-gap-charts-2017-3.
61. Thomas, R., Cooper, M., Cardazone, G., Urban, K., Bohrer, A., Long, M., Yee, L., Krivkovich, A., Huang, J., Prince, S., Kumar, A., & Coury, S. (2020). *Women in the workplace 2020: Corporate America is at a critical crossroads.* McKinsey & Company and Lean In. https://wiw-report.s3.amazonaws.com/Women_in_the_Workplace_2022.pdf.
62. A., R. S. (2020, August 25). I am a Black woman and my mentor is a privileged White man. *Illumination.* https://medium.com/illumination-curated/i-am-a-black-woman-and-my-mentor-is-a-privileged-white-man-7ebd08086cd9. Quote appears in para. 3.
63. Ten things everyone should know about race. (2003). *Race—The power of an illusion.* California Newsreel, Public Broadcasting System. https://newsreel.org/guides/race/10things.htm
64. Bonam, C. M., & Shih, M. (2009). Exploring multiracial individual's comfort with intimate interracial relationships. *Journal of Social Issues, 65,* 87–103.
65. Anti-social media: 10,000 racial slurs a day on Twitter, finds Demos. (2014, February 7). Demos. https://demos.co.uk/press-release/anti-social-media-10000-racial-slurs-a-day-on-twitter-finds-demos-2/.
66. Tynes, B. M. (2015, December). *Online racial discrimination: A growing problem for adolescents.* American Psychological Association. https://www.apa.org/science/about/psa/2015/12/online-racial-discrimination.
67. Perrin, A. (2020, October 15). *23 percent of users in U.S. say social media led them to change views on an issue; some cite Black Lives Matter.* Pew Research Center. https://www.pewresearch.org/fact-tank/2020/10/15/23-of-users-in-us-say-social-media-led-them-to-change-views-on-issue-some-cite-black-lives-matter/.
68. Vasilogambros, M., & National Journal. (2015, April 10). Why is it so hard to talk about race? *The Atlantic.* https://www.theatlantic.com/politics/archive/2015/04/why-is-it-so-hard-to-talk-about-race/431934/. Quote appears in para. 3.
69. Wilson, B. L. (2020, June 8). I'm your Black friend, but I won't educate you about racism. That's on you. *The Washington Post.* https://www.washingtonpost.com/outlook/2020/06/08/black-friends-educate-racism/.
70. Campt, D. (2020, April 27). *Message to White allies from a Black anti-racism expert: You're doing it wrong.* Medium. https://medium.com/progressively-speaking/message-to-white-allies-from-a-black-racial-dialogue-expert-youre-doing-it-wrong-39c09b3908a5.
71. Oluo, I. (2018). *So you want to talk about race.* Seal Press. Quote appears on p. 221.
72. Eddo-Lodge, R. (2017, May 30). Why I'm no longer talking to White people about race. *The Guardian.* https://www.theguardian.com/world/2017/may/30/why-im-no-longer-talking-to-white-people-about-race. Quote appears in para. 4.
73. Oluo, I. (2019). *So you want to talk about race.* Seal Press. Quotes in this paragraph appear on p. 40.
74. Meta, J. (2017). *On allies asking to be taught about race.* Medium. https://medium.com/@johnmetta/when-you-walk-into-the-valley-933ff8c079c0.
75. Oluo, I. (2019). *So you want to talk about race.* Seal Press. Quotes in this paragraph appear on p. 40.
76. DiAngelo, R. (2018). *White fragility: Why it's so hard for White people to talk about racism.* Beacon Press.
77. Acho, E. (2020). *Uncomfortable conversations with a Black man.* Flatiron Books.
78. Tatum, B. D. (2017). *Why are all the Black kids sitting together in the cafeteria? And other conversations about race.* Basic Books.

79. Oluo, I. (2019). *So you want to talk about race.* Seal Press. Quotes appears on pp. 45 and 168, respectively.
80. Williams, P. S. (2017, December 19). *I've lived as a man & a woman—here's what I learned* [Video]. TED. https://www.youtube.com/watch?v=lrYx7HaUlMY. Quotes begin at 2:30 mins and 7:55 mins, respectively.
81. Andreas, O., Eleni, K., Kimmo, S., & Ivanka, S. (2016). Testosterone and estrogen impact social evaluations and vicarious emotions: A double-blind placebo-controlled study. *Emotion, 4,* 515–523.
82. *Hyperprolactinemia diagnosis and treatment: A patient's guide.* (2011, February). Hormone Health Network. https://academic.oup.com/jcem/article/96/2/35A/2709499.
83. PFLAG [Parents, Families, and Friends of Lesbians and Gays]. (2021, January). *PFLAG national glossary of terms.*https://pflag.org/glossary#:~:text=Gender%20Expansive%3A%20An%20umbrella%20term,or%20expected%20societal%20gender%20norms.
84. Weiss, S. (2018, February 15). 9 things people get wrong about being nonbinary. *Teen Vogue.* https://www.teenvogue.com/story/9-things-people-get-wrong-about-being-non-binary. Quote appears in para. 3.
85. Wamsley, L. (2021, June 2). *A guide to gender identity terms.* NPR. https://www.npr.org/2021/06/02/996319297/gender-identity-pronouns-expression-guide-lgbtq.
86. GenderGP. (2021, April 22). *My first hand experience of discovering I'm agender.* https://www.gendergp.com/discovering-im-agender-first-hand-experience/. Quote appears in para. 9.
87. Abrams, M. (2019, December 20). *64 terms that describe gender identity and expression.* Healthline. https://www.healthline.com/health/different-genders.
88. Hancox, L. (n.d.). *Top 8 tips for coming out as trans.* Ditch the Label. http://www.ditchthelabel.org/8-tips-for-coming-out-as-trans/. Quotes appear in tip 4 and tip 1, respectively.
89. *Pronouns matter.* (n.d.). Lionel Cantú Queer Center, University of California Santa Cruz. https://queer.ucsc.edu/education-resources/pronouns-matter.html. Quote appears in fifth bulleted point in "Why Is It Important" section.
90. *Pronouns matter.* (n.d.).
91. *Transgender teen on why "deadnaming" Elliot Page is harmful.* (2020, December 4). CBC Kids News. https://www.cbc.ca/kidsnews/post/transgender-teen-on-why-deadnaming-elliot-page-is-harmful. Quote appears in para. 13.
92. Scharrer, E., & Blackburn, G. (2018). Cultivating conceptions of masculinity: Television and perceptions of masculine gender role norms. *Mass Communication & Society, 21*(2), 149–177.
93. Blackburn, G., & Scharrer, E. (2019). Video game playing and beliefs about masculinity among male and female emerging adults. *Sex Roles, 80*(5/6), 310–324.
94. Weinberg, F. J., Treviño, L. J., & Cleveland, A. O. (2019). Gendered communication and career outcomes: A construct validation and prediction of hierarchical advancement and non-hierarchical rewards. *Communication Research, 46*(4), 456–502.
95. Baldoni, J. (2017). *Why I'm done trying to be "man enough"* [Video]. TED. https://www.ted.com/talks/justin_baldoni_why_i_m_done_trying_to_be_man_enough/up-next?referrer=playlist-how_masculinity_is_evolving#t-117,247. Quotes begin at 16:20 mins and 15:45 mins, respectively.
96. Sandberg, S. (2013). *Lean in: Women, work, and the will to lead.* Knopf.
97. Staley, O. (2016, September 27). Here's new ammunition from McKinsey for women fighting for equality in the workplace. *Quartz.* https://qz.com/793109/a-mckinsey-and-lean-in-report-on-women-in-the-workplace-study-shows-women-are-still-trailing-men-in-opportunities/.
98. Lean In. (2014, March 9). *Ban bossy—I'm not bossy. I'm the boss* [Video]. YouTube. https://www.youtube.com/watch?v=6dynbzMlCcw&list=UUgjlS2OBBggUw92bqzjAaVw. Quote begins at 1:07 mins.
99. Murphy, C. (2016, May 17). 6 people share what it took to make their interfaith relationships work. *Women's Health.* https://www.womenshealthmag.com/relationships/a19969422/interfaith-relationships/.
100. Hussein, Y. (2015, December 3). *Are you afraid to be Muslim in America?* Huffington Religion. http://www.huffingtonpost.com/yasmin-hussein/are-you-afraid-to-be-muslim-in-america_b_8710826.html. Quote appears in para. 8.
101. Ellis, S. (2020, March 25). *Here's the right time to talk about religion on dates.* Elite Daily. https://www.elitedaily.com/p/talking-about-religion-on-dates-can-be-tricky-so-heres-when-to-do-it-22654330. Quote appears in para. 7.
102. Charoensap-Kelly, P., Mestayer, C., & Knight, G. B. (2020). Religious talk at work: Religious identity management in the United States workplace. *Journal of Communication & Religion, 43*(1), 55–74. Quote appears on p. 61.
103. Riggio, M. (n.d.). What's it like to have Down syndrome. *National Geographic Kids.* https://kids.nationalgeographic.com/pages/article/down-syndrome. Quote appears in para. 5.
104. Basingstoke and District Disability Forum. (2013, October 29). *Talk to me. Physical disability awareness* [Video]. YouTube. https://www.youtube.com/watch?v=CL8GMxRW_5Y. Quotes begin at 0:46 mins and 2:32 mins, respectively.
105. Ladua, E. (n.d.). 4 disability euphemisms that need to bite the dust. *Center for Disability Rights.* http://cdrnys.org/blog/disability-dialogue/the-disability-dialogue-4-disability-euphemisms-that-need-to-bite-the-dust/.
106. Wright, E. (2020, February 11). *Whatever you do don't call me differently abled.* The Startup. https://medium.com/swlh/whatever-you-do-dont-call-me-differently-abled-d947ac029801.
107. Ladau, E. (n.d.). 4 disability euphemisms that need to bite the dust. *Center for Disability Rights.* http://cdrnys.org/blog/disability-dialogue/the-disability-dialogue-4-disability-euphemisms-that-need-to-bite-the-dust/. Quote appears in para. 2.
108. Koester, N. (2019, January 19). *Choosing the right words.* National Center on Disability and Journalism. https://ncdj.org/wp-content/uploads/2019/01/Choosing-the-Right-Words.pdf.

109. Koester (2019). Quote appears on slide 20.
110. Raine, C. (2017, October 22). *6 tips—how to talk to a Deaf/HOH person* [Video]. YouTube. https://www.youtube.com/watch?v=d-MHRa0LKG0. Quote begins at 1:10 mins.
111. Seven things you should stop saying and doing to disabled people. (2017). *The Guardian.* https://www.theguardian.com/inequality/2017/nov/15/seven-things-you-should-stop-saying-doing-to-disabled-people.
112. Twardowski, B. (2020, January 9). *Opinion: What this wheelchair user wants you to know.* Next Avenue. https://www.nextavenue.org/what-this-wheelchair-user-wants-you-to-know/. Quotes appears in paras. 6 and 12, respectively.
113. Riggio, M. (n.d.). What's it like to have Down syndrome? *National Geographic Kids.* https://kids.nationalgeographic.com/pages/article/down-syndrome. Quote appears in last two paragraphs.
114. Miller, G. (2015, January 11). *There's only one real difference between liberals and conservatives.* Huffington Post. http://www.huffingtonpost.com/galanty-miller/theres-only-one-real-diff_b_6135184.html. Quote appears in final paragraph.
115. CATO Institute. (2020, July 22). *Poll: 62 percent of Americans say they have political views they're afraid to share.* https://www.cato.org/survey-reports/poll-62-americans-say-they-have-political-views-theyre-afraid-share#introduction.
116. Cohen, D. (2021, January 13). One-half of online harassment victims attributed their experiences to political differences. *ADWeek.* https://www.adweek.com/media/one-half-of-online-harassment-victims-attributed-their-experiences-to-political-differences/.
117. Anderson, M., & Auxier, B. (2020, August 19). *55 percent of U.S. social media users say they are "worn out" by political posts and discussions.* Pew Research Center. https://www.pewresearch.org/fact-tank/2020/08/19/55-of-u-s-social-media-users-say-they-are-worn-out-by-political-posts-and-discussions/.
118. Blotky, A. (2020, August 31). Talking about politics constructively in 2020.[Personal blog.] https://andrewblotky.medium.com/talking-about-politics-constructively-in-2020-1141c21921f1. Quote appears in para. 3.
119. Rainie, L., Anderson, J., & Albright, J. (2017, March 29). *The future of free speech.* Pew Research Center: Internet & Technology. http://www.pewinternet.org/2017/03/29/the-future-of-free-speech-trolls-anonymity-and-fake-news-online/.
120. Editorial: Don't be a troll. (2017, April 9). *Iowa State Daily.* http://www.iowastatedaily.com/opinion/editorials/article_f2edd1e8-1d45-11e7-9ccf-ebde4f9f215c.html. Quote appears in para. 3.
121. Kelly, C. (2019, April 12). *Keeping it civil: How to talk politics without letting things turn ugly.* NPR. https://www.npr.org/2019/04/12/712277890/keeping-it-civil-how-to-talk-politics-without-letting-things-turn-ugly. Quote appears in para. 13.
122. Kelly (2019). Quote appears in para. 10.
123. Kelly (2019). Quote appears in para. 10.
124. Corwin, A. I. (2018). Overcoming elderspeak: A qualitative study of three alternatives. *The Gerontologist, 58*(4), 724–729.
125. Ryan, E. B., & Butler, R. N. (1996). Communication, aging, and health: Toward understanding health provider relationships with older clients. *Health Communication, 8,* 191–197.
126. Dukes, J. (2016, July 21). Self-imposed age discrimination: Being your own worst enemy. *Forbes.* https://www.forbes.com/sites/nextavenue/2015/07/21/self-imposed-age-discrimination-being-your-own-worst-enemy/#6b878e6e456c.
127. Strom, R., & Strom, P. (2015). Assessment of intergenerational communication and relationships. *Educational Gerontology, 41*(1), 41–52.
128. Kroger, J., Martinussen, M., & Marcia, J. E. (2010). Identity status change during adolescence and young adulthood: A meta-analysis. *Journal of Adolescence, 33,* 683–698.
129. Galanaki, E. P. (2012). The imaginary audience and the personal fable: A test of Elkind's theory of adolescent egocentrism. *Psychology, 3,* 457–466.
130. Weeks, K. P., & Schaffert, C. (2019). Generational differences in definitions of meaningful work: A mixed methods study. *Journal of Business Ethics, 156*(4), 1045–1061.
131. Stevanin, S., Palese, A., Bressan, V., Vehviläinen-Julkunen, K., & Kvist, T. (2018). Workplace-related generational characteristics of nurses: A mixed-method systematic review. *Journal of Advanced Nursing, 74*(6), 1245–1263.
132. Eschlemann, K. J., King, M., Mast, D., Ornellas, R., & Hunter, D. (2017). The effects of stereotype activation on generational differences. *Work, Aging & Retirement, 3*(2), 200–208.
133. Rudolph, C. W., Rauvola, R. S., Costanza, D. P., & Zacher, H. (2020). Generations and generational differences: Debunking myths in organizational science and practice and paving new paths forward. *Journal of Business and Psychology,* 1–23.
134. Trubela, T. B. (2020, June 2). There are huge benefits to having much older friends (and much younger ones too). *Good Housekeeping.* https://www.goodhousekeeping.com/life/relationships/a32669687/older-friends/. Quote appears in para. 21.
135. Vasadi, E. (n.d.). My 6 culture shocks with moving to America. [Personal blog.] https://editvasadi.com/american-culture-shock/. Quote appears in "Culture Shock 2" subheading.
136. Vasadi (n.d.). Quote appears in para. 2 of "Culture Shock 4."
137. Sue, D. W. (2001). Multidimensional facets of cultural competence. *Counseling Psychologist, 26*(6), 790–821.
138. Spitzberg, B. H., & Chagnon, G. (2011). Conceptualizing intercultural competence. In D. K. Deardorff (Ed.), *The Sage Handbook of Intercultural Competence* (pp. 2–52). Sage.
139. Hook, J. N., Davis, D. E., Owen, J., Worthington, E. L., Jr., & Utsey, S. O. (2013). Cultural humility: Measuring openness to culturally diverse clients. *Journal of Counseling Psychology, 60*(3), 353–366.
140. Tervalon, M., & Murray-García, J. (2010). Cultural humility versus cultural competence: A critical distinction in defining physician training outcomes in multicultural education. *Journal of Health Care for the Poor and Underserved, 9*(2), 117–125.

141. Pettigrew, T. F., & Tropp, L. R. (2000). Does intergroup contact reduce prejudice? Recent meta-analytic findings. In S. Oskamp (Ed.), *Reducing prejudice and discrimination: Social psychological perspectives* (pp. 93–114). Erlbaum.
142. Paluck, E. L., Green, S. A., & Green, D. P. (2018, July 10). The contact hypothesis re-evaluated. *Behavioural Public Policy.* Online article made available by Cambridge University Press. https://osf.io/preprints/socarxiv/w2jkf.
143. Broockman, D., & Kalla, J. (2016). Durably reducing transphobia: A field experiment on door-to-door canvassing. *Science, 352*(6282), 220–224.
144. Collins, L. (2016, October 2). I fell in love with a Frenchman—but didn't speak the language. *The Guardian.* https://www.theguardian.com/lifeandstyle/2016/oct/02/pardon-my-french-conscious-coupling-through-a-language-barrier. Quotes appear in paras. 14, 15, 15, and 19, respectively.
145. Ward, M. (2020, June 4). Top 5 myths about India. BreatheDreamGo. https://breathedreamgo.com/top-5-myths-about-india/.
146. Gudykunst, W. B., & Nishida, T. (2001). Anxiety, uncertainty, and perceived effectiveness of communication across relationships and cultures. *Journal of Intercultural Relations, 25*, 55–71.
147. Oberg, K. (1960). Cultural shock: Adjustment to new cultural environments. *Practical Anthropology, 7*, 177–182.
148. Oberg (1960).
149. Vasadi, E. (n.d.). Moving countless times. [Personal blog.] https://editvasadi.com/meet-edit. Quote appears in para. 3.
150. Chang, L. C.-N. (2011). My culture shock experience. *ETC: A Review of General Semantics, 68*(4), 403–405.
151. Oberg, K. (1960). Cultural shock: Adjustment to new cultural environments. *Practical Anthropology, 7*, 177–182.
152. Oberg (1960).
153. Kim, Y. Y. (2008). Intercultural personhood: Globalization and a way of being. *International Journal of Intercultural Relations, 32*, 359–368.
154. Kim (2008).
155. Kim, Y. Y. (2005). Adapting to a new culture: An integrative communication theory. In W. B. Gudykunst (Ed.), *Theorizing about intercultural communication* (pp. 375–400). Sage.
156. Decker, B., (2016, January 19). *The key to making friends in college.* Odyssey. https://www.theodysseyonline.com/the-key-to-making-friends-in-college. Quotes appear in para. 3 and 6, respectively.

CHAPTER 5

1. Lovato, D. (2020, September 1). Demi Lovato's deeply personal letter on the pandemic, mental health and Black Lives Matter. *Vogue.* https://www.vogue.com/article/demi-lovato-letter-of-hope-pandemic-mental-health-black-lives-matter. Quote appears in para. 7.
2. Chiu, M. (2021, May 19). Demi Lovato comes out as non-binary: "I'll be changing my pronouns to they/them.'" *People.* https://people.com/music/demi-lovato-comes-out-non-binary-changing-pronouns-they-them/. Quote appears in para. 6.
3. Chiu, M. (2021, May 19). Quote appears in para. 12.
4. Anders, C. (2021, June 22). More than 1 million non-binary adults live in the U.S., a pioneering study finds. *Washington Post.* https://www.washingtonpost.com/dc-md-va/2021/06/22/first-population-estimate-lgbtq-non-binary-adults-us-is-out-heres-why-that-matters/.
5. Rice, N. (2021). Demi Lovato says their family has done an "incredible job" of using they/them pronouns. *People.* https://people.com/music/demi-lovatos-family-has-done-incredible-job-of-using-they-them-pronouns/. Quote appears in para. 7.
6. Wilson, R. (2013, December 2). What dialect do you speak? A map of American English. *Washington Post.* https://www.washingtonpost.com/blogs/govbeat/wp/2013/12/02/what-dialect-to-do-you-speak-a-map-of-american-english/.
7. What does air mean in Malay? (n.d.). Dictionary. *Educalingo.* https://educalingo.com/en/dic-ms/air.
8. Sacks, O. (1989). *Seeing voices: A journey into the world of the deaf.* University of California Press, p. 17.
9. *What is LGBTQ?* (n.d.). The Lesbian, Gay, Bisexual & Transgender Community Center. https://gaycenter.org/about/lgbtq/.
10. Braidwood, E. (2018, April 19). What does queer mean? *PinkNews.* https://www.pinknews.co.uk/2018/04/19/what-does-queer-mean/. Quote appears in screen shot of tweet by love, shauna.
11. Pearce, W. B. & Cronen, V. E. (1980). *Communication, action and meaning: The creation of social realities.* Praeger.
12. People wash their clothing in Barf every day. (2009, June 17). *Adweek.* https://www.adweek.com/creativity/people-wash-their-clothing-barf-every-day-14028/.
13. Pous, T. (2017, February 22). Do you know how to pronounce these common words? *BuzzFeed.* https://www.buzzfeed.com/terripous/are-you-pronouncing-these-common-words-correctly?utm_term=.dybPWXYK4#.psAAn1kx6.
14. Assuage. (n.d.). Dictionary.com. http://www.dictionary.com/browse/assuage.
15. Pearce, W. B., & Cronen, V. (1980). *Communication, action, and meaning.* Praeger. See also Barge, J. K. (2004). Articulating CMM as a practical theory. *Human Systems: The Journal of Systemic Consultation and Management, 15*, 193–204; and Griffin, E. M. (2006). *A first look at communication theory* (6th ed.). McGraw-Hill.
16. Associated Press National Opinion Research Center (NORC) for Public Affairs Research. (2016, April 15). *New survey shows Americans believe civility is on the decline.* http://www.apnorc.org/PDFs/Rudeness/APNORC%20Rude%20Behavior%20Report%20%20PRESS%20RELEASE.pdf.
17. Andrews, C. G. (2018, May 8). "Fernweh": A farsickness of longing for unseen places. *Good Nature Travel.* https://www.nathab.com/blog/fernweh-a-farsickness-or-longing-for-unseen-places/
18. For a summary of scholarship supporting the notion of linguistic determinism, see Boroditsky, L. (2010, July 23). Lost in translation. *Wall Street Journal Online.* http://www.wsj.com/articles/SB10001424052748703467304575383131592767868.
19. Vedantham, S. [host]. (2018, January 29). *Lost in translation: The power of language to shape how we view the world.* NPR. https://www.npr.org/templates/transcript/transcript.

php?storyId=581657754. Quote appears in para. 40 of transcript.
20. Boroditsky, L. (2012). How the languages we speak shape the way we think: The FAQs. In M. J. Spivey, K. McRae, & M. F. Joanisse (Eds.) *The Cambridge Handbook of Psycholinguistics* (pp. 615–631). Cambridge University Press.
21. Vervecken, D., & Hannover, B. (2015). Yes I can! Effects of gender fair job descriptions on children's perceptions of job status, job difficulty, and vocational self-efficacy. *Social Psychology, 46*(2), 76–92.
22. Lee, J. K. (2015). "Chairperson" or "chairman"?—A study of Chinese EFL teachers' gender inclusivity. *Australian Review of Applied Linguistics, 38*(1), 24–49.
23. Prewitt-Freilino, J. L., Caswell, T. A., & Laakso, E. K. (2012). The gendering of language: A comparison of gender equality in countries with gendered, natural gender, and genderless languages. *Sex Roles, 66*(3/4), 268–281.
24. *Gendered languages may play a role in limiting women's opportunities, new research finds.* (2019, January 24). The World Bank. https://www.worldbank.org/en/news/feature/2019/01/24/gendered-languages-may-play-a-role-in-limiting-womens-opportunities-new-research-finds. Quote appears in para. 10.
25. Jakiela, P., & Ozier, O. (2018, June). *Gendered language.* Policy Research Working Paper 8464. World Bank Group. https://documents1.worldbank.org/curated/en/405621528167411253/pdf/WPS8464.pdf.
26. Is there a language without gender in third person pronouns? (2014). Linguistics. https://linguistics.stackexchange.com/questions/7164/is-there-a-language-without-gender-in-third-person-pronouns#:~:text=The%20World%20Atlas%20of%20Language,pronouns%20(Iraqw%20and%20Burunge).
27. *Gendered languages may play a role in limiting women's opportunities, new research finds.* (2019, January 24). The World Bank. https://www.worldbank.org/en/news/feature/2019/01/24/gendered-languages-may-play-a-role-in-limiting-womens-opportunities-new-research-finds. Quote appears in para. 10.
28. DePino, L. (2019, August 9). How I came to my own name. *New York Times.* https://www.nytimes.com/2019/08/09/well/family/how-i-came-to-own-my-name.html. Quotes appear in paras. 21 and 22.
29. Borget, J. (2012, November 9). Biracial names for biracial babies. *Mom Stories.* http://blogs.babycenter.com/mom_stories/biracial-baby-names-110912/.
30. Edelman, B., Luca, M., & Svirsky, D. (2017.) Racial discrimination in the sharing economy: Evidence from a field experiment. *American Economic Journal: Applied Economics, 9*(2), 1–22.
31. Derous, E., Ryan, A. M., & Nguyen, H. D. (2012). Multiple categorization in resume screening: Examining effects on hiring discrimination against Arab applicants in field and lab settings. *Journal of Organizational Behavior, 33*(4), 544–570.
32. Bertrand, M., & Mullainathan, S. (2004). Are Emily and Greg more employable than Lakisha and Jamal? A field experiment on labor market discrimination. *The American Economic Review, 4,* 991–1013.
33. No names, no bias? (2015, October 31). *The Economist.* http://www.economist.com/news/business/21677214-anonymising-job-applications-eliminate-discrimination-not-easy-no-names-no-bias.
34. Derwing, T. M., & Munro, M. J. (2009). Putting accent in its place: Rethinking obstacles to communication. *Language Teaching, 42*(4), 476–490.
35. Agudo, R.R. (July 14, 2018). *New York Times.* https://www.nytimes.com/2018/07/14/opinion/sunday/everyone-has-an-accent.html.
36. Waxman, O. B. (2015, February 10). This the world's hottest accent. *Time.* http://time.com/3702961/worlds-hottest-accent/.
37. Hansen, K., & Dovidio, J. F. (2016). Social dominance orientation, nonnative accents, and hiring recommendations. *Cultural Diversity and Ethnic Minority Psychology, 22*(4), 544–551.
38. Grant, A. (2020, December 7). *The power of powerless communication: Adam Grant* (transcript of May 2013 TEDxEast talk). https://singjupost.com/the-power-of-powerless-communication-adam-grant-transcript/?singlepage=1. Quotes in this paragraph appear in transcript para. 13 and 18, respectively.
39. Association of American Colleges & Universities. (2021). *2021 report on employer views of higher education.* https://www.aacu.org/2021-report-employer-views-higher-education.
40. Tang, C.-C., Draucker, C., Tejani, M. A., & Von Ah, D. (2018). Patterns of interactions among patients with advanced pancreatic cancer, their caregivers, and healthcare providers during symptom discussions. *Supportive Care in Cancer, 26*(10), 3497–3506.
41. Fragale, A. R. (2006, November). The power of powerless speech: The effects of speech style and task interdependence on status conferral. *Organizational Behavior and Human Decision Processes, 101*(2), 243–261.
42. Blankenship, K. L., & Craig, T. Y. (2007). Language and persuasion: Tag questions as powerless speech or as interpreted in context. *Science Direct, 43,* 112–118. https://citeseerx.ist.psu.edu/viewdoc/download?doi=10.1.1.1087.5453&rep=rep1&type=pdf.
43. Cain, S. (2013). *Quiet: The power of introverts in a world that can't stop talking.* Crown.
44. Cain, S. (n.d.). *7 ways to use the power of powerless communication.* Quiet Revolution. https://www.quietrev.com/7-ways-to-use-powerless-communication/. Quotes appear in para. 7.
45. Sidani, S., Reeves, S., Hurlock-Chorostecki, C., van Soeren, M., Fox, M., & Collins, L. (2018). Exploring differences in patient-centered practices among healthcare professionals in acute care settings. *Health Communication, 33*(6), 716–723.
46. How collaboration wins: Leadership, benefits, and best practices. (2018, January 12). *Harvard Business Review.* https://hbr.org/sponsored/2018/01/how-collaboration-wins#:~:text=More%20than%20a%20touchstone%20for,promote%20internal%20and%20external%20cooperation.
47. Hayakawa, S. I. (1964). *Language in thought and action.* Harcourt Brace.

48. Labov, T. (1992). Social and language boundaries among adolescents. *American Speech, 4*, 339–366.
49. Peters, M. (2017, January 27). *The hidden dangers of euphemisms*. BBC News. http://www.bbc.com/capital/story/20170126-the-hidden-danger-of-euphemisms.
50. Mitchell, A., Gottfried, J., Barthel, M., & Sumida, N. (2018, June 18). *Distinguishing between factual and opinion statements in the news*. Pew Research Center. https://www.journalism.org/2018/06/18/distinguishing-between-factual-and-opinion-statements-in-the-news/.
51. Epstein, D. (2019, July 20). Chances are, you're not as open-minded as you think. *Washington Post*. https://www.washingtonpost.com/opinions/chances-are-youre-not-as-open-minded-as-you-think/2019/07/20/0319d308-aa4f-11e9-9214-246e594de5d5_story.html. Quote appears in the title.
52. Obermeyer, Z., Powers, B., Vogeli, C., & Mullainathan, S. (2019, October 25). Dissecting racial bias in an algorithm used to manage the health of populations. *Science*. https://blog.ed.ted.com/2017/01/12/how-to-tell-fake-news-from-real-news/.
53. McClure, L. (2017, January 12). *How to tell fake news from real news*. TEDEd. http://blog.ed.ted.com/2017/01/12/how-to-tell-fake-news-from-real-news/.
54. *Difference between facts and opinions*. (n.d.). Difference Between. http://www.differencebetween.info/difference-between-facts-and-opinions.
55. Serota, K. B., Levine, T. R., & Boster, F. J. (2010). The prevalence of lying in America: Three studies of self-reported lies. *Human Communication Research, 36*(1), 2–25.
56. McCornack, S. A., & Levine, T. R. (1990). When lies are uncovered: Emotional and relational outcomes of discovered deception. *Communication Monographs, 57*, 119–138.
57. Guthrie, J., & Kunkel, A. (2013). Tell me sweet (and not-so-sweet) little lies: Deception in romantic relationships. *Communication Studies, 64*(2), 141–157.
58. Kaplar, M. E., & Gordon, A. K. (2004). The enigma of altruistic lying: Perspective differences in what motivates and justifies lie telling within romantic relationships. *Personal Relationships, 11*, 489–507.
59. Bryant, E. (2008). Real lies, white lies and gray lies: Towards a typology of deception. *Kaleidoscope: A Graduate Journal of Qualitative Communication Research, 7*, 723–748.
60. Gunderson, P. R., & Ferrari, J. R. (2008). Forgiveness of sexual cheating in romantic relationships: Effects of discovery method, frequency of offense, and presence of apology. *North American Journal of Psychology, 10*, 1–14.
61. Heitler, S. (2012, October 2). The problem with over-emotional political rhetoric. *Psychology Today*. https://www.psychologytoday.com/blog/resolution-not-conflict/201210/the-problem-over-emotional-political-rhetoric. Quote appears in para. 2.
62. For a discussion of racist language, see Bosmajian, H. A. (1983). *The language of oppression*. University Press of America.
63. Garcia-Navarro, L. (2020, October 18). *The consequences of dehumanizing language in politics*. NPR. https://www.npr.org/2020/10/18/925069809/the-consequences-of-politics-dehumanizing-language.
64. Abarcar, S. (2018). *Have you ever called someone a racial/ethnic slur?* Quora. https://www.quora.com/Have-you-ever-called-someone-a-racial-ethnic-slur.
65. Saleem, H. M., Dillon, K. P., Benesch, S., & Ruths, D. (2016). *A web of hate: Tackling hateful speech in online social spaces*. Workshop on Text Analytics for Cybersecurity and Online Safety. https://dangerousspeech.org/a-web-of-hate-tackling-hateful-speech-in-online-social-spaces/. Quote appears in title.
66. Connley, C. (2018, April 25). *4 workplace microaggressions that can kill your confidence—and what to do about them*. CNBC Make It. https://www.cnbc.com/2018/04/25/workplace-microaggressions-can-kill-your-confidence-heres-what-to-do.html. Quote appears in para. 2 under item 1.
67. Sue, D. W., Capodilupo, C. M., Torino, G. C., Bucceri, J. M., Holder, A. M. B., Nadal, K. L., & Esquilin, M. (2007). Racial microaggressions in everyday life: Implications for clinical practice. *American Psychologist, 62*, 271–286.
68. Sue, D. W. (2010). *Microaggressions in everyday life: Race, gender, and sexual orientation*. John Wiley & Sons.
69. Nigatu, H. (2013, December 9). *21 racial microaggressions you hear on a daily basis*. BuzzFeed. https://www.buzzfeed.com/hnigatu/racial-microaggressions-you-hear-on-a-daily-basis. Quote appears in photo 18.
70. Mason, S. A. (2013, November 6). *Microaggressions: The little things people say* [Video]. YouTube. https://www.youtube.com/watch?v=ScOA-_tsi-Y. Quote is at 39 seconds.
71. Montana, S. (2017, April 14). A micro-list of microaggressions against women in the workplace. *CRN*. https://wotc.crn.com/blog/a-micro-list-of-microaggressions-against-women-in-the-workplace. Quote appears in para. 8.
72. Mason (2013). *Microaggressions: The little things people say* [Video]. YouTube. https://www.youtube.com/watch?v=ScOA-_tsi-Y. Quote is at 2:42 mins.
73. Mason (2013). Quote is at 3 mins.
74. Limbong, A. (2020, June 9). *Microaggressions are a big deal: How to talk about them and when to walk away*. NPR. https://www.npr.org/2020/06/08/872371063/microaggressions-are-a-big-deal-how-to-talk-them-out-and-when-to-walk-away. Quote appears in para. 5.
75. Parsons, A. (2017). The effect of microaggressions, predominantly White institutions, and support service on the academic success of minority students. *Perspectives* (University of New Hampshire), 1–10.
76. Nadal, K. L., Wong, Y., Griffin, K. E., Davidoff, K., & Sriken, J. (2014). The adverse impact of racial microaggressions on college students' self-esteem. *Journal of College Student Development, 55*, 461–474.
77. Hollingsworth, D. W., Cole, A. B., O'Keefe, V. M., Tucker, R. P., Story, C. R., & Wingate, L. R. R. (2017). Experiencing racial microaggressions influences suicide ideation through perceived burdensomeness in African Americans. *Journal of Counseling Psychology, 64*, 104–111.
78. Joll, C., & Sunstein, C. R. (2006). The law of implicit bias. *California Law Review, 94*(4), 969–996.
79. Connley, C. (2018, April 25). *4 workplace microaggressions that can kill your confidence—and what to do about them*. CNBC Make It. https://www.cnbc.com/2018/04/25/

workplace-microaggressions-can-kill-your-confidence-heres-what-to-do.html. Quote appears in para. 4 under item 1.
80. Broido, E. M. (2000, January/February). The development of social justice allies during college: A phenomenological investigation. *Journal of College Student Development, 41*, 3–18.
81. Sue, D. W., Alsaidi, S., Awad, M. N., Glaeser, E., Calle, C. Z., & Mendez, N. (2019). Disarming racial microaggressions: Microintervention strategies for targets, white allies, and bystanders. *American Psychologist, 1*, 128–142.
82. Limbong, A. (2020, June 9). *Microaggressions are a big deal: How to talk about them and when to walk away.* NPR. https://www.npr.org/2020/06/08/872371063/microaggressions-are-a-big-deal-how-to-talk-them-out-and-when-to-walk-away. Quote appears in para. 1 under "Is there a risk . . .".
83. Limbong (2020). Quote appears in para. 3 under "If someone says . . .".
84. Kothari, P. (n.d.). *How to fix a microaggression you didn't mean to commit.* Ellevate. https://www.ellevatenetwork.com/articles/8034-how-to-fix-a-microaggression-you-didn-t-mean-to-commit. Quote appears in Item 3.
85. Hyatt, M. (2016, February 26). *How a small shift in your vocabulary can instantly change your attitude.* Michael Hyatt & Co. https://michaelhyatt.com/how-a-shift-in-your-vocabulary-can-instantly-change-your-attitude/. Quote appears in fourth to last paragraph.

CHAPTER 6

1. TEDx Talks. (2017, December 15). *Chanel Lewis: Listening is radical* [Video]. YouTube. https://www.youtube.com/watch?v=pfppBsJDrpA. Quote appears at 1:24 mins.
2. TEDx Talks. (2017, December 15). Quote appears at 6:44 mins.
3. TEDx Talks. (2017, December 15). Quote appears at 6:50 mins.
4. Covey, S. (1989). *The 7 habits of highly effective people.* Simon & Schuster.
5. Kalargyrou, V., & Woods, R. H. (2011). Wanted: Training competencies for the twenty-first century. *International Journal of Contemporary Hospitality Management, 23*(3), 361–376.
6. Kochhar, R. (2020, March 23). *New, emerging jobs and the green economy are boosting demand for analytical skills.* Pew Research Center. https://www.pewresearch.org/fact-tank/2020/03/23/new-emerging-jobs-and-the-green-economy-are-boostingdemand-for-analytical-skills/.
7. Kalargyrou, V., & Woods, R. H. (2011). Wanted: Training competencies for the twenty-first century. *International Journal of Contemporary Hospitality Management, 23*(3), 361–376.
8. Davis, J., Foley, A., Crigger, N., & Brannigan, M. C. (2008). Healthcare and listening: A relationship for caring. *International Journal of Listening, 22*(2), 168–175.
9. Pryor, S., Malshe, A., & Paradise, K. (2013). Salesperson listening in the extended sales relationship: An exploration of cognitive, affective, and temporal dimensions. *Journal of Personal Selling & Sales Management, 33*(2), 185–196.
10. Brockner, J., & Ames, D. (2010, December 1). Not just holding forth: The effect of listening on leadership effectiveness. *Social Science Electronic Publishing.* http://papers.ssrn.com/sol3/papers.cfm?abstract_id=1916263.
11. Jonsdottir, I. J. & Fridriksdottir, K. (2019). Active listening: Is it the forgotten dimension in managerial communication? *International Journal of Listening. 34*(3), 178–188.
12. Johnston, M. K., & Reed, K. (2017). Listening environment and the bottom line: How a positive environment can improve financial outcomes. *International Journal of Listening, 31*(2), 71–79.
13. Gordon, P., James Allan, C., Nathaniel, B., Derek, J. K., & Jonathan, A. F. (2015). On the reception and detection of pseudo-profound bullshit. *Judgment and Decision Making, 10*(6), 549–563.
14. Bodie, G. D., Vickery, A. J., Cannava, K., & Jones, S. M. (2015). The role of "active listening" in informal helping conversations: Impact on perceptions of listener helpfulness, sensitivity, and supportiveness and discloser emotional improvement. *Western Journal of Communication, 79*(2), 151–173.
15. Mallory, T. E., Kluger, A. N., Martin, J., & Pery, S. (2021). Women listening to women at zero-acquaintance: Interpersonal befriending at the individual and dyadic levels. *International Journal of Listening.* Advanced online publication.
16. Lopez-Rosenfeld, M., Calero, C. I., Fernandez Slezak, D., Garbulsky, G., Bergman, M., Trevisan, M., & Sigman, M. (2015). Neglect in human communication: Quantifying the cost of cell-phone interruptions in face to face dialogs. *PLOS ONE, 10*(6).
17. Bodie, G. D., Vickery, A. J., & Gearhart, C. C. (2013). The nature of supportive listening, I: Exploring the relation between supportive listeners and supportive people. *International Journal of Listening, 27*, 39–49.
18. Fletcher, G. O., Kerr, P. G., Li, N. P., & Valentine, K. A. (2014). Predicting romantic interest and decisions in the very early stages of mate selection: Standards, accuracy, and sex differences. *Personality & Social Psychology Bulletin, 4*, 540–550.
19. What women want from men: A good listener. (2012, October 3.) *Ingenio Advisor.* http://www.ingenio.com/CommunityServer/UserBlogPosts/Advisor_Louise_PhD/What-Women-Want-from-Men—A-Good-Listener/630187.aspx. Quote appears in para. 3.
20. TEDx Talks. (2017, April 7). *Stephen O'Keefe: How to listen better–tips from a deaf guy* [Video]. YouTube. https://www.youtube.com/watch?v=dx70vvOSlNY. Quote appears at 3:00 mins.
21. TEDx Talks. (2017, April 7). Quote appears at 3:19 mins.
22. TEDx Talks. (2017, April 7). Quote appears at 6:22 mins.
23. TEDx Talks. (2017, April 7). Quote appears at 6:43 mins.
24. TEDx Talks. (2017, April 7). Quote appears at 10:20 mins.
25. TEDx Talks. (2017, April 7). Quote appears at 6:37 mins.
26. Powers, W. G., & Witt, P. L. (2008). Expanding the theoretical framework of communication fidelity. *Communication Quarterly, 56*, 247–267; Fitch-Hauser, M., Powers, W. G., O'Brien, K., & Hanson, S. (2007). Extending the conceptualization of listening fidelity. *International Journal of Listening, 21*, 81–91; Powers, W. G., & Bodie, G. D. (2003). Listening fidelity: Seeking congruence between cognitions of the listener and the sender. *International Journal of Listening, 17*, 19–31.
27. Kim, Y. G. (2016). Direct and mediated effects of language and cognitive skills on comprehension of oral narrative

28. texts (listening comprehension) for children. *Journal of Experimental Child Psychology, 141*, 101–120.
28. Fontana, P. C., Cohen, S. D., & Wolvin, A. D. (2015). Understanding listening competency: A systematic review of research scales. *International Journal of Listening, 29*(3), 148–176.
29. Ames, D., Maissen, L. B., & Brockner, J. (2012, June). The role of listening in interpersonal influence. *Journal of Research in Personality, 46*, 345–349.
30. Ames, D., Maissen, L. B., & Brockner, J. (2012, June).
31. *Accenture research finds listening more difficult in today's digital workplace.* (2015, February 26). Accenture. Retrieved from https://newsroom.accenture.com/industries/global-media-industry-analyst-relations/accenture-research-finds-listening-more-difficult-in-todays-digital-workplace.htm.
32. *Accenture research finds listening more difficult in today's digital workplace.* (2015, February 26).
33. rurounikenji. (2012, July 8). *Girlfriend literally never listens to me* [message board post]. The Student Room. http://www.thestudentroom.co.uk/showthread.php?t=2075372. Quote appears in first paragraph.
34. Chapman, S. G. (2012). *The five keys to mindful communication: Using deep listening and mindful speech to strengthen relationships, heal conflicts, and accomplish your goals.* Shambhala Publications.
35. Robertson, R. R. (2016, October 12). Normani Kordei opens up about her struggle with cyberbullies and racist trolls. *Essence.* http://www.essence.com/celebrity/normani-kordei-cyberbullies-racist-trolls.
36. Bodie, G. D. (2011). The Revised Listening Concepts Inventory (LCI-R): Assessing individual and situational differences in the conceptualization of listening. *Imagination, Cognition and Personality, 30*(3), 301–339.
37. Gearhart, C. C., Denham, J. P., & Bodie, G. D. (2014). Listening as a goal-directed activity. *Western Journal of Communication, 78*(5), 668–684.
38. Parker, K., Horowitz, J., & Stepler, R. (2017, December 5). *On gender differences, no consensus on nature vs. nurture: Americans say society places a higher premium on masculinity than on femininity.* Pew Research Center. https://www.pewresearch.org/social-trends/2017/12/05/on-gender-differences-no-consensus-on-nature-vs-nurture/.
39. Pence, M. E., & James, T. A. (2015). The role of sex differences in the examination of personality and active empathic listening: An initial exploration. *International Journal of Listening, 29*(2), 85–94.
40. Welch, S. A., & Mickelson, W. (2020). Listening across the life span: A listening environment comparison. *International Journal of Listening, 34*(2), 97–109.
41. Welch, S. A., & Mickelson, W. (2020).
42. Sargent, S. L., & Weaver, J. B, III. (2003). Listening styles: Sex differences in perceptions of self and others. *International Journal of Listening, 17*(1), 5–18.
43. Janusik, L., & Imhof, M. (2017). Intercultural listening: Measuring listening concepts with the LCI-R. *International Journal of Listening, 31*(2), 80–97.
44. Stewart, M. C., & Arnold, C. L. (2018). Defining social listening: Recognizing an emerging dimension of listening. *International Journal of Listening, 32*(2), 85–100.
45. Vickery, A. (2018). "Listening enables me to connect with others": Exploring college students' (mediated) listening metaphors. *International Journal of Listening, 32*(2), 69–84.
46. Stewart, M. C., & Arnold, C. L. (2018). Defining social listening: Recognizing an emerging dimension of listening. *International Journal of Listening, 32*(2), 85–100.
47. Datareportal. (2021, July). *Global social media stats.* https://datareportal.com/social-media-users.
48. Trujillo, J. (2020, November 20). Viral TikTok star doggface208 buys a new house in cash: Meet the Native-Mexican that skated into the world's heart. *Hola!* https://us.hola.com/celebrities/20201119ft7yi5eht5/tiktok-star-doggface208-nathan-apodaca-buys-house.
49. Bariso, J. (2019, October 16). A Panera employee says she was fired after her video went viral: Her response is a lesson in emotional intelligence. *Inc.* https://www.inc.com/justin-bariso/a-panera-employee-says-she-was-fired-after-her-video-went-viral-her-response-is-a-lesson-in-emotional-intelligence.html.
50. Stewart, M. C., & Arnold, C. L. (2018). Defining social listening: Recognizing an emerging dimension of listening. *International Journal of Listening, 32*(2), 85–100.
51. Morrison, C. (2021, April 1). The 20 social listening statistics your company must know. *Everyonesocial.* https://everyonesocial.com/blog/social-listening-statistics/.
52. Gearhart, C. C., & Maben, S. K. (2019). Active and empathic listening in social media: What do stakeholders really expect. *International Journal of Listening, 0,* 1–22.
53. Segarra, L. M. (2017, May 9). Nugget boy just broke Twitter's retweet record, and he got a year of free nuggets. *Fortune.* https://fortune.com/2017/05/09/nugget-boy-retweet-record-twitter-wendys/.
54. *Accenture research finds listening more difficult in today's digital workplace.* (2015, February 26). Accenture. https://newsroom.accenture.com/industries/global-media-industry-analyst-relations/accenture-research-finds-listening-more-difficult-in-todays-digital-workplace.htm.
55. Hansen, J. (2007). *24/7: How cell phones and the Internet change the way we live, work, and play.* Praeger. See also Turner, J. W., & Reinsch, N. L. (2007). The business communicator as presence allocator: Multicommunicating, equivocality, and status at work. *Journal of Business Communication, 44,* 36–58.
56. Stewart, M. C., & Arnold, C. L. (2018). Defining social listening: Recognizing an emerging dimension of listening. *International Journal of Listening, 32*(2), 85–100.
57. Storch, S. L., & Juarez-Paz, A. V. O. (2018). Family communication: Exploring the dynamics of listening with mobile devices. *International Journal of Listening, 32*(2), 115–126. Quote appears on p. 120.
58. Storch, S. L., & Juarez-Paz, A. V. O. (2018). Quote appears on p. 120.
59. Gazzaley, A., & Rosen, L. D. (2016). *The distracted mind: Ancient brains in a high-tech world.* MIT Press.
60. Strayer, D., & Watson, J. (2012, March). Top multitaskers help explain how brain juggles thoughts.

Scientific American. https://www.scientificamerican.com/article/supertaskers-and-the-multitasking-brain/.

61. Spence, A., Beasley, K., Gravenkemper, H., Hoefler, A., Ngo, A., Ortiz, D., & Campisi, J. (2020). Social media use while listening to new material negatively affects short-term memory in college students. *Physiology & Behavior, 227*(1).
62. Hemp, P. (2009, December 4). Death by information overload. *Harvard Business Review.* http://www.ocvets4pets.com/archive17/Death_by_Information_Overload_-_HBR.org.pdf.
63. Bryant, A. (n.d.). How to be a better listener. *New York Times.* https://www.nytimes.com/guides/smarterliving/be-a-better-listener. Quote appears in para. 10.
64. Bryant, A. (n.d.). Quote appears in para. 10.
65. Imhof, M. (2003). The social construction of the listener: Listening behaviors across situations, perceived listener status, and cultures. *Communication Research Reports, 20,* 357–366.
66. Zohoori, A. (2013). A cross-cultural comparison of the HURIER Listening Profile among Iranian and U.S. students. *International Journal of Listening, 27,* 50–60.
67. Shirota, N. (2021, February 28). Nagara listening: Japanese listeners' behavior in multiactivity settings. *International Journal of Listening.*
68. Parks, E. S. (2020, April 6). Listening across the ages: Measuring generational listening differences with the LCI-R. *International Journal of Listening.*
69. Cole, J. (2020, September 23). *Googling, Ubering and Xeroxing: How Zooming became a verb in six months.* Center for the Digital Future. https://www.digitalcenter.org/columns/zooming/. Quote appears in para 31.
70. Zoom. (2020, April 22). 90-day security plan progress report: April 22. *Zoom Blog.* https://blog.zoom.us/90-day-security-plan-progress-report-april-22/.
71. Sadler, M. (2021, July 1). 84 current video conferencing statistics for the 2021 market. *TrustRadius.* https://www.trustradius.com/vendor-blog/web-conferencing-statistics-trends.
72. Prossack, A. (2021, February 10). 5 statistics employers need to know about the remote workforce. *Forbes.* https://www.forbes.com/sites/ashiraprossack1/2021/02/10/5-statistics-employersneed-to-know-about-the-remote-workforce/?sh=4360c198655d.
73. Steward, J. (2021, August 6). *The ultimate list of remote work statistics for 2021.* Findstack. https://findstack.com/remote-work-statistics/.
74. Prossack, A. (2021, February 10). 5 statistics employers need to know about the remote workforce. *Forbes.* https://www.forbes.com/sites/ashiraprossack1/2021/02/10/5-statistics-employersneed-to-know-about-the-remote-workforce/?sh=4360c198655d.
75. ThinkImpact. (2021). eLearning statistics. https://www.thinkimpact.com/elearning-statistics/.
76. Lindsay, K. (2021). The art of listening in virtual teams. *Aptimore.* https://www.aptimore.com/article/the-art-of-listening-in-virtual-teams/.
77. Lindsay, K. (2021). Quote appears in para. 7.
78. Lindsay, K. (2021). Quote appears in para. 8.
79. Rayome, A. D. (2020, June 22). *6 Zoom rules you probably broke today.* CNET. https://www.cnet.com/tech/services-and-software/6-zoom-rules-you-probably-broke-at-work-today/. Quote appears in para. 8.
80. Feldman, J. (2020). Listening and falling silent: Towards *technics* of collectivity [Special feature]. *Sociologica, 14*(2).
81. Callahan, G. (2020, July 27). How to fight Zoom fatigue: Five practical steps. *Rev.* https://www.rev.com/blog/how-to-fight-zoom-fatigue-five-practical-steps.
82. Victor, D. (2021, May 6). "I'm not a cat," says lawyer having Zoom difficulties. *New York Times.* https://www.nytimes.com/2021/02/09/style/cat-lawyer-zoom.html.
83. Gershman, S. (2020, May 4). Stop zoning out in Zoom meetings. *Harvard Business Review.* https://hbr.org/2020/05/stop-zoning-out-in-zoom-meetings.
84. Callahan, G. (2020, July 27). How to fight Zoom fatigue: Five practical steps. *Rev.* https://www.rev.com/blog/how-to-fight-zoom-fatigue-five-practical-steps. Quote appears in para. 8.
85. Gershman, S. (2020, May 4). Stop zoning out in Zoom meetings. *Harvard Business Review.* https://hbr.org/2020/05/stop-zoning-out-in-zoom-meetings. Quote appears in para. 12.
86. Supiano, B. (2020, April 23). Why is Zoom so exhausting? *The Chronicle of Higher Education.* https://www.chronicle.com/article/Why-Is-Zoom-So-Exhausting-/248619.
87. u/kaleighhealy (2019, January 11). *Always the listener, never the listened to . . .* [Online forum post]. Reddit. https://www.reddit.com/r/infp/comments/aes73o/always_the_listener_never_the_listened_to/. Quote appears in para. 1.
88. u/kaleighhealy (2019, January 11). Quote appears in para. 1.
89. Sahay, S. (2021). Organizational listening during organizational change: Perspectives of employees and executives. *International Journal of Listening.* Advance online publication.
90. Spence, A., Beasley, K., Gravenkemper, H., Hoefler, A., Ngo, A., Ortiz, D., & Campisi, J. (2020). Social media use while listening to new material negatively affects short-term memory in college students. *Physiology & Behavior, 227*(1).
91. Taylor, A. (2019, October 31). Living for social media vs. being present in the moment. *Northeast Ohio Parent.* https://www.northeastohioparent.com/blogger/living-for-social-media-vs-being-present-in-the-moment/. Quote appears in para. 1.
92. Taylor, A. (2019, October 31) Quote appears in para. 1.
93. Vangelisti, A. L., Knapp, M. L., & Daly, J. A. (1990). Conversational narcissism. *Communication Monographs,57,* 251–274.
94. Walshe, I. (2018, April 24). Know a conversational narcissist? *Mayo News.* https://www.mayonews.ie/living/nurturing/31980-know-a-conversational-narcissist.
95. Ward, A. F. (2013, July 16). The neuroscience of everybody's favorite topic. *Scientific American.* https://www.scientificamerican.com/article/the-neuroscience-of-everybody-favorite-topic-themselves/.
96. Bryant, A. (n.d.) How to be a better listener. *New York Times.* https://www.nytimes.com/guides/smarterliving/be-a-better-listener. Quote appears in para. 17.
97. Headlee, C. (2016, July 19). *How a great conversation is like a game of catch* [Video]. TED Ideas. https://ideas.ted.com/how-a-great-conversation-is-like-a-game-of-catch/.

98. Talking to your parents—or other adults. (2015, February). *Nemours*. https://kidshealth.org/en/teens/talk-to-parents.html?WT.ac=ctg/. Quote appears in para. 7 of "Raising Difficult Topics" section.
99. Kochhar, R. (2020, March 23). *New, emerging jobs and the green economy are boosting demand for analytical skills*. Pew Research Center. https://www.pewresearch.org/fact-tank/2020/03/23/new-emerging-jobs-and-the-green-economy-are-boostingdemand-for-analytical-skills/.
100. Eggenberger, A. L. B. (2019). Active listening skills as predictors of success in community college students. *Community College Journal of Research and Practice, 45*(5), 324–333.
101. Gearhart, C. G., & Bodie, G. D. (2011). Active-empathic listening as a general social skill: Evidence from bivariate and canonical correlations. *Communication Reports, 24*, 86–98.
102. Bodie, G. D., Vickery, A. J., & Gearhart, C. C. (2013). The nature of supportive listening, I: Exploring the relation between supportive listeners and supportive people. *International Journal of Listening, 27*, 39–49.
103. Wolbe, S. (2016, April 29). *Active listening to improve . . . everything*. HuffPost. https://www.huffpost.com/entry/active-listening-to-impro_b_9810266. Quote appears in para. 1.
104. Rogers, C. R. (1959). *A theory of therapy, personality, and interpersonal relationships: As developed in the client-centered framework*. McGraw-Hill. Quote appears on p. 210.
105. Rogers, C. R. (1957/2007). The necessary and sufficient conditions of therapeutic personality change. *Psychotherapy: Theory, Research, Practice, Training, 44*(3), 240–248. (Reprinted from "The necessary and sufficient conditions of therapeutic personality change," 1957, Journal of Consulting Psychology, 21(2), 95–103.)
106. Rogers, C. R. (1959). *A theory of therapy, personality, and interpersonal relationships: As developed in the client-centered framework*. McGraw-Hill.
107. Bodie, G. D., Vickery, A. J., Cannava, K., & Jones, S. M. (2015). The role of "active listening" in informal helping conversations: Impact on perceptions of listener helpfulness, sensitivity, and supportiveness and discloser emotional improvement. *Western Journal of Communication, 79*(2), 151–173.
108. Bodie, G. D., Vickery, A. J., & Gearhart, C. C. (2013). The nature of supportive listening, I: Exploring the relation between supportive listeners and supportive people. *International Journal of Listening, 27*, 39–49.
109. Kochhar, R. (2020, March 23). *New, emerging jobs and the green economy are boosting demand for analytical skills*. Pew Research Center. https://www.pewresearch.org/fact-tank/2020/03/23/new-emerging-jobs-and-the-green-economy-are-boostingdemand-for-analytical-skills/.
110. Jonsdottir, I. J., & Fridriksdottir, K. (2019). Active listening: is it the forgotten dimension in managerial communication? *International Journal of Listening, 34*(3), 178–188.
111. Sharifirad, M. S. (2013). Transformational leadership, innovative work behavior, and employee well-being. *Global Business Perspectives, 1*, 198–225.
112. Listening Ears. (n.d.) *Our work*. https://listeningears.org/our-work/.
113. Chia, H. L. (2009). Exploring facets of a social network to explicate the status of social support and its effects on stress. *Social Behavior & Personality: An International Journal, 37*(5), 701–710. See also Segrin, C., & Domschke, T. (2011). Social support, loneliness, recuperative processes, and their direct and indirect effects on health. *Health Communication, 26*, 221–232.
114. Shirota, N. (2021, February 28). Nagara listening: Japanese listeners' behavior in multiactivity settings. *International Journal of Listening*.
115. Robinson, J. D., & Tian, Y. (2009). Cancer patients and the provision of informational social support. *Health Communication, 24*, 381–390.
116. Gearhart, C. G., & Bodie, G. D. (2011). Active-empathic listening as a general social skill: Evidence from bivariate and canonical correlations. *Communication Reports, 24*, 86–98.
117. Michelson, T., & Kluger, A. (2021). Can listening hurt you? A meta-analysis of the effects of exposure to trauma on listener's stress. *International Journal of Listening*. Advance online publication.
118. Bodie, G. D., Vickery, A. J., Cannava, K., & Jones, S. M. (2015). The role of "active listening" in informal helping conversations: Impact on perceptions of listener helpfulness, sensitivity, and supportiveness and discloser emotional improvement. *Western Journal of Communication, 79*(2), 151–173.
119. Singh, S. (2021, August 3). *22 customer stories that are so wholesome, they'll cheer you right up*. BuzzFeed. https://www.buzzfeed.com/simrinsingh/people-are-sharing-the-nicest-things-customers-have-done. Quote appears in para. 21.
120. Singh, S. (2021, August 3). Quote appears in para. 21.
121. Paraschos, S. (2013). Unconventional doctoring: A medical student's reflections on total suffering. *Journal of Palliative Medicine, 16*, 325.
122. Burleson, B. (2008). What counts as effective emotional support? In M. T. Motley (Ed.), *Studies in Applied Interpersonal Communication* (pp. 207–227). Sage.
123. Guo, J., & Turan, B. (2016). Preferences for social support during social evaluation in men: The role of worry about a relationship partner's negative evaluation. *Journal of Social Psychology, 156*(1), 122–129.
124. Olson, R. (2014). A time-sovereignty approach to understanding carers of cancer patients' experiences and support preferences. *European Journal of Cancer Care, 23*(2), 239–248.
125. Young, R. W., & Cates, C. M. (2004). Emotional and directive listening in peer mentoring. *International Journal of Listening, 18*, 21–33.
126. Huerta-Wong, J. E., & Schoech, R. (2010). Experiential learning and learning environments: The case of active listening skills. *Journal of Social Work Education, 46*, 85–101.
127. Emanuel, D. (2021, August 5). *Men are sharing the "guy secrets" they don't usually share with women, and they range from gross to really sad*. BuzzFeed. https://www.buzzfeed.com/daniellaemanuel/men-embarrassed-tell-women. Quote appears in para. 12.
128. Giddens, A. (1984). *The constitution of society: Outline of the theory of structuration*. University of California Press.
129. Headley, J. (2013). *It's not about the nail* [Video]. YouTube. https://www.youtube.com/watch?v=-4EDhdAHrOg. Quote appears at 0:45.

130. Tannen, D. (2010). He said, she said. *Scientific American Mind, 21*(2), 55–59.
131. Emanuel, D. (2021, August 5). *Men are sharing the "guy secrets" they don't usually share with women, and they range from gross to really sad*. BuzzFeed. https://www.buzzfeed.com/daniellaemanuel/men-embarrassed-tell-women. Quote appears in para. 12.
132. Goldsmith, D. (2000). Soliciting advice: The role of sequential placement in mitigating face threat. *Communication Monographs, 67*, 1–19.
133. 10 things to say (and not to say) to someone with depression. (2022, June 27). *Health*. https://www.health.com/condition/depression/10-things-to-say-and-10-not-to-say-to-someone-with-depression.
134. Stewart, M., Letourneau, N., Masuda, J. R., Anderson, S., Cicutto, L., McGhan, S., & Watt, S. (2012). Support needs and preferences of young adolescents with asthma and allergies: "Just no one really seems to understand." *Journal of Pediatric Nursing, 27*(5), 479–490.
135. Helping adults, children cope with grief. (2001, September 13). *Washington Post*.
136. Eisenberg, E. M., Murphy, A. G., Sutcliffe, K., Wears, R., Schenkel, S., Perry, S., & Vanderhoef, M. (2005). Communication in emergency medicine: Implications for patient safety. *Communication Monographs, 72*, 390–413. Quote appears on p. 401.
137. TEDx Talks. (2017, April 7). *Stephen O'Keefe: How to listen better–tips from a deaf guy* [Video]. YouTube. https://www.youtube.com/watch?v=dx70vvOSlNY. Quote appears at 9:27 mins.
138. TEDx Talks. (2017, April 7). Quote appears at 9:55 mins.
139. Learn anything with the Feynman Technique. (2016, December 17). *The Science Explorer*. http://thescienceexplorer.com/universe/learn-anything-feynman-technique.
140. Wood, Z. R. (2018, April). *Why it's worth listening to people you disagree with* [Video]. TED. https://www.ted.com/talks/zachary_r_wood_why_it_s_worth_listening_to_people_we_disagree_with. Quote appears in title.
141. Luedtke, K. (1987, January 7). What good is free speech if no one listens? *Los Angeles Times*. http://articles.latimes.com/1987-01-07/local/me-2347_1_free-speech.
142. TEDx Talks. (2017, December 15). *Chanel Lewis: Listening is radical* [Video]. YouTube. https://www.youtube.com/watch?v=pfppBsJDrpA. Quote appears at 3:19 mins.
143. TEDx Talks. (2017, December 15). Quote appears at 2:35 mins.

CHAPTER 7

1. Karimi, F., & Mezzofiore, G. (2021, October 2). *He's the most popular man on TikTok. And he doesn't say a word*. CNN. https://www.cnn.com/2021/10/02/world/tiktok-khaby-lame-cec/index.html.
2. Horowitz, J., & Lorenz, T. (2021, June 2). Khaby Lame, the everyman of the internet. *New York Times*. https://www.nytimes.com/2021/06/02/style/khaby-lame-tiktok.html.
3. Horowitz, J., & Lorenz, T. (2021, June 2). Quote appears in para. 8.
4. Karimi, F., & Mezzofiore, G. (2021, October 2). *He's the most popular man on TikTok. And he doesn't say a word*. CNN. https://www.cnn.com/2021/10/02/world/tiktok-khaby-lame-cec/index.html.
5. As quoted in Karimi, F., & Mezzofiore, G. (2021, October 2). Quote appears in para 14.
6. Research supporting these claims is cited in Burgoon, J. K., & Hoobler, G. D. (2002). Nonverbal signals. In M. L. Knapp & J. A. Daly (Eds.), *Handbook of interpersonal communication* (3rd ed., pp. 240–299). Sage.
7. Jones, S. E., & LeBaron, C. D. (2002). Research on the relationship between verbal and nonverbal communication: Emerging interactions. *Journal of Communication, 52*, 499–521.
8. For a survey of the issues surrounding the definition of nonverbal communication, see Knapp, M., & Hall, J. A. (2010). *Nonverbal communication in human interaction* (6th ed.). Wadsworth.
9. Knapp, M., & Hall, J. A. (2010). *Nonverbal communication in human interaction* (6th ed.). Wadsworth.
10. For a discussion of intentionality, see Knapp & Hall (2010), pp. 9–12.
11. Tracy, J. L., & Matsumoto, D. (2008, August 19). The spontaneous expression of pride and shame: Evidence for biologically innate nonverbal displays. *Proceedings from the National Academy of Science, 105*(33), 11, 655–11, 660.
12. Palmer, M. T., & Simmons, K. B. (1995). Communicating intentions through nonverbal behaviors: Conscious and nonconscious encoding of liking. *Human Communication Research, 22*, 128–160.
13. Schroeder, J., & Epley, N. (2015). The sound of intellect: Speech reveals a thoughtful mind, increasing a job candidate's appeal. *Psychological Science, 26*(6), 877–891.
14. Korte, K. (2021, March 14). Resting bitch face in the office: The fallout & "what if" scenarios. *LinkedIn*. https://www.linkedin.com/pulse/resting-bitch-face-office-fallout-what-scenarios-kim-korte. Quote appears in para. 5.
15. Horel, R. (2020, March 1). 16 surprising reasons you should smile today. *LinkedIn*. https://www.linkedin.com/pulse/16-surprising-reasons-you-should-smile-today-rob-horel.
16. Horel, R. (2020, March 1).
17. Wenner Moyer, M. (2016). Eye contact: How long is too long? *Scientific American Mind, 27*(1), 8.
18. Wenner Moyer (2016).
19. Glass, L. (2012). *The body language advantage*. Fair Winds Press.
20. Riggio, R. (2021, October 10). Secret body language cues that make people more likable. *Psychology Today*. https://www.psychologytoday.com/us/blog/cutting-edge-leadership/202110/secret-body-language-cues-make-people-more-likable.
21. Gurney, D. J., Howlett, N., Pine, K., Tracey, M., & Moggridge, R. (2017). Dressing up posture: The interactive effects of posture and clothing on competency judgements. *British Journal of Psychology, 108*, 436–451.
22. Ritchel, M. (2015, June 27). The mouth is mightier than the pen. *New York Times*. https://www.nytimes.com/2015/06/28/business/the-mouth-is-mightier-than-the-pen.html/. Quote appears in para. 4.
23. Meyer, K. (2016, July 17). *The four dimensions of tone of voice*. Nielsen Norman Group. https://www.nngroup.com/articles/tone-of-voice-dimensions/.

24. Guéguen, N., & Vion, M. (2009). The effect of a practitioner's touch on a patient's medication compliance. *Psychology, Health and Medicine, 14,* 689–694.
25. Segrin, C. (1993). The effects of nonverbal behavior on outcomes of compliance gaining attempts. *Communication Studies, 11,* 169–187.
26. Hornik, J. (1992). Effects of physical contact on customers' shopping time and behavior. *Marketing Letters, 3,* 49–55.
27. Smith, D. E., Gier, J. A., & Willis, F. N. (1982). Interpersonal touch and compliance with a marketing request. *Basic and Applied Social Psychology, 3,* 35–38. See also Soars, B. (2009). Driving sales through shoppers' sense of sound, sight, smell and touch. *International Journal of Retail & Distribution Management, 37*(3), 286–298.
28. Carlson, E. N. (2013). Overcoming barriers to self-knowledge: Mindfulness as a path to seeing yourself as you really are. *Perspectives on Psychological Science, 8,* 173–186.
29. TEDx Talk. (2020, March 31). *Joe Navarro: the power of nonverbal communication* [Video]. YouTube. https://www.youtube.com/watch?v=fLaslONQAKM. Quote appears at 10:05 mins.
30. TEDx Talk. (2020, March 31). Quote appears at 10:16 mins.
31. Miller, H. J., Kluver, D., Thebault-Spieker, J., Terveen, L. G., & Hecht, B. J. (2017). Understanding emoji ambiguity in context: The role of text in emoji-related miscommunication. *Proceedings of the Eleventh International Conference on Web and Social Media,* ICWSM 2017, pp. 152–161.
32. Safeway clerks object to "service with a smile." (1998, September 2). *San Francisco Chronicle.*
33. Burgoon, J. K., & Hale, J. L. (1988). Nonverbal expectancy violations: Model elaboration and application to immediacy behaviors. *Communication Monographs, 55,* 58–79.
34. Guerrero, L. K., & Bachman, G. F. (2008). Relational quality and relationships: An expectancy violations analysis. *Journal of Social and Personal Relationships, 23*(6), 943–963.
35. Miller-Ott, A., & Kelly, L. (2015). The presence of cell phones in romantic partner face-to-face interactions: An expectancy violation theory approach. *Southern Communication Journal, 4,* 253–270.
36. Amy, J. C. C., Caroline, A. W., Andy, J. Y., & Dana, R. C. (2015). Preparatory power posing affects nonverbal presence and job interview performance. *Journal of Applied Psychology, 4,* 1286–1295.
37. Ekman, P. (1992). An argument for basic emotions. *Cognition and Emotion, 6,* 169–200. Quote appears on page 177.
38. Ekman, P. (1992).
39. Bogart, K. R. (2020). Socioemotional functioning with facial paralysis: Is there a congenital or acquired advantage? *Health Psychology, 39*(4), 345–354.
40. Bogart, K. R. (2021, March 1). Look beyond face value: The psychology of facial paralysis. *Psychology Today.* https://www.psychologytoday.com/us/blog/disability-is-diversity/202103/look-beyond-face-value-the-psychology-facial-paralysis. Quote appears in para. 11.
41. Bogart, K. R., Tickle-Degnen, L., & Joffe, M. (2012). Social interaction experiences of adults with Moebius syndrome: A focus group. *Journal of Health Psychology, 17*(8), 1212–1222.
42. Bogart, K. R., Tickle-Degnen, L., & Ambady, N. (2014). Communicating without the face: Holistic perception of emotions of people with facial paralysis. *Basic and Applied Social Psychology, 36*(4), 309–320.
43. Reilly, M., & Parsa, K. (2019, December 17). Botox and emotional expressivity. *Psychology Today.* https://www.psychologytoday.com/us/blog/dissecting-plastic-surgery/201912/botox-and-emotional-expressivity.
44. Wyffels, M. L., Ray, B. B., Laurita, J. T., Zbib, N., Bachour, K., Glass, G. E., & Stotland, M. A. (2020). Impact of glabellar paralysis on facial expression of emotion. *Aesthetic Surgery Journal, Volume 40*(4), 430–436.
45. Reilly, M., & Parsa, K. (2019, December 17). Botox and emotional expressivity. *Psychology Today.* https://www.psychologytoday.com/us/blog/dissecting-plastic-surgery/201912/botox-and-emotional-expressivity.
46. Roberts, J. (2020, October 9). Celebrities who've been refreshingly candid about Botox: These stars didn't hold back. *Marie Claire.* https://www.marieclaire.com/beauty/news/g4289/celebrities-who-have-spoken-about-botox/?slide=16. Quote appears in para.1.
47. Cohen, S., Koegel, R., Koegel, L. K., Engstrom, E., Young, K., & Quach, A. (2021). Using self-management and visual cues to improve responses to nonverbal social cues in adults with autism spectrum disorder. *Behavior Modification.* Advance online publication.
48. Rourke, B. P. (1989). *Nonverbal learning disabilities: The syndrome and the model.* Guilford Press.
49. Denworth, L. (2020, April 1). *How people with autism forge friendships.* Spectrum News. https://www.spectrumnews.org/features/deep-dive/how-people-with-autism-forge-friendships/.
50. Cross, E. S., & Franz, E. A. (2003, March 30–April 1). *Talking hands: Observation of bimanual gestures as a facilitative working memory mechanism.* Paper presented at the 10th annual meeting of the Cognitive Neuroscience Society, New York.
51. Burnett, S. (2014, August 4). Have you ever wondered why Asians spontaneously make V-signs in photos? *Time.* http://time.com/2980357/asia-photos-peace-sign-v-janet-lynn-konica-jun-inoue/.
52. Cosgrove, B. (2014, July 4). V for victory: Celebrating a gesture of solidarity and defiance. *Time.* http://time.com/3880345/v-for-victory-a-gesture-of-solidarity-and-defiance/.
53. Shittu, H., & Query, C. (2006). *Absurdities, scandals & stupidities in politics.* Genix Press.
54. Kleinke, C. R. (1977). Compliance to requests made by gazing and touching experimenters in field settings. *Journal of Experimental Social Psychology, 13,* 218–233.
55. Patel, S., & Scherer, K. (2013). Vocal behavior. In M. L. Knapp & J. A. Hall (Eds.), *Nonverbal communication* (pp. 167–204). De Gruyter Mouton.
56. Argyle, M. F., Alkema, F., & Gilmour, R. (1971). The communication of friendly and hostile attitudes: Verbal and nonverbal signals. *European Journal of Social Psychology, 1,* 385–402.
57. Bond, C. F., Jr., & DePaulo, B. (2006). Accuracy of deception judgments. *Personality and Social Psychology Review, 10,* 214–234.
58. Mattes, K., Popova, V., & Evans, J. R. (2021). Deception detection in politics: Can voters tell when politicians are lying? *Political Behavior.* Advance online publication.

59. Wired. (2020, January 27). *Former FBI agent explains how to detect lying and deception* [Video]. YouTube. https://www.youtube.com/watch?v=tpJcBozuF6A. Quote appears at 0:25 mins.
60. Wired. (2019, May 21). *Former FBI agent explains how to read body language* [Video]. YouTube. https://www.youtube.com/watch?v=4jwUXV4QaTw. Quote appears at 2:13 mins.
61. Knapp, M. L. (2006). *Lying and deception in close relationships.* Cambridge University Press.
62. Morris, W. L., Sternglanz, R. W., Ansfield, M. E., Anderson, D. E., Snyder, J. H., & DePaulo, B. M. (2016). A longitudinal study of the development of emotional deception detection within new same-sex friendships. *Personality & Social Psychology Bulletin, 2,* 204–218.
63. Levine, T. R. (2014). Truth-default theory (TDT): A theory of human deception and deception detection. *Journal of Language & Social Psychology, 33*(4), 378–392.
64. Ein-Dor, T. T., Perry-Paldi, A., Zohar-Cohen, K., Efrati, Y., & Hirschberger, G. (2017). It takes an insecure liar to catch a liar: The link between attachment insecurity, deception, and detection of deception. *Personality & Individual Differences, 113,* 81–87.
65. Denault, V. (2020). Misconceptions about nonverbal cues to deception: A covert threat to the justice system? *Frontiers in Psychology, 11.* Article 73,460.
66. Denault, V. (2020).
67. Denault, V., Plusquellec, P., Jupe, L. M., St-Yves, M., Dunbar, N. E., Hartwig, M., . . . van Koppen, P. J. (2020). The analysis of nonverbal communication: The dangers of pseudoscience in security and justice contexts. *Anuario de Psicología Jurídica, 30,* 1–12.
68. American Civil Liberties Union. (2017). *Bad trip: Debunking the TSA's "behavior detection" program.* https://www.aclu.org/report/bad-trip-debunkingtsas-behavior-detection-program.
69. Hart, C. (2020, October 28). What is a white lie? *Psychology Today.* https://www.psychologytoday.com/us/blog/the-nature-deception/202010/what-is-white-lie.
70. Mayle, W. (2021, April 22). The secret trick for catching a liar every time, say psychologists. *Eat This, Not That!* https://www.eatthis.com/secret-trick-catching-liar-psychologists/.
71. As quoted in Feintzeig, R. (2021, January 10). The lies we tell during job interviews. *The Wall Street Journal.* https://www.wsj.com/articles/the-lies-we-tell-during-job-interviews-11,610,326,800. Quote appears in para. 3.
72. Chen, M., Zhang, T., Zhang, R., Wang, N., Yin, Q., Li, Y., Liu, J., Liu, T., & Li, X. (2020). Neural alignment during face-to-face spontaneous deception: Does gender make a difference? *Human Brain Mapping, 41,* 4964–4981.
73. Vasconcellos, S. J. L., Rizzatti, M., Barbosa, T. P., Schmitz, B. S., Coelho, V. C. N., & Machado, A. (2019). Understanding lies based on evolutionary psychology: A critical review. *Trends in Psychology, 27,* 141–153.
74. Feintzeig, R. (2021, January 10). The lies we tell during job interviews. *The Wall Street Journal.* https://www.wsj.com/articles/the-lies-we-tell-during-job-interviews-11,610,326,800.
75. Feintzeig, R. (2021, January 10).
76. Khan, W., Crockett, K., O'Shea, J., Hussain, A., & Khan, B. M. (2021). Deception in the eyes of deceiver: A computer vision and machine learning based automated deception detection. *Expert Systems with Applications, 169.* Article 114,341.
77. Kleinberg, B., Mozes, M., Arntz, A., & Verschuere, B. (2018). Using named entities for computer-automated verbal deception detection. *Journal of Forensic Science, 63*(3), 714–723.
78. Vrij, A., Granhag, P. A., Mann, S., & Leal, S. (2011). Outsmarting the liars: Toward a cognitive lie detection approach. *Current Directions in Psychological Science, 20,* 28–32.
79. Kleinberg, B., Mozes, M., Arntz, A., & Verschuere, B. (2018). Using named entities for computer-automated verbal deception detection. *Journal of Forensic Science, 63*(3), 714–723.
80. Mattes, K., Popova, V., & Evans, J. R. (2021). Deception detection in politics: Can voters tell when politicians are lying? *Political Behavior.* Advance online publication.
81. Luke, T. (2019). Lessons from Pinocchio: Cues to deception may be highly exaggerated. *Perspectives on Psychological Science, 14,* 646–671.
82. Wired. (2020, January 27). *Former FBI agent explains how to detect lying and deception* [Video]. YouTube. https://www.youtube.com/watch?v=tpJcBozuF6A. Quote appears at 5:46 mins.
83. Wired. (2020, January 27).
84. DePaulo, B. M., Lindsay, J. J., Malone, B. E., Muhlenbruck, L., Charlton, K., & Cooper, H. (2003). Cues to deception. *Psychological Bulletin, 129*(1), 74–118.
85. Abouelenien, M., P'erez-Rosas, V., Mihalcea, R., & Burzo, M. (2017). Detecting deceptive behavior via integration of discriminative features from multiple modalities. *IEEE Transactions on Information Forensics and Security, 12*(5), 1042–1055.
86. Abouelenien, M., P'erez-Rosas, V., Mihalcea, R., & Burzo, M. (2017).
87. Wired. (2020, January 27). *Former FBI agent explains how to detect lying and deception* [Video]. YouTube. https://www.youtube.com/watch?v=tpJcBozuF6A.
88. Wired. (2020, January 27). Quote appears at 4:31 mins.
89. Vrij, A. (2019). Deception and truth detection when analyzing nonverbal and verbal cues. *Applied Cognitive Psychology, 33,* 160–167.
90. Hartwig, M., & Granhag, P. A. (2015). Exploring the nature and origin of beliefs about deception: Implicit and explicit knowledge among lay people and resumed experts. In P. A. Granhag, A. Vrij, & B. Verschuere (Eds.), *Detecting Deception: current challenges and cognitive approaches.* Wiley.
91. DePaulo, B. M., Lindsay, J., Malone, B. E., Muhlenbruck, L., Charlton, K., & Cooper, H. (2003). Cues to deception. *Psychological Bulletin, 129,* 74–118.
92. Mann, S., Ewens, S., Shaw, D., Vrij, A., Leal, S., & Hillman, J. (2013). Lying eyes: Why liars seek deliberate eye contact. *Psychiatry, Psychology and Law, 20*(3), 452–461.
93. Khan, W., Crockett, K., O'Shea, J., Hussain, A., & Khan, B. M. (2021). Deception in the eyes of deceiver: A computer vision and machine learning based automated deception

detection. *Expert Systems with Applications, 169.* Article 114, 341.
94. Wired. (2020, January 27). *Former FBI agent explains how to detect lying and deception* [Video]. YouTube. https://www.youtube.com/watch?v=tpJcBozuF6A.
95. Mattes, K., Popova, V., & Evans, J. R. (2021). Deception detection in politics: Can voters tell when politicians are lying? *Political Behavior.* Advance online publication.
96. Brinke, L., & Weisbuch, M. (2020). How verbal-nonverbal consistency shapes the truth. *Journal of Experimental Social Psychology, 89.* Article 103,978.
97. Bond, C. F., Jr., & DePaulo, B. (2006). Accuracy of deception judgments. *Personality and Social Psychology Review, 10,* 214–234.
98. Glass, L. (2012). *The body language advantage.* Fair Winds Press.
99. Riggio, R. (2021, October 10). Secret body language cues that make people more likable. *Psychology Today.* https://www.psychologytoday.com/us/blog/cutting-edge-leadership/202110/secret-body-language-cues-make-people-more-likable.
100. Waters, H. (2013, December 13). Fake it 'til you become it: Amy Cuddy's power poses, visualized. *TED Blog.* http://blog.ted.com/fake-it-til-you-become-it-amy-cuddys-power-poses-visualized/. Quote appears in title.
101. Zloteanu, M., Krumhuber, E. G., & Richardson, D. C. (2021). Sitting in judgment: How body posture influences deception detection and gazing behavior. *Behavioral Sciences, 11*(6). Article 85.
102. Maurer, R. E., & Tindall, J. H. (1983). Effect of postural congruence on client's perception of counselor empathy. *Journal of Counseling Psychology, 30,* 158–163. See also Hustmyre, C., & Dixit, J. (2009, January 1). Marked for mayhem. *Psychology Today.* https://www.psychologytoday.com/articles/200901/marked-mayhem.
103. Ray, G., & Floyd, K. (2006). Nonverbal expressions of liking and disliking in initial interaction: Encoding and decoding perspectives. *Southern Communication Journal, 71*(1), 45–65.
104. Myers, M. B., Templer, D., & Brown, R. (1984). Coping ability of women who become victims of rape. *Journal of Consulting and Clinical Psychology, 52,* 73–78.
105. Ekman, P. (1985). *Telling lies: Clues to deceit in the marketplace, politics, and marriage.* Norton.
106. Farley, J., Risk, E., & Kingstone, A. (2013). Everyday attention and lecture retention: The effects of time, fidgeting, and mind wandering. *Frontiers in Psychology, 4.* Article 619.
107. Aspiranti, K. B., & Hulac, D.M. (2021). Using fidget spinners to improve on-task classroom behavior for students with ADHD. *Behavior Analysis Practice.* Advance online publication.
108. Nicholls, E. (2020, January 30). Everything you need to know about fidgeting. *Healthline.* https://www.healthline.com/health/fidgeting#causes.
109. Lee-Falcon, M. (2021, October 12). Why so many people are turning to fidget toys. *Seen in the City.* https://seeninthecity.co.uk/2021/10/12/why-so-manypeople-are-turning-to-fidget-toys/.
110. Hulac, D. M., Aspiranti, K., Kriescher, S., Briesch, A. M., & Athanasiou, M. (2020). A multisite study of the effect of fidget spinners on academic performance. *Contemporary School Psychology.* Advance online publication.
111. Musicus, A., Tal, A., & Wansink, B. (2014). Eyes in the aisles: Why is Cap'n Crunch looking down at my child? *Environment and Behavior, 47*(7), 715–733.
112. Murphy, K. (2014, May 18). Psst. Look over here . . . *New York Times,* p. SR6.
113. Farroni, T., Csibra, G., Simion, F. & Johnson, M. H. (2002). Eye contact detection in humans from birth. *Proceedings of the National Academy of Sciences of the United States of America, 99*(14), 9602–9605.
114. Papagiannopoulou, E. A., Chitty, K. M., Hermens, D. F., Hickie, I. B., & Lagopoulos, J. (2014). A systematic review and meta-analysis of eye-tracking studies in children with autism spectrum disorders. *Social Neuroscience, 9*(6), 610–632.
115. Akechi, H., Senju, A., Uibo, H., Kikuchi, Y., Hasegawa, T., & Hietanen, J. K. (2013). Attention to eye contact in the West and East: Autonomic responses and evaluative ratings. *Plos ONE, 8*(3), 1–10.
116. Ekman, P. (1992). An argument for basic emotions. *Cognition and Emotion, 6,* 169–200.
117. Ekman, P. (2016). What scientists who study emotion agree about. *Perspectives on Psychological Science, 11*(1), 31–34.
118. Ekman, P. (1992). An argument for basic emotions. *Cognition and Emotion, 6,* 169–200.
119. Gendron, M., Roberson, D. van der Vyver, J. M., & Feldman, L. (2014). Perceptions of emotion from facial expressions are not culturally universal: Evidence from a remote culture. *Emotion, 14*(2), 251–262.
120. Ekman, P. (1999). Facial expressions. In T. Dalgleish and M. Power (Eds.), *Handbook of Cognition and Emotion* (pp. 316–320). John Wiley & Sons.
121. Ekman, P., Friesen, W. V., O'Sullivan, M., Diacoyanni-Tarlatzis, I., Krause, R. Pitcarin, T., Scherer, K., Chan, A., Heider, K., LeCompte, W.A., Ricci-Bitti, P. E., Tomita, M., & Tzavaras, A. (1987). Universals and cultural differences in the judgments of facial expressions of emotion. *Journal of Personality and Social Psychology, 53*(4), 712–717.
122. Ekman, P. (1999). Facial expressions. In T. Dalgleish and M. Power (Eds.), *Handbook of Cognition and Emotion* (pp. 316–320). John Wiley & Sons.
123. Ze, W., Huifang, M., Jessica, L., & Fan, L. (2016). The insidious effects of smiles on social judgments. *Advances in Consumer Research, 44,* 665–669.
124. Arapova, M. A. (2017). Cultural differences in Russian and Western smiling. *Russian Journal of Communication, 1,* 34–52.
125. Arapova (2017).
126. Laird, J. D. (1974). Self-attribution of emotion: The effects of expressive behavior on the quality of emotional experience. *Journal of Personality and Social Psychology, 29,* 475–486.
127. Kuehne, M., Siwy, I., Zaehle, T., Heinze, H., & Lobmaier, J. S. (2019). Out of focus: Facial feedback manipulation modulates automatic processing of unattended emotional faces. *Journal of Cognitive Neuroscience, 31*(11), 1631–1640.
128. Kuehne, M., Zaehle, T. & Lobmaier, J.S. (2021). Effects of posed smiling on memory for happy and sad facial expressions. *Scientific Reports* 11. Article 10,477.

129. Starkweather, J. A. (1961). Vocal communication of personality and human feeling. *Journal of Communication, 11*(2), 63–72; and Scherer, K. R., Koiwunaki, J., & Rosenthal, R. (1972). Minimal cues in the vocal communication of affect: Judging emotions from content-masked speech. *Journal of Psycholinguistic Speech, 1*(3), 269–285. See also Cox, F. S., & Olney, C. (1985). *Vocalic communication of relational messages*. Paper presented at the annual meeting of the Speech Communication Association, Denver.
130. Burns, K. L., & Beier, E. G. (1973). Significance of vocal and visual channels for the decoding of emotional meaning. *Journal of Communication, 23*, 118–130. See also Hegstrom, T. G. (1979). Message impact: What percentage is nonverbal? *Western Journal of Speech Communication, 43*, 134–143; and McMahan, E. M. (1976). Nonverbal communication as a function of attribution in impression formation. *Communication Monographs, 43*, 287–294.
131. Kimble, C. E., & Seidel, S. D. (1991). Vocal signs of confidence. *Journal of Nonverbal Behavior, 15*, 99–105.
132. Cartei, V., Oakhill, J., Garnham A., Banerjee, R., & Reby, D. (2021). Voice cues influence children's assessment of adults' occupational competence. *Journal of Nonverbal Behavior, 45*(2), 281–296.
133. Sorokowski, P., Puts, D., Johnson, J., Żółkiewicz, O., Oleszkiewicz, A., Sorokowska, A., Kowal, M., Borkowska, B., & Pisanski, K. (2019). Voice of authority: Professionals lower their vocal frequencies when giving expert advice. *Journal of Nonverbal Behavior, 43*, 257–269.
134. Pisanski, K., Olesziewicz, A., Plachetka, J., Gmiterek, M., & Reby, D. (2018). Voice pitch modulation in human mate choice. *Proceedings of the Royal Society B: Biological Sciences, 285*(1893), 1–8.
135. Hardy, T. L. D., Boliek, C. A., Aalto, D., Lewicke, J., Wells, K., & Reiger, J. M. (2020). Contributions of voice and nonverbal communication to perceived masculinity-femininity for cisgender and transgender communicators. *Journal of Speech, Language, and Hearing Research, 63*, 931–947.
136. Hardy, T. L. D., Boliek, C. A., Aalto, D., Lewicke, J., Wells, K., & Reiger, J. M. (2020).
137. Hardy, T. L. D., Reiger, J. M., Wells, C., & Boliek, C. A. (2021). Associations between voice and gestural characteristics of transgender women and self-rated femininity, satisfaction, and quality of life. *American Journal of Speech Language Pathology, 30*, 663–672.
138. Gupta, N. D., Etcoff, N. L., & Jaeger, M. M. (2015, June 14). Beauty in mind: The effects of physical attractiveness on psychological well-being and distress. *Journal of Happiness Studies, 17*(3), 1313–1325.
139. Milazzo, C., & Mattes, K. (2016). Looking good for election day: Does attractiveness predict electoral success in Britain? *British Journal of Politics & International Relations, 18*(1), 161–179.
140. Haas, A., & Gregory, S. W., Jr. (2005). The impact of physical attractiveness on women's social status and interactional power. *Sociological Forum 20*(3), 449–471.
141. Gunnell, J. J., & Ceci, S. J. (2010). When emotionality trumps reason: A study of individual processing style and juror bias. *Behavioral Sciences & the Law, 28*(6), 850–877.
142. Rice, H., Murphy, C., Nolan, C., & Kelly, M. (2020). Measuring implicit attractiveness bias in the context of innocence and guilt evaluations. *International Journal of Psychology & Psychological Therapy, 20*(3) 273–275.
143. Willis, J., & Todorov, A. (2006). First impressions: Making up your mind after a 100-ms exposure to a face. *Psychological Science, 17*, 592–598.
144. Bennett, J. (2010, July 19). The beauty advantage: How looks affect your work, your career, your life. *Newsweek*. http://www.newsweek.com/2010/07/19/the-beauty-advantage.html.
145. Behrend, T., Toaddy, S., Thompson, L. F., & Sharek, D. J. (2012). The effects of avatar appearance on interviewer ratings in virtual employment interviews. *Computers in Human Behavior 28*(6), 2128–2133.
146. Mobius, M. M., & Rosenblat, T. S. (2005, June 24). Why beauty matters. *American Economic Review, 96*(1), 222–235.
147. Golle, J., Mast, F. W., & Lobmaier, J. S. (2014). Something to smile about: The interrelationship between attractiveness and emotional expression. *Cognition & Emotion, 28*(2), 298–310.
148. Abdala, K. F., Knapp, M. L., & Theune, K. E. (2002). Interaction appearance theory: Changing perceptions of physical attractiveness through social interaction. *Communication Theory, 12*, 8–40.
149. Agthe, M., Sporrle, M., & Maner, J. K. (2011). Does being attractive always help? Positive and negative effects of attractiveness on social decision making. *Personality and Social Psychology Bulletin, 37*, 1042–1054.
150. Frevert, T. K., & Walker, L. S. (2014). Physical attractiveness and social status. *Sociology Compass, 8*, 313–323.
151. Gurney, D. J., Howlett, N., Pine, K., Tracey, M., & Moggridge, R. (2017). Dressing up posture: The interactive effects of posture and clothing on competency judgements. *British Journal of Psychology, 108*, 436–451.
152. Adam, H., & Galinsky, A. D. (2012). Enclothed cognition. *Journal of Experimental Social Psychology, 48*, 918–925.
153. Sarda-Joshi, G. (2016, February 29). 7 ways clothes change the way you think. *Brain Fodder*. https://brainfodder.org/psychology-clothes-enclothed-cognition/.
154. Mendoza S. A., & Parks-Stamm, E. J. (2020). Embodying the police: The effects of enclothed cognition on shooting decisions. *Psychological Reports, 23*(6), 2353–2371.
155. Wang, Y., John, D. R., & Griskevicious, V. (2019). Does the devil wear Prada? Luxury product experiences can affect prosocial behavior. *International Journal of Research in Marketing*. Article 01357.
156. Wang, Y., Stoner, J., & John, D. R. (2019). Counterfeit luxury consumption in a social context: The effects on females' moral disengagement and behavior. *Journal of Consumer Psychology, 29*(2), 207–225.
157. Kraus, M., & Mendes, W. B. (2014). Sartorial symbols of social class elicit class-consistent behavioral and physiological responses: A dyadic approach. *Journal of Experimental Psychology, 143*(6), 2330–2340.
158. Roberts, S., Owen, R. C., & Havlicek, J. (2010). Distinguishing between perceiver and wearer effects in clothing color-associated attributions. *Evolutionary Psychology, 8*(3), 350–364.
159. McCall, T. (2013, August 1.) Why is it so difficult for the average American woman to shop for clothes? *Fashionista*.

https://fashionista.com/2013/08/why-is-it-so-difficult-for-the-average-american-woman-to-shop-for-clothes.
160. Bell, E. (2016, May 13). *Wearing heels to work is a game women have been losing for decades.* The Conversation. https://www.thenewsminute.com/article/wearing-heels-work-game-women-have-been-losing-decades-43294.
161. Gurung, R. A. R., Brickner, M., Leet, M., & Punke, E. (2018). Dressing "in code": Clothing rules, propriety, and perceptions. *The Journal of Social Psychology, 158*(5), 553–557.
162. Howlett, N., Pine, K. J., Cahill, N., Orakçıoğlu, I., & Fletcher, B. C. (2015). Unbuttoned: The interaction between provocativeness of female work attire and occupational status. *Sex Roles, 72*, 105–116.
163. Gurney, D. J., Howlett, N., Pine, K., Tracey, M., & Moggridge, R. (2017). Dressing up posture: The interactive effects of posture and clothing on competency judgements. *British Journal of Psychology, 108*, 436–451.
164. Statista. (2021, September 6). *Americans with at least one tattoo 2021, by generation.* https://www.statista.com/statistics/259601/share-of-americans-with-at-least-one-tattoo-by-age/.
165. Hunter, D. (2021, January 1). *Which country's residents have the most tattoos?* https://authoritytattoo.com/most-tattooed-country/.
166. Jackson, C. (2019, August 29). *More Americans have tattoos today than seven years ago.* Ipsos. https://www.ipsos.com/en-us/news-polls/more-americans-have-tattoos-today.
167. Michalak, J. (2021, May 26). Why has the popularity of tattoos grown? Byrdie. https://www.byrdie.com/why-are-tattoos-popular-3,189,518.
168. Mull, A. (2019, October 29). Tattoos now have an exit strategy. *The Atlantic.* https://www.theatlantic.com/health/archive/2019/10/semi-permanent-tattoos/601012/.
169. Michalak, J. (2021, May 26). Why has the popularity of tattoos grown? Byrdie. https://www.byrdie.com/why-are-tattoos-popular-3,189,518.
170. Brown, K. A., McKimmie, B. M., & Zarkadi, T. (2018) The defendant with the prison tattoo: The effect of tattoos on mock jurors' perceptions. *Psychiatry, Psychology and Law, 25*(3), 386–403.
171. Statista. (2021, September 6). *Americans with at least one tattoo 2021, by generation.* https://www.statista.com/statistics/259601/share-of-americans-with-at-least-one-tattoo-by-age/.
172. Palmisano, K. (2020, December 10). More physicians have tattoos, and how that makes them more genuine at the bedside. *Kevin MD.* https://www.kevinmd.com/blog/2020/12/more-physicians-have-tattoos-and-how-that-makes-them-more-genuine-at-the-bedside.html. Quote appears in para. 3.
173. Musambira, G. W., Raymond, L., & Hastings, S. O. (2016). A comparison of college students' perceptions of older and younger tattooed women. *Journal of Women & Aging, 28*(1), 9–23.
174. Skoda, K., Oswald, F., Brown, K., Hesse, C., & Pedersen, C. L. (2020). Showing skin: Tattoo visibility status, egalitarianism, and personality are predictors of sexual openness among women. *Sexuality and Culture, 24*, 1935–1956.
175. Broussard, K. A., & Harton, H. C. (2018). Tattoo or taboo? Tattoo stigma and negative attitudes toward tattooed individuals. *The Journal of Social Psychology, 158*(5), 521–540.
176. Baumann, C., Timming, A. R., & Gollan, P. J. (2016). Taboo tattoos? A study of the gendered effects of body art on consumers' attitudes toward visibly tattooed front-line staff. *Journal of Retailing and Consumer Services, 29*, 31–39.
177. Skoda, K., Oswald, F., Brown, K., Hesse, C., & Pedersen, C. L. (2020). Showing skin: Tattoo visibility status, egalitarianism, and personality are predictors of sexual openness among women. *Sexuality and Culture, 24*, 1935–1956.
178. Galbarczyk, A., & Ziomkiewicz, A. (2017). Tattooed men: Healthy bad boys and good-looking competitors. *Personality and Individual Differences, 106*, 122–125.
179. Workopolis. (2014, October 6). *Research reveals how your tattoos affect your chances of getting the job.* https://careers.workopolis.com/advice/research-reveals-how-your-tattoos-affect-your-chances-of-getting-the-job/.
180. French, M. T., Maclean, J. C., Robins, P. K., Sayed, B., & Shiferaw, L. (2016). Tattoos, employment, and labor market earnings: Is there a link in the ink? *Southern Economic Journal, 82*(4), 1212–1246.
181. Mull, A. (2019, October 29). Tattoos now have an exit strategy. *The Atlantic.* https://www.theatlantic.com/health/archive/2019/10/semi-permanent-tattoos/601012/. Quote appears in para.14.
182. Hart, S., Field, T., Hernandez-Reif, M., & Lundy, B. (1998). Preschoolers' cognitive performance improves following massage. *Early Child Development and Care, 143*, 59–64. For more about the role of touch in relationships, see Keltner, D. (2009). *Born to be good: The science of a meaningful life.* Norton, pp. 173–198.
183. Field, T. (2010). Touch for socioemotional and physical well-being: A review. *Developmental Review, 30*(4), 367–383.
184. Montagu, A. (1972). *Touching: The human significance of the skin.* Harper & Row, p. 93.
185. Feldman, R. (2011). Maternal touch and the developing infant. In M. Hertenstein & S. Weiss (Eds.), *Handbook of touch* (pp. 373–407). Springer.
186. Saarinen, A., Harjunen, V., Jasinskaja-Lahti, I., Jääskeläinen, I. P., & Ravaja, N. (2021). Social touch experience in different contexts: A review. *Neuroscience & Biobehavioral Reviews, 131*, 360–372.
187. Kelly, M., Svrcek, C., King, N., Scherpbier, A., & Dornan, T. (2020). Embodying empathy: A phenomenological study of physician touch. *Medical Education, 54*, 400–407.
188. Wanko Keutchafo, E. L., Kerr, J., & Jarvis, M. A. (2020). Evidence of nonverbal communication between nurses and older adults: A scoping review. *BMC Nursing, 19*. Article 53.
189. Wanko Keutchafo, E. L., Kerr, J., & Jarvis, M. A. (2020).
190. Saarinen, A., Harjunen, V., Jasinskaja-Lahti, I., Jääskeläinen, I. P., & Ravaja, N. (2021). Social touch experience in different contexts: A review. *Neuroscience & Biobehavioral Reviews, 131*, 360–372.
191. Gulledge, N., & Fischer-Lokou, J. (2003). Another evaluation of touch and helping behaviour. *Psychological Reports, 92*, 62–64.
192. Saarinen, A., Harjunen, V., Jasinskaja-Lahti, I., Jääskeläinen, I. P., & Ravaja, N. (2021). Social touch experience

193. Willis, F. N., & Hamm, H. K. (1980). The use of interpersonal touch in securing compliance. *Journal of Nonverbal Behavior, 5*, 49–55.
194. Jacob, C., & Guéguen, N. (2014). The effect of compliments on customers' compliance with a food server's suggestion. *International Journal of Hospitality Management, 40*, 59–61.
195. Luangrath, A. W., Peck, J., & Gustafsson, A. (2020). Should I touch the customer? Rethinking interpersonal touch effects from the perspective of the touch initiator. *Journal of Consumer Research, 47*(4), 588–607.
196. Kraus, M. W., Huang, C., & Keltner, D. (2010). Tactile communication, cooperation, and performance: An ethological study of the NBA. *Emotion, 10*(5), 745–749.
197. Saarinen, A., Harjunen, V., Jasinskaja-Lahti, I., Jääskeläinen, I. P., & Ravaja, N. (2021). Social touch experience in different contexts: A review. *Neuroscience & Biobehavioral Reviews, 131*, 360–372.
198. Williams, J. C., & Lebsock, S. (2018). Now what? Social media has created a remarkable moment for women, but is this really the end of the harassment culture? *Harvard Business Review*. https://hbr.org/coverstory/2018/01/now-what.
199. See Hall, E. (1969). *The hidden dimension*. Anchor Books.
200. LeFebvre, L., & Allen, M. (2014). Teacher immediacy and student learning: An examination of lecture/laboratory and self-contained course sections. *Journal of the Scholarship of Teaching and Learning, 14*(2), 29–45.
201. Wouda, J. C., & van de Wiel, H. B. (2013). Education in patient–physician communication: How to improve effectiveness? *Patient Education and Counseling, 90*(1), 46–53.
202. Mumm, J., & Mutlu, B. (2011, March). Human-robot proxemics: Physical and psychological distancing in human-robot interaction. *Proceedings of the 6th International Conference on Human-Robot Interaction in Lausanne, Switzerland*. http://www.cs.cmu.edu/~illah/CLASSDOCS/p331-mumm.pdf.
203. Samani, S. A., Rasid, S. Z. A., Sofian, S. (2015, October). Perceived level of personal control over the work environment and employee satisfaction and work performance. *Performance Improvement, 54*(9), 28–35.
204. Bernstein, B., & Waber, B. (2019, November–December). The truth about open offices. *Harvard Business Review*. https://hbr.org/2019/11/the-truth-about-open-offices.
205. Bernstein, B., & Waber, B. (2019, November–December).
206. Sander, E. (2021, July 7). Science confirms it: Open offices are a nightmare. *Fast Company*. https://www.fastcompany.com/90652947/science-confirms-it-open-offices-are-a-nightmare.
207. Sparks, S. D. (2018, November 5). Sounding an alarm: Background noise can hurt student achievement. *Education Week*. https://www.edweek.org/teaching-learning/sounding-an-alarm-background-noise-can-hurt-student-achievement/2018/11.
208. Barrett, P., Davies, F., Zhang, Y., & Barrett, L. (2017). The holistic impact of classroom spaces on learning in specific subjects. *Environment & Behavior, 49*(4), 425–451.
209. Bruneau, T. J. (2012). Chronemics: Time-binding and the construction of personal time. *ETC: A Review of General Semantics, 69*(1), 72.
210. Ballard, D. I., & Seibold, D. R. (2000). Time orientation and temporal variation across work groups: Implications for group and organizational communication. *Western Journal of Communication, 64*, 218–242.
211. Levine, R. (1997). *A geography of time: The temporal misadventures of a social psychologist*. Basic Books.
212. See, for example, Hill, O. W., Block, R. A., & Buggie, S. E. (2000). Culture and beliefs about time: Comparisons among black Americans, black Africans, and white Americans. *Journal of Psychology, 134*, 443–457.
213. Levine, R., & Wolff, E. (1985, March). Social time: The heartbeat of culture. *Psychology Today, 19*, 28–35. See also Levine, R. (1987, April). Waiting is a power game. *Psychology Today, 21*, 24–33.
214. Burgoon, J. K., Buller, D. B., & Woodall, W. G. (1996). *Nonverbal communication*. McGraw-Hill, p. 148. See also White, L. T., Valk, R., & Dialmy, A. (2011). What is the meaning of "on time"? The sociocultural nature of punctuality. *Journal of Cross-Cultural Psychology, 42*(3), 482–493.
215. Matsumoto, D. (2006). Culture and nonverbal behavior. In V. Manusov & M. L. Patterson (Eds.), *Sage handbook of nonverbal communication* (pp. 219–235). Sage.
216. Hall, E. (1969). *The hidden dimension*. Anchor Books.
217. Kelly, D. J., Liu, S., Rodger, H., Miellet, S., Ge, L., & Caldara, R. (2011). Developing cultural differences in face processing. *Developmental Science, 14*(5), 1176–1184.
218. Yuki, M., Maddux, W. W., & Masuda, T. (2007). Are the windows to the soul the same in the East and West? Cultural differences in using the eyes and mouth as cues to recognize emotions in Japan and the United States. *Journal of Experimental Social Psychology, 43*, 303–311.
219. Linneman, T. J. (2013). Gender in *Jeopardy!* Intonation variation on a television game show. *Gender & Society, 27*, 82–105; Wolk, L., Abdelli-Beruh, N. B., & Slavin, D. (2012). Habitual use of vocal fry in young adult female speakers. *Journal of Voice, 26*, 111–116.
220. Anderson, R. C., Klofstad, C. A., Mayew, W. J., & Venkatachalam, M. (2014). Vocal fry may undermine the success of young women in the labor market. *PLoS ONE, 9*, e97506.
221. Yuasa, I. P. (2010). Creaky voice: A new feminine voice quality for young urban-oriented upwardly mobile American women? *American Speech, 85*, 315–337.
222. Booth-Butterfield, M., & Jordan, F. (1988). *"Act like us": Communication adaptation among racially homogeneous and heterogeneous groups*. Paper presented at the Speech Communication Association meeting, New Orleans.
223. Warnecke, A. M., Masters, R. D., & Kempter, G. (1992). The roots of nationalism: Nonverbal behavior and xenophobia. *Ethnology and Sociobiology, 13*, 267–282.
224. Hall, J. A. (2006). Women and men's nonverbal communication. In V. Manusov & M. L. Patterson (Eds.), *The Sage handbook of nonverbal communication* (pp. 201–218). Sage.
225. Hadiani, D., & Ariyani, E. D. (2020) Students' verbal and nonverbal communication patterns: A gender perspective. *Proceedings of the International Conference on Applied Science and Technology on Social Science, 544*, 282–286.
226. Lin, Y., Ding, H., & Zhang, Y. (2021). Gender differences in identifying facial, prosodic, and semantic emotions show category and channel-specific effects mediated by encoder's gender. *Journal of Speech, Language, and Hearing Research, 64*, 2941–2955.

227. Zand, S., Baradaran, M., Najafi, R., & Maleki, A., & Mahdipour, A. (2020). Culture and gender in nonverbal communication. *Journal of Behavioral Public Administration, 8*, 123–130.
228. Gurney, D. J., Howlett, N., Pine, K., Tracey, M., & Moggridge, R. (2017). Dressing up posture: The interactive effects of posture and clothing on competency judgements. *British Journal of Psychology, 108*, 436–451.
229. Gurney, D. J., Howlett, N., Pine, K., Tracey, M., & Moggridge, R. (2017).
230. Holden, M. (2017, August 15). Being a woman in the workplace means getting pressured to wear makeup on the job. *Allure*. https://www.allure.com/story/women-pressured-to-wear-makeup-at-work-as-a-double-standard.
231. Denton, E. (2017, March 29). How much the average woman spends on makeup in her life. *Allure*. https://www.allure.com/story/average-woman-spends-on-makeup.
232. Guerin, L. (2021). *Can my employer require me to wear makeup?* Nolo. https://www.employmentlawfirms.com/resources/can-my-employer-require-me-wear-makeup.htm.
233. Bernard, P., Content, J., Servais, L., Wollast, R., & Gervais, S. (2020). An initial test of the cosmetics dehumanization hypothesis: Heavy makeup diminishes attributions of humanness-related traits to women. *Sex Roles, 83*, 315–327.
234. Bernard, P., Content, J., Servais, L., Wollast, R., & Gervais, S. (2020).
235. Elan, P. (2020, February 23). Beauty and the bloke: Why more men are wearing makeup. *The Guardian*. https://www.theguardian.com/fashion/2020/feb/23/beauty-and-the-bloke-why-more-men-are-chooisng-to-wear-makeup-warpaint.
236. O'Grady, M. (2021, May 10). Makeup is for everyone. *New York Times*. https://www.nytimes.com/2021/05/10/t-magazine/men-makeup-gender-norms.html.
237. Elan, P. (2020, February 23). Beauty and the bloke: Why more men are wearing makeup. *The Guardian*. https://www.theguardian.com/fashion/2020/feb/23/beauty-and-the-bloke-why-more-men-are-chooisng-to-wear-makeup-warpaint.
238. Imam, J. (2016). "Resting bitch face" is real, scientists say. CNN. https://www.cnn.com/2016/02/03/health/resting-bitch-face-research-irpt/index.html.
239. Gibson, C. (2016, February 2). Scientists have discovered what causes resting bitch face. *The Washington Post*. https://www.washingtonpost.com/news/arts-and-entertainment/wp/2016/02/02/scientists-have-discovered-the-source-of-your-resting-bitch-face/.
240. Riggio, R. (2021, October 10). Secret body language cues that make people more likable. *Psychology Today*. https://www.psychologytoday.com/us/blog/cutting-edge-leadership/202110/secret-body-language-cues-make-people-more-likable.
241. Gibson, C. (2016, February 2). Scientists have discovered what causes resting bitch face. *The Washington Post*. https://www.washingtonpost.com/news/arts-and-entertainment/wp/2016/02/02/scientists-have-discovered-the-source-of-your-resting-bitch-face/. Quote appears in para. 13.
242. Mayo, C., & Henley, N. M. (Eds.). (2012). *Gender and nonverbal behavior*. Springer Science & Business Media.

CHAPTER 8

1. Eidell, L. (2021, March 15). *Bachelor* and *Bachelorette* couples still together: The complete list. *Glamour*. https://www.glamour.com/story/bachelor-bachelorette-couples-history.
2. Boardman, M. (2015, February 7). Sean Lowe, Catherine Giudici explain why so few *Bachelor* couples get married. *US Weekly*. http://www.usmagazine.com/entertainment/news/sean-lowe-catherine-giudici-explain-why-so-few-bachelor-couples-wed-201572. Quote appears in para. 6.
3. Lim, T. S., & Bowers, J. W. (1991). Facework: Solidarity, approbation, and tact. *Human Communication Research,17*, 415–450.
4. Frei, J. R., & Shaver, P. R. (2002). Respect in close relationships: Prototype, definition, self-report assessment, and initial correlates. *Personal Relationships, 9*, 121–139.
5. Marano, H. E., (2014, January 1). Love and power. *Psychology Today*. https://www.psychologytoday.com/articles/201401/love-and-power.
6. Hashim, I. M., Mohd-Zaharim, N., & Khodarahimi, S. (2012). Perceived similarities and satisfaction among friends of the same and different ethnicity and sex at workplace. *Psychology, 3*, 621–625.
7. Nelson, P. A., Thorne, A., & Shapiro, L. A. (2011). I'm outgoing and she's reserved: The reciprocal dynamics of personality in close friendships in young adulthood. *Journal of Personality, 79*(5), 1113–1148.
8. Lin, R., & Utz, S. (2017). Self-disclosure on SNS: Do disclosure intimacy and narrativity influence interpersonal closeness and social attraction? *Computers in Human Behavior, 70*, 426–436.
9. Batool, S., & Malik, N. I. (2010, June). Role of attitude similarity and proximity in interpersonal attraction among friends. *International Journal of Innovation, Management and Technology, 1*(2), 142–146.
10. See, for example, Roloff, M. E. (1981). *Interpersonal communication: The social exchange approach*. Sage.
11. Duck, S. W. (2011). Similarity and perceived similarity of personal constructs as influences on friendship choice. *British Journal of Clinical Psychology, 12*, 1–6.
12. Hughes, B. T., Flournoy, J. C., & Srivastava, S. (2021). Is perceived similarity more than assumed similarity? An interpersonal path to seeing similarity between self and others. *Journal of Personality and Social Psychology, 121*(1), 184–200.
13. Sias, P. M., Drzewiecka, J. A., Meares, M., Bent, R., Konomi, Y., Ortega, M., & White, C. (2008). Intercultural friendship development. *Communication Reports, 21*, 1–13.
14. Hashim, I. M., Mohd-Zaharim, N., & Khodarahimi, S. (2012). Perceived similarities and satisfaction among friends of the same and different ethnicity and sex at workplace. *Psychology, 3*, 621–625.
15. Finkel, E. J., Eastwick, P. W., Karney, B. R., Reis, H. T., & Sprecher, S. (2012). Online dating: A critical analysis from the perspective of psychological science. *Psychological Science in the Public Interest, 13*(1), 3–66.
16. See Rossiter, C. M., Jr. (1974). Instruction in metacommunication. *Central States Speech Journal, 25*, 36–42; and Wilmot, W. W. (1980). Metacommunication: A reexamination and extension. In D. Nimmo (Ed.), *Communication yearbook 4*. Transaction.

17. Grey, J. (2017, December 29). *Don't let the little things turn into big things in your relationship*. The Good Men Project. https://goodmenproject.com/sex-relationships/dont-let-little-things-become-big-things-relationship-cmtt/.
18. Stone, D., Patton, B., & Heen, S. (2010). *Difficult conversations: How to discuss what matters most*. Penguin.
19. Tamir, D. I., & Mitchell, J. P. (2012). Disclosing information about the self is intrinsically rewarding. *Proceedings of the National Academy of Science, 109*(21), 8038–8043.
20. Altman, I., & Taylor, D. A. (1973). *Social penetration: The development of interpersonal relationships*. Holt, Rinehart and Winston.
21. Whitbourne, S. K. (2014, April 1). The secret to revealing your secrets. *Psychology Today*. https://www.psychologytoday.com/blog/fulfillment-any-age/201404/the-secret-revealing-your-secrets. Quote appears in para. 2.
22. Luft, J. (1969). *Of human interaction*. National Press.
23. Summarized in Pearson, J. (1989). *Communication in the family*. Harper & Row, pp. 252–257.
24. Chen, Y., & Nakazawa, M. (2009). Influences of culture on self-disclosure as relationally situated in intercultural and interracial friendships from a social penetration perspective. *Journal of Intercultural Communication Research, 38*(2), 77–98.
25. Rosenfeld, L. B., & Gilbert, J. R. (1989). The measurement of cohesion and its relationship to dimensions of self-disclosure in classroom settings. *Small Group Behavior, 20*, 291–301.
26. The friends I've never met. (2014, December 15). *Femsplain*. https://femsplain.com/the-friend-i-ve-never-met-7ae521269047. Quotes appear in paras. 4 and 5. (Maya and Jad are pseudonyms. The post does not include names.)
27. Patton, B. R., & Giffin, K. (1974). *Interpersonal communication: Basic text and readings*. Harper & Row.
28. Ledbetter, A. M. (2014). The past and future of technology in interpersonal communication theory and research. *Communication Studies, 65*(4), 456–459.
29. Ishii, K. (2017). Online communication with strong ties and subjective well-being in Japan. *Computers in Human Behavior, 66*, 129–137.
30. Quan-Haase, A., Mo, G. Y., & Wellman, B. (2017). Connected seniors: How older adults in East York exchange social support online and offline. *Information, Communication & Society, 20*(7), 967–983.
31. Juvonen, J., Schacter, H. L., & Lessard, L. M. (2021). Connecting electronically with friends to cope with isolation during COVID-19 pandemic. *Journal of Social and Personal Relationships, 38*(6), 1782–1799.
32. These are my "real" friends: Removing the stigma of online friendships. (2013, January 29). *Persephone*. http://persephonemagazine.com/2013/01/these-are-my-real-friends-removing-the-stigma-of-online-friendships/. Quote appears in para. 4.
33. Durrotul, M. (2017). The use of social media in intercultural friendship development. *Profetik, 10*(1), 5–20.
34. The phubbing truth. (2013, October 8). *Wordability*. http://wordability.net/2013/10/08/the-phubbing-truth/.
35. Przybylski, A. K., & Weinstein, N. (2013). Can you connect with me now? How the presence of mobile communication technology influences face-to-face conversation quality. *Journal of Social and Personal Relationships, 30*, 237–246.
36. Kerner, I. (2017, May 16). *What counts as "cheating" in the digital age?* CNN. https://www.cnn.com/2017/05/16/health/cheating-internet-sex-kerner/index.html.
37. Yao, M. Z., & Zhong, Z. (2014). Loneliness, social contacts and Internet addiction: A cross-lagged panel study. *Computers in Human Behavior, 30*, 164–170.
38. Donnelly, M. (2017, May 3). *Here's to the best friends who feel like family*. Thought Catalog. https://thoughtcatalog.com/marisa-donnelly/2017/05/heres-to-the-best-friends-who-feel-like-family/. Quote appears in para. 8.
39. Donnelly (2017). Quote appears in para. 8.
40. Deci, E., La Guardia, J., Moller, A., Scheiner, M., & Ryan, R. (2006). On the benefits of giving as well as receiving autonomy support: Mutuality in close friendships. *Personality & Social Psychology Bulletin, 32*, 313–327.
41. Demir, M., & Özdemir, M. (2010). Friendship, need satisfaction and happiness. *Journal of Happiness Studies, 11*, 243–259.
42. Elmer, T., Boda, Z., & Stadtfeld, C. (2017). The co-evolution of emotional well-being with weak and strong friendship ties. *Network Science* (Cambridge University Press), *5*(3), 278–307.
43. Graber, R., Turner, R., & Madill, A. (2016). Best friends and better coping: Facilitating psychological resilience through boys' and girls' closest friendships. *The British Journal of Psychology, 107*(2), 338–358.
44. Demir, M., Özdemir, M., & Marum, K. (2011). Perceived autonomy support, friendship maintenance, and happiness. *Journal of Psychology, 145*, 537–571.
45. Salinas, J., O'Donnell, A., Kojis, D. J., Pase, M. P., DeCarli, C., Rentz, D. M., Berkman, L. F., Beiser, A., & Seshadri, S. (2021). Association of social support with brain volume and cognition. *JAMA Network Open, 4*(8), e2121122–e2121122.
46. Hall, J. A. (2019). How many hours does it take to make a friend? *Journal of Social and Personal Relationships, 36*, 1278–1296.
47. Harding, N. (2021, March 6). "My long distance friendship is also my strongest." *Cosmopolitan*. https://www.cosmopolitan.com/uk/love-sex/a35709493/long-distance-friendship/. Quote appears in last paragraph.
48. Charleston, A. (2017, January 16). To those who overshare on social media: Don't. *Odyssey*. https://www.theodysseyonline.com/open-letter-those-overshare-social-media.
49. Bello, R. S., Brandau-Brown, F. E., Zhang, S., & Ragsdale, J. (2010). Verbal and nonverbal methods for expressing appreciation in friendships and romantic relationships: A cross-cultural comparison. *International Journal of Intercultural Relations, 34*, 294–302.
50. Guerrero, L. K., Farinelli, L., & McEwan, B. (2009). Attachment and relational satisfaction: The mediating effect of emotional communication. *Communication Monographs, 76*, 487–514.
51. Davis, J. R., & Gold, G. J. (2011). An examination of emotional empathy, attributions of stability, and the link between perceived remorse and forgiveness. *Personality & Individual Differences, 50*, 392–397.

52. Deci, E., La Guardia, J., Moller, A., Scheiner, M., & Ryan, R. (2006). On the benefits of giving as well as receiving autonomy support: Mutuality in close friendships. *Personality & Social Psychology Bulletin, 32,* 313–327.
53. van der Horst, M., & Coffe, H. (2012). How friendship network characteristics influence subjective well-being. *Social Indicators Research, 107,* 509–529.
54. Gold, S. S. (2008, September 10). Do *you* have friends of other races? *Glamour.* https://www.glamour.com/story/do-you-have-friends-of-other-races. Quote appears in para. 5.
55. Allport, G. W. (1954). *The nature of prejudice.* Perseus.
56. Abrams, J. R., McGaughey, K. J., & Haghighat, H. (2018). Attitudes toward Muslims: A test of the parasocial contact hypothesis and contact theory. *Journal of Intercultural Communication Research, 47*(4), 276–292.
57. Shook, N. J., Hopkins, P. D., & Koech, J. M. (2016). The effect of intergroup contact on secondary group attitudes and social dominance orientation. *Group Processes & Intergroup Relations, 19*(3), 328–342.
58. Phillips, B. A., Fortney, S., & Swafford, L. (2019). College students' social perceptions toward individuals with intellectual disability. *Journal of Disability Policy Studies, 30*(1), 3–10.
59. Lytle, A., & Levy, S. R. (2019). Reducing ageism: Education about aging and extended contact with older adults. *Gerontologist, 59*(3), 580–588.
60. Pekerti, A. A., van de Vijver, F. J. R., Moeller, M., & Okimoto, T. G. (2020, March). Intercultural contacts and acculturation resources among international students in Australia: A mixed-methods study. *International Journal of Intercultural Relations, 75,* 56–81.
61. Bukhari, S., Mushtaq, H., & Aurangzaib, S. (2016). Attitudes towards transgender: A study of gender and influencing factors. *Journal of Gender & Social Issues, 15*(2), 93–112.
62. Dovidio, J. F., Love, A., Schellhaas, F. M. H., & Hewstone, M. (2017). Reducing intergroup bias through intergroup contact: Twenty years of progress and future directions. *Group Processes & Intergroup Relations, 20*(5), 606–620.
63. Tucker, J. A. (2016, December 28). *Actually, friendship is a powerful antidote to racism.* Foundation for Economic Education. https://fee.org/articles/actually-friendship-is-a-powerful-antidote-to-racism/.
64. *The single best antidote to prejudice and racism? Cross-race friendship.* (2010, November 23). The Leadership Conference Education Fund. https://civilrights.org/edfund/resource/the-single-best-antidote-to-prejudice-and-racism-cross-race-friendship/.
65. Davis, D. (n.d.) *What do you do when someone just doesn't like you?* TEDxCharlottesville. https://www.ted.com/talks/daryl_davis_what_do_you_do_when_someone_just_doesn_t_like_you/transcript?language=en.
66. Mancini, T., & Imperato, C. (2020). Can social networks make us more sensitive to social discrimination? E-contact, identity processes and perception of online sexual discrimination in a sample of Facebook users. *Social Sciences, 9*(4), 1–11.
67. Dahlberg, L. (2015, August 10). *10 great things I learned from my best friend who is quadriplegic.* Life Beyond Numbers. https://lifebeyondnumbers.com/10-great-things-learned-best-friend/.
68. Wölfer, R., Schmid, K., Hewstone, M., & Zalk, M. (2016). Developmental dynamics of intergroup contact and intergroup attitudes: Long-term effects in adolescence and early adulthood. *Child Development, 87*(5), 1466–1478.
69. McDonnall, M. C., & Antonelli, K. (2020). The impact of a brief meeting on employer attitudes, knowledge, and intent to hire. *Rehabilitation Counseling Bulletin, 63*(3), 131.
70. Quinton, W. J. (2019). Unwelcome on campus? Predictors of prejudice against international students. *Journal of Diversity in Higher Education, 12*(2), 156–169.
71. Tucker, J. A. (2016, December 28). *Why do people poke fun at the claim that friendship is evidence of equanimity?* Foundation for Economic Education. https://fee.org/articles/actually-friendship-is-a-powerful-antidote-to-racism/. Quote appears in para. 9.
72. Ansberry, C. (2021, March 2). The power of friendship across generations. *The Wall Street Journal.* https://www.wsj.com/articles/the-power-of-friendship-across-generations-11614717112.
73. Furlan, J. (2019, August 19). *Accept the awkward: How to make friends (and keep them).* NPR. https://www.npr.org/2019/08/15/751479810/make-new-friends-and-keep-the-old. Quote appears in para. 4.
74. Zhang, A. (n.d.) Why intergenerational friendships matter—and how to form them. *The Good Trade.* https://www.thegoodtrade.com/features/intergenerational-friendships. Quote appears in para. 6.
75. Minow, M. (1998). Redefining families: Who's in and who's out? In K. V. Hansen & A. I. Garey (Eds.), *Families in the U.S.: Kinship and domestic policy* (pp. 7–19). Temple University Press. (Originally published in the *University of Colorado Law Review,* 1991, *62,* 269–285.)
76. Galvin, K. M. (2006). Diversity's impact of defining the family: Discourse-dependence and identity. In R. L. West & L. H. Turner (Eds.), *The family communication sourcebook* (pp. 3–20). Sage.
77. For background on this theory, see Baumrind, D. (1991). The influence of parenting styles on adolescent competence and substance use. *The Journal of Early Adolescence, 11,* 56–95.
78. Koerner, A. F., & Fitzpatrick, M. A. (2006). Family communication patterns theory: A social cognitive approach. In D. O. Braithwaite & L. A. Baxter (Eds.), *Engaging theories in family communication: Multiple perspectives* (pp. 50–65). Sage.
79. Janik McErlean, A. B., & Lim, L. X. C. (2020). Relationship between parenting style, alexithymia and aggression in emerging adults. *Journal of Family Issues, 41*(6), 853–874.
80. Cameron, M., Cramer, K. M., & Manning, D. (2020). Relating parenting styles to adult emotional intelligence: A retrospective study. *Athens Journal of Social Sciences, 7*(3), 185–198.
81. Hamon, J. D., & Schrodt, P. (2012). Do parenting styles moderate the association between family conformity orientation and young adults' mental well-being? *Journal of Family Communication, 12,* 151–166. Quote appears on p. 162.
82. Edwards, R., Hadfield, L., Lucey, H., & Mauthner, M. (2006). *Sibling identity and relationships: Brothers and sisters.* Routledge. Quote appears on p. 4.

83. Epstein, L. (2014, August 4). *16 things that only half-siblings understand*. BuzzFeed. https://www.buzzfeed.com/leonoraepstein/things-only-half-siblings-understand?utm_term=.gp1jQpMB1#.lmZROVGWY. Quote appears in item 7.
84. Stewart, R. B., Kozak, A. L., Tingley, L. M., Goddard, J. M., Blake, E. M., & Cassel, W. A. (2001). Adult sibling relationships: Validation of a typology. *Personal Relationships, 8*, 299–324.
85. Riggio, H. (2006). Structural features of sibling dyads and attitudes toward sibling relationships in young adulthood. *Journal of Family Issues, 27*, 1233–1254.
86. So . . . my brother just moved out. (2009, January 16). Grasscity. https://forum.grasscity.com/threads/so-my-brother-just-moved-out.322445/. Quote appears in para. 7.
87. Scharf, M., Shulman, S., & Avigad-Spitz, L. (2005). Sibling relationships in emerging adulthood and in adolescence. *Journal of Adolescent Research, 20*, 64–90.
88. Riggio, H. (2006). Structural features of sibling dyads and attitudes toward sibling relationships in young adulthood. *Journal of Family Issues, 27*, 1233–1254.
89. Hamwey, M. K., & Whiteman, S. D. (2021). Jealousy links comparisons with siblings to adjustment among emerging adults. *Family Relations, 70*(2), 483–497.
90. Myers, S. A., & Goodboy, A. K. (2010). Relational maintenance behaviors and communication channel use among adult siblings. *North American Journal of Psychology, 12*, 103–116.
91. Duke, M. P. (2013, March 23). *The stories that bind us: What are the twenty questions?* HuffPost. http://www.huffingtonpost.com/marshall-p-duke/the-stories-that-bind-us-_b_2918975.html.
92. Szczesniak, M., & Tulecka, M. (2020). Family functioning and life satisfaction: The mediatory role of emotional intelligence. *Psychology Research and Behavior Management, 13*, 223–232.
93. Ledbetter, A. M. (2019). Parent–child privacy boundary conflict patterns during the first year of college: Mediating family communication patterns, predicting psychosocial distress. *Human Communication Research, 45*(3), 255–285.
94. Ritu Modi. (2017). Parental encouragement and mental health among students. *Indian Journal of Health and Wellbeing, 8*(5), 398–401.
95. Feiler, B. (2013). *The secrets of happy families: Improve your mornings, rethink family dinner, fight smarter, go out and play, and much more*. HarperCollins.
96. Ahmetoglu, G., Swami, V., & Chamorro-Premuzic, T. (2010). The relationship between dimensions of love, personality, and relationship length. *Archives of Sexual Behavior, 34*, 1181–1190.
97. Malouff, J. M., Schutte, N. S., & Thorsteinsson, E. B. (2013). Trait emotional intelligence and romantic relationship satisfaction: A meta-analysis. *American Journal of Family Therapy, 42*, 53–66.
98. Knapp, M. L., & Vangelisti, A. L. (2009). *Interpersonal communication and human relationships* (6th ed.). Allyn and Bacon.
99. Canary, D. J., & Stafford, L. (Eds.). (1994). *Communication and relational maintenance*. Academic Press. See also Lee, J. (1998). Effective maintenance communication in superior-subordinate relationships. *Western Journal of Communication, 62*, 181–208.
100. Quittner, E. (2019, December 9). *When the DM slide actually works: 4 couples who met on Instragram*. Man Repeller. https://www.manrepeller.com/2019/12/instagram-couples-dms.html.
101. Sprecher, S., & Hampton, A. J. (2017). Liking and other reactions after a get-acquainted interaction: A comparison of continuous face-to-face interaction versus interaction that progresses from text messages to face-to-face. *Communication Quarterly, 65*(3), 333–353.
102. Floyd, K., Hess, J. A., Miczo, L. A., Halone, K. K., Mikkelson, A. C., & Tusing, K. (2005). Human affection exchange: VIII. Further evidence of the benefits of expressed affection. *Communication Quarterly, 53*, 285–303.
103. Wilson, S. R., Kunkel, A. D., Robson, S. J., Olufowote, J. O., & Soliz, J. (2009). Identity implications of relationship (re)definition goals: An analysis of face threats and facework as young adults initiate, intensify, and disengage from romantic relationships. *Journal of Language and Social Psychology, 28*, 32–61.
104. Duran, R. L., & Kelly, L. (2017). Knapp's model of relational development in the digital age. *Iowa Journal of Communication, 49*(1/2), 22–45.
105. Burgess, B. (n.d.). *What to do if it gets awkward on an internet date*. Synonym. http://classroom.synonym.com/gets-awkward-internet-date-23312.html. Quote appears in para. 1.
106. Caughlin, J. P., & Sharabi, L. L. (2013). A communicative interdependence perspective of close relationships: The connections between mediated and unmediated interactions matter. *Journal of Communication, 63*(5), 873–893.
107. Flaa, J. (2013, October 29). *I met my spouse online: 9 online dating lessons learned the hard way*. HuffPost. http://www.huffingtonpost.com/jennifer-flaa/9-online-dating-lessons_b_4174334.html. Quote appears in para. 13.
108. Aslay, J. (2012, November 3). *You lost me at hello, how to get past the awkward first meeting*. Understand Men Now. http://www.jonathonaslay.com/2012/11/03/you-lost-me-at-hello-how-to-get-past-the-awkward-first-meeting/. Quote appears in para. 11.
109. Flaa, J. (2013, October 29). *I met my spouse online: 9 online dating lessons learned the hard way*. HuffPost. http://www.huffingtonpost.com/jennifer-flaa/9-online-dating-lessons_b_4174334.html. Quote appears in paragraph 13.
110. Meyers, S. (2014, December 23.) 5 ways to put your date at ease (and alleviate awkward tension). *Fox News Magazine*. http://magazine.foxnews.com/love/5-ways-put-your-date-at-ease-and-alleviate-awkward-tension.
111. Meyers (2014). Quote appears in the last paragraph.
112. Weiner, Z. (2015, July 24). *20 unexpected ways to tell your new relationship is getting serious*. Bustle. https://www.bustle.com/articles/99437-20-unexpected-ways-to-tell-your-new-relationship-is-getting-serious.
113. Dunleavy, K., & Booth-Butterfield, M. (2009). Idiomatic communication in the stages of coming together and falling apart. *Communication Quarterly, 57*, 416–432.

114. Sarah. (2021, August 30). *Breaking up over text—how cool is it?* Bonobology. https://www.bonobology.com/breaking-up-over-text/. Quotes appear in para. 5 under "Is It Rude?"
115. D'Agata, M. T., Kwantes, P. J., & Holden, R. R. (2021). Psychological factors related to self-disclosure and relationship formation in the online environment. *Personal Relationships, 28*(2), 230–250.
116. Chapman, G. (2010). *The five love languages: The secret to love that lasts.* Northfield Publishing.
117. Frisby, B. N., & Booth-Butterfield, M. (2012). The "how" and "why" of flirtatious communication between marital partners. *Communication Quarterly, 60,* 465–480.
118. Merolla, A. J. (2010). Relational maintenance during military deployment: Perspectives of wives of deployed US soldiers. *Journal of Applied Communication Research, 38*(1), 4–26.
119. Geiger, A. W. (2016, November 30). *Sharing chores a key to good marriage, say majority of married adults.* Pew Research Center. https://www.pewresearch.org/fact-tank/2016/11/30/sharing-chores-a-key-to-good-marriage-say-majority-of-married-adults/.
120. Papp, L. M. (2018). Topics of marital conflict in the everyday lives of empty nest couples and their implications for conflict resolution. *Journal of Couple & Relationship Therapy, 17*(1), 1–24.
121. Soin, R. (2011). Romantic gift giving as chore or pleasure: The effects of attachment orientations on gift giving perceptions. *Journal of Business Research, 64,* 113–118.
122. Van Raalte, L. J. (2021). The effects of cuddling on relational quality for married couples: A longitudinal investigation. *Western Journal of Communication, 85*(1), 61–82.
123. Egbert, N., & Polk, D. (2006). Speaking the language of relational maintenance: A validity test of Chapman's (1992) *Five Love Languages. Communication Research Reports, 23*(1), 19–26.
124. Bland, A. M., & McQueen, K. S. (2018). The distribution of Chapman's love languages in couples: An exploratory cluster analysis. *Couple and Family Psychology: Research and Practice, 7*(2), 103–126.
125. Hughes, J. L., & Camden, A. A. (2020). Using Chapman's five love languages theory to predict love and relationship satisfaction. *Psi Chi Journal of Psychological Research, 25,* 234–244.
126. Chapman, G. (2010). *The five love languages: The secret to love that lasts.* Northfield Publishing.
127. See, for example, Baxter, L. A., & Montgomery, B. M. (1998). A guide to dialectical approaches to studying personal relationships. In B. M. Montgomery & L. A. Baxter (Eds.), *Dialectical approaches to studying personal relationships* (pp. 1–16). Erlbaum; and Ebert, L. A., & Duck, S. W. (1997). Rethinking satisfaction in personal relationships from a dialectical perspective. In R. J. Sternberg & M. Hojjatr (Eds.), *Satisfaction in close relationships* (pp. 190–216). Guilford.
128. Summarized by Baxter, L. A. (1994). A dialogic approach to relationship maintenance. In D. J. Canary & L. Stafford (Eds.), *Communication and relational maintenance* (pp. 233–254). Academic Press.
129. Baxter (1994).
130. Morris, D. (1971). *Intimate behavior.* Kodansha Globe, pp. 21–29.
131. Adapted from Baxter, L. A., & Montgomery, B. M. (1998). A guide to dialectical approaches to studying personal relationships. In B. M. Montgomery & L. A. Baxter (Eds.), *Dialectical approaches to studying personal relationships* (pp. 1–16). Erlbaum.
132. Siffert, A., & Schwarz, B. (2011). Spouses' demand and withdrawal during marital conflict in relation to their subjective well-being. *Journal of Social and Personal Relationships, 28,* 262–277.
133. Lee, T. (2016, March 17). Jillian Harri's advice for new "Bachelorette" JoJo Fletcher: "It sounds cliché, but be yourself." *Hello!* http://us.hellomagazine.com/health-and-beauty/12016031712860/jillian-harris-bachelorette-advice-jojo-fletcher-motherhood/. Quote appears in headline.

CHAPTER 9

1. Brown, B. (2015). *Rising strong.* Spiegel & Grau. Quotes appear on p. 31.
2. Brown (2015). Quotes appear on p. 31.
3. Bernard Meltzer Quotes. (n.d.). AZ Quotes. https://www.azquotes.com/author/9957-Bernard-Meltzer. Quote appears seventh.
4. Brown, B. (2015). *Rising strong.* Spiegel & Grau. Quote appears on p. 32.
5. Sillars, A. L. (2009). Interpersonal conflict. In C. Berger, M. Roloff, & D. R. Roskos-Ewoldsen (Eds.), *Handbook of communication science* (2nd ed., pp. 273–289). Sage.
6. Kelly, J. (2021, May 21). Survey asks employees at top U.S. companies if they'd give up $30,000 to work from home: The answers may surprise you. *Forbes.* https://www.forbes.com/sites/jackkelly/2021/05/21/survey-asks-employees-at-top-us-companies-if-theyd-give-up-30000-to-work-from-home-the-answers-may-surprise-you/?sh=546ef438330f.
7. McKenzie, L. (2021, April 27). Students want online learning options post-pandemic. *Inside Higher Ed.* https://www.insidehighered.com/news/2021/04/27/survey-reveals-positive-outlook-online-instruction-post-pandemic.
8. Ong, J., De Santo, R., Heir, J., Siu, E., Nirmalan, N, Ofori, M. B., Awotide, A., Hassan, O., Ramos, R., Badaoui, T., Ogley, V., Saad, C., Sabbatasso E., Wyse, S. M., & Wu, C. (2020, December 2). 7 missing pieces: Why students prefer in-person over online classes. *University Affairs.* https://www.universityaffairs.ca/features/feature-article/7-missing-pieces-why-students-prefer-in-person-over-online-classes/.
9. Brownlee, D. (2019, May 7). Instead of emailing while angry, do this. *Forbes.* https://www.forbes.com/sites/danabrownlee/2019/03/07/instead-of-emailing-while-angry-do-this/?sh=19c9231b4617. Quote appears in para. 2.
10. 5 ways to effectively handle conflict in virtual teams. (2018, October 22). *The Virtual Hub.* https://www.thevirtualhub.com/blog/5-ways-to-effectively-handle-conflict-in-virtual-teams/. Quote appears in para. 1 under item 3.
11. Virtual work skills have become a must. (2020). *Great People Inside.* https://greatpeopleinside.com/virtual-work-skills-have-become-a-must/. Quotes appear in para. 7 and the first subheading, respectively.

12. Brown, B. (2015). *Rising strong*. Spiegel & Grau. Quote appears on p. 33.
13. For a discussion of reactions to disconfirming responses, see Vangelisti, A. L., & Crumley, L. P. (1998). Reactions to messages that hurt: The influence of relational contexts. *Communication Monographs, 64*, 173–196. See also Cortina, L. M., Magley, V. J., Williams, J. H., & Langhout, R. D. (2001). Incivility in the workplace: Incidence and impact. *Journal of Occupational Health Psychology, 6*, 64–80.
14. Gibb, J. (1961). Defensive communication. *Journal of Communication, 11*, 141–148. See also Eadie, W. F. (1982). Defensive communication revisited: A critical examination of Gibb's theory. *Southern Speech Communication Journal, 47*, 163–177.
15. Gottman, J. M., & Levenson, R. W. (2002). A two-factor model for predicting when a couple will divorce: Exploratory analyses using 14-year longitudinal data. *Family Process, 41*(1), 83–96; Gottman, J. M., Coan, J., Carrere, S., & Swanson, C. (1998). Predicting marital happiness and stability from newlywed interactions. *Journal of Marriage and the Family, 60*(1), 5–22. http://www.jstor.org/pss/353438; Carrere, S., Buehlman, K. T., Gottman, J. M., Coan, J. A., & Ruckstuhl, L. (2000). Predicting marital stability and divorce in newlywed couples. *Journal of Family Psychology, 14*(1), 42–58; Gottman, J. M. (1991). Predicting the longitudinal course of marriages. *Journal of Marital and Family Therapy, 17*(1), 3–7; Gottman, J. M., & Krokoff, L. J. (1989). The relationship between marital interaction and marital satisfaction: A longitudinal view. *Journal of Consulting and Clinical Psychology, 57*, 47–52; Carrere, S., & Gottman, J. M. (1999). Predicting divorce among newlyweds from the first three minutes of a marital conflict discussion. *Family Process, 38*(3), 293–301.
16. Gottman, J. (1994). *Why marriages succeed or fail: And how you can make yours last*. Simon & Schuster.
17. Gottman, J., & Gottman, J. (2017, March). The natural principles of love. *Journal of Family Theory & Review, 9*(1), 7–26.
18. Elium, D. (n.d.). What is the difference between a complaint and a criticism? https://donelium.wordpress.com/2011/10/10/what-is-the-difference-between-a-complaint-and-a-criticism/.
19. Gottman, J. M. (2009). *The marriage clinic*. Norton.
20. Cissna, K. N. L., & Seiburg, E. (1995). Patterns of interactional confirmation and disconfirmation. In M. V. Redmond (Ed.), *Interpersonal communication: Readings in theory and research* (pp. 301–317). Harcourt Brace.
21. Brown, B. (2015). *Rising strong*. Spiegel & Grau. Quote appears on p. 32.
22. Cissna, K. N. L., & Seiburg, E. (1995). Patterns of interactional confirmation and disconfirmation. In J. Stewart (Ed.), *Bridges not walls: A book about interpersonal communication* (6th ed., pp. 237–246). McGraw-Hill.
23. Brown, B. (2015). *Rising strong*. Spiegel & Grau. Quote appears on p. 32.
24. De Vries, R. E., Bakker-Pieper, A., & Oostenveld, W. (2010). Leadership = communication? The relations of leaders' communication styles with leadership styles, knowledge sharing and leadership outcomes. *Journal of Business & Psychology, 25*, 367–380.
25. Yu, P.-L. (2017). Innovative culture and professional skills: The use of supportive leadership and individual power distance orientation in IT industry. *International Journal of Manpower, 38*(2), 198–214.
26. Malloy, T. E. (2017, May). Interpersonal attraction in dyads and groups: Effects of the hearts of the beholder and the beheld. *European Journal of Social Psychology, 48*(3), 285–302.
27. Montoya, R. M., Kershaw, C., & Prosser, J. L. (2018). A meta-analytic investigation of the relation between interpersonal attraction and enacted behavior. *Psychological Bulletin, 144*(7), 673–709.
28. Burns, M. E., & Pearson, J. C. (2011). An exploration of family communication environment, everyday talk, and family satisfaction. *Communication Studies, 62*(2), 171–185.
29. LaBelle, S., & Johnson, Z. D. (2020). The relationship of student-to-student confirmation and student engagement. *Communication Research Reports, 37*(5), 234–242.
30. See Wilmot, W. W. (1987). *Dyadic communication*. Random House, pp. 149–158; and Andersson, L. M., & Pearson, C. M. (1999). Tit for tat? The spiraling effect of incivility in the workplace. *Academy of Management Review, 24*, 452–471. See also Olson, L. N., & Braithwaite, D. O. (2004). "If you hit me again, I'll hit you back": Conflict management strategies of individuals experiencing aggression during conflicts. *Communication Studies, 55*, 271–286.
31. Wilmot, W. W., & Hocker, J. L. (2007). *Interpersonal conflict* (7th ed., pp. 21–22). McGraw-Hill.
32. Wilmot & Hocker (2007), pp. 23–24.
33. Brown, B. (2015). *Rising strong*. Spiegel & Grau. Quote appears on p. 33.
34. Riter, T., & Riter, S. (2005). Why can't women just come out and say what they mean? *Family Life*. https://www.familylife.com/articles/topics/marriage/staying-married/husbands/why-cant-women-just-come-out-and-say-what-they-mean/. Quotes appear in paras. 6 and 8, respectively.
35. Pikiewicz, K. (2015, April 24). How trying to make everyone happy can make you miserable. *Psychology Today*. https://www.psychologytoday.com/us/blog/meaningful-you/201504/how-trying-make-everyone-happy-can-make-you-miserable. Quote appears in title.
36. Bach, G. R., & Goldberg, H. (1974). *Creative aggression*. Doubleday.
37. Meyer, J. R. (2004). Effect of verbal aggressiveness on the perceived importance of secondary goals in messages. *Communication Studies, 55*, 168–184; New Mexico Commission on the Status of Women. (2002). *Dealing with sexual harassment*. https://www.tandfonline.com/doi/abs/10.1080/10510970409388611?journalCode=rcst20.
38. Warning signs of abuse. What to look for. (n.d.). National Domestic Violence Hotline. https://www.thehotline.org/identify-abuse/domestic-abuse-warning-signs/.
39. Yan, H. (2019, November 26). *A 911 call with a fake pizza order helped stop a possible attack. But what if you can't speak to 911?* CNN. https://www.cnn.com/2019/11/26/us/what-to-do-if-you-cant-speak-to-911/index.html.
40. Why people abuse. Abuse is never okay. Learn why it continues. (n.d.). National Domestic Violence Hotline. https://www.thehotline.org/identify-abuse/why-do-people-abuse/.
41. Brandon, J. (n.d.). 37 more quotes on handling workplace conflict. *Inc*. https://www.inc.com/john-brandon/37-more-quotes-on-handling-workplace-conflict.html. Quote appears in item 30. (It must be acknowledged that, although this quote is commonly attributed to Churchill, he may not

42. Cekic, Ö. (2018, September). *Why I have coffee with people who send me hate mail* [Video]. TED. https://www.ted.com/talks/ozlem_cekic_why_i_have_coffee_with_people_who_send_me_hate_mail?language=en. Quote is in opening statement.
43. Cekic (2018). Quote is at 7:29 minutes.
44. Cekic (2018).
45. Cekic, O. (2020). *Overcoming hate through dialogue: Confronting prejudice, racism, and bigotry with conversation—and coffee*. Mango Media.
46. Zacchilli, T. L., Hendrick, C., & Hendrick, S. S. (2009). The romantic partner conflict scale: A new scale to measure relationship conflict. *Journal of Social and Personal Relationships, 26*, 1073–1096.
47. Todorov, E.-H., Paradis, A., & Godbout, N. (2021). Teen dating relationships: How daily disagreements are associated with relationship satisfaction. *Journal of Youth and Adolescence, 50*(8), 1510–1520.
48. Shell, G. R. (2006). *Bargaining for advantage: Negotiation strategies for reasonable people*. Penguin. Quote appears on p. 6.
49. Dickinson, A. (2017, May 18). How to negotiate with a bully. *Forbes*. https://www.forbes.com/sites/alexandradickinson/2017/05/18/how-to-negotiate-with-a-bully/#271b5e2e42bc. Quote appears in para. 1.
50. Dickinson (2017). Quote appears in para. 9.
51. Leonhardt, T. (2017, May 2). *Five tips for winning business negotiations with bullies*. Fast Company. https://www.fastcompany.com/40416298/five-tips-for-winning-business-negotiations-with-bullies. Quote appears in para. 7.
52. Fisher, R., & Ury, W. (1981). *Getting to yes: Negotiating agreement without giving in*. Houghton Mifflin.
53. Brown, B. (2015). *Rising strong*. Spiegel & Grau. Quote appears on p. 56.
54. Brown (2015). Quote appears on p. 34.
55. Brown (2015). Quote appears on p. 35.
56. Brown (2015). Quote appears on p. 37.
57. Niemiec, E. (2010, September 27). *Emotions and Italians*. Life in Italy. http://www.lifeinitaly.com/italian/emotions. Quote appears in para. 1.
58. Kim-Jo, T., Benet-Martinez, V., & Ozer, D. J. (2010). Culture and interpersonal conflict resolution styles: Role of acculturation. *Journal of Cross-Cultural Psychology, 41*, 264–269.
59. Hammer, M. R. (2009). Solving problems and resolving conflict using the Intercultural Style Model and Inventory. In M. A. Moodian (Ed.), *Contemporary leadership and intercultural competence* (pp. 219–232). Sage.
60. Hammer (2009).
61. *Gallup global emotions*. (2021). Gallup. https://www.gallup.com/analytics/349280/gallup-global-emotions-report.aspx.

CHAPTER 10

1. Travae, M. (2002, February 8). Nathália Rodrigues runs a YouTube channel focused on financial education. *Black Brazil Today*. https://blackbraziltoday.com/nathalia-rodrigues-runs-a-youtube-channel-focused-on-financial-education/.
2. World's 50 greatest leaders: Nathália Rodrigues. (2021). *Fortune*. https://fortune.com/worlds-greatest-leaders/2021/nathalia-rodrigues/.
3. Ruge, E. (2021, March 31). Meet Brazil's financial guru for a new generation. *Americas Quarterly*. https://www.americasquarterly.org/article/meet-brazils-financial-guru-for-a-new-generation/.
4. World's 50 greatest leaders: Nathália Rodrigues. (2021). *Fortune*. https://fortune.com/worlds-greatest-leaders/2021/nathalia-rodrigues/.
5. *The key attributes employers seek on college graduates' resumes*. (2021, April 13). National Association of Colleges and Employers. https://www.naceweb.org/about-us/press/the-key-attributes-employers-seek-on-college-graduates-resumes/.
6. Half, R. (2017, April 26). The benefits of teamwork in the workplace. *The Robert Half Blog*. https://www.roberthalf.com/blog/management-tips/the-value-of-teamwork-in-the-workplace. Quote is second subtitle.
7. Warrick, D. (2016). What leaders can learn about teamwork and developing high performance teams from organization development practitioners. *Performance Improvement, 3*, 13–21.
8. Amstrup, E. (2019, November 21). The global state of customer service. *Microsoft Dynamics 365 Blog*. https://cloudblogs.microsoft.com/dynamics365/bdm/2019/11/21/the-global-state-of-customer-service/.
9. *Improved communication essential to enhance customer satisfaction with after-sale service*. (2017, April 27). J.D. Power. http://india.jdpower.com/sites/default/files/2017042in.pdf.
10. *Inside Edition*. (2019, August 25). *Trader Joe's staff stop toddler mid-meltdown by dancing* [Video]. YouTube. https://www.youtube.com/watch?v=opAUyY8Wxx8.
11. Myatt, M. (2012, April 4). 10 communication secrets of great leaders. *Forbes*. https://www.forbes.com/sites/mikemyatt/2012/04/04/10-communication-secrets-of-great-leaders/#6cc9808a22fe/. Quote appears in para. 3.
12. *Communication barriers in the modern workplace*. (2018). The Economist Intelligence Unit Limited. https://impact.economist.com/perspectives/sites/default/files/EIU_Lucidchart-Communication%20barriers%20in%20the%20modern%20workplace.pdf.
13. Deutschendorf, H. (2018, December 6). 5 reasons empathy is the most important leadership skill. *Fast Company*. https://www.fastcompany.com/90272895/5-reasons-empathy-is-the-most-important-leadership-skill.
14. Hicks, J. (2020, October 12). Leader communication styles and organizational health. *The Health Care Manager*. https://journals.lww.com/healthcaremanagerjournal/fulltext/2020/10000/leader_communication_styles_and_organizational.6.aspx?casa_token=XQDoj8QEsR8AAAAA:td7DfqpPAWGQJnkWL5edKLvfl8hb1_h5JiuHR9P5pUiP6cWtP1NX-2_XZtZg8npViv-p-e7Ae3OFesF9YQrFP-1k.
15. Shores, M. (2021, February 16). Three habits of highly effective communicators. *Forbes*. https://www.forbes.com/sites/forbesbusinesscouncil/2021/02/16/three-habits-of-highly-effective-communicators/?sh=574d66454ce3.
16. Profita, M. (2018, October 2). 8 tips for starting a college senior job search. *The Balance Careers*. https://www.thebalancecareers.com/starting-a-college-senior-job-search-2059879. Quote appears in para. 2.

17. Kellam, J. (2015, January 9). I highly recommend this student . . . *Saint Vincent College Faculty Blog.* http://info.stvincent.edu/faculty-blog/i-highly-recommend-this-student. Quote appears in third to last paragraph.
18. Jain, P. (2021, May 10). *Building a personal brand on social media.* House of Scholars. https://houseofscholars.net/2021/05/10/building-a-personal-brand-on-social-media/. Quote appears in para. 3.
19. Tarpey, M. (2018, August 9). *Not getting job offers? Your social media could be the reason.* CareerBuilder. https://www.careerbuilder.com/advice/not-getting-job-offers-your-social-media-could-be-the-reason.
20. St. John, A. (2017, August 1). Looking for a job? First, clean up your social media presence. *Consumer Reports.* https://www.consumerreports.org/employment-careers/clean-up-social-media-presence-when-looking-for-a-job/.
21. Hiring managers "spell out" the biggest deal-breakers regarding job candidates' resumes. (2018, February 4). *Talent Inc.* https://www.talentinc.com/press-2018-02-14.
22. *Number of employers using social media to screen candidates at all-time high, find latest CareerBuilder study.* (2017, June 15). CISION PR Newswire. http://www.prnewswire.com/news-releases/number-of-employers-using-social-media-to-screen-candidates-at-all-time-high-finds-latest-careerbuilder-study-300474228.html.
23. Tarpey, M. (2018, August 9). *Not getting job offers? Your social media could be the reason.* CareerBuilder. https://www.careerbuilder.com/advice/not-getting-job-offers-your-social-media-could-be-the-reason.
24. *Social media etiquette in the workplace.* (2014, July 29). The Human Resource. https://www.toolbox.com/hr/hr-compliance/blogs/social-media-etiquette-in-the-workplace-072914/. Quote appears in para. 5.
25. Patrik, A. (n.d.) *Have you ever gotten a job purely through networking?* Quora. https://www.quora.com/Have-you-ever-gotten-a-job-purely-through-networking.
26. Akinmade, O. (n.d.) *Have you ever gotten a job purely through networking?* Quora. https://www.quora.com/Have-you-ever-gotten-a-job-purely-through-networking.
27. Han, E. (n.d.) *Have you ever gotten a job purely through networking?* Quora. https://www.quora.com/Have-you-ever-gotten-a-job-purely-through-networking. Quote appears in para. 2 of Ed Han post.
28. Flynn, J. (2021, September 13). *24+ important networking statistics [2021]: The power of connections in the workplace.* Zippia. https://www.zippia.com/advice/networking-statistics/.
29. Kaufman, W. (2011, February 3). *A successful job search: It's all about networking.* National Public Radio. https://www.npr.org/2011/02/08/133474431/a-successful-job-search-its-all-about-networking.
30. Dodds, P. S., Muhamad, R., & Watts, D. J. (2003). An experimental study of search in global social networks. *Science, 301,* 827–829.
31. Brustein, D. (2014, July 22). 17 tips to survive your next networking event. *Forbes.* https://www.forbes.com/sites/yec/2014/07/22/17-tips-to-survive-your-next-networking-event/#59f1c18c7cd4. Quote appears in tip 2.
32. Bridges, J. (2020, July 31). 13 things you need to include in your career portfolio. *Reputation Defender.* https://www.reputationdefender.com/blog/job-seekers/13-things-you-need-to-include-in-your-career-portfolio. Quote appears in tip 12.
33. Tullier, M. (2002). *The art and science of writing cover letters: The best way to make a first impression.* Monster.com. http://resume.monster.com/coverletter/coverletters.
34. Burnison, G. (2019, January 26). *6 things I loved about the most impressive resume I've ever seen—based on 20 years of hiring.* CNBC Make It. https://www.cnbc.com/2019/06/26/most-impressive-resume-ever-based-on-20-years-of-hiring-and-interviewing.html. Quotes appear in para. 1 and 4, respectively.
35. Kurtuy, A. (n.d.). Show contact information on your resume—how-to & examples. *CareerBlog.* https://novoresume.com/career-blog/contact-information-on-resume.
36. Smith, A. (n.d.). *3 reasons you should ditch that resume objective—and 3 things you can do instead.* The Muse. https://www.themuse.com/advice/ditch-resume-objective-what-to-do-instead.
37. Severt, N. (2021, November 17). 40 best resume tips 2021: Great tricks and writing advice. *Zety.* https://zety.com/blog/resume-tips.
38. *10 resume writing tips to help you land a job.* (2021, June 25). Career Guide. https://www.indeed.com/career-advice/resumes-cover-letters/10-resume-writing-tips. Quote appears in tip 5.
39. 30 resume accomplishments examples to demonstrate your value. (2021, April 20). *Jobscan.* https://www.jobscan.co/blog/resume-accomplishments-examples/.
40. 30 resume accomplishments . . . (2021, April 20). Quote is paraphrased version of "Resume Accomplishments" example 1.
41. Burnison, G. (2019, January 26). *6 things I loved about the most impressive resume I've ever seen—based on 20 years of hiring.* CNBC Make It. https://www.cnbc.com/2019/06/26/most-impressive-resume-ever-based-on-20-years-of-hiring-and-interviewing.html. Quotes appears tip 4, paras. 4 and 2, respectively.
42. *How to send your resume: As a PDF or Word document?* (2021, November 30). Resume Coach. https://www.resumecoach.com/send-resume-word-or-pdf/.
43. *Settling the debate: One or two page resumes.* (n.d.). ResumeGo. https://www.resumego.net/research/one-or-two-page-resumes/.
44. Graham, A. (2011, January 14). You won't land a job if you can't follow directions. *Forbes.* http://www.forbes.com/sites/work-in-progress/2011/01/14/you-wont-land-a-job-if-you-cant-follow-directions/.
45. Zhang, L. (n.d) *The ultimate guide to researching a company pre-interview.* The Muse. https://www.themuse.com/advice/the-ultimate-guide-to-researching-a-company-preinterview.
46. Moss, C. (2013, September 28). 14 weird, open-ended job interview questions asked at Apple, Amazon and Google. *Business Insider.* http://www.businessinsider.com/weird-interview-questions-from-apple-google-amazon-2013-9?op=1#ixzz37PdzyCTk. See also: *Top 25 oddball interview questions for 2014.* Glassdoor. http://www.glassdoor.com/Top-25-Oddball-Interview-Questions-LST_KQ0,34.htm.

47. Smith, A. (2012, April 20). *5 illegal interview questions and how to dodge them.* Forbes. https://www.forbes.com/sites/dailymuse/2012/04/20/5-illegal-interview-questions-and-how-to-dodge-them/#6072984e191f. Quotes appear in para. 1.
48. Smith (2012). Quote appears in para. 8.
49. Smith, A. (n.d.). *5 steps to acing your interview presentation.* The Muse. https://www.themuse.com/advice/5-steps-to-acing-your-interview-presentation.
50. *Do's and don'ts of PowerPoint presentations.* (n.d.). TotalJobs. https://www.totaljobs.com/careers-advice/job-interview-advice/powerpoint-pitfalls.
51. Mitchell, N. R. (n.d.). *Top 10 interview tips from an etiquette professional.* Experience. http://www.experience.com/entry-level-jobs/jobs-and-careers/interview-resources/top-10-interview-tips-from-an-etiquette-professional/. Quote appears in para. 7.
52. How to answer a salary requirements interview question (with example answers). (2021, June 29). *Glassdoor Blog.* https://www.glassdoor.com/blog/guide/salary-requirements/. Quote appears under "Example answer 1."
53. *51 great questions to ask in an interview.* (n.d.). The Muse. https://www.themuse.com/advice/51-interview-questions-you-should-be-asking.
54. Joyce, S. P. (n.d.). *Questions not to ask in an interview (40+ examples).* Job Hunt. https://www.job-hunt.org/questions-not-to-ask-an-interviewer/.
55. DePaul, K. (2020, December 23). Negotiating a job offer? Here's how to get what you want. *Harvard Business Review.* https://hbr.org/2020/12/negotiating-a-job-offer-heres-how-to-get-what-you-want. Quote appears in para. 3 under "Start from a Place" subheading.
56. Berger, J. (2020). *The catalyst: How to change anyone's mind.* Simon & Schuster.
57. DePaul, K. (2020, December 23). Negotiating a job offer? Here's how to get what you want. *Harvard Business Review.* https://hbr.org/2020/12/negotiating-a-job-offer-heres-how-to-get-what-you-want. Quote appears in final paragraph.
58. *8 tips for success in a long distance interview.* (2016, May 14). The Everygirl. http://theeverygirl.com/8-tips-for-success-in-a-long-distance-interview/. Quote appears in tip 2.
59. Nguyen, T. V. (n.d.). Inside Google's company culture: Creative and gratifying. *Grove Blog.* https://blog.grovehr.com/google-company-culture. Quote appears in item 1 under "Their Unique Approach."
60. Consumers expect the brands they support to be socially responsible. (2019, October 2). *Business Wire.* https://www.businesswire.com/news/home/20191002005697/en/Consumers-Expect-the-Brands-they-Support-to-be-Socially-Responsible.
61. Laforet, S. (2016). Effects of organisational culture on organisational innovation performance in family firms. *Journal of Small Business and Enterprise Development, 2,* 379–407.
62. Ward, C. (1996). Acculturation. In D. Landis & R. S. Bhagat (Eds.), *Handbook of intercultural training* (pp. 124–147). Sage.
63. *Do you feel culture shock in your new job?* (n.d.) Hire Imaging. https://hireimaging.com/articles/career-tips/do-you-feel-culture-shock-in-your-new-job/.
64. Belgium, H. (2017). *Five unexpected adjustments you will need to make in your new job.* Hays. https://social.hays.com/2017/09/19/five-adjustments-new-job/.
65. *How to adapt to change in the workplace.* (2021, March 10). Indeed. https://www.indeed.com/career-advice/career-development/adapting-to-change. Quote appears in para. 2 under "Stay Positive."
66. Richards, E. (2020). *How to survive workplace culture shock.* A Top Career. https://atopcareer.com/survive-workplace-culture-shock/. Quote appears in item 6.
67. Deloitte Millennial Survey 2018. (2018). Deloitte. https://www2.deloitte.com/global/en/pages/about-deloitte/articles/millennialsurvey.html.
68. Hoang, X. M. (n.d.). *7 most wanted work benefits to attract Millennials.* The Undercover Recruiter. https://theundercoverrecruiter.com/benefits-attract-millennial-talent/.
69. Rivers, T. B. (2015, November 16). 11 ways to make your workplace appealing to Millennials. *iOffice Blog.* https://www.iofficecorp.com/blog/11-ways-to-make-your-workplace-appealing-to-millennials.
70. Richinick, M. (2020, April 17). Workplace etiquette: 21 dos and don'ts of the workplace. *Northeastern University Graduate Programs Blog.* https://www.northeastern.edu/graduate/blog/workplace-etiquette/.
71. Richinick (2020, April 17). Quotes appears in paras. 26 and 20, respectively.
72. Richinick (2020, April 17). Quotes appears in item 8.
73. Rosh, L., & Offermann, L. (2013, October). Be yourself, but carefully. *Harvard Business Review.* http://hbr.org/2013/10/be-yourself-but-carefully/ar/1.
74. Parker, T. (2013, October 25). *30 non-Americans on the American norms they find weird.* Thought Catalogue. http://thoughtcatalog.com/timmy-parker/2013/10/30-non-americans-on-the-weirdest-things-that-are-norms-to-americans/.
75. Zimmerman, M. (2010, December 22). *Losing your temper at work: How to survive it.* CBS News. https://www.cbsnews.com/news/losing-your-temper-at-work-how-to-survive-it/#. Quote appears in para. 5.
76. Chandler, N. (n.d.). *10 tips for managing conflict in the workplace.* HowStuffWorks. http://money.howstuffworks.com/business/starting-a-job/10-tips-for-managing-conflict-in-the-workplace1.htm#page=1.
77. *Foot-in-mouth stories . . . post your shame.* (2008, August 14). The Straight Dope message board. https://boards.straightdope.com/sdmb/archive/index.php/t-479416.html. Quote appears in the post by Mister Rik.
78. Zaslow, J. (2010, January 6). *Before you gossip, ask yourself this. . . .* Moving On. http://online.wsj.com/article/SB10001424052748704160504574640111681307026.html.
79. Calero-Holmes, B. (2021, November 18). 10 distractions that kill workplace productivity. *Business News Daily.* https://www.businessnewsdaily.com/8098-distractions-kiling-productivity.html.
80. *Leaping lizards! OfficeTeam survey reveals managers' most embarrassing moment at work.* (2011, January 18). OfficeTeam. http://rh-us.mediaroom.com/2011-01-18-LEAPING-LIZARDS.

81. Glazer, R. (2021, April 19). I asked 2,000 people about their remote work experience—here's what they shared. *Forbes*. https://www.forbes.com/sites/robertglazer/2021/04/19/i-asked-2000-people-about-their-remote-work-experience-heres-what-they-shared/?sh=27bf9a7c7b4a. Quote appears as first subheading.
82. Glazer, R. (2021, April 19).
83. Glazer, R. (2021, April 19).
84. Borsellino, R. (2020). *7 essential tips for working from home during the coronavirus pandemic*. The Muse. https://www.themuse.com/advice/coronavirus-work-from-home-tips. Quote appears in para. 4 of item 2.
85. Borsellino, R. (2020). Quote appears in para. 2 of item 3.
86. *SHRM research reveals negative perceptions of remote work*. (2021, July 26). Society of Human Resource Management. https://www.shrm.org/about-shrm/press-room/press-releases/pages/-shrm-research-reveals-negative-perceptions-of-remote-work.aspx.
87. Griffin, D., & Berry, D. (2018, December 12). *Out of sight, out of mind? Managing the remote worker*. Gartner Research. https://www.gartner.com/en/documents/3895668/out-of-sight-out-of-mind-managing-the-remote-worker.
88. Dempsey, B. (2010, February 10). The seven most universal job skills. *Forbes*. https://www.forbes.com/2010/02/18/most-important-job-skills-personal-finance-universal.html#78e44e375db2. Quote appears in Item 1.

CHAPTER 11

1. Zhuo, J. (2019). *The making of a manager: What to do when everyone looks to you*. Penguin. Quote appears on p. 23.
2. Zhuo, J. (2019).
3. Ralon, A., Rothenberg, J., Odeh, G., Turney, M., & Wu, Y. (2021). How followers contribute to team success, leadership transformation and organizational excellence. *Journal of International Business and Management, 4*(12), 1–7.
4. 100 best companies to work for. (2021). *Fortune*. https://fortune.com/best-companies/2021/.
5. *Great place to work profile series*. (2018). Hospitality for all: How Hilton treats employees at all levels like guests; and how that "for all" culture drives its success. https://www.greatplacetowork.com/images/reports/2018-GPTW-Profile-Series-_Hilton_Hospitality_For_All.pdf.
6. *How to use your initiative at work*. (n.d.). Success at School. https://successatschool.org/advicedetails/703/How-to-use-your-initiative-at-work. Quote appears in para. 4.
7. Suda, L. (2013). *In praise of followers*. Paper presented at PMI® Global Congress 2013—North America, New Orleans, LA. Newtown Square, PA: Project Management Institute. https://www.pmi.org/learning/library/importance-of-effective-followers-5887. Quote appears in para. 2 under "The Effective Follower."
8. Smith, J., & Lebowitz, S. (2019, October 7). 14 signs you're secretly the boss' favorite. *Business Insider*. https://www.businessinsider.com/signs-youre-the-bosss-favorite-2016-6. Quote appears in para. 1 of "You Are Their Go-To Person."
9. Rockwell, D. (2017, January 29). How to practice feedback-seeking and take your career to the next level. *Leadership Freak*. https://leadershipfreak.blog/2017/01/29/how-to-practice-feedback-seeking-and-take-your-career-to-the-next-level/.
10. *10 ways to help and support colleagues at work*. (2021, April 22). Indeed. https://www.indeed.com/career-advice/career-development/helping-and-supporting-others-at-work?aceid=. Quote appears in para. 1.
11. Leung, L, A., Brindley, P., Anderson, S., Park, J., Vergis, A., & Gillman, L. M. (2018). Followership: A review of the literature in healthcare and beyond. *Journal of Critical Care, 46,* 99–104. Quote appears in para. 2 of Discussion.
12. Chaleff, I. (2016, August 25). *No, you probably shouldn't follow every order from your boss*. Government Executive. https://www.govexec.com/management/2016/08/no-you-probably-shouldnt-follow-every-order-your-boss/131050/. Quote appears in second to last paragraph.
13. Chaleff, I. (2016).
14. Esthersanni. (2021, April 10). *How to deal with difficult people*. Medium. https://medium.com/illuminations-mirror/how-to-deal-with-difficult-people-cca8b97c9630. Quote appears in para. 6.
15. Esthersanni. (2021, April 10). Quote appears under "Listen."
16. Markway, B. (2015, March 3). 20 expert tactics for dealing with difficult people. *Psychology Today*. https://www.psychologytoday.com/us/blog/living-the-questions/201503/20-expert-tactics-dealing-difficult-people/. Quote appears in tip 14.
17. Lapin, R. (2008). *Working with difficult people*. Dorling Kindersley.
18. *Leaving a job professionally: Wrapping up your current position before moving on*. (2010). Claros Group. http://www.claros-group.com/leavingjob.pdf.
19. Duhigg, C. (2016, February 25). The work issue: What Google learned from its quest to build the perfect team. *New York Times Magazine*. https://www.nytimes.com/2016/02/28/magazine/what-google-learned-from-its-quest-to-build-the-perfect-team.html.
20. Robles, M. M. (2012). Executive perceptions of the top 10 soft skills needed in today's workplace. *Business Communication Quarterly, 75,* 453–465.
21. Lutgen-Sandvik, P., Riforgiate, S., & Fletcher, C. (2011). Work as a source of positive emotional experiences and the discourses informing positive assessment. *Western Journal of Communication, 75,* 2–27.
22. Conley, M. (2021, March 17). 45 quotes that celebrate teamwork, hard work, and collaboration. *Hubspot*. https://blog.hubspot.com/marketing/teamwork-quotes. Quote appears in item 5 under "Teamwork Quotes to Inspire Collaboration."
23. Marby, E. A. (1999). The systems metaphor in group communication. In L. R. Frey (Ed.), *Handbook of group communication theory and research* (pp. 71–91). Sage.
24. Is your team too big? Too small? What's the right number? (2006, June 14). *Knowledge@Wharton*. http://knowledge.wharton.upenn.edu/article.cfm?articleid=1501.
25. Lowry, P., Roberts, T. L., Romano, N. C., Jr., Cheney, P. D., & Hightower, R. T. (2006). The impact of group size and social presence on small-group communication. *Small Group Research, 37,* 631–661.

26. Hackman, J. (1987). The design of work teams. In J. Lorsch (Ed.), *Handbook of organizational behavior* (pp. 315–342). Prentice-Hall.
27. LaFasto, F., & Carson, C. (2001). *When teams work best: 6,000 team members and leaders tell what it takes to succeed.* Sage; Larson, C. E., & LaFasto, F. M. J. (1989). *Teamwork: What must go right, what can go wrong.* Sage.
28. Simms, A., & Nichols, T. (2014). Social loafing: A review of the literature. *Journal of Management Policy and Practice, 15*(1), 58–67.
29. Wagner, R., & Harter, J. K. (2006). When there's a freeloader on your team. Excerpt from *The elements of great managing.* Gallup Press. https://stybelpeabody.com/pdf/Freeloader_on_Your_Team.pdf.
30. Paknad, D. (n.d.). The 5 dynamics of low performing teams. Don't let freeloaders or fear undermine your team. *WorkBoard.* http://www.workboard.com/blog/dynamics-of-low-performing-teams.php.
31. Duhigg, C. (2016, February 25). The work issue: What Google learned from its quest to build the perfect team. *New York Times Magazine.* https://www.nytimes.com/2016/02/28/magazine/what-google-learned-from-its-quest-to-build-the-perfect-team.html.
32. Duhigg, C. (2016, February 25). Quote appears in para. 44.
33. Gouran, D. S., Hirokawa, R. Y., Julian, K. M., & Leatham, G. B. (1992). The evolution and current status of the functional perspective on communication in decision-making and problem-solving groups. In S. A. Deetz (Ed.), *Communication yearbook* 16 (pp. 573–600). Sage. See also Wittenbaum, G. M., Hollingshead, A. B., Paulus, P. B., Hirokawa, R. Y., Ancona, D. G., Peterson, R. S., Jehn, K. A., & Yoon, K. (2004). The functional perspective as a lens for understanding groups. *Small Group Research, 35,* 17–43.
34. Mayer, M. E. (1998). Behaviors leading to more effective decisions in small groups embedded in organizations. *Communication Reports, 11,* 123–132.
35. Murphy, K. (2020, April 29). Why Zoom is terrible. *New York Times.* https://www.nytimes.com/2020/04/29/sunday-review/zoom-video-conference.html. Quote appears in para. 6.
36. Lauritsen, J. (2021, January 15). Top 5 virtual meetings WORST practices (and how to fix them). [Personal blog.] https://jasonlauritsen.com/2021/01/top-5-virtual-meetings-worst-practices-and-how-to-fix/. Quote appears in para. 2 of item 1.
37. Fisher, B. A. (1970). Decision emergence: Phases in group decision making. *Speech Monographs, 37,* 53–66.
38. This terminology originated with Tuckman, B. W (1965). Developmental sequence in small groups. *Psychological Bulletin, 63*(6), 384–399.
39. Hülsheger, U. R., Anderson, N., & Salgado, J. F. (2009). Team-level predictors of innovation at work: A comprehensive meta-analysis spanning three decades of research. *Journal of Applied Psychology, 94,* 1128–1145.
40. Clark, D. (2012, May 23). How to deal with difficult co-workers. *Forbes.* http://www.forbes.com/sites/dorieclark/2012/05/23/how-to-deal-with-difficult-co-workers/#6c21476a191d.
41. Dugan, D. (n.d.). *Co-workers from hell: Dealing with difficult colleagues.* Salary.com. http://www.salary.com/co-workers-from-hell-dealing-with-difficult-colleagues/.
42. Waller, B. M., Hope, L., Burrowes, M., & Morrison, E. R. (2011). Twelve (not so) angry men: Managing conversational group size increases perceived contribution by decision makers. *Group Processes and Intergroup Relations, 14,* 835–843.
43. Rothwell, J. D. (2013). *In mixed company* (8th ed.). Wadsworth-Cengage.
44. Janis, I. (1982). *Groupthink: Psychological studies of policy decisions and fiascoes.* Houghton Mifflin. See also Baron, R. S. (2005). So right it's wrong: Groupthink and the ubiquitous nature of polarized group decision making. In M. P. Zanna (Ed.), *Advances in experimental social psychology* (Vol. 37, pp. 219–253). Elsevier Academic Press.
45. Janis, I. L. (1972). *Victims of groupthink: A psychological study of foreign-policy decisions and fiascoes.* Houghton Mifflin.
46. Challenger explosion. (n.d.) *History.* https://www.history.com/topics/challenger-disaster.
47. *Penn State scandal fast facts.* (2018, March 28). CNN. https://www.cnn.com/2013/10/28/us/penn-state-scandal-fast-facts/index.html.
48. Decker, B. M. (2018, August 28). In a Catholic church where even the pope covers for sexual abuse, everywhere is as bad as Boston. *USA Today.* https://www.usatoday.com/story/opinion/2018/08/28/pope-francis-knew-cardinal-mccarrick-sexual-abuse-catholic-churchcolumn/1109251002/.
49. Chavez, N. (2018, January 25). *What others knew: Culture of denial protected Nassar for years.* CNN. https://www.cnn.com/2018/01/23/us/nassar-sexual-abuse-who-knew/index.html.
50. Adapted from Rothwell, J. D. (2013). *In mixed company* (8th ed.). Wadsworth-Cengage, pp. 139–142, see note 49.
51. Zhuo, J. (2019). *The making of a manager: What to do when everyone looks to you.* Penguin. Information paraphrased here appears on p. 126.
52. Love them or hate them, virtual meetings are here to stay. (2021, April 10). *The Economist.* https://www.economist.com/international/2021/04/10/love-them-or-hate-them-virtual-meetings-are-here-to-stay. Quote appears in headline.
53. *25 percent of all professional jobs in North America will be remote by the end of next year.* (2021, December 7). Ladders. https://www.theladders.com/press/25-of-all-professional-jobs-in-north-america-will-be-remote-by-end-of-next-year.
54. Capdeferro, N., & Romero, M. (2012). Are online learners frustrated with collaborative learning experiences? *International Review of Research in Open & Distance Learning, 13,* 26–44.
55. Bond, S. (2020, April 3). *A must for millions, Zoom has a dark side—and an FBI warning.* NPR. https://www.npr.org/2020/04/03/826129520/a-must-for-millions-zoom-has-a-dark-side-and-an-fbi-warning.
56. Brown, N. (2020, April 4). Best practices for video conferencing security. *PaloAlto Networks.* https://blog.paloaltonetworks.com/2020/04/network-video-conferencing-security/.
57. Adams, S. (2020, April 22). Zoom meeting etiquette: 15 tips and best practices for online video conference

meetings. *Penn Live*. https://www.pennlive.com/coronavirus/2020/04/zoom-meeting-etiquette-15-tips-and-best-practices-for-online-video-conference-meetings.html. Quote appears in para. 8.
58. Gupta, A. H. (2020, April 14). It's not just you: In online meetings, many women can't get a word in. *New York Times*. https://www.nytimes.com/2020/04/14/us/zoom-meetings-gender.html. Quote appears in para. 2.
59. Bailenson, J. N. (2021, February 23). Nonverbal overload: A theoretical argument for the causes of Zoom fatigue. *Technology, Mind, and Behavior, 2*(1), nonpaginated online. https://tmb.apaopen.org/pub/nonverbal-overload/release/2.
60. Fadilpašić, S. (2021, November 4). *Remote working is actually increasing the amount of time we spend in meetings*. Tech Radar. https://www.techradar.com/news/remote-working-is-actually-increasing-the-amount-of-time-we-all-spend-in-meetings.
61. Snow, S. (2014, September 11). *How Matt's machine works*. Fast Company. http://www.fastcompany.com/3035463/how-matts-machine-works. Quotes appear in para. 7 of "A Secret Sauce."
62. *Your brainstorming invitee list: Why diversity is the mother of innovation*. SmartStorming. https://www.smartstorming.com/your-brainstorming-invitee-list-why-diversity-is-the-mother-of-innovation/. Quote appears in the title.
63. Markman, A. (2017, May 18). Your team is brainstorming all wrong. *Harvard Business Review*. https://hbr.org/2017/05/your-team-is-brainstorming-all-wrong. Quote appears in para. 3.
64. 10 brainstorming strategies that work. (2018, April 10). *Forbes*. https://www.forbes.com/sites/forbesagencycouncil/2018/04/10/10-brainstorming-strategies-that-work/#70db12aa5da7/
65. Athuraliya, A. (2018, June 1). The ultimate list of essential visual brainstorming techniques. *Creately Blog*. https://creately.com/blog/diagrams/visual-brainstorming-techniques/.
66. Bolger, M. (n.d.). *Anonymous Q&A on Zoom*. Virtual. https://virtual.facilitator.cards/anonymous-qa-on-zoom-using-google-slides-meg-bolger.
67. Bohm, D. (1996). *On dialogue* (L. Nichol, Ed.). London: Routledge & Kegan Paul.
68. *How communities are banding together to help small businesses*. (2021, June 22). Waze. https://medium.com/waze/how-communities-are-banding-together-to-help-small-businesses-a9715326ef06.
69. Houck, B. (2020, October 7). A restaurant made entirely from geodesic domes is a perfect metaphor for dining in 2020. *Detroit Eater*. https://detroit.eater.com/2020/10/7/21506040/east-eats-detroit-new-restaurant-geodesic-domes-outdoor-seating-jefferson-chalmers-covid-19-dining.
70. Sherman, A. (2021, February 18). *A restaurant made entirely from geodesic domes is a perfect metaphor for dining in 2020*. EATER Detroit. https://detroit.eater.com/2020/10/7/21506040/east-eats-detroit-new-restaurant-geodesic-domes-outdoor-seating-jefferson-chalmers-covid-19-dining. Quote appears in para. 7.
71. See, for example, Pavitt, C. (2003). Do interacting groups perform better than aggregates of individuals? *Human Communication Research, 29*, 592–599; Wittenbaum, G. M. (2004). Putting communication into the study of group memory. *Human Communication Research, 29*, 616–623; and Frank, M. G., Feely, T. H., Paolantonio, N., & Servoss, T. J. (2004). Individual and small group accuracy in judging truthful and deceptive communication. *Group Decision and Negotiation, 13*, 45–54.
72. Rae-Dupree, J. (2008, December 7). Innovation is a team sport. *New York Times*. http://www.nytimes.com/2008/12/07/business/worldbusiness/07iht-innovate.1.18456109.html?_r=0.
73. Information drawn from García, M., & Cañado, M. (2011). Multicultural teamwork as a source of experiential learning and intercultural development. *Journal of English Studies, 9*, 145–163; and van Knippenberg, D., van Ginkel, W. P., & Homan, A. C. (2013, July). Diversity mindsets and the performance of diverse teams. *Organizational Behavior and Human Decision Processes, 121*, 183–193.
74. Adler, R. B., & Elmhorst, J. M. (2010). *Communicating at work: Principles and practices for business and the professions* (10th ed.). McGraw-Hill, pp. 278–279.
75. van Knippenberg, D., van Ginkel, W. P., & Homan, A. C. (2013, July). Diversity mindsets and the performance of diverse teams. *Organizational Behavior and Human Decision Processes, 121*, 183–193.
76. Dewey, J. (1910). *How we think*. Heath.
77. Poole, M. S. (1991). Procedures for managing meetings: Social and technological innovation. In R. A. Swanson & B. O. Knapp (Eds.), *Innovative meeting management* (pp. 53–109). 3M Meeting Management Institute. See also Poole, M. S., & Holmes, M. E. (1995). Decision development in computer-assisted group decision making. *Human Communication Research, 22*, 90–127.
78. Lewin, K. (1951). *Field theory in social science*. Harper & Row, pp. 30–59.
79. Hastle, R. (1983). *Inside the jury*. Harvard University Press.
80. Zhuo, J. (2019). *The making of a manager: What to do when everyone looks to you*. Penguin. Quote appears on p. 12.
81. Gallo, A. (2013, May 2). Act like a leader before you are one. *Harvard Business Review*. https://hbr.org/2013/05/act-like-a-leader-before-you-a. Quote appears in para. 1.
82. Aristotle. (1958). *Politics, Book 7*. Oxford University Press.
83. Sethuraman, K., & Suresh, J. (2014, August 25). Effective leadership styles. *International Business Research, 7*(9), 165–172.
84. Van Wart, M. (2013). Lessons from leadership theory and the contemporary challenges of leaders. *Public Administration Review, 73*(4), 553–565.
85. The following types of power are based on the categories developed by French, J. R., & Raven, B. (1968). The basis of social power. In D. Cartright & A. Zander (Eds.), *Group dynamics*. Harper & Row, p. 565.
86. Rothwell, J. D. (2013). *In mixed company: Communicating in small groups* (8th ed.). Cengage Learning.
87. Rothwell, J. D. (2004). *In mixed company: Small group communication* (5th ed.). Wadsworth, pp. 247–282.

88. Greenleaf, R. K., & Spears, L. C. (2002). *Servant leadership: A journey into the nature of legitimate power and greatness.* Paulist Press.
89. Alonderiene, R., & Majauskaite, M. (2016). Leadership style and job satisfaction in higher education institutions. *International Journal of Educational Management, 30*(1), 140–164.
90. Bhatti, N., Maitlo, G. M., Shaikh, N., Hashmi, M. A., & Shaikh, F. M. (2012). The impact of autocratic and democratic leadership style on job satisfaction. *International Business Research, 5*(2), 192–201.
91. Skogstad, A., Hetland, J., Glasø, L., & Einarsen, S. (2014). Is avoidant leadership a root cause of subordinate stress? Longitudinal relationships between laissez-faire leadership and role ambiguity. *Work and Stress, 28*(4), 323–341.
92. Asplund, J., & Blacksmith, N. (2012, March 6). https://news.gallup.com/businessjournal/152981/strengths-based-goal-setting.aspx.
93. Tischler, L., Giambatista, R., McKeage, R., & McCormick, D. (2016). Servant leadership and its relationships with core self-evaluation and job satisfaction. *Journal of Values-Based Leadership, 9*(1), 1–20. http://scholar.valpo.edu/cgi/viewcontent.cgi?article=1148&context=jvbl.
94. Crwys-Williams, J. (Ed.). (2010). *In the words of Nelson Mandela.* Walker. Quote appears on p. 62.
95. Fiedler, F. E. (1967). *A theory of leadership effectiveness.* McGraw-Hill.
96. Hersey, P., & Blanchard, K. (2001). *Management of organizational behavior: Utilizing human resources* (8th ed.). Prentice-Hall.
97. Blake, R., & Mouton, J. (1964). *The Managerial Grid: The key to leadership excellence.* Gulf Publishing Co.
98. Blake, R., & Mouton, J. (1985). *The Managerial Grid III: The key to leadership excellence.* Gulf Publishing Co.
99. Kuhnert, K. W., & Lewis, P. (1987). Transactional and transformational leadership: A constructive/developmental analysis. *Academy of Management Review, 12,* 648–657.
100. Boies, K., Fiset, J., & Gill, H. (2015). Communication and trust are key: Unlocking the relationship between leadership and team performance and creativity. *The Leadership Quarterly, 26,* 1080–1094.
101. Amin, M., Tatlah, I. A., & Khan, A. M. (2013). Which leadership style to use? An investigation of conducive and non-conducive leadership style(s) to faculty job satisfaction. *International Research Journal of Art an Humanities, 41,* 229–253.
102. Weinstein, B. (2007, August 17). How to deal with bully bosses. CIO. https://www.cio.com/article/274669/how-to-deal-with-bully-bosses.html. Quotes appear in last two paragraphs on p. 1.
103. Stefanyk, A. (2020, March 5). How to avoid the swoop and poop. *Kanopi Blog.* https://kanopi.com/blog/how-to-avoid-the-swoop-and-poop/. Paraphrased passage is based on para. 2.
104. Costanzo, L. A., & Di Domenico, M. (2015). A multi-level dialectical-paradox lens for top management team strategic decision-making in a corporate venture. *British Journal of Management, 3,* 484–506.
105. Kuhnert, K. W., & Lewis, P. (1987). Transactional and transformational leadership: A constructive/developmental analysis. *Academy of Management Review, 12,* 648–657.
106. Bass, B. M. (1990). From transactional to transformational leadership: Learning to share the vision. *Organizational Dynamics, 3,* 19–31.
107. Pierro, A., Raven, B. H., Amato, C., & Bélanger, J. J. (2013). Bases of social power, leadership styles, and organizational commitment. *International Journal of Psychology, 48*(6), 1122–1134.
108. Van Wart, M. (2013). Lessons from leadership theory and the contemporary challenges of leaders. *Public Administration Review, 73*(4), 553–565.
109. U.S. Bureau of Labor Statistics. (2021, August 31). Table 1. Number of jobs held by individuals ages 18 to 54 in 1978–2018 by educational attainment, sex, race, Hispanic or Latino ethnicity, and age. Author: Washington, DC. https://www.bls.gov/news.release/nlsoy.t01.htm.
110. Zhuo, J. (2019). Quote appears on p. 215.
111. Zhuo, J. (2019). Quote appears on p. 206.

CHAPTER 12

1. Hardcastle, H. (2020). *The case for student mental health days* [Video]. TEDxSalem. https://www.ted.com/talks/hailey_hardcastle_the_case_for_student_mental_health_days.
2. Dwyer, K. K., & Davidson, M. M. (2012, April–June). Is public speaking really more feared than death? *Communication Research Reports, 29,* 99–107. This study found that public speaking was selected more often as a common fear than any other fear, including death. However, when students were asked to select a top fear, students selected death most often.
3. For an example of how demographics have been taken into consideration in great speeches, see Stephens, G. (1997, Fall). Frederick Douglass's multiracial abolitionism: "Antagonistic cooperation" and "redeemable ideals" in the July 5 speech. *Communication Studies, 48,* 175–194. On July 5, 1852, Douglass gave a speech titled "What to the Slave Is the 4th of July?," attacking the hypocrisy of Independence Day in a slaveholding republic. It was one of the greatest antislavery speeches ever given, and part of its success stemmed from the way Douglass sought common ground with his multiracial audience.
4. A tutorial on the different types of college students is available at Lumen Learning, College Success. Available at https://courses.lumenlearning.com/suny-collegesuccess-lumen1/chapter/types-of-students/ Accessed September 1, 2021.
5. Ody, B. (2018). Neglected patients: Deliberate indifference in state prisons. In L. Schnoor (Ed.), *Winning orations, 2018.* Interstate Oratorical Association. Britton was coached by Hope Willoughby and Matt Delzer.
6. Prof. Reich posts his videos on Facebook's Inequality Media page and on YouTube. A small portion of this transcript was edited to reduce potential political reaction. That excision is as follows: "Trump has intentionally cleaved America into two warring camps: pro-Trump or anti-Trump. Now most Americans aren't passionate conservatives or liberals, Republicans or Democrats. But they have become impassioned for or against Trump. As a result . . ."

7. Different analyses might use these terms in slightly different ways. See, for example, Kumar, M. (2022, June 25). *The relationship between beliefs, values, attitudes and behaviours.* https://owlcation.com/social-sciences/Teaching-and-Assessing-Attitudes.
8. Stutman, R. K., & Newell, S. E. (1984, Fall). Beliefs versus values: Silent beliefs in designing a persuasive message. *Western Journal of Speech Communication, 48*(4), 364.
9. Stutman & Newell (1984).
10. Mahoney, B. (2015). They're not a burden: The inhumanity of anti-homeless legislation. In L. Schnoor (Ed.), *Winning orations, 2015* (pp. 20–22). Interstate Oratorical Association. Brianna was coached by Kellie Roberts.
11. Mahoney (2015).
12. In information science parlance, these are referred to as Boolean terms.
13. McKibban, A. (2017). Criminalizing survivors. In L. Schnoor (Ed.), *Winning orations.* Interstate Oratorical Association, p. 163.
14. Some recent works specifically refer to public speaking anxiety, or PSA. See, for example, Bodie, G. D. (2010, January). A racing heart, rattling knees, and ruminative thoughts: Defining, explaining, and treating public speaking anxiety. *Communication Education, 59*(1), 70–105.
15. See, for example, Clark, A. H. (2018, October 16). How to harness your anxiety. *New York Times.* https://www.nytimes.com/2018/10/16/well/mind/how-to-harness-your-anxiety.html.
16. See, for example, Borhis, J., & Allen, M. (1992, January). Metaanalysis of the relationship between communication apprehension and cognitive performance. *Communication Education, 41*(1), 68–76.
17. Daly, J. A., Vangelisti, A. L., & Weber, D. J. (1995, December). Speech anxiety affects how people prepare speeches: A protocol analysis of the preparation process of speakers. *Communication Monographs, 62,* 123–134.
18. Researchers generally agree that communication apprehension has three causes: genetics, social learning, and inadequate skills acquisition. See, for example, Finn, A. N. (2009). Public speaking: What causes some to panic? *Communication Currents, 4*(4), 1–2.
19. See, for example, Sawyer, C. R., & Behnke, R. R. (1997, Summer). Communication apprehension and implicit memories of public speaking state anxiety. *Communication Quarterly, 45*(3), 211–222.
20. Adapted from Ellis, A. (1977). *A new guide to rational living.* North Hollywood, CA: Wilshire Books. G. M. Philips listed a different set of beliefs that he believes contributes to reticence. The beliefs are as follows: (1) an exaggerated sense of self-importance (reticent people tend to see themselves as more important to others than others see them); (2) effective speakers are born, not made; (3) skillful speaking is manipulative; (4) speaking is not that important; (5) I can speak whenever I want to; I just choose not to; (6) it is better to be quiet and let people think you are a fool than prove it by talking (they assume they will be evaluated negatively); and (7) what is wrong with me requires a (quick) cure. See Keaten, J. A., Kelly, L., & Finch, C. (2000). Effectiveness of the Penn State Program in changing beliefs associated with reticence. *Communication Education, 49*(2), 134–145.
21. Behnke, R. R., Sawyer, C. R., & King, P. E. (1987, April). The communication of public speaking anxiety. *Communication Education, 36,* 138–141.
22. Honeycutt, J. M., Choi, C. W., & DeBerry, J. R. (2009, July). Communication apprehension and imagined interactions. *Communication Research Reports, 26*(2), 228–236.
23. Behnke, R. R., & Sawyer, C. R. (1999, April). Milestones of anticipatory public speaking anxiety. *Communication Education, 48*(2), 165.
24. See, for example, Goldhill, O. (2017, April 22). *Rhetoric scholars pinpoint why Trump's inarticulate speaking style is so persuasive.* Quartz. https://qz.com/965004/rhetoric-scholars-pinpoint-why-trumps-inarticulate-speaking-style-is-so-persuasive.
25. Tamir, D., Templeton, E., Ward, A., & Zaki, J. (2018, May). Media usage diminishes memory for experiences. *Journal of Experimental Social Psychology 76,* 161–168. https://www.sciencedirect.com/science/article/abs/pii/S002210311730505X. See also Majid, F. (2020, January 21). *Memory loss happens to healthy young people too.* Neurogrow. https://neurogrow.com/memory-loss-happens-to-healthy-young-people-too/ .
26. Hinton, J. S., & Kramer, M. W. (1998, April). The impact of self-directed videotape feedback on students' self-reported levels of communication competence and apprehension. *Communication Education, 47*(2), 151–161. Significant increases in competency and decreases in apprehension were found using this method.
27. Economy, P. (2017, May 11). *These three body language mistakes make millennials look really unprofessional.* Inc. https://www.inc.com/peter-economy/these-3-body-language-mistakes-make-millennials-look-really-unprofessional.html.
28. See, for example, Rosenfeld, L. R., & Civikly, J. M. (1976). *With words unspoken.* New York: Holt, Rinehart and Winston, p. 62. Also see Chaiken, S. (1979). Communicator physical attractiveness and persuasion. *Journal of Personality and Social Psychology, 37,* 1387–1397.
29. Research has confirmed that speeches practiced in front of other people tend to be more successful. See, for example, Smith, T. E., & Frymier, A. B. (2006, February). Get "real": Does practicing speeches before an audience improve performance? *Communication Quarterly, 54,* 111–125.
30. A study demonstrating this stereotype is Street, R. L., Jr., & Brady, R. M. (1982, December). Speech rate acceptance ranges as a function of evaluative domain, listener speech rate, and communication context. *Speech Monographs, 49,* 290–308.
31. See, for example, Mulac, A., & Rudd, M. J. (1977). Effects of selected American regional dialects upon regional audience members. *Communication Monographs, 44,* 184–195. Some research, however, suggests that nonstandard dialects do not have the detrimental effects on listeners that were once believed. See, for example, Johnson, F. L., & Buttny, R. (1982, March). White listeners' responses to "sounding black" and "sounding white": The effect of message content on judgments about language. *Communication Monographs, 49,* 33–39. See also Monteserín, M. L., & Zevin, J.

(2016, September 8). Investigating the impact of dialect prestige on lexical decision. *Proceedings of Interspeech 2016* (pp. 2214–2218). https://www.isca-speech.org/archive_v0/Interspeech_2016/pdfs/1549.PDF.

32. Smith, V., Siltanen, S. A., & Hosman, L. A. (1998, Fall). The effects of powerful and powerless speech styles and speaker expertise on impression formation and attitude change. *Communication Research Reports, 15*(1), 27–35. In this study, a powerful speech style was defined as one without hedges and hesitations such as *uh* and *anda*.

33. Hardcastle, H. (2019, July 22). Quoted in Matias, D., Feeling blue? Oregon students allowed to take "mental health days." NPR. https://www.npr.org/2019/07/22/744074390/feeling-blue-oregon-students-allowed-to-take-mental-health-days.

34. Hardcastle (2020). *The case for student mental health days* [Video]. TEDxSalem. https://www.ted.com/talks/hailey_hardcastle_the_case_for_student_mental_health_days.

CHAPTER 13

1. See, for example, Stern, L. (1985). *The structures and strategies of human memory*. Dorsey Press. See also Turner, C. (1987, June 15). Organizing information: Principles and practices. *Library Journal, 112*(11), 58.
2. Booth, W. C., Colomb, G. C., & Williams, J. M. (2003). *The craft of research*. University of Chicago Press.
3. Koehler, C. (1998, June 15). Mending the body by lending an ear: The healing power of listening. *Vital Speeches of the Day,* 543.
4. Hallum, E. (1998). Untitled. In L. Schnoor (Ed.), *Winning orations, 1998* (pp. 4–6). Interstate Oratorical Association, p. 4. Hallum was coached by Clark Olson.
5. Monroe, A. (1935). *Principles and types of speech*. Scott, Foresman.
6. Adapted from http://vaughnkohler.com/wp-content/uploads/2013/01/Monroe-Motivated-Sequence-Outline-Handout1.pdf, accessed May 23, 2013.
7. Graham, K. (1976, April 15). The press and its responsibilities. *Vital Speeches of the Day,* 42.
8. Tucker, A. C. (2018). Bad medicine: $765 billion of medical waste. In L. Schnoor (Ed.), *Winning Orations, 2018* (pp. 15–16). Interstate Oratorical Association, p. 15. Tucker was coached by Hope Willoughby and Matt Delzer.
9. Tucker, A. C. (2018).
10. Anderson, G. (2009). Don't reject my homoglobin. In L. Schnoor (Ed.), *Winning Orations, 2009* (pp. 33–35). Interstate Oratorical Association, p. 33. Anderson was coached by Leah White.
11. Chakrabarty, P. (2018, April 10). *Four billion years of evolution in six minutes* [Video]. TED. https://www.ted.com/talks/prosanta_chakrabarty_four_billion_years_of_evolution_in_six_minutes.
12. Robbins, A. (2019, January 26). A frat boy and a gentleman. *New York Times*. https://www.nytimes.com/2019/01/26/opinion/sunday/fraternity-sexual-assault-college.html.
13. Eggleston, T. S. (n.d.). The key steps to an effective presentation. [Personal blog.] https://tseggleston.com/key-steps/.
14. Cook, E. (n.d.). *Making business presentations work*. Business Know-How. www.businessknowhow.com/manage/presentation101.htm.
15. Kotelnikov, V. (n.d.). Effective presentations. Retrieved September 12, 2021 from http://www.1000ventures.com/business_guide/crosscuttings/presentations_main.html.
16. Toastmasters International, Inc.'s Communication and Leadership Program. Retrieved May 19, 2010 from www.toastmasters.org.
17. Mehta, A. (2018). *Where do your online returns go?* [Video]. TED@UPS. https://www.ted.com/talks/aparna_mehta_where_do_your_online_returns_go.
18. Hill, N. (2018). Shut up or get out: How nuisance ordinances fail our communities. In L. Schnoor (Ed.), *Winning orations, 2018* (pp. 82–84). Interstate Oratorical Association, p. 82. Hill was coached by Mark Rittenour.
19. Williams, R. (2017). Asking for it: The lies of the rape culture. In L. Schnoor (Ed.), *Winning orations, 2017* (pp. 22–26). Interstate Oratorical Association. Williams was coached by Michael Gray and Baker Weilert.
20. See Williams, R. (2017).
21. Wideman, S. (2006). Planning for peak oil: Legislation and conservation. In L. Schnoor (Ed.), *Winning orations, 2006* (pp. 7–9). Interstate Oratorical Association, p. 7. Wideman was coached by Brendan Kelly.
22. Herbert, B. (2007, April 26). Hooked on violence. *New York Times*. http://www.nytimes.com/2007/04/26/opinion/26herbert.html.
23. Sherriff, D. (1998, April 1). *Bill Gates too rich* [Online forum comment]. CRTNET.
24. Elsesser, K. (2010, March 4). And the gender-neutral Oscar goes to . . . *New York Times*. http://www.nytimes.com/2010/03/04/opinion/04elsesser.html.
25. Gieseck, A. (2015). Problem of police brutality with regard to the deaf community. In L. Schnoor (Ed.), *Winning orations, 2015* (pp. 82–84). Interstate Oratorical Association, p. 83.
26. Boyle, J. (2015). Spyware stalking. In L. Schnoor (Ed.), *Winning orations, 2015* (pp. 32–34). Interstate Oratorical Association, p. 32. Julia was coached by Judy Santacaterina and Lisa Roth.
27. Notebaert, R. C. (1998, November 1). Leveraging diversity: Adding value to the bottom line. *Vital Speeches of the Day,* 47.
28. McCarley, E. (2009). On the importance of drug courts. In L. Schnoor (Ed.), *Winning orations, 2009* (pp. 36–38). Interstate Oratorical Association, p. 36.
29. See McCarley, E. (2009). On the importance of drug courts.
30. Teresa Fishman, director of the Center for Academic Integrity at Clemson University, quoted in Gabriel, T. (2010, August 1). Plagiarism lines blur for students in digital age. *New York Times*. https://www.nytimes.com/2010/08/02/education/02cheat.html.

CHAPTER 14

1. Sigler, M. G. (2010, August 4). *Eric Schmidt: Every 2 days we create as much information as we did up to 2003*. TechCrunch. http://techcrunch.com/2010/08/04/schmidt-data.

2. See, for example, Wurman, R. S. (2000). *Information anxiety 2*. Que.
3. Buffett, W. E. (2012, November 25). A minimum tax for the wealthy. *New York Times*. http://www.nytimes.com/2012/11/26/opinion/buffett-a-minimum-tax-for-the-wealthy.html.
4. See Fransden, K. D., & Clement, D. A. (1984). The functions of human communication in informing: Communicating and processing information. In C. C. Arnold & J. W. Bowers (Eds.), *Handbook of rhetorical and communication theory* (pp. 338–399). Allyn and Bacon.
5. Allocca, K. (2011, November). *Why videos go viral* [Video]. TEDYouth. https://www.ted.com/talks/kevin_allocca_why_videos_go_viral?language=en.
6. Cabalan, S. (2018). Eleven and engaged: On America's unseen child marriage crisis. In L. Schnoor (Ed.), *Winning orations, 2018* (pp. 31–33). Interstate Oratorical Association (p. 31).
7. Cacioppo, J. T., & Petty, R. E. (1979). Effects of message repetition and position on cognitive response, recall, and persuasion. *Journal of Personality and Social Psychology, 37*, 97–109.
8. Schulz, K. (2011, March). *On being wrong* [Video]. TED2011. https://www.ted.com/talks/kathryn_schulz_on_being_wrong?language=en.
9. MacPherson-Krutsky, C. (2020, July 24). *3 questions to ask yourself next time you see a graph, chart or map*. The Conversation. https://theconversation.com/3-questions-to-ask-yourself-next-time-you-see-a-graph-chart-or-map-141348.
10. Grant, A. (2021). *What frogs in hot water can teach us about thinking again*. Ted@BCG.https://www.ted.com/talks/adam_grant_what_frogs_in_hot_water_can_teach_us_about_thinking_again.

CHAPTER 15

1. Malami, S. (2022). Zoom meeting with George Rodman, January 15, 2022.
2. For an explanation of social judgment theory, see Griffin, E. (2012). *A first look at communication theory* (8th ed.). McGraw-Hill.
3. See, for example, Jaska, J. A., & Pritchard, M. S. (1994). *Communication ethics: Methods of analysis* (2nd ed.). Wadsworth.
4. Some research suggests that audiences may perceive a direct strategy as a threat to their freedom to form their own opinions. This perception hampers persuasion. See Brehm, J. W. (1966). *A theory of psychological reactance*. Academic Press.
5. Golden, J. L., Berquist, G. F., Coleman, W. E., Golden, R., & Sproule, J. M., (Eds.). (2011). *The rhetoric of Western thought: From the Mediterranean world to the global setting* (10th ed.). Kendall/Hunt, p. 72.
6. Heath, D., & Heath, C. (2010). *Switch: How to change things when change is hard*. Currency/Random House, p. 117.
7. Haidt, J. (2012). *The righteous mind*. Pantheon, p. 50.
8. Haidt (2012), p. 68.
9. Anthon, J. (2020); Yocum, R. (2015). Maternal mortality in modern America. In L. Schnoor (Ed.), *Winning orations 2020* (pp. 17–19). Interstate Oratorical Association. Julia was coached by Chas Womelsdorf.
10. Yocum, R. (2015). A deadly miscalculation. In L. Schnoor (Ed.), *Winning orations 2015* (pp. 97–99). Interstate Oratorical Association. Rebeka was coached by Mark Hickman.
11. For an excellent review of the effects of evidence, see Reinard, J. C. (1988, Fall). The empirical study of persuasive effects of evidence: The status after fifty years of research. *Human Communication Research, 15*(1), 3–59.
12. There are, of course, other classifications of logical fallacies than those presented here. See, for example, Warnick, B., & Inch, E. (1994). *Critical thinking and communication: The use of reason in argument* (2nd ed.). Macmillan, pp. 137–161.
13. Sprague, J., & Stuart, D. (1992). *The speaker's handbook* (3rd ed.). Harcourt Brace Jovanovich, p. 172.
14. Myers, L. J. (1981). The nature of pluralism and the African-American case. *Theory into Practice, 20*, 3–4. Cited in Samovar, L. A., & Porter, R. E. (1995). *Communication between cultures* (2nd ed.). Wadsworth, p. 251.
15. Samovar & Porter, *Communication between cultures*, pp. 154–155.
16. For an example of how one politician failed to adapt to his audience's attitudes, see Hostetler, M. J. (1998, Winter). Gov. Al Smith confronts the Catholic question: The rhetorical legacy of the 1928 campaign. *Communication Quarterly, 46*(1), 12–24. Smith was reluctant to discuss religion, attributed bigotry to anyone who brought it up, and was impatient with the whole issue. He lost the election. Many years later, John F. Kennedy dealt with "the Catholic question" more reasonably and won.
17. Reich, R. (2018, August 12). 10 steps to finding common ground. [Personal blog.] https://robertreich.org/post/176921589945.
18. DeVito, J. A. (1986). *The communication handbook: A dictionary*. Harper & Row, pp. 84–86.
19. Rodriguez, G. (2010). *How to develop a winning sales plan*. PowerHomeBiz.com. www.powerhomebiz.com/062006/salesplan.htm.
20. Sanfilippo, B. (2010). *Winning sales strategies of top performers*. SelfGrowth.com. www.selfgrowth.com/articles/Sanfilippo2.html.
21. Kraidin, D. (2019, May 9). Saeed Malami '20 wins national forensics championship, first Interstate Oratorical Association Championship. *The Lafayette*.
22. Personal email correspondence with George Rodman, March 2021.
23. Malami, S. (2019). Child slavery in the chocolate industry. In *Winning orations, 2019*. Interstate Oratorical Association. Edited for classroom use with visual aids.

Glossary

abstract language Wording that makes vague references to people, objects, events, and experiences.

abstraction ladder A conceptualization in which a vague statement is modified to become increasingly more specific.

accent Pronunciation perceived as different from other people's.

acculturation The process of adapting to a culture.

active listening Listening to understand contextual and emotional components of a message while empathizing with the speaker.

actuate To move members of an audience toward a specific behavior.

***ad hominem* fallacy** A fallacious argument that attacks the integrity of a person to weaken the person's position.

addition The articulation error that involves adding extra parts to words.

advising response A response in which the receiver offers suggestions about how the speaker should deal with a problem.

affect blend The combination of two or more expressions, each showing a different emotion.

affect displays Facial expressions, body movements, and vocal traits that reveal emotional states.

affinity The degree to which people like or appreciate one another, whether they display that outwardly or not.

all-channel network A communication network pattern in which all group members share information with one another.

ally Someone from a dominant social group (e.g., White, male, cisgender, management) who actively advocates for fair treatment and social justice for others.

altruistic lies Deception intended to be unmalicious, or even helpful, to the person to whom it is told.

analogies An extended comparison that can be used as supporting material in a speech.

analytical listening A listening style in which the primary goal is to understand a message.

anchor The position supported by audience members before a persuasion attempt.

anecdote A brief, personal story used to illustrate or support a point in a speech.

apathetic siblings Those who are relatively indifferent toward one another and seldom or never communicate.

***argumentum ad populum* fallacy** Fallacious reasoning based on the dubious notion that because many people favor an idea, you should, too.

***argumentum ad verecundiam* fallacy** Fallacious reasoning that tries to support a belief by relying on the testimony of someone who is not an authority on the issue being argued.

articulation The process of pronouncing all the necessary parts of a word, with all its distinct syllables.

assertive communication A style of communicating that directly expresses the sender's needs, thoughts, or feelings, delivered in a way that does not attack the receiver.

assertive language Wording that is clear and direct.

asynchronous communication Communication that occurs when there is a lag between the creation and reception of a message.

attending The process of focusing on certain stimuli in the environment.

attitude The predisposition to respond to an idea, person, or thing favorably or unfavorably.

attribution The process of attaching meaning to behavior.

audience analysis A consideration of characteristics, including the type, goals, demographics, beliefs, attitudes, and values of listeners.

audience involvement The level of commitment and attention that listeners devote to a speech.

audience participation Listener activity during a speech; a technique to increase audience involvement.

authoritarian An approach in which parents (or other types of leaders) are strict and expect unquestioning obedience.

authoritarian leadership A style in which a designated leader uses coercive and reward power to dictate the group's actions.

authoritative An approach in which parents (or other types of leaders) are firm, clear, and strict, while encouraging open communication.

avoidance spiral Occurs when relational partners reduce their involvement with one another, withdraw, and become less invested in the relationship.

bar chart A visual aid that compares two or more values by showing them as elongated horizontal rectangles.

basic speech structure The division of a speech into introduction, body, and conclusion.

behavioral interview A question and answer session that focuses on an applicant's past performance (behavior) as it relates to the job at hand.

belief An underlying conviction about the truth of an idea, often based on cultural training.

brainstorming A group activity in which members share as many ideas about a topic as possible without stopping to judge or rule anything out.

breakout groups A strategy used when the number of members is too large for effective discussion. Subgroups simultaneously address an issue and then report back to the group at large.

cause-effect pattern An organizing plan for a speech that demonstrates how one or more events result in another event or events.

chain network A communication network in which information passes sequentially from one member to another.

channel The medium through which a message passes from one person to another.

chronemics The study of how humans use and structure time.

citation A brief statement of supporting material in a speech.

climax pattern An organizing plan for a speech that builds ideas to the point of maximum interest or tension.

clip intro A brief explanation or comment before a visual aid is used.

clip outro A brief summary or conclusion after a visual aid has been used.

clip wraparound A brief introduction before a visual aid is presented, accompanied by a brief conclusion afterward.

coculture A group that is part of an encompassing culture.

coercive power The ability to influence others by threatening or imposing unpleasant consequences.

cohesiveness The degree to which members feel part of a group and want to remain in it.

collaborative language A communication approach that encourages people to think together without treating any one person's opinion as dominant.

collectivistic culture A group in which members focus on the welfare of the group as a whole more than on individual identity.

column chart A visual aid that compares two or more values by showing them as elongated vertical rectangles.

comforting A response style in which a listener reassures, supports, encourages, or distracts the person seeking help.

communication climate The emotional tone of a relationship as it is expressed in the messages that the partners send and receive.

communication competence The ability to achieve one's goals through communication and, ideally, maintain healthy relationships.

communication The process of creating meaning through symbolic interaction.

communicative coherence Consistency of a person's verbal and nonverbal communication.

competitive siblings Those who consider themselves to be rivals vying for scarce resources such as their parents' time and respect.

compromise An agreement that gives both parties at least some of what they wanted, although both sacrifice part of their goals.

conclusion (of a speech) The final part of a speech, in which the main points are reviewed and final remarks are made to motivate the audience or help listeners remember key ideas.

concrete language Wording that refers to specific people, behaviors, objects, or events.

confirmation bias The emotional tendency to interpret new information as reinforcing of one's existing beliefs.

confirming messages Actions and words that express respect for another person.

conflict (storming) stage When group members openly defend their positions and question those of others.

conflict An expressed struggle between at least two interdependent parties who perceive incompatible goals, scarce rewards, and/or interference from the other party in achieving their goals.

connection power The ability to influence others based on having relationships that might help the group reach its goal.

connotative meanings Informal, implied interpretations for words, often positive or negative. See denotative meanings.

contempt Verbal and nonverbal messages that ridicule or belittle another person.

content message A message that communicates information about the subject being discussed.

control The amount of influence one has over others. See relational message.

conversational narcissists People who focus on themselves and their interests instead of listening to and encouraging others.

convincing A speech goal that aims at changing audience members' beliefs, values, or attitudes.

coordinated management of meaning (CMM) The perspective that people co-create meaning in the process of communicating with each other.

counterfeit question A question that is not truly a request for new information.

credibility The believability of a speaker or other source of information.

critical listening Listening in which the goal is to evaluate the quality or accuracy of the speaker's remarks.

criticism A message that is personal, all-encompassing, and accusatory.

cultural humility An approach that considers effective intercultural communication to be an ongoing process in which people continually learn about others and themselves in an environment that is empowering, respectful, and adaptive.

culture The language, values, beliefs, traditions, and customs shared by a group of people.

cyberbullying A malicious act in which one or more parties aggressively harasses a victim online, often in public forums.

cyberstalking Ongoing obsessive and/or malicious monitoring of the social media presence of a person.

database A computerized collection of information that can be searched in a variety of ways to locate information that the user is seeking.

debilitative communication apprehension An intense level of anxiety about speaking before an audience, resulting in poor performance.

deception bias The tendency to assume that others are lying.

defensive listening A response style in which the receiver perceives a speaker's comments as an attack.

defensiveness Protecting oneself by counterattacking the other person or justifying one's own behavior.

deletion An articulation error that involves leaving off parts of spoken words.

democratic leadership A style in which a leader invites the group's participation in decision making.

demographics Audience characteristics that can be analyzed statistically, such as age, gender, education, and group membership.

denotative meanings Formally recognized definitions of words. See connotative meanings.

description A type of speech that uses details to create a "word picture" of something's essential factors.

developmental model (of relational maintenance) Theoretical framework based on the idea that communication patterns are different in various stages of interpersonal relationships.

diagram A line drawing that shows the most important components of an object.

dialect A version of the same language that includes substantially different words and meanings.

dialogue A process in which people let go of the notion that their ideas are more correct or superior to others' and instead seek to understand an issue from many different perspectives.

digital dirt Unflattering information (whether true or not) that has been posted about a person online.

direct aggressive message Actions and words that attack the position and perhaps the dignity of another person. See passive aggression.

direct persuasion Persuasion that does not try to hide or disguise the speaker's persuasive purpose.

disconfirming messages Actions and words that imply a lack of agreement or respect for another person.

disfluencies Vocal interruptions such as stammering and use of "uh," "um," and "er."

disinhibition The tendency to transmit messages without considering their consequences.

dyadic communication Message exchange between two people.

either–or fallacy Fallacious reasoning that sets up false alternatives, suggesting that if the inferior one must be rejected, then the other must be accepted.

emblems Deliberate gestures with precise meanings, known to virtually all members of a cultural group.

emergence (norming) stage When a group moves from conflict toward a single solution.

emergent leader A member who assumes leadership without being appointed by higher-ups.

emotional intelligence (EI) A person's ability to understand and manage their own emotions and to deal effectively with the emotions of others.

emotive language Opinion statements meant to stir up strong emotional reactions.

enclothed cognition The phenomenon of adopting traits or enacting behaviors associated with articles of clothing.

environment Both the physical setting in which communication occurs and the personal perspectives of the parties involved.

eportfolio A personal website on which a person presents portfolio information, such as work samples, awards, and letters of commendation.

equivocal language Words that have more than one dictionary definition.

equivocation A deliberately vague statement that can be interpreted in more than one way.

escalatory conflict spiral A reciprocal pattern of communication in which messages between two or more communicators reinforce one another in an increasingly negative manner.

ethical persuasion Persuasion in an audience's best interest that does not depend on false or misleading information to induce change in that audience.

ethnicity A social construct that refers to the degree to which a person identifies with a particular group, usually on the basis of nationality, culture, religion, or some other perspective.

ethnocentrism The attitude that one's own culture is superior to others'.

ethos Appeals based on the credibility of the speaker.

euphemism Mild or indirect word choices used in place of more direct but less pleasant ones.

evasion A deliberately vague statement.

expectancy violation theory The proposition that nonverbal cues cause physical and/or emotional arousal, especially if they deviate from what is considered normal.

expert power The ability to influence others by virtue of one's perceived expertise on the subject in question.

explanations Speeches or presentations that clarify ideas and concepts already known but not understood by an audience.

extemporaneous speech A speech that is planned in advance but presented in a direct, conversational manner.

face The socially approved identity that a communicator tries to present.

facework Verbal and nonverbal behavior designed to create and maintain a communicator's face and the face of others; synonymous with identity management.

facilitative communication apprehension A moderate level of anxiety about speaking before an audience that helps improve the speaker's performance

factual example A true, specific case that is used to demonstrate a general idea.

factual statement An assertion that can be verified as being true or false.

fallacy An error in logic.

fallacy of ad hominem A statement that attacks a person's character rather than debating the issues at hand.

fallacy of approval The irrational belief that it is vital to win the approval of virtually every person a communicator deals with.

fallacy of catastrophic failure The irrational belief that the worst possible outcome will probably occur.

fallacy of overgeneralization Irrational beliefs in which conclusions (usually negative) are based on limited evidence or communicators exaggerate their shortcomings.

fallacy of perfection The irrational belief that a worthwhile communicator should be able to handle every situation with complete confidence and skill.

family People who share affection and resources and who think of themselves and present themselves as a family, regardless of genetics.

Feynman Technique A process proposed by physicist Richard Feynman to make sense of complicated information.

flow chart A diagram that depicts the steps in a process with shapes and arrows.

formal outline A consistent format and set of symbols used to identify the structure of ideas.

formal role A role assigned to a person by group members or an organization, usually to establish order.

frame switching Adopting the perspectives of different cultures.

gatekeeper A person who acts as a hub for receiving messages and sharing (or not sharing) them with others.

gender A socially constructed set of expectations about what it means to be masculine, feminine, and intermediate points along a masculine-feminine continuum.

general purpose One of three basic ways a speaker seeks to affect an audience: to entertain, inform, or persuade.

groupthink The tendency in some groups to support ideas without challenging them or providing alternatives.

halo/horns effect A form of bias that overgeneralizes positive or negative traits.

haptics The study of touch.

hearing The process wherein sound waves strike the eardrum and cause vibrations that are transmitted to the brain. See listening.

hegemony The dominance of one culture over another.

high self-monitors People who pay close attention to their own behavior and to others' reactions, adjusting their communication to create the desired impression.

high-context culture A group that relies heavily on subtle, often nonverbal cues to convey meaning and maintain social harmony.

hostile siblings Those who harbor animosity toward each other and often stop communicating.

hyperpersonal communication The phenomenon in which digital interaction creates deeper relationships than those which arise through face-to-face interaction.

hypothetical example An example that asks an audience to imagine an object or event.

identity management Strategies used by communicators to influence the way others view them.

illustrators Nonverbal behaviors that accompany and support verbal messages.

imaginary audience A heightened self-consciousness that makes it seem as if people are always observing and judging you.

immediacy Expression of interest and attraction communicated verbally and/or nonverbally.

implicit bias A prejudice and stereotype that someone harbors without consciously thinking about it.

impromptu speech A speech given "off the top of one's head," without preparation.

indirect communication Hinting at a message instead of expressing thoughts and feelings directly.

indirect persuasion Persuasion that disguises or deemphasizes the speaker's persuasive goal.

individualistic culture A group in which members focus on the value and welfare of individual members more than on the group as a whole.

inferential statement A conclusion (accurate or not) that someone arrives at after considering evidence.

informal role A role usually not explicitly recognized by a group that describes functions of group members rather than their positions. These are sometimes called "functional roles."

information anxiety The psychological stress of dealing with too much information.

information hunger Audience desire, created by a speaker, to learn information.

informational interview A structured meeting in which a person seeks answers from someone whose knowledge might help them succeed.

informative purpose statement A complete statement of the objective of a speech, worded to stress audience knowledge and/or ability.

in-group Members of a social group with which a person identifies.

insensitive listening The failure to recognize the thoughts or feelings that are not directly expressed by a speaker, and instead accepting the speaker's words at face value.

instructions Remarks that teach something to an audience in a logical, step-by-step manner.

insulated listening A style in which the receiver ignores or is oblivious to undesirable information.

intercultural communication competence The ability to engage effectively with people from different cultures based on understanding of, and respect for, different perspectives.

intergroup contact hypothesis A proposition based on evidence that prejudice tends to diminish when people have personal contact with those they might otherwise stereotype.

interpersonal communication Two-way interactions between people who share emotional closeness and treat each other as unique individuals.

interpretation The perceptual process of attaching meaning to stimuli that have previously been selected and organized.

intersectionality The idea that people are influenced in unique ways by the complex overlap and interaction of multiple identities.

intimate distance One of four distance zones, ranging from skin contact to 18 inches.

intrapersonal communication Communication that occurs as internal dialogue.

introduction (of a speech) The first structural unit of a speech, in which the speaker captures the audience's attention and previews the main points to be covered.

irrational thinking Beliefs that have no basis in reality or logic; one source of debilitative communication apprehension.

jargon Specialized vocabulary used by people with common backgrounds and experience (more formal and enduring than slang).

Johari Window A model that describes the relationship between self-disclosure and self-awareness.

judging response A response that evaluates the sender's thoughts or behaviors in some way. The evaluation may be favorable or unfavorable.

kinesics The study of body movement, facial expression, gesture, and posture.

laissez-faire leadership A style in which a leader takes a hands-off approach, imposing few rules on a team.

language A collection of symbols governed by rules and used to convey messages between individuals.

latitude of acceptance In social judgment theory, statements that a receiver would not reject.

latitude of noncommitment In social judgment theory, statements that a receiver would not care strongly about one way or another.

latitude of rejection In social judgment theory, statements that a receiver would not accept.

leanness The lack of nonverbal cues to clarify a message.

legitimate power The ability to influence group members based on one's official position. Synonymous with being a nominal leader.

line chart A visual aid consisting of a grid that maps out the direction of a trend by plotting a series of points.

linguistic relativism The notion that language influences the way users experience the world.

listening fidelity The degree of congruence between what a listener understands and what the message sender was attempting to communicate.

listening The process wherein the brain recognizes impulses as sound and gives them meaning. See hearing.

logos Appeals based on logical reasoning

longing siblings Those who admire and respect one another but interact less frequently and with less depth than they would like.

lose–lose problem solving An approach to conflict resolution in which neither party achieves their goals.

low self-monitors People who express what they are thinking and feeling without much attention to the impression their behavior creates.

low-context culture A group that uses language primarily to express thoughts, feelings, and ideas as directly as possible.

managerial grid A two-dimensional model that identifies leadership styles as a combination of concern (in varying degrees) for people and/or for the task at hand.

manipulators Movements in which one part of the body (usually the hands) massages, rubs, holds, pinches, picks, or otherwise involves an object or body part.

manuscript speech A speech that is read word for word from a prepared text.

mass communication The transmission of messages to large, usually widespread audiences via TV, internet, movies, magazines, and other forms of mass media.

masspersonal communication Messages that, to varying degrees, combine elements of mass and interpersonal communication.

memorized speech A speech learned and delivered by rote without a written text.

message Information shared (intentionally or not) via words and nonverbal behaviors.

metacommunication Messages that refer to other messages; communication about communication.

microaggressive language Subtle, everyday messages that (intentionally or not) stereotype or demean people on the basis of sex, race, gender, appearance, or some other factor.

microresistance Everyday behaviors that call attention to hurtful language and stereotypes.

mindful listening Being fully present with people—paying close attention to their gestures, manner, and silences, as well as to what they say.

model (in speeches and presentations) A replica of an object being discussed. It is usually used when it would be difficult or impossible to use the actual object.

monochronic The use of time that emphasizes punctuality, schedules, and completing one task at a time.

narration The presentation of speech supporting material as a story with a beginning, middle, and end.

negativity bias The perceptual tendency to focus more on negative indicators than on positive ones.

negotiation An interactive process meant to help people reach agreement that best suits their goals.

networking The process of creating and maintaining relationships that further one's goals.

noise External, physiological, and psychological distractions that interfere with the accurate transmission and reception of a message.

nominal leader The ability to influence group members based on one's official position. Synonymous with legitimate power.

nonassertion The inability or unwillingness to express one's thoughts or feelings.

nonverbal communication Messages expressed without words, as through body movements, facial expressions, eye contact, tone of voice, and so on.

norms Typically unspoken shared values, beliefs, behaviors, and processes that influence group members.

number chart A visual aid that lists numbers in tabular form to clarify information.

online surveillance Discreet monitoring of the social media presence of unknowing targets.

opinion statement An assertion based on the speaker's beliefs.

organization The perceptual process of mentally grouping stimuli into patterns.

organizational communication Interaction among members of a relatively large, permanent structure (such as a nonprofit agency or business) in order to pursue shared goals.

orientation (forming) stage When group members become familiar with one another's positions and tentatively volunteer their own.

out-group People who do not belong to the social groups with which a person identifies.

paralanguage Nonlinguistic means of vocal expression: rate, pitch, tone, and so on.

paraphrasing Feedback in which the receiver rewords the speaker's thoughts and feelings to confirm understanding and/or express empathy.

passive aggression An indirect expression of aggression, delivered in a way that allows the sender to maintain a facade of innocence. See direct aggressive message.

pathos Appeals based on emotion

perceived self The person we believe ourselves to be in moments of candor.

perception A process in which people reach conclusions about others and the world around them.

perception checking A three-part method for verifying the accuracy of interpretations, including an objective description of the behavior, two possible interpretations, and a request for confirmation of the interpretations.

performance norms Shared expectations that influence how group members handle the job at hand.

permissive An approach in which parents do not require children to follow many rules.

personal distance One of four distance zones, ranging from 18 inches to 4 feet.

personal fable The sense, common in adolescence, that you are different from everybody else.

personality Characteristic ways that a person tends to think and behave in a variety of situations.

persuasion The act of motivating a listener, through communication, to change a particular belief, attitude, value, or behavior.

phonological rules Linguistic guidelines that govern how sounds are combined to form words.

phubbing A mixture of the words *phoning* and *snubbing*, used to describe episodes in which people pay attention to their devices rather than to the people around them.

pictogram A visual aid that conveys its meaning through an image of an actual object.

pie chart A visual aid that divides a circle into wedges, representing percentages of the whole.

pitch The highness or lowness of one's voice.

polychronic The use of time that emphasizes flexible schedules in which multiple tasks are pursued at the same time.

polymediation The range of communication channel options available to communicators.

portfolio A collection of a person's best work along with awards, letters of commendation, and other materials that provide evidence of their credentials.

positive spiral Occurs when one person's confirming message leads to a similar or more confirming response from the other person.

post hoc **fallacy** Fallacious reasoning that mistakenly assumes that one event causes another because they occur sequentially.

power distance The perceived gap between those with substantial power and resources and those with less.

power The ability to influence others' thoughts and/or actions.

pragmatic rules Guidelines that govern how people use language in everyday interaction.

prejudice An unfairly biased and intolerant attitude toward a group of people.

presenting self The image a person presents to others.

problem census A technique in which members write their ideas on cards, which are then posted and grouped by a leader to generate key ideas the group can then discuss.

problem-solution pattern An organizing pattern for a speech that describes an unsatisfactory state of affairs and then proposes a plan to remedy the problem.

procedural norms Shared expectations that influence how a group operates or reaches decisions.

prompting Using silence and brief statements to encourage a speaker to continue talking.

proposition of fact A claim bearing on issue in which there are two or more sides of conflicting factual evidence.

proposition of policy A claim bearing on an issue that involves adopting or rejecting a specific course of action.

proposition of value A claim bearing on an issue involving the worth of some idea, person, or object.

proxemics The study of how people and animals use space.

pseudolistening An imitation of true listening.

public communication Communication that occurs when a group becomes too large for all members to contribute; characterized by an unequal amount of speaking and by limited verbal feedback.

public distance One of four distance zones, ranging from 12 feet or more.

purpose statement A complete sentence that describes precisely what a speaker wants to accomplish.

race A construct originally created to explain differences between people whose ancestors originated in different regions of the world—Africa, Asia, Europe, and so on.

rate The speed at which a speaker utters words.

reductio ad absurdum **fallacy** Fallacious reasoning that unfairly attacks an argument by extending it to such extreme lengths that it looks ridiculous.

referent power The ability to influence others by virtue of being liked or respected.

reflected appraisal The influence of others on one's self-concept.

reflecting Listening that helps the person speaking hear and think about the words they have just spoken.

reflective thinking method A systematic approach to solving problems. Also known as Dewey's reflective thinking method.

reinforcement (performing) stage When group members endorse the decision they have made.

relational dialectics The perspective that partners in interpersonal relationships deal with simultaneous and opposing forces of connection versus autonomy, predictability versus novelty, and openness versus privacy.

relational listening A listening style that is driven primarily by the desire to build emotional closeness with the speaker.

relational message A message that expresses the social relationship between two or more individuals. See content message.

relational spiral A reciprocal communication pattern in which each person's message reinforces the other's emotional tone.

relative words Terms that gain their meaning by comparison.

remembering The act of recalling previously introduced information. The amount of recall drops off in two phases: short term and long term.

residual message The part of a message a receiver can recall after short- and long-term memory loss.

respect The degree to which a person holds another in esteem, whether or not they like them.

responding Providing observable feedback to another person's behavior or speech.

reward power The ability to influence others by granting or promising desirable consequences.

richness The degree to which nonverbal cues clarify and/or reinforce a verbal message.

role Pattern of behavior enacted by particular members.

round robin Session in which members each address the group for specified amount of time.

rules Explicit, official guidelines that govern group functions and member behavior.

salience How much weight people attach to cultural characteristics in a particular situation.

selection interview A conversation aimed at evaluating a candidate for a job or other opportunity.

selection The perceptual act of attending to some stimuli in the environment and ignoring others.

selective listening A listening style in which the receiver responds only to messages that interest them.

self-concept A set of largely stable perceptions about oneself.

self-disclosure The process of deliberately revealing information about oneself that is significant and that would not normally be known by others.

self-esteem The part of the self-concept that involves evaluations of self-worth.

self-fulfilling prophecy A prediction or expectation of an event that makes the outcome more likely to occur than would otherwise have been the case.

self-monitoring Paying close attention to situational cues and adapting one's behavior accordingly.

self-serving bias The tendency to judge other harshly but cast oneself in a favorable light.

semantic rules Guidelines that govern the meaning of language as opposed to its structure.

servant leadership A style based on the idea that a leader's job is mostly to recruit outstanding team members and provide the support they need to do a good job.

sex A biological category such as male or female. See gender.

signpost A phrase that emphasizes the importance of upcoming material in a speech.

sincere question A question posed with the genuine desire to learn from another person.

situational leadership A theory that argues that the most effective leadership style varies according to the circumstances involved.

slang Language used by a group of people whose members belong to a similar coculture or other group. (Less formal and enduring than jargon).

slurring The articulation error that involves overlapping the end of one word with the beginning of the next.

small group A limited number of interdependent people who interact with one another over time to pursue shared goals.

small-group communication Communication that occurs within a group in which every member can participate actively with the other members.

social comparison Evaluating oneself in comparison to others.

social distance One of four distance zones, ranging from 4 to 12 feet.

social exchange theory The idea that relationships are worth maintaining if the rewards are greater than or equal to the costs involved.

social judgment theory The theory that opinions will change only in small increments, and only when the target opinions lie within the receiver's latitudes of acceptance and noncommitment.

social listening The process of attending to, observing, understanding, and responding to stimuli through mediated, electronic, and social channels.

social loafing The tendency of group members to do as little work as possible.

social media Dynamic websites and applications that enable individual users to create and share content or to participate in personal networking.

social media snarks People who post insulting comments about others.

social media trolls Individuals whose principal goal is to disrupt public discourse by posting false claims and prejudiced remarks, usually anonymously.

social norms Shared expectations that influence how group members relate to one another.

social penetration model A theory that describes how intimacy can be achieved via the breadth and depth of self-disclosure.

sound bite A brief recorded excerpt from a longer statement.

space pattern An organizing plan in a speech that arranges points according to their physical location.

specific purpose The precise effect that the speaker wants to have on an audience. It is expressed in the form of a purpose statement.

statistics Numbers arranged or organized to show how a fact or principle is true for a large percentage of cases.

stereotype A widely held but oversimplified or inaccurate idea about a group of people.

stonewalling Refusing to engage with the other person.

substitution The articulation error that involves replacing part of a word with an incorrect sound.

supportive listening Listening with the goal of helping a speaker deal with personal dilemmas.

supportive siblings Those who talk regularly and are accessible and emotionally close to one another.

survey research Information gathering in which the responses of a population sample are collected to disclose information about the larger group.

symbol An arbitrary sign used to represent a thing, person, idea, or event in ways that make communication possible.

synchronous communication Communication that occurs in real time.

syntactic rules Guidelines that govern how symbols can be arranged (e.g., sentence structure).

target audience That part of an audience that must be influenced to achieve a persuasive goal.

task-oriented listening A listening style in which the goal is to secure information necessary to get a job done.

teams Groups of people who are successful at reaching goals.

territoriality The tendency to claim spaces or things as one's own, at least temporarily.

testimony Supporting material that proves or illustrates a point by citing an authoritative source.

thesis statement A complete sentence describing the central idea of a speech.

time pattern An organizing plan for a speech based on chronology.

topic pattern An organizing plan for a speech that arranges points according to logical types or categories.

trait theories of leadership A school of thought based on the belief that some people are born to be leaders and others are not.

transactional communication model Proposes that communicators co-create meaning while simultaneously exchanging messages within the context of social, relational, and cultural expectations.

transformational leadership Defined by a leader's devotion to help a team fulfill an important mission.

transition A phrase that connects ideas in a speech by showing how one relates to the other.

trolling Attacking others via online channels.

truth bias The tendency to assume that others are being honest.

uncertainty avoidance The cultural tendency to seek stability and honor tradition instead of welcoming risk, uncertainty, and change.

understanding The act of interpreting a message by following syntactic, semantic, and pragmatic rules.

uses and gratification theory An approach used to understand why and how people actively seek out specific media to satisfy their particular needs, including providing information, facilitating personal relationships, defining personal identity, and entertainment.

value A deeply rooted belief about a concept's inherent worth.

visual aids Graphic devices used in a speech to illustrate or support ideas.

visualization A technique for rehearsal using a mental visualization of the successful completion of a speech.

wheel network A communication network in which a gatekeeper regulates the flow of information to and from all other members. See gatekeeper.

win–lose problem solving An approach to conflict resolution in which one party strives to achieve their goal at the expense of the other.

win–win problem solving An approach to conflict resolution in which the parties work together in an attempt to satisfy all their goals.

word chart A visual aid that lists words or terms in tabular form in order to clarify information.

working outline A constantly changing organizational aid used in planning a speech.

Credits

PHOTOGRAPHS
Page 2 © LightField Studios/Shutterstock.
9 Media Res/ Echo Films/ Hello Sunshine / Album.
12 Lisa Werner / Alamy Stock Photo.
16 © Prostock-studio/Shutterstock.
20 © Ground Picture/Shutterstock.
23 © Willrow Hood/Shutterstock/ © Shutterstock.
29 © Koon MadzCat/Shutterstock.
31 © Koshiro K/Shutterstock.
33 © View Apart/Shutterstock.
35 © Kaspars Grinvalds/Shutterstock.
37 © mundissima/Shutterstock.
39 © LightField Studios/Shutterstock.
42 © asiandelight/Shutterstock.
44 © Jacob Lund/Shutterstock.
48 © Dragon Images/Shutterstock.
51 Erik Pendzich / Alamy Stock Photo.
53 © iStock/VladimirFloyd.
60 Album / Alamy Stock Photo.
68 Album / Alamy Stock Photo.
74 © daizuoxin/Shutterstock.
77 Abaca Press / Alamy Stock Photo.
80 © Sky Cinema/Shutterstock.
86 © Jacob Lund/Shutterstock.
88 © Rena Schild/Shutterstock.
93 Cal Sport Media / Alamy Stock Photo.
97 © iStock/Ridofranz.
104 © Daisy Daisy/Shutterstock.
107 © jfmdesign/iStockphoto.
111 ZUMA Press, Inc. / Alamy Stock Photo.
112 © Fabio Pagani/Shutterstock.
115 © James Steidl/Shutterstock.
120 © iStock/Wavebreakmedia.
122 Pictorial Press Ltd / Alamy Stock Photo.
128 © Abraham_stockero/Shutterstock.
131 © Jamie Lamor Thompson/Shutterstock.
136 © leungchopan/Shutterstock.
139 © Hananeko_Studio/Shutterstock.
142 © ian routledge/Shutterstock.
143 © Drazen Zigic/Shutterstock.
145 © Dmytro Zinkevych/Shutterstock.
148 © Pixel-Shot/Shutterstock.
151 © fizkes/Shutterstock.
156 © fizkes/Shutterstock.
160 © Jovica Varga/Shutterstock.
162 © GIO_LE/Shutterstock.
163 © fizkes/Shutterstock.
167 © wavebreakmedia/Shutterstock.
168 © Odua Images/Shutterstock.
169 © Denis Makarenko/Shutterstock.
173 Moviestore Collection Ltd / Alamy Stock Photo.
177 Album / Alamy Stock Photo.
180 © fizkes/Shutterstock.
182 © BAZA Production/Shutterstock.
188 REUTERS / Alamy Stock Photo.
188 © Kathy Hutchins/Shutterstock.
192 © iStock/Leks_Laputin.
195 © iStock/jacoblund.
197 © iStock/FilippoBacci.
202 © iStock/Chainarong Prasertthai.
205 PictureLux / The Hollywood Archive / Alamy Stock Photo.
209 © 123RF/Pei Ting Hung.
212 Moviestore Collection Ltd / Alamy Stock Photo.
218 © 123RF/Dmytro.
226 © Oliver Marquardt/123RF.
229 © iStock/Mixmike.
247 © Cristian Storto/Shutterstock.
249 Photo by Danny Clark.
251 © 123RF/lightfieldstudios.
256 © iStock/fizkes.
260 © iStock/gremlin.
262 © iStock/Rawpixel.
265 © 123RF/Wavebreak Media Ltd.
273 © iStock/Thomas Northcut.
282 © Paul Bradbury/iStockphoto.
283 © iStock/~UserGI15613517.
288 © iStock/liorpt.
292 © iStock/FG Trade.
302 © iStock/cagkanasyin.
304 © iStock/Kateryna Onyshchuk.
311 © Jeff Chiu/AP/Shutterstock.
322 © Ground Picture/Shutterstock.
324 Courtesy of Hailey Hardcastle.
328 © fizkes/Shutterstock.
332 © ReeldealHD on Offset/Shutterstock.
334 Reeldeal Images / Alamy Stock Photo.
336 © Andrea Raffin/Shutterstock.

342 © fizkes/Shutterstock.
350 Cavan Images / Alamy Stock Photo.
352 Courtesy of Dan O'Connor.
366 © Pixelvario/Shutterstock.
382 © Ground Pictures/Shutterstock.
384 Courtesy of Adam Grant.
395 © artemisphoto/Shutterstock.
412 © fizkes/Shutterstock.
414 Courtesy of Saeed Salmi.
419 © zitrik/iStockphoto.
424 Pictorial Press Ltd / Alamy Stock Photo.
435 Allstar Picture Library Ltd / Alamy Stock Photo.
437 Courtesy of Saeed Salmi.

CARTOONS

Page 8 Warren Miller/The New Yorker Collection/The Cartoon Bank.
13 CALVIN and HOBBES © 1994 Watterson. Distributed by UNIVERSAL PRESS SYNDICATE. Reprinted with permission. All rights reserved.
15 Mike Twohy/The New Yorker Collection/The Cartoon Bank.
34 Rina Piccolo Cartoon used with permission of Rina Piccolo and the Cartoonist Group. All rights reserved.
41 Rina Piccolo Cartoon used with permission of Rina Piccolo and the Cartoonist Group. All rights reserved.
55 Edward Frascino/The New Yorker Collection/The Cartoon Bank.
62 William Steig/The New Yorker Collection/The Cartoon Bank.
117 Leo Cullum/Cartoonstock.
134 Zitz © 2003 Zitz Partnership distributed by King Features Syndicate Inc.
138 DILBERT 2009 Scott Adams. Used by permission of ANDREWS MCMEEL SYNDICATION. All rights reserved.
141 CALVIN AND HOBBES © 1995 Watterson. Reprinted with permission of ANDREWS MCMEEL SYNDICATION. All rights reserved.
155 Courtesy of Ted Goff.
170 © Dmytro Onopko/Shutterstock.
172 DILBERT © 2006 Scott Adams. Used by permission of ANDREWS MCMEEL SYNDICATION. All rights reserved.
178 Alex Gregory The New Yorker Collection/The Cartoon Bank.
186 Alex Gregory/Cartoon Collections.
195 Leo Cullum The New Yorker Collection/The Cartoon Bank.
200 Leo Cullum The New Yorker Collection/The Cartoon Bank.
221 Zitz © 2006 Zitz Partnership distributed by King Features Syndicate Inc.
232 Crock © 2019 North American Syndicate. World rights reserved.
245 Leo Cullum The New Yorker Collection/The Cartoon Bank.
294 © Shutterstock.
299 Sizemore, Jim/Cartoonstock.
372 DILBERT © 2006 Scott Adams. Used by permission of ANDREWS MCMEEL SYNDICATION. All rights reserved.
394 By permission of Leigh Rubin and Creators Syndicate, Inc.

FIGURES

Page 57 © 123RF/Evgeny Gromov.
356 Source: Adapted from "Outlining the Speech," Speech Program, University of Colorado at Colorado Springs.
399 Source: Kim Goodwin Twitter post July 19, 2018 at https://twitter.com/kimgoodwin/status/1020029572266438657?lang=en

Index

Abarcar, Sonya, 121
ABC, as news source, 39
Abstraction ladder, 115, 116
Abstract language, 115, 117
Abusive partners, 241
Accenting behavior, 170
Accents, 112
Acceptance, 415–16
Accommodation, 237–38
Accomplishments, 285
Acculturation, 279
Acknowledgment, 235
Acronyms, 116
Action zone, 341
Active listening, 143–44, 174–75
Active strategies, for learning, 99
Acts of service, 218
Actuating, 420
Adams, Sean, 303
Addition, 344
Ad hominem fallacy, 426
Adjustment shock, 99
Admiration, 198
Adoption, 420
Advising responses, 149–50
Advocacy, with social media, 33
Affect blends, 178
Affect displays, 168, 186
Affinity, 196
Affirming words, 218
Age:
   of audience members, 328
   as coculture, 95–96
   listening and, 138–39
   mediated communication use and, 40–41
Agender, 90
Agreement, 235
Akinmade, Oyinkan, 261
All-channel networks, 280

Allocca, Kevin, 391–92
AllSides, 39
Ally, 123, 124
#ALSIceBucketChallenge, 33
Alternation strategy, 222
Altman, Irwin, 200
Altruistic lies, 120
Ambiguity:
   in mediated messages, 28, 42
   in nonverbal communication, 164, 166–67
   tolerance for, 98
   uncertainty avoidance and, 83–84
American Psychological Association, 384
Analogies, 372–73
Analytical listening, 154–56
Anderson, Grant, 366–67
Anecdotes, 364, 373
Anonymity, in virtual meetings, 306
Anthon, Julia, 423
Anticlimactic organization, 361
Anxiety, about job interview, 273
Apathetic siblings, 213
Apodaca, Nathan, 24, 136
Apologies:
   in friendships, 207
   for microaggressions, 123
   at work, 283
Appearance:
   for job interview, 273
   nonverbal communication related to, 179–82
   for remote work, 284
   during speech delivery, 341
   in virtual meetings, 304
Appreciation, 197–98, 207, 263
Appropriateness, 29, 43, 401, 402
Approval, 337
Argumentum ad populum fallacy, 427

Argumentum ad verecundiam fallacy, 427
Aristotle, 294, 311, 421, 433
Articulation, 343–44
Aslay, Jonathan, 216
Assertive communication, 240–45
   defined, 240
   leadership potential and, 312
   process of, 242–43
   self-assessment, 244–45
Assertive language, 112–14
Associated Press, 39
Assumptions, 123, 153
Asynchronous communication, 29
Attending, 132
Attention:
   capturing, 364
   distancing behavior and, 234
   fidgeting and, 176
   in virtual meetings, 304
   when listening, 141–42, 143–44
Attitudes:
   of audience members, 329–31
   defined, 329
   positive, 338
Attractiveness:
   physical, 179–80
   of visual aids, 401
Attribution, 58
Audience(s), 327–31
   action by, 425
   capturing attention of, 364
   demographics of, 327–29
   demonstrating importance of topic to, 365–66
   desired response from, 425
   effect of problem on, 423–24
   expectations of, 332
   imaginary, 96
   for job interview presentation, 272

Audience(s) (*Cont.*)
  outlines for, 359
  for persuasive speeches, 428–30
  purpose for, 327
  referring to, 364
  for social media, 23
  target, 428–29
Audience analysis, 327–31, 432
Audience involvement, 393–95
Audience participation, 394
Audio files (for speeches), 395–96
Authenticity, 54
Authoritarian leadership, 313–14
Authoritarian parents, 211
Authoritative parents, 211
Authority, appeal to, 427
Authority–obedience managers, 315
Authority rule, 310
Authors, of fake stories, 38
Autism, 169
Automattic, 304–5
Avoidance, 234, 237
Avoidance spirals, 235–36
Avoiding stage of relationships, 217

Babe, Ann, 80
Baby Boomers, 96
Bach, George, 239–40
Bachelor, The (television program), 194, 202
Bachelorette, The (television program), 194, 202, 223
Baldoni, Justin, 91
"Ban Bossy" movement, 92
Bandwagon appeals, 427
Bar charts, 398, 400
Barrett, Lisa Feldman, 62
Basic speech structure, 355
Beals, Jennifer, 173
Beauty premium, 179–80
Begging the question, 428
Behavior(s):
  expectations and, 55
  focus on, 117
  interpretations of, 242
  listening as, 135
Behavioral descriptions, 242
Behavioral interviews, 274
Being-oriented friendships, 206
Being Wrong (Schulz), 394
Beliefs:
  of audience members, 329–31
  defined, 330

Bell, W. Kamau, 68
Bias(es):
  confirmation, 118, 433
  deception, 173
  implicit, 123, 124
  and misinformation, 39
  negativity, 60
  reflection on own, 123
  self-serving, 61
  truth, 173
Bibliography (speech), 355–56, 376
Big picture, of questions, 394
Blake, Robert, 314, 315
Blind area (Johari Window), 201
Blotky, Andrew, 95
Body (speech), 391–92
Body art, 181–82
Body movements, 175–78, 341
Bogart, Kathleen, 169
Bona fide occupational qualifications, 271
Bonding stage of relationships, 216
Booth, Wayne, 355
Borsellino, Regina, 284
Bosses, difficult, 293 (*See also* Leaders)
Boston Consulting Group, 402
Bosworth, Andrew, 45
Botox, 169
Boyle, Julia, 373
Bradberry, Travis, 63
Brainstorming, 305
Breakout groups, 306
Brown, Brené, 54, 228, 231, 233–34, 248–49, 252
Brown, Steve, 228, 231, 234, 249, 252
Brownlee, Dana, 230
Bryant, Adam, 142
BTS, 51
Buffett, Warren, 390, 391
Bullies and bullying, 36–37, 51–52, 246
Burnison, Gary, 265
Business presentations, organization of, 369

Cadman, Joseph, 260
Cain, Susan, 113
Calaban, Shayla, 393
Call to Courage, The (television program), 249
Calmness, 292–93
Camaraderie, developing, 303
Cameras, in videoconferencing, 303–4
Campt, David, 89

Career success, 257–87
  culture and, 83
  gender and, 92
  and identity management, 66
  LinkedIn use and, 32
  listening skills and, 131
  setting stage for, 259–63
  social media and, 32
  work lessons from *Undercover Boss*, 70
CARE mnemonic, 362
"Case for Student Mental Health Days, The" (Hardcastle), 344–47
Catastrophic failure, fallacy of, 337
Categorization, 58–59
Catfishing, 35–36, 44
Cause-effect patterns, 362, 424
CBS, as news source, 39
Cekic, Özlem, 241
Chain networks, 280
Chakrabarty, Prosanta, 368–69
Chaleff, Ira, 292
Chalkboards, 400
Challenger explosion, 302
Chang, Lynn Chih-Ning, 100
Channels, 5, 28–29
Chapman, Gary, 218
Character, 421, 431
Charisma, 432–33
Charts, 396–400
Cheaper by the Dozen (film), 212
"Child Slavery in the Chocolate Industry" (Malami), 433–37
Chopra, Priyanka, 30
Chronemics, 185
Churchill, Winston, 240
Circumscribing stage of relationships, 217
Cisgender, 90
Citations, 355–56, 374, 375, 425
Civility, 42–43
Clarification, supporting material for, 368–69
Clarity, 390–92, 422
Clemente, Jim, 173, 174, 175
Climax patterns, 361
Clip intros, 396
Clip outros, 396
Clip wraparounds, 396
Clothing:
  for job interview, 273
  as nonverbal communication, 180–81

for remote work, 284
for virtual meetings, 304
CMM (coordinated management of meaning), 108
Co-created meaning, 7
Cocultures:
defined, 76
types of, 86–96
understanding, 76–79
Coercion, 415
Coercive power, 313
#coffeedialogues, 241
Cognitive complexity, 12
Coherence, 175
Cohesiveness, 300–301
Collaboration:
communication as, 5
culture and, 81
in identity management, 68–69
and social media, 23–24
Collaborative language, 112–14
Collectivistic cultures, 80–82
assertive and collaborative language in, 114
conflict communication in, 251
Collins, Lauren, 98
Collins, Oliver, 98
Column charts, 398
Comfort, offering, 150–51
Commitment, 13, 15
Common ground, 42, 95, 429
Common interests, 197
Communication:
characteristics of, 4–7
competence in, 11–15
culture and anxiety about, 81–82
defined, 4–5
many approaches to, 11
misconceptions about, 15–16
models of, 5–7
types of, 8–11
unethical behaviors, 417
Communication apprehension, 335–39
overcoming, 338, 339
sources of, 336–37
Communication climates:
confirming and disconfirming messages in, 231–35
conflict and, 231–37
defined, 231
development of, 235–37

Communication competence, 11–15, 41–45
Communications, defined, 4
Communication strategies:
for active listening, 143–44
adapting to speaking situations, 329
for assertive and collaborative language, 114
for brainstorming sessions, 305
for business presentations, 369
for conclusions, 367
for conflict management, 230
cultural differences in persuasion, 427
for dealing with abusive partners, 241
for dealing with difficult team members, 300
for dealing with sexual harassment, 239
to demonstrate leadership potential, 312
to detect deception, 174–75
to distinguish between facts and opinions, 119
to evaluate misinformation, 28–29
for expressing yourself clearly, 117
for family relationships, 213
for fighting fair, 232
for friendships, 207, 210
humblebragging, 67
individuality vs. stereotypes, 59
informative speaking techniques, 390
for job interview presentations, 272
for job interviews, 270, 275
for managing dialectical tensions, 222
for meeting online dates, 216
for microresistance, 122–23
for nonverbal communication at work, 165–66
persuasion skills in sales, 432
practicing presentations, 340
presentation software, 401
for professional networks, 263
for questioning, 153
ranking of ethical faults, 418
for reaching group decisions, 310
for remote interviews, 278
for self-disclosing, 203
for social media, 10
social media for career success, 32
for talking about race, 89
for working with a difficult boss, 293

for working with slackers, 296
work lessons, 70
Communication technology:
conflict management and, 230
ending relationships on, 217
engaging with diverse people with, 97
job interviews with, 278
learning about, 303
listening in virtual spaces with, 140
plagiarism and, 374–75
See also Mediated communication; specific types
Communicative coherence, 175
Communicators:
competence of, 11–12
self-monitoring by, 13
in transactional communication model, 6
Comparison-contrast statements, 372–73
Competence, 114, 156, 312, 431 (See also Communication competence)
Competition:
culture and, 81
minimizing, in teams, 301
Competitive siblings, 212
Complementary differences, 197
Complementing behavior, 170
Comprehension checking, 395
Compromise, 247
Concepts, speeches about, 386
Conclusion (speech), 366–67, 392
Concrete language, 115
Confidence, 114, 366
Confidentiality, 207
Confirmation bias, 118, 433
Confirming messages, 233–35
Conflict, 227–55
choosing methods for management of, 250
communication climates and, 231–37
cultural approaches to, 251–52
culture and, 83
defined, 229
in family relationships, 213
in groups, 302
negotiation strategies, 245–51
styles of expressing, 237–45
understanding, 228–31

Conflict communication styles, 237–45
  assertiveness, 240–45
  direct aggression, 240
  indirect communication, 238–39
  nonassertiveness, 237–38
  passive aggression, 239–40
Conflict (storming) stage of groups, 299
Conformity, 302–3
Confusion:
  about facts, inferences, and opinions, 118–19
  abstract language and, 117
Connection:
  with diverse people, 100
  social media and, 31, 34, 35
Connection–autonomy dialectic, 220–21
Connection power, 312–13
Connotative meaning, 107
Consensus, 310
Consequence statements, 243
Contact frequency, in friendships, 209
Contempt, 233
Content:
  in active listening, 144
  informative speaking by, 386
  user-generated, 23
Content message, 196
Context:
  of leadership, 314
  for listening, 151
  for nonverbal communication, 164
  in transactional communications model, 7
Contradicting, 172
Control, 196, 232
Conversational narcissists, 142
Convincing, 419–20
Cook, Ethel, 369
Coordinated management of meaning (CMM), 108
Cornell University, 176
Cosmetics dehumanization hypothesis, 188
Counterfeit questions, 153, 154
Country club managers, 315
Cover letters, 265, 266
Covey, Stephen, 131
COVID-19 pandemic:
  fundraising during, 24
  group problem-solving during, 307
  misinformation during, 37
  remote work and, 283
  rise of mediated communication due to, 202
  and use of masspersonal communication, 25
  value of teamwork in, 292
  virtual meetings and, 303
Craft of Research, The (Booth et al.), 355
Credentials, of sources, 425
Credibility:
  defined, 430
  establishing, 366
  persuasive speaking and, 430–33
  of websites, 333
Credibility, of speaker, 156
Critical listening, 156–57
Criticism, 233, 305
Cuddy, Amy, 168
Cultural humility, 97
Culture, 75–103
  of audience members, 327
  conflict and, 251–52
  in diverse groups, 308
  expectations of friends and, 207
  facial expression and, 178
  frame switching and, 69
  intercultural communication competence, 96–100
  listening and, 138–39
  nonverbal communication and, 186–87
  persuasion and, 427
  respecting differences in, 281
  understanding, 76–79
  use of time and, 185
  values and norms in, 79–86
  in workplaces, 279–80
  See also Cocultures; Diversity
Culture shock, 99, 100
Currency, of evidence, 333
Customer satisfaction, 258–59
Cyberbullying, 36–37, 51
Cybersecurity, 303
Cyberstalking, 36–37

Daeley, Spike, 81
Databases, 333, 334
Davis, David, 209–10
Deadnaming, 91
"Deaf Justice" (O'Connor), 375–79
Debilitative communication apprehension, 336–39
Deception:
  as distancing tactic, 234
  lies and evasion, 120
  in nonverbal communication, 172–75
  via social media, 35–36
  See also Honesty
Deception bias, 173
Decker, Benjamin, 100
Deepfakes, 39
Defensive listening, 142–43
Defensiveness, 233, 395
Definitions, 371
Dehumanization, with name calling, 121
Deletion, 343
Delivery (speech), 341–44
  auditory aspects of, 342–44
  types of, 339–40
  visual aspects of, 341–42
Demand–withdraw pattern, 222
Democratic leadership, 314
Demographics, audience, 327–29
Demonstrations, 425
Denial, 222
Denotative meaning, 107
DePaul, Kristi, 277
Derived credibility, 431
Descriptions, 386, 423–25
Detachment, 234
"Detox," from social media, 44
Developmental model (of relation maintenance), 214–18
Devil's advocate, 303
Dewey, John, 308
Diagrams, 396
Dialects, 106, 343
Dialogue, 306–7
DiCaprio, Leonardo, 424
Dickinson, Alexandra, 246
Dictionary.com, 390
Differentiating stage of relationships, 216
Difficult Conversations, 199
Digital communication technology, 5 (*See also* Communication technology)
Digital dirt, 260
Digital infidelity, 204
Directly aggressive messages, 240
Direct persuasion, 420
Direct rewards, 425
Direct speech, 80–81
Disability, 92–94
Disagreement, respectful, 303

Disclosures (*See* Self-disclosure)
Disconfirming messages, 231–33
Discontinuance, 420
Discounting, 234
Discrimination, 121–22
Discussion formats, 305–7
Disfluencies, 178
Disinhibition, 42
Disrespect, 234
Distance, 183–84
Distancing behavior, 234
Distractions, 204
Diversity:
    as asset, in groups, 307
    of audience members, 327
    in brainstorming sessions, 305
    contact with diverse people, 97
    cultural differences in persuasion, 427
    finding online, 202–3
    in friendships, 209–10
    in groups, 308
    See also Culture
Division, rule of, 359
Dobbs, Lou, 398, 400
Doing-oriented friendships, 206
Donnelly, Marisa, 205
Don't Look Up (film), 424
Downward social comparisons, 52
Dozing listening, 138
"Draw back and leap" pattern, 100
Dunbar, Robin, 34
Dunbar's number, 34
Dydadic communication, 8
Dynamism, 432
Dysfunctional team roles, 298

East Eats, 307
Eddo-Lodge, Renni, 89
EEOC (Equal Employment Opportunity Commission), 271
Effort, communication competence and, 13
Eggleston, T. Stephen, 369
eHarmony, 199
EI (emotional intelligence), 62–64, 162
Either-or fallacy, 426
Ekman, Paul, 177
Elderspeak, 95
Elford, Megan, 57
Ellis, Albert, 336
Emails, 217
Emblems, 170
Emergence (norming) stage of groups, 299

Emergencies, planning for, 310
Emergent leaders, 311
Emoticons, 170
Emotion(s):
    in active listening, 144
    basic, 177
    conveyed with nonverbal cues, 167–68
    describing, 242–43
    emotional tone of relationship (*see* Communication climate)
    listening for unexpressed, 146–47
    while talking about race, 89
Emotional appeals, 157, 417, 421
Emotional expressiveness, 252
Emotional Intelligence 2.0 (Bradberry), 63
Emotional intelligence (EI), 62–64, 162
Emotive language, 121
Empathy:
    communication competence and, 12
    in conflicts, 232
    as dimension of emotional intelligence, 62–63
    when dealing with challenges, 293
Empowerment, 316
Encanto (film), 60
Enclothed cognition, 180
Encouragement, 144
Engagement, 312
Entertainment, social media for, 24
Enthusiasm, 432–33
Environments:
    nonverbal communication and, 184–85
    in transactional communications model, 7
    for virtual meetings, 303
    in virtual spaces, 140
Epley, Nicholas, 166
Eportfolio, 264
Equal Employment Opportunity Commission (EEOC), 271
Equivocal language, 115–16
Equivocation, 115–17, 120, 174
Escalatory conflict spirals, 235
Ethical persuasion, 416–17, 428
Ethics:
    adapting to speaking situations, 329
    chart and graph use, 398, 400
    communicating vs. not communicating, 16
    indirect persuasion, 421

    unethical communication behaviors, 417
Ethnicity, 87–89
Ethnocentrism, 98
Ethos, 421
Euphemisms, 93, 116–18
Evasions, 119, 120
Events, speeches about, 386
Evidence, 156, 425
Examples, 371
Expectancy violation theory, 166–67
Expectations:
    about gender roles, 91–92
    of audience, 332
    behavior and, 55
    cultural, 98
    and identity management, 66–67
    organizing speech based on, 429
Experimenting stage of relationships, 214
Expert opinion, 310
Expert power, 312
Explanations, 386, 424, 425
"Exposing Panera" TikTok video, 24
Expressed struggle, 229
Extemporaneous speeches, 340
External noise, 6
Eye contact, 165, 175–77, 186, 342

Fabrications, 120
Face, 66, 303–4
Facebook, 34, 290
Facebook Addiction Disorder (FAD), 34–35
Face-to-face communication:
    with diverse people, 97
    listening during, 141–42
    mediated and, 27–30
    meeting online dates, 216
    time spent in, 42–43
Facework, 66
Facial expressions, 165, 168–69, 177, 188–89, 342
Facilitative communication apprehension, 335
Fact(s):
    citing startling, 364
    inferences, opinions, and, 118–19
    propositions of, 418–19
Fact checkers, 38
Factual examples, 371
Factual statements, 118

FAD (Facebook Addiction Disorder), 34–35
Fallacies:
  with communication apprehension, 336–37
  in persuasive speaking, 426–28
Fallacy of ad hominem, 121
Fallacy of approval, 337
Fallacy of catastrophic failure, 337
Fallacy of overgeneralization, 337
Fallacy of perfection, 337
False cause fallacy, 426, 427
False information, 417
Familiarity:
  capturing attention with familiar references, 364
  gravitating to, 61
  in informative speeches, 389
  and organization of speech, 361
Family, 211
Family relationships, 211–13
Farr, Christina, 22, 45
Faulty analogies, 428
Fears, rational vs. irrational, 338
Feedback, 134, 291
Feelings (*See* Emotions)
Feldman, Robert, 174
Ferraz, Christina, 162
Fey, Tina, 77
Feynman, Richard, 154
Feynman Technique, 154
Fidgeting, 175, 176
Figure-ground principle, 57
First impressions, 59–60, 198
Fisher, Roger, 247
Flaa, Jennifer, 216
Flip pads, 400
Flow charts, 398, 399
Follower, 290–93
Follow up:
  after interviews, 276
  in group problem solving, 310
Forcefield analysis, 309
Formal outlines, 355–56
Formal roles, 297
Forming stage of groups, 299
Four Horsemen of the Apocalypse, 233
Frame switching, 69
#fridaysforfuture, 33
Friends (television program), 205
Friendship, 205–10
  diversity in, 97
  listening and, 131
  making, 210
  mediated communication and, 28, 31, 34
Fuller, Mark, 138
Full-sentence outlines, 356
Functionality, of websites, 333
Fundamental attribution error, 61
Fundraising, 33

Galileo Galilei, 427
Gallo, Amy, 311
Gandhi, Mahatma, 50, 311
Gardner, John E., 5
Gatekeepers, 22, 280
Gates, Bill, 311, 372
Gender:
  of audience members, 328
  clothing and, 181
  as coculture, 90–92
  listening and, 135–36
  mediated communication and, 35–36
  nonverbal communication and, 187–89
  references to, in language, 110–11
Gender nonconforming people, 188
Genderqueer, 90
Gender roles, 91–92
Generalizations, 428
General purpose, 325
Generation:
  as coculture, 95–96
  listening and, 138–39
  mediated communication use and, 40–41
Generation Xers, 96, 138–39
Generation Zers, 138–39
Gershman, Sarah, 140
Gestures, 247
Gieseck, Alyssa, 373
Gifts, 218
Goals:
  focus on shared, 301
  group, 295
  individual, 295
  perceived incompatibility of, 229
  personal, 54, 57, 66
  setting realistic, 56
Goffman, Erving, 66
Gofundme, 33
Goldilocks principle, 201
Goleman, Daniel, 62
Good, inherent, of communication, 15

Google, 279, 294, 297, 332, 374
Google Scholar, 333
Gossip, 282
Gottman, John, 233
Graham, Katharine, 365, 367
Grant, Adam, 112, 114, 384, 387, 390, 392, 393, 396, 402–10
Graphs, 398, 400
Greenwell, Lucy, 10
Griffin, Shaquem, 93
Grimes, 80
Group(s), 290–310
  audience membership in, 328
  communicating as a follower, 290–93
  communicating in, 294–98
  individual roles in, 297–98
  interaction in, 298–307
  leadership in (*see* Leadership)
  motivational factors in, 295
  multicultural, 308
  power distributed in, 313
  problem solving in, 307–10
  reaching decisions in, 310
  rules and norms in, 295–97
Group goals, 295
Group problem solving, 307–10
  advantages of, 307
  decision-making methods in, 310
  structured approach to, 308–10
Groupthink, 302
Guiltmakers, 239

Haas, Susan Biali, 56
Hahn, Melissa, 83
Haidt, Jonathan, 421
Half, Robert, 258
Hall, Edward T., 183, 184, 186
Hallum, Elizabeth, 362
Halo effect, 61
Hambleton, Julie, 10
Hamilton, Angel (zilianOP), 27
Han, Ed, 261
Hancox, Lewis, 90–91
Handouts, 400
Haptics, 166, 182
Harassment, 36–37, 238–39
Hardcastle, Hailey, 324, 325, 344–47
Harris, Jillian, 223
Harris, Kamala, 311
Harvard Business Review, 113
Hasty generalizations, 428

Hay, Louise, 56
Hearing, 132
Hegemony, 98
Heitler, Susan, 121
Hidden agendas, 153
Hidden area (Johari Window), 201
High-context communication, 84
High-context cultures, 82–83, 251–52
High-disclosure friendships, 206
High-obligation friendships, 207
High self-monitors, 69–70
Hill, Nathan, 370
Hilton, 291
Hinting, 120
Hirsch, Becca, 92
Hoge, Rachel, 58
Homesickness, 100
Honesty, 36, 67–68, 232, 267, 431
(*See also* Deception; Lies)
Hopkins, Tom, 130
Horns effect, 61
Hostile siblings, 213
Hostility, neutralizing, 429–30
Humblebragging, 67
Humility:
    balancing competence with, 114
    culture and, 81
Humor:
    capturing attention, 364, 365
    in microresistance, 123
    personalizing speeches with, 393
    in speeches, 332
    in workplace, 281–82
Humoring, 234
Hussein, Yasmin, 92
Hyperpersonal communication, 28
Hypothetical examples, 371

Ice Bucket challenge, 33
Identity(-ies):
    mistaken, 260–61
    multiple, 68
    presenting professional, 278
Identity management, 65–71
    characteristics of, 68–71
    and honesty, 67–68
    with mediated communication, 28, 259–61
    with nonverbal cues, 168
    public and private selves in, 65–66
    purposes of, 66–67
Identity needs, 389

Illustrators, 170
Imaginary audiences, 96
Immediacy, 196
Impartiality, 156, 431
Impersonal demeanor, 234
Implicit bias, 123, 124
Impoverished managers, 314–15
Impromptu speeches, 340
Inappropriate content, posts with, 29, 43
Inattention, 234
Incremental persuasion, 415–16
Indirect communication, 238–39
Indirect persuasion, 420–21
Indirect speech, 80–81
Individual goals, 295
Individualistic cultures, 80–82
    assertive and collaborative language in, 114
    conflict communication in, 251
Individuality, focusing on, 59
Individual responsibilities, 310
Inferences, 118–19
Inferential statements, 118
Inflammatory language, 119, 121
Informal roles, 297
Information:
    evaluating, 38–39, 95
    gathering, for speech, 332–35
    for group problem solving, 309
    knowledge vs., 384
    listening for, 154–55
Informational interviews, 262–63
Information anxiety, 384, 389
Information hunger, 389
Information overload, 302, 384, 385, 389
Information underload, 302
Informative purpose statement, 388–89
Informative speaking, 383–411
    indirect persuasion with, 420–21
    persuasive speaking vs., 387
    sample speech, 402–10
    supporting material for, 392–402
    techniques of, 388–92
    types of, 386–87
In-group, 77, 79
Initial credibility, 431
Initiating stage of relationships, 214
Inner critics, 56
Insensitive listening, 141
Inside Out (film), 177

Insincerity, 417
Instagram, 182
Instructions, 386–87
Insulated listening, 142
Integrating stage of relationships, 215
Integrity, 329
Intensifying stage of relationships, 214–15
Intention statements, 243
Interaction:
    in groups, 294, 298–307
    keeping records of, 269
    patterns of, 280
    persuasion and, 416
    with relational partners, 198
    shortening of, 234
    on social media, 23
Intercultural communication competence, 96–100
Interdependence, 229, 295, 301
Interest(s):
    focusing on, in negotiations, 247
    supporting material to add, 370
Intergroup contact hypothesis, 209
Internal motivation, 62
Internet:
    evaluation of websites, 333
    social isolation with overuse of, 35
    speech research on, 332–33
    See also Communication technology; Mediated communication
Interpersonal communication, 8–9, 193–225
    characteristics of, 194–96
    choosing relational partners for, 197–200
    communication climates in (*see* Communication climates)
    conflict in (*see* Conflict)
    defined, 194
    in family relationships, 211–13
    in friendships, 205–10
    and masspersonal communication, 24, 25, 27
    online, 202–5
    and relational dialectics, 220–23
    in romantic relationships, 213–20
    self-disclosure and, 200–203
    self-disclosure in (*see* Self-disclosure)
Interpretation, in perception process, 57–58
Intersectionality theory, 87

Interviews:
  behavioral, 274
  gathering information for speeches with, 334
  informational, 262-63
  job (See Job interviews)
Intimacy, 28, 35
Intimate Behavior (Morris), 220
Intimate distance, 183
Intrapersonal communication, 8
Introduction (speech), 364-66, 391
Inverted pyramid format, 369
Irrational fears, 338
Irrational thinking, 336-37
Isolation, 35, 79
It's Not About the Nail (video), 150

Jackson, Jesse, Sr., 373
James, William, 56
Jargon, 116, 117
Jeffries, Mark, 281
Job applicants, 263-69
Job interview presentations, 272
Job interviews, 269-78
  humblebragging in, 67
  lies in, 174
  participating in, 272-77
  by phone or video, 278
  preparing for, 269-72
Jobs, resigning, 316, 318
Jobs, Steve, 30
Johari Window, 201
Johnson, Dennis, 303
Johnson, Kara, 59
Jokers, 240
Jokes (See Humor)
Jonas, Joe, 30
Jonas, Nick, 30
Joyce, Susan P., 276
Judging responses, 149
Judgments, 149, 232

Kailey, Matt, 59
Kellam, Jim, 259
Kelley, Robert E., 290
Keynote (software), 400
Kim, Young Yum, 100
Kinesics, 175
King, Gayle, 12
KKK (Ku Klux Klan), 209-10
Knapp, Mark, 214
Knowledge:
  information vs., 384
  in intercultural communication competence, 99
Koehler, Carol, 362
Konnikova, Maria, 42
Kordei, Normani, 135
Korte, Kim, 165
Kotelnikov, Vadim, 369
Kothari, Pooja, 123
Ku Klux Klan (KKK), 209-10

Ladau, Emily, 93
Laissez-faire leadership, 314
Lame, Khaby, 162, 177, 189
Language, 105-27
  defined, 106
  detecting deception in, 174
  in informative speaking, 390, 391
  misunderstandings and, 114, 116-17
  nature of, 106-10
  power of, 110-14
  troublesome, 117-24
Latitudes of acceptance, 415-16
Latitudes of noncommitment, 415-16
Latitudes of rejection, 415-16
Law, for job interviews, 271
Lawrence, Jennifer, 336, 424
Leaders:
  behaving like, 70
  communication and, 259
  emergent, 311
  nominal, 312
Leadership, 311-17
  approaches to, 313-14
  demonstrating potential for, 312
  as learned skill, 311-12
  listening and, 131
  power and, 312-13
  self-assessment, 317
  situational, 314-15
  trait theories of, 311-12
  transformational, 315-16
Leading questions, 153
Lean In (Sandberg), 91-92
Leanness, message, 27-28
Leaving jobs, 316, 318
Legitimate power, 312
Leonard, Rachel, 54
Levine, Robert, 185
Lewis, Chanel, 130, 135, 143, 157
LexisNexis, 333
Library research, 333-34
Lies, 119-20, 174 (See also Deception; Honesty)

Lie to me* (television program), 172-73
Likability, 432, 433
Lincoln, Abraham, 42, 337
Line charts, 398, 399
Linguistic relativism, 110
LinkedIn, 32, 260, 263
Lionel Cantú Queer Center, 91
Listening, 129-59
  analytical, 154-56
  critical, 156-57
  defined, 132
  to detect deception, 174-75
  to difficult people, 292
  during discussions about race, 89
  in family relationships, 213
  faulty habits in, 140-44
  in friendships, 207
  to informative speeches, 389-90
  misconceptions about, 132-37
  as natural process, 134-35
  overcoming challenges to effective, 137-40
  radical, 130
  relational, 144-51
  social, 136-37
  supportive, 135-36, 144-51
  task-oriented, 134, 151-54
  value of, 130-32
  in virtual spaces, 139-40
Listening Ears, 144-45
Listening fidelity, 132
Livestreaming, 25, 27
Location, interview, 272
Lodhia, Devarshi, 94
Logos (logical reasoning), 422
Long-distance romantic relationships, 30
Longing siblings, 212
Long-term friendships, 206
Lose-lose problem solving, 246
Lovato, Demi, 106, 111
Love languages, 218-20
Low-context communication, 84
Low-context cultures, 82-83, 251-52
Low-disclosure friendships, 206
Lowe, Sean, 194
Low-obligation friendships, 207
Low self-monitors, 69-70
Loyalty, 207

Mahoney, Brianna, 330-31
Main points of speech, 359-60, 363, 365

Majority control, 310
Makeup, 188
Making of a Manager, The (Zhuo), 290
Malami, Saeed, 414, 422, 433–37
Managerial grid, 314–15
Mandela, Nelson, 314
Manipulators, 176
Manners, 273–74, 280–83
Manry, Bo, 10
Manuscript speeches, 340
Marital satisfaction, 35
Markman, Art, 305
Martin, Noelle, 37
Mass communication, 10, 22–27
Mass media, 22–27
Masspersonal communication, 24, 25, 27
McCarley, Evan, 374
McElvey, Amanda, 54
McKibban, Allison, 334
Mean Girls (musical), 77
Meaning(s):
　co-created, 7
　in communication, 16
　in language, 107–8
Media, 22–27
Mediated communication:
　conflict management and, 230
　face-to-face vs., 27–30
　influences on, 40–41
　interpersonal relationships in, 202–5
　job interviews with, 278
　networking with, 261–62
　preference for, 35
　time spent on, 42–43
　See also Communication technology; *specific types*
Meetings, 301–7
Mehta, Aparna, 370
Meltzer, Bernand, 228
Memorable points, supporting materials for, 370
Memorized speeches, 340
Mental health, social media use and, 34–35
Message(s):
　conveyed via channels, 5
　for persuasive speaking, 422–28
　richness of, 27–29
　send nonverbally, 163
　separating speaker from, 155
　in transactional communication model, 6

Message overload, 137
Meta, 39
Meta, John, 89
Metacommunication, 199–200
Metaphors, 372
#MeToo movement, 183
Meyers, Seth, 216
Microaggressions, 121–23
Microaggressive language, 121–22
Microresistance, 122–23
Middle-of-the-road managers, 315
Millennials, 96
Minari (film), 122
Mindful courage, 89
Mindful listening, 131, 135
Minions, 107
Miranda, Lin-Manuel, 311
Mirroring, 176
Misinformation, on social media, 37–39, 44
Misrepresentation, 36
Mission, 316
Mistaken identities, 260–61
Misunderstandings, 114, 116–17, 150
Mnemonics, 362
Models, as visual aids, 395
Moderation strategy, 222
Molinsky, Andy, 83
Monetary compensation, 277
Monochronic, 185
Monroe, Alan, 363
Mood, 166, 365
Morale, 314
Morning Show, The (television program), 9
Morris, Desmond, 220
Motivated Sequence, 363, 364
Motivation, 62, 295
Mouton, Jane, 314, 315
Mull, Amanda, 182
Multicultural teams, 308
Murphy's Law, 310
Musk, Elon, 80
Musk, Exa Dark Sideræl, 80
Musk, X Æ A-12, 80
Muslims, 92
Mute, in videoconferencing, 304
Myatt, Matt, 259
My Fair Lady (musical), 112

Nadal, Kevin, 122, 123
Name calling, 121

Names(s):
　culture and, 80
　language and, 111–12
Narration, 374
Nary, Dot, 93
NASA, 302
Nath Finanças, 258
National Basketball Association, 183
National Science Foundation, 384
Navarro, Joe, 164, 173
NBC, as news source, 39
Needs, 248–49, 389
Negative experience, previous, 336
Negativity bias, 60
Negotiation strategies, 245–51
　compromise, 247
　for job offers, 277
　lose–lose problem solving, 246
　for use with bullies, 246
　win–lose problem solving, 245–46
　win–win problem solving, 247–51
Nelson, Shasta, 96
Nervous energy, 338
Networking, 261–63
Never Have I Ever (television program), 68
Noise, 6–7, 137–38
Nominal leaders, 312
Nonassertion, 237–38
Nonbinary, 90, 106
Noncommitment, latitude of, 415–16
Nonimmediacy, 234
Nonverbal communication, 161–91
　in active listening, 144
　asking for help with, 247
　characteristics of, 163–69
　defined, 162–63
　demonstrating charisma with, 432–33
　functions of, 169–75
　influences on, 186–89
　and message richness, 27–29
　types of, 175–86
Nonverbal learning disorder (NVLD), 169
Norming stage of groups, 299
Norms:
　cultural, 79–86
　in groups, 295–97
Notes for speaking, 352–54, 356–57, 359
Note-taking, 140, 154
#NotInMyName, 92
#NuggsForCarter, 136–37

Number charts, 396, 397
NVLD (nonverbal learning disorder), 169

Objectivity, 333
Objects, 386, 395
Observation, passive, 99
Occasion, for speech, 331–32, 364
O'Connor, Dan, 352–55, 375–79
Ody, Britton, 328
O'Keefe, Stephen, 132, 134, 135, 153
Oluo, Ijeoma, 89
Omissions, 120
Onion, The, 38
Online classes, 230
Online communication (*See* Mediated communication)
Online dating, 30–31, 215, 216
Online research, 332–33
Online support groups, 32–33
Online surveillance, 36
Online teams, 230
Open area (Johari Window), 201
Open-mindedness:
  to different opinions, 95
  in intercultural communication competence, 98–99
  in multicultural teams, 308
Openness, 198, 221
Openness–privacy dialectic, 221
Open office plans, 185
Opinion(s), 364
Opinions:
  expert, 310
  framed as questions, 153
  inferences, facts, and, 118–19
Opinion statements, 118
Organization:
  of business presentations, 369
  of informative speeches, 391–92
  of outlines, 360–64
  in perception process, 57
  of persuasive speech, 429
Organizational communication, 9
Organizational communication networks, 280
Organizational culture, 279–80
Orientation (forming) stage of groups, 299
Originality, 374–75
Osaka, Naomi, 88
Outcome categorization of persuasion, 419–20

Out-group, 77, 79
Outlines, 359–64
  formal, 355–56
  full-sentence, 356
  organization of, 360–64
  principles of, 359–60
  samples, 356, 358, 375–76, 434
  standard format for, 359
  standard symbols used in, 359
  working, 354–55
Overcoming Hate Through Dialogue (Cekic), 241
Overcommunicating, 16
Over-correction, 291
Overgeneralization, 61, 337
Oversharing, 282

Palmisano, Katherine, 182
Paralanguage, 178, 186
Parallel wording, rule of, 360
Paraphrasing:
  of audience questions, 394
  in negotiations, 249
  in supportive listening, 146–47
  in task-oriented listening, 152, 154
Parents and parenting relationships, 211
Parrish, Jenny, 65
Participation, in meetings, 301
Passive observation, 99
Pathos, 421–22
Patience:
  in intercultural communication competence, 99
  in overcoming culture shock, 100
Pauses, in online communication, 204
Perceived self, 65
Perception, 56–65
  and emotional intelligence, 62–64
  and perception checking, 35, 63
  stages of, 56–58
  tendencies in, 58–61
Perception checking, 35, 63
Perception management, 175
Perfection, 337
Performance norms, 296, 297
Performing stage of groups, 299–300
Permanence, of communication, 5, 29–30
Permissive parents, 211
Perry, Tyler, 424
Perseverance, 99
Personal attacks, 121

Personal communications, at work, 282–83
Personal distance, 183–84
Personal experience, 336, 366, 431
Personal fables, 96
Personal factors, in listening, 145–46
Personal identity (*See* Identity management)
Personality, 52
Personalization, 393
Personal relationships (*See* Relationship(s))
Persuasion, 414–18
Persuasive appeals, types of, 421–22
Persuasive speaking, 413–40
  adapting, to audience, 428–30
  categorizing types of, 418–22
  characteristics of, 414–18
  creating the message, 422–28
  credibility in, 430–33
  informative speaking vs., 387
  patterns in, 362
  sample speech, 433–37
Petrik, Andrei, 261
Phone interviews, 278
Phonological rules, 108
Photos (visual aids), 395
Phubbing, 42–43, 204
Physical attractiveness, 179–80
Physical needs, 389
Physical noise, 138
Physical touch (*See* Touch)
Physiological noise, 6, 138
Pictograms, 397
Pie charts, 397
Pink slime journalism, 39
Pisano, Vincent, 59
Pitch, of voice, 179
Pitch, voice, 343
Place, for speech, 331–32
Placke, Scott, 433
Plagiarism, 374–75, 417
Polarization strategy, 222
Policy, propositions of, 419
Political viewpoints, 94–95
Polychronic, 185
Polymediation, 28
Polymer marking surfaces, 400
Ponton, Rod, 140
Pope, Alexander, 373
Porter, Richard, 427
Portfolio, work, 264, 273
Positive spirals, 236

Poster boards, 400
Post hoc fallacy, 426, 427
Posture, 165, 175–76, 304, 341–42
Poundstone, Paula, 142
Power:
   of language, 110–14
   leadership and, 312–13
   in win–lose problem solving, 245
Power distance, 84–85
Powerful language, 267
PowerPoint, 400, 401
Practical needs, 389
Practice, for speech, 340
Pragmatic rules, 108, 110
Predictability–novelty dialectic, 221–22
Preference for online social interaction, 35
Prejudice, 97
Preliminary notes, for speaking, 352–54
Preparation, for speech, 336, 338, 339, 366
Presence, 10
Presentation (speech), 339–40
Presentations:
   for job interviews, 272
   organizing, 369
Presentation software, 400, 401
Presenting self, 65
Previous negative experience, 336
Prezi, 400
Priorities, 54
Privacy, 43, 213, 221
Private self, 65–66
Proactivity, 291
Problem(s):
   addressing, as a follower, 291–92
   analyzing, 309
   communication limits in solving, 15
   demonstrating, 423–24
   describing, 423
   identifying, 248, 309
Problem census, 306
Problem-solution patterns, 362–64
Problem solving:
   metacommunication for, 199–200
   mutually beneficial, 232
   negotiation strategies for, 245–51
   See also Group problem solving
Procedural norms, 296, 297
Processes, 4, 360, 386
Professional environments, 277, 279–85
Professional identity, 260

Professional networks, 261–63
Professional social media, 32
Project Aristotle, 294, 297
Prompting, 147
Pronouns:
   in communications about gender, 91
   culture and, 80
   power of language and, 106
Proof, 369, 371
Propositions, 418–19
Propositions of fact, 418–19
Propositions of policy, 419
Propositions of value, 419
Proxemics, 183
Psuedoaccomodators, 239
Psuedolistening, 141
Psychological noise, 7, 137–38
Psychological safety, 297
Public awareness, building, 258
Public communication, 9–10, 323–49, 351–81
   analyzing the situation, 327–32
   building, 352–59
   choosing the topic, 325
   communication apprehension, 335–39
   conclusion, 366–67
   defining the purpose, 325–26
   delivery guidelines, 341–44
   gathering information for, 332–35
   introduction, 364–66
   outlining, 359–64
   presenting, 339–40
   sample speeches, 344–47, 375–79, 402–10, 433–37
   stating the thesis, 326–27
   supporting material for, 368–75
   transitions in, 368
   See also Informative speaking; Persuasive speaking
Public distance, 184
Public self, 65–66
Punctuation, 41
Purpose:
   audience, 327
   of informative speaking, 386–89
   of persuasive speaking, 387
   of public communication, 325–26
Purpose statement, 325–26, 388–89, 422

Quality time, 218
Question(s):
   asking open-ended, 95

   for audience, 364
   counterfeit, 153, 154
   to detect deception, 174
   in group problem solving, 309
   to illicit feedback, 291
   in job interviews, 270–71, 274–76
   leading, 153
   sincere, 153, 154
   in supportive listening, 146
   tag, 344
   in task-oriented listening, 152, 153
Question-and-answer period, 394–95
Quiet: The Power of Introverts (Cain), 113
Quotations, 364, 369, 370, 373

Race and racism:
   coculture and, 87–89
   defining, 76
   discussions about, 89
   friendship as antidote to, 209–10
   microaggressions and, 121–22
   on social media, 88
Radical listening, 130
Raine, Charlie, 94
Ramirez, Brianna, 24, 25, 136
Ratatouille (film), 24
"Ratatouille: The TikTok Musical," 24
Rate, speech, 336, 343
Rational fears, 338
Rayome, Alison DeNisco, 140
RBF (resting blank face), 188–89
Realistic purpose statements, 325
Reasoning, examining, 156
Receivers, 338
Recognition, 234–35
Red herrings, 428
Redirection, 271
Reductio ad absurdum fallacy, 426
Referent power, 313
Referrals, 262
Reflected appraisal, 52
Reflecting, 144, 147–48
Reflective thinking method, 308–9
Reframing, 222, 273
Regionalisms, 116
Regulating, 170, 172
Reich, Robert, 329, 429
Reinforcement (performing) stage of groups, 299–300
Rejection, latitude of, 415–16
Relational dialectics, 220–23
Relational intimacy, 28, 35
Relational listening, 144–51

Relational messages, 196
Relational partners, choosing, 197–200
Relational spirals, 235
Relationship(s):
   building, in teams, 301
   cultivating strong, 281
   listening and, 131
   mediated communication competence and, 42–43
   nonverbal cues effect on, 168–69
   references to, 364
   social media use and, 24, 30–35
   touch in, 183
Relative words, 116
Reliability, of visual aids, 402
Religion, 92
Remembering, 134
Remote work, 139, 283–85, 303
Repeating, 170
Repetition, 392–93
Reputation, 259
ReputationDefender.com, 260
Research, 38, 269–70, 277, 332–35
Reserve, 234
Residual messages, 134
Resigning, 316, 318
Resources:
   in groups, 307
   perceived as scarce, 230–31
   for solutions, 310
Respect, 196, 303
Responding:
   defined, 134
   describing desired response, 425
   during remote work, 284
Resting blank face (RBF), 188–89
Restraining forces, 309
Restraint, 234
Restriction of topics, 234
Result orientation, 325–26
Resumes, 265, 267–69, 273
Reuters, 39
Revenge porn, 37
Reverse image searches, 38–39
Reward power, 313
Rewards, 425
Rhetoric (Aristotle), 421
Rhetorical Triad, 421–22
Richardson, Danny, 79
Richness, message, 27–29
Ridgeway, Erin "Big Debo," 79
Riggio, Melissa, 92–94
Riley, Connor, 11

Robbins, Mel, 56
Roberts, Julia, 169
Robots, 184
Rockwell, Dan, 291
Rodrigues, Nathália, 258
Rodriguez, George, 432
Roles, in groups, 297–98
Romantic relationships, 213–20
   abusive partners, 241
   digital infidelity in, 204
   love languages and, 218–20
   stages of, 214–18
Round robins, 306
Rule(s):
   in groups, 295–97
   of language, 108, 110
Rule of division, 359
Rule of parallel wording, 360
Rule of seven, 401

Saban, Nick, 131
Safety:
   cybersecurity, 303
   psychological, 297
   on social media, 43–45
Safeway, 166
Sales plan development, 432
Salience, 76–77
Samovar, Larry A., 427
Samson, Jaclyn, 79
Sandberg, Sheryl, 91–92
Sanfilippo, Barbara, 432
Santos-Díaz, Stephanie, 59
Sapir-Whorf hypothesis, 110
Schadenfreude, 178
School Strike 4 Climate, 33
Schulman, Nev, 35
Schulz, Kathryn, 394
Scuderi, Royale, 56
Secondary traumatic stress, 145
Selection, in perception process, 56–57
Selection interviews (*See* Job interviews)
Selection strategy, 222
Selective listening, 141
Self-assessment:
   Are You Overloaded?, 385
   How Assertive Are You?, 244–45
   How Do You Use Language?, 109
   How Emotionally Intelligent Are You?, 64
   How Good a Follower Are You?, 290
   How Much Do You Know About Other Cultures?, 78

   How Worldly Are Your Nonverbal Communication Skills?, 171
   Main Points and Subpoints, 363
   Persuasive Speech, 430
   Speech Anxiety Symptoms, 339
   What Are Your Listening Strengths?, 133
   What Is Your Love Language?, 219
   What Kind of Friendship Do You Have?, 208
   What's the Forecast for Your Communication Climate?, 236
   What's Your Leadership Style?, 317
   What Type of Communicator Are You?, 14
   What Type of Social Media Communicator Are You?, 26–27
Self-awareness, 62
Self-centeredness, 142
Self-concept, 50–51
Self-defeating thinking, 56
Self-disclosure, 200–203
   with diverse people, 99
   in friendships, 206, 207
   via mediated communication, 28
   at work, 282
Self-esteem, 35, 51–52
Self-fulfilling prophecies, 53–56, 336, 338
Self-monitoring, 13, 69–71
Self-promotion, 81
Self-protection, on social media, 43–45
Self-regulation, 62
Self-serving bias, 61
Self-serving lies, 120
Semantic rules, 108
Servant leadership, 314
Service learning, 263
Seven, rule of, 401
7 Habits of Highly Effective People, The (Covey), 131
Sex, 90–92
Sexual harassment, 238–39
Shores, Mary, 259
Shortening of interactions, 234
Short-term friendships, 206
Siblings, 211–13
Sign language, 107
Signposts, 393
Silence, 16, 86
Similes, 372
Simplicity, 16, 389–91, 401
Sincere questions, 153, 154

Sincerity, 329
Situational factors:
   in communication competence, 11–12
   for deception, 174
   high-context cultures and, 82
   in relational and supportive listening, 145–46
Situational leadership, 314–15
Size:
   of audience, 23, 328
   of small groups, 295
   of visual aids, 401
Slackers (*See* Social loafing)
Slang, 116, 117
Slurring, 344
Slurs, 121
Small-group communication, 9
Small groups, 294–95, 301 (*See also* Group(s))
Smiling, 178
Smith, Angela, 271
Smith, Hanley, 91
Snapchat, 29–30
Snoop Dogg, 24, 136
Social comparison, 52–53
Social connection, 4
Social distance, 184
Social exchange theory, 198
Social expectations, 91–92
Social isolation, 35
Socialization, 279
Social judgment theory, 415–16
Social listening, 136–37, 141–42
Social loafing, 295
Social media, 21–47
   benefits of, 30–33
   characteristics of, 23–24
   communication competence with, 41–45
   communications on, 10–11
   defined, 22–23
   discussing politics on, 95
   drawbacks of, 22, 34–39
   engaging with diverse people on, 97
   face-to-face communication vs., 27–30
   as glamorized version of reality, 53, 54, 56
   healthy relationship with, 10
   influences on, 40–41
   limiting time on, 44–45
   listening and, 136–37
   message overload and, 137
   as news source, 39
   personal attacks on, 121
   plagiarism and, 374
   power of language on, 106
   racism on, 88
   role of mass media and, 22–27
   types of content, 25
   uses of, 24
   validation on, 204
   See also Mediated communication
Social media snarks, 94, 95
Social media trolls, 94, 95
Social needs, 389
Social norms, 296
Social penetration model, 200–201
Social roles, 298
Social skills, 35, 63
Social support, 32–33
Software, presentation, 400, 401
Solution(s):
   advantages of, 424
   demonstrating, 425
   describing, 424
   developing, 309–10
   following up on, 310
   negotiating, 249–50
   in problem-solution patterns, 362
   providing, 312
Sound bites, 395
Sources, of information, 38, 39
So You Want to Talk About Race (Oluo), 89
Space, 186
Space patterns, 361
Speaking notes, 356–57, 359
Specificity, 117
Specific purpose, 325, 326
Speech anxiety (*see* Communication apprehension)
Speeches (*See* Public communication)
Spokespersons, minorities as, 123
Stage fright (*see* Communication apprehension)
Stagnating stage of relationships, 217
Stalking, via social media, 36–37
Statistics, 371–72, 397
Statue of Liberty, 419
Status, 302–3
Stereotypes:
   friendships and, 209
   gender-related, 40
   and perception, 58–60
Stereotyping, 327
Stevens A., Rebecca, 87
Stewart, Kristen, 188
Stimuli, intensity of, 57
Stonewalling, 233
StopNCII.org, 37
Storming stage of groups, 299
Straw man arguments, 428
Strengths, personal, 260
Structure (speech), 355
   basic, 355
   of informative speeches, 391–92
   of persuasive speeches, 423–25
Subpoints of speech, 359, 360, 363
Substituting, 170
Substitution, 344
Sub-subpoints of speech, 359
Success criteria, 309
Sue, Derald Wing, 121, 123
Superficial relationships, 34
Supporting forces, 309
Supporting material, 368–75
   alternative media for presenting, 400
   emphasizing important points with, 392–93
   functions of, 368–70
   generating audience involvement in, 393–95
   for informative speeches, 392–402
   and plagiarism vs. originality, 374–75
   presentation software for, 400
   styles of, 373–74
   types of, 371–73
   visual aids, 395–400
Supportive followers, 291
Supportive listening, 135–36, 144–51
Supportive siblings, 212
Suppression, 417
Surveillance, online, 36
Survey research, 334–35
Swearing, 110
Sweetman, Kate, 84–85
Symbols and symbolism, 4–5, 106–7, 359
Synchronous communication, 29
Syntactic rules, 108

Tag questions, 344
Talking, 86
Talley, Lloyd M., 307
Tan, Tiffany, 56
Tannen, Deborah, 150
Target audience, 428–29

Task goals, 309
Task-oriented listening, 134, 151–54
Task roles, 297, 298
Tattoos, 181–82
Taylor, Aisha, 141–42
Taylor, Dalmas, 200
Team managers, 315
Team members, difficult, 300
Teams:
    communicating in, 294–98
    development of, 299–300
    multicultural, 308
    online, 230
    See also Group(s)
Teamwork, 258
    in COVID-19 pandemic, 292
    as valued skill, 294
Technology (*See* Communication technology)
TED Talks, 130, 241, 332, 344, 384, 402
Te'O, Manti, 36, 44
Terminal credibility, 431
Terminating stage of relationships, 217
Territoriality, 184
Testimony, 373
Text messaging, 29, 35, 217
Theodoridis, Alexander, 121
Thesis statement, 326–27, 388–89
Thoughts, unexpressed, 146–47
Thunberg, Greta, 33, 311
TikTok, 23–24, 162
Time:
    listening and, 145
    mediated vs. face-to-face communication and, 42–43
    for multicultural teams, 308
    needed for groups, 295
    nonverbal communication and, 185–86
    as perceived scarce resource, 231
    for speech, 331
Time patterns, 360–61
Tinder, 30
Tinder trap, 179
Ting, 135
Toastmasters International, 369
Tone, of speech, 166, 365
Topic, 234, 301, 325, 365–66, 387
Topic patterns, 361–62
Touch, 166, 182–83, 197, 218–19
Trait theories of leadership, 311
Transactional communication model, 5–7
Transformational leadership, 315–16

Transgender, 90
Transitions, 368
Transmediation, 28–29
Transparency, 316
Transportations Security Administration, 173
Triple delivery, 401
Triumph, Tim, 70
Trivial tyrannizers, 240
Trolling, 42
Trump, Donald, 340, 398, 400
Trust:
    in friendships, 207, 209–10
    in online classes and teams, 230
Truth bias, 173
Tucker, Anna Claire, 366
Tucker, Jeffrey, 210
Tufte, Edward R., 401
Tuiasosopo, Ronaiah, 36
Turkle, Sherry, 43
Turner, Sophie, 30
Twardowski, Barbara, 94
Twitter, 29, 107

Uncensored (Wood), 154
Uncertainty avoidance, 83–84
Undercover Boss (television program), 70
Understanding, 15, 42, 132, 210, 249
Undivided attention, 42–43
Unexpressed thoughts, 146–47
University of Colorado at Colorado Springs, 357
Unknown area (Johari Window), 201
Unmet needs, 248–49
Uptalk, 187
Upward social comparisons, 52–53
URLs, providing, 265
Ury, William, 247
User-generated content, 23
Uses and gratifications theory, 24

Validation, 204, 207
Value(s):
    of audience members, 329–31
    cultural, 79–86
    defined, 330
    propositions of, 419
Vasadi, Edit, 76, 82, 96–97, 99
Vasilogambros, Matt, 89
Venting, 150
Verbal communication, 162–63
Videoconferencing platforms, 303
Video interviews, 278

Videos, as visual aids, 395–96
Video sharing services, 24
Viral, going, 24, 25
Virtual communities or groups, 31
Virtual meetings, 303–5
Visual aids, 395–402
    defined, 395
    ethical use of, 398, 400
    rules for using, 401, 402
    software/media for presenting, 400
    speaking notes as, 359
    types of, 395–98
Visualization, 338
Vocal fry, 187
Voice, 166, 178–79
Volume, voice, 343
Volunteerism, 263
Volunteers, 394

Wall Street Journal, The, 39
Walther, Joseph, 28
Wansink, Brian, 176
Ward, Mariellen, 98–99
Warning signs, groupthink, 302
Weakness, personal, 270
Websites, evaluating, 333
"Webs of hate," 121
Webster, Daniel, 343
Weinstein, Bob, 315
Wendy's, 136–37
"What Frogs in Hot Water Can Teach Us About Thinking Again" (Grant), 402–10
Wheel networks, 280
Whistleblowers.gov, 292
Whiteboards, 400
White House Coronavirus Task Force, 398
White lies, 120, 174
"Why Videos Go Viral" (Allocca), 391–92
Wideman, Stephanie, 371
Widisto, Aushaf, 39
Wikipedia, 333, 374
Wilkerson, Carter, 136–37
Williams, Paula Stone, 90
Williams, Reagan, 371
Winfrey, Oprah, 4, 12, 16, 311
Win–lose problem solving, 245–46
Win–win problem solving, 247–51
Withholding information, 417
Withholders, 240
Wood, Zachary R., 154

Word charts, 396, 397
Working outlines, 354–55
Working remotely, 283–85
Worklife with Adam Grant (podcast), 384
Workplace communication, 257–87
    age and, 96
    environment and, 184–85
    essential skills for, 258–59
    for job applicants, 263–69
    in job interviews, 269–78
    nonverbal, 165–66
    organizing business presentations, 369
    persuasion skills in sales, 432
    presentation software for, 401
    in professional environments, 277, 279–85
    sexual harassment and, 238–39

Ye, 188
Yocum, Rebecca, 424
YouTube, 23, 24, 391

Zhu, 79
Zhuo, Julie, 290, 303, 311, 318
zilianOP (Angel Hamilton), 27
Zoom, 139, 140, 306
Zoom fatigue, 304